T0332058

The Analysis of
Irregular Shaped
Structures

About the International Code Council

The International Code Council® is the leading global source of model codes and standards and building safety solutions that include product evaluation, accreditation, technology, codification, training, and certification. The Code Council's codes, standards, and solutions are used to ensure safe, affordable, and sustainable communities and buildings worldwide. The International Code Council family of solutions includes the ICC Evaluation Service, the International Accreditation Service, General Code, S. K. Ghosh Associates, NTA Inc., Progressive Engineering Inc., ICC Community Development Solutions, and the Alliance for National & Community Resilience. The Code Council is the largest international association of building safety professionals and is the trusted source of model codes and standards, establishing the baseline for building safety globally and creating a level playing field for designers, builders, and manufacturers.

Washington, DC Headquarters
500 New Jersey Avenue, NW, 6th Floor, Washington, DC 20001

Regional Offices
Eastern Regional Office (Birmingham, AL)
Central Regional Office (Chicago, IL)
Western Regional Office (Los Angeles, CA)

Distribution Center (Lenexa, KS)

888-ICC-SAFE (888-422-7233)
www.iccsafe.org

The Analysis of Irregular Shaped Structures

Wood Diaphragms and Shear Walls

R. Terry Malone, P.E., S.E.

Scott E. Breneman, Ph.D., P.E., S.E.

Robert W. Rice, CBO

Second Edition

INTERNATIONAL
CODE COUNCIL®

S. K. GHOSH
ASSOCIATES

New York Chicago San Francisco
Athens London Madrid
Mexico City Milan New Delhi
Singapore Sydney Toronto

Library of Congress Cataloging-in-Publication Data

Names: Malone, R. Terry, author. | Breneman, Scott E., author. | Rice, Robert W., author.
Title: The analysis of irregular shaped structures : wood diaphragms and shear walls /
    R. Terry Malone, P.E., S.E., Scott E. Breneman, Ph.D., P.E., S.E., Robert W. Rice, CBO.
Description: Second edition. | New York : McGraw Hill, [2022] | Includes bibliographical references
    and index. | Summary: "This is a second edition of a popular guide for professionals in need of
    solutions for lateral load path problems in wood building construction and design. This edition
    is updated to include newly developed structure types and new design solutions. It is co-branded
    with The International Code Council (ICC) and incorporates up-to-date structural codes, as well
    all other requirements and standards"—Provided by publisher.
Identifiers: LCCN 2022014767 (print) | LCCN 2022014768 (ebook) |
    ISBN 9781264278824 (hardcover) | ISBN 9781264278831 (ebook)
Subjects: LCSH: Structural design. | Shear walls—Design and construction. |
    Diaphragms (Structural engineering)—Design and construction.
Classification: LCC TA658.4 .M35 2022 (print) | LCC TA658.4 (ebook) |
    DDC 624.1/7—dc23/eng/20220509
LC record available at https://lccn.loc.gov/2022014767
LC ebook record available at https://lccn.loc.gov/2022014768

The Analysis of Irregular Shaped Structures: Wood Diaphragms and Shear Walls, Second Edition

1 2 3 4 5 6 7 8 9   LHN   27 26 25 24 23 22

ISBN 978-1-264-27882-4
MHID 1-264-27882-9

This book is printed on acid-free paper.

Sponsoring Editor Ania Levinson	**Project Manager** Rishabh Gupta, MPS Limited	**Art Director, Cover** Jeff Weeks
Editing Supervisor Stephen M. Smith	**Copy Editor** Yashoda Rawat, MPS Limited	**Composition** MPS Limited
Production Supervisor Lynn M. Messina	**Proofreader** Alekha C. Jena	
Acquisitions Coordinator Elizabeth M. Houde	**Indexer** Ariel Tuplano	

Dedicated to, and in appreciation of, those who inspire us.

To our families, especially our wives:

Jerri
Courtney
Lisa

About the Authors

R. Terry Malone, P.E., S.E., is a licensed structural engineer in Washington, Oregon, and Arizona and senior technical director of the Project Resources and Solutions Division of WoodWorks. He provides project support and education to architects, engineers, and design professionals regarding designing and constructing wood buildings, oversees the development of technical educational content, and is involved in professional standards committees through the American Wood Council. Prior to joining WoodWorks, Mr. Malone was a principal in consulting structural engineering firms in Washington and Oregon and also conducted third-party structural plan reviews. He previously served as a faculty member at St. Martin's College in Lacey, Washington. He has over 45 years of wood design experience and experience with lateral-force-resisting systems.

Scott E. Breneman, Ph.D., P.E., S.E., is a licensed structural engineer and professional engineer in the State of California, is a licensed professional engineer in the State of Washington, and is experienced with structural design, seismic rehabilitation, and peer review of building projects of all shapes and materials. He received bachelor's and master's degrees from the University of Florida and a doctorate from Stanford University. As a senior technical director at WoodWorks, Dr. Breneman currently provides project support and education to AEC professionals constructing wood buildings, oversees the development of technical educational content, and generally tries to stay knowledgeable on the rapidly progressing field of mass timber. He is a past president of the Structural Engineers Association of Central California and is involved in numerous professional standards committees through the American Society of Civil Engineers, American Wood Council, and APA—The Engineered Wood Association.

Robert W. Rice, CBO, is operations manager/chief building official for Northwest Code Professionals, which provides building department services for jurisdictions throughout Oregon and Washington. Following 10 years of work as a structural designer for engineering firms in Southern Oregon, and a developing interest in building codes, he worked as building official for Josephine County. Oregon and assistant state building official for the State of Oregon. He is also actively involved in code development with ICC and the State of Oregon and is a past member of ICC's Building Code Action Committee. Mr. Rice is a past president of Oregon Building Officials Association and the Southern Oregon Chapter of ICC. He was also a part-time instruction for the construction technology program at Rogue Community College.

For instructors of classes using this book as a text, a solutions manual for the end-of-chapter problems is available at www.mhprofessional.com/AnalysisofIrregularShapedStructures2E.

Table of Contents

Preface

Residential and commercial buildings have become more complex than structures built only a few decades ago. To create architecturally appealing structures, horizontal and vertical offsets in the diaphragms, multiple reentrant corners, multiple irregularities, and fewer vertical lateral-force-resisting elements have become commonplace. The structural configurations of many modern buildings require very complex lateral load paths. Most texts and publications available only address simple rectangular diaphragms and shear walls. Methods of analysis for these simpler diaphragms and shear walls do not easily adapt to complex diaphragms and shear wall layouts in irregular shaped structures.

Calculating the forces that are to be transferred across multiple discontinuities and detailing the design requirements on the construction documents can be very challenging and time consuming. Various methods of analyzing the distribution of lateral loads in complex structures were developed in the early 1980s, based largely on work done by the Applied Technology Council (ATC-7),[1] the APA—The Engineered Wood Association,[2] and by Edward F. Diekmann[3] among others. But the distribution of this information has been limited, making some of the material hard to find. Innovations in wood construction have also introduced cross-laminated timber (CLT) into wood roof, floor, and wall constructions. While basic load path analysis is material independent, the use of CLT in lateral-force-resisting diaphragms and shear walls brings in new design considerations for practicing engineers.

The purpose of this publication is to consolidate information into one source to provide a comprehensive coverage of the analysis of modern irregular shaped structures through numerous step-by-step examples, and to bring it to the forefront of the engineering and code official communities. A secondary objective is to demonstrate how to achieve the *necessary* complete lateral load paths through shear wall and diaphragm discontinuities. The complex diaphragm, shear wall, and load path issues addressed in this book are representative of today's demand on design professionals and code officials. Most of the examples in this book are based on light-frame wood construction using diaphragms that can be idealized as flexible. Shear walls are typically considered to be rigid bodies using wood or cold-formed steel framing with wood sheathing but vary in stiffness due to larger openings.

The information presented in this book is intended to serve as a guideline for recognizing irregularities and developing the procedures necessary to resolve the forces along complicated load paths. The examples provide a progressive coverage of basic to very complex illustrations of load paths in the complicated structures. The benefits of the methods presented herein allow creation of complete lateral load paths when none appear to be possible. Most of the examples presented throughout the book and in the solutions manual show shear wall and diaphragm configurations that would be considered minimal lateral-force-resisting systems, without redundancy and under maximum demand. This has been done to simplify the examples. Reducing the number of vertical lateral-force-resisting elements, combined with multiple complicated load paths, and then designing to the maximum element capacity is neither suggested nor encouraged by the authors. In most cases, more direct, conservative, and simpler solutions to load paths are available. The methods and examples included are intended to provide the design professional with reasonable and rational analytical tools that can be used to solve complex problems, but do not represent the only methods available.

It has been the authors' experience from private design practice, teaching, and plan reviews that the knowledge in the engineering and code administration communities regarding the analysis of wood diaphragms and shear walls varies greatly. Design professionals need to learn and mentor the art of understanding and establishing complete load paths. This is increasingly important due to an increased reliance on structural analysis programs in the design process. Although it is helpful to have a basic understanding of simple shear walls and diaphragms prior to reading this book, enough fundamental information is provided for the laymen to follow the complex examples.

This book is based on the 2021 IBC,[4] ASCE7-16,[5] the 2018 NDS,[8] and the 2021 edition of the Special Design Provisions for Wind and Seismic (SDPWS).[9] It is assumed that the reader has a working understanding of and access to these design codes and standards, including the applicable loads, load combinations, allowable stresses, and adjustment factors. Publications covering the basic concepts and methods of addressing analysis and design of wood structures can be referenced in *The Design of Wood Structures*[6] and SEAOC's *Structural/Seismic Design Manual, Vol. 2*,[7] which provide a comprehensive coverage of fundamentals of wood lateral-force-resisting system analysis and design. The opinions and interpretations are those of the authors, based on experience, and are intended to reflect current structural practice. Engineering judgment and experience has been used in establishing the procedures presented in this book when there was an absence of documentation or well-established procedures available. Although every attempt has been made to eliminate errors and to provide complete accuracy in this publication, it is the responsibility of the design professional or individual using these procedures to verify the results. Users of this information assume all liability arising from such use.

Comments or questions about the text, examples, or problems may be addressed to any of the authors through this address: malone.breneman.rice@gmail.com.

R. Terry Malone, P.E., S.E.
Senior Technical Director,
WoodWorks
Prescott Valley, Arizona

Scott Breneman, Ph.D., P.E., S.E.
Senior Technical Director,
WoodWorks
Deer Park, Washington,

Robert W. Rice, CBO
Operations Manager/
Building Official
Northwest Code
Professionals
Grants Pass, Oregon

References

1. Applied Technology Council (ATC), *Guidelines for Design of Horizontal Wood Diaphragms, ATC-7*, Applied Technology Council, Redwood, CA, 1981.
2. APA—The Engineered Wood Association, *APA Research Report 138, Plywood Diaphragms, APA Form E315H*, APA—The Engineered Wood Association, Engineering Wood Systems, Tacoma, WA, 2000.
3. Diekmann, E. F., "Design of Wood Diaphragms," *Journal of Materials Education*, Fourth Clark C. Heritage Memorial Workshop, Wood Engineering Design Concepts, University of Wisconsin, WI, 1982.
4. International Code Council (ICC), *International Building Code, 2021 with Commentary*, ICC, Brea, CA, 2021.
5. American Society of Civil Engineers (ASCE), *ASCE/SEI 7-16 Minimum Design Loads for Buildings and Other Structures*, ASCE, New York, 2016.
6. Breyer, D. E., Martin, Z., and Cobeen, K. E., *Design of Wood Structures ASD*, 8th ed., McGraw-Hill, New York, 2020.
7. Structural Engineers Association of California (SEAOC), *IBC Structural/Seismic Design Manual*, Vol. 2, SEAOC, CA, 2018.
8. American Wood Council, *National Design Specification for Wood Construction and Supplement*, Leesburg, VA, 2018.
9. American Wood Council, *Special Design Provisions for Wind and Seismic with Commentary*, Leesburg, VA, 2021.

Nomenclature

Organizations

AF&PA
American Forest and Paper Association
1111 19th St., NW
Suite 800
Washington, District of Columbia 20036

APA
APA—The Engineered Wood Association
PO Box 11700
Tacoma, Washington 98411-0700

ASCE
American Society of Civil Engineers
1801 Alexander Bell Dr.
Reston, Virginia 20191

ATC
Applied Technology Council
2471 E. Bayshore Rd.
Suite 512
Palo Alto, California 94303

Building Seismic Safety Council (a council
  of the National Institute of Building
  Safety)
Washington, District of Columbia 20005

ICC
International Codes Council
3060 Saturn Street,
Suite 100, Brea, California 92821

AWC
American Wood Council
22 Catoctin Circle, SE, Suite 201
Leesburg, Virginia 20175

SEAOC
Structural Engineers Association
  of California
555 University Ave., Suite 126
Sacramento, California 95825

USDA
US Department of Agriculture
Forest Products Laboratory
Madison, Wisconsin 53726

WPC
Wood Products Council
WoodWorks
1101 K St NW Ste 700
Washington, District of Columbia 20005

Abbreviations

Allow.	allowable	MWFRS	main wind force-resisting system
ASD	allowable stress design	N.A.	neutral axis
Bm.	beam	N.G.	no good
Blk'g.	blocking	o.c.	on center
CLT	cross laminated timber	o.k.	okay
Discont.	discontinuous	OSB	oriented strand board
Diaph.	diaphragm	PW	plywood
Ecc.	eccentricity, eccentric	req'd.	required
FS	factor of safety	SDC	seismic design category
Flr.	floor	SDS	short period design spectral acceleration parameter
Hdr.	header		
I.P.	inflection point	Shr.	shear
Lds.	loads	SFRS	seismic force-resisting system
LRFD	load and resistance factor design	Sht'g.	sheathing
		STR	strength, strength design
LFRS	lateral force resisting system	SW	shear wall
max.	maximum	Trib.	tributary
min.	minimum	UNO	unless noted otherwise
MLFRS	main lateral-force-resting system	Unif.	uniform
MSFRS	main seismic force-resisting system	WSP	wood structural panels

Units

ft	foot, feet	ksf	kips per square foot
ft^2	square foot, square feet	pcf	pounds per cubic foot
in	inch, inches	plf	pounds per lineal foot
in^2	square inch, square inches	psf	pounds per square foot
k	kip, kips, 1000 lb	psi	pounds per square inch
ksi	kips per square inch		

Symbols

A	area (in^2, ft^2)
A_{net}	net area (in^2, ft^2)
A.R.	aspect ratio (length to width or length to depth)
ATS	automatic tensioning system anchor, shrinkage compensating
A_x	torsional amplification factor

b	length of shear wall parallel to lateral force, distance between chords of shear wall (in, or ft)
b_{eff}	effective width of moment-resisting arm between centerline of hold-down rod and centerline of compression boundary member of the shear wall used to determine the overturning force (ft)
b_i	individual full height section of perforated shear wall
b_s	width (breadth) of a CLT shear wall panel
b', d'	shallower width or depth of diaphragm (ft)
C	compression force (lb or kips)
C.M.	center of mass
C.R.	center of rigidity
C_b	bearing length (in)
C_D	load duration factor
C_{di}	diaphragm factor for nail connections
C_{eg}	end grain factor for wood connections
C_f	size factor for sawn lumber
C_G	CLT shear wall capacity adjustment factor for specific gravity
CL	distance from face of hold-down to the centerline of the anchor bolt (in)
C.L.	centerline
C_s	seismic response coefficient
D	dead load (lb, k, plf, klf, psf, ksf)
D	depth (ft)
d	depth of solid wood section (in)
d_a	vertical elongation of overturning anchorage (in)
d_e	depth of member less the distance from the connector to the unloaded edge (in)
Diaph 2	diaphragm 2
DL	dead load (lb, k, plf, klf)
$d_{req,d}$	depth required (ft)
E	modulus of elasticity (psi, ksi)
$EI_{eff,f}$	flatwise effective bending stiffness of CLT (lb-in²/ft)
e	eccentricity (in, ft)
e_n	nail deformation (in)
e_f	fastener deformation (in)
$e_{f\parallel}$	fastener deformation on panel edges parallel to applied load (in)
$e_{f\perp}$	fastener deformation on panel edges perpendicular to applied load (in)
f_a	axial stress (psi, ksi)
f_b	bending stress (psi, ksi)
F_b'	allowable bending stress, adjusted (psi, ksi)
$F_b S_{eff}$	flatwise reference flexural design capacity of CLT (lb-ft/ft)

F_{9B}	the force at grid line 9B (lb, k)
f_c	compression stress (psi, ksi)
F_{CL}	force at centerline (lb, k)
$F'_{c\perp}$	allowable bearing stress perpendicular to the grain, adjusted (psi, ksi)
F_{chord}	chord force (lb, k)
$F_{collector}, F_{coll.}$	collector force (lb, k)
F_{max}	maximum force (lb, k)
$F_{o/t}$	overturning force (lb, k)
F_{strut}	strut force (lb, k)
F_T	torsional force (lb, kips)
ft	feet
ft-lb	foot-pounds
ft-k	foot-kips
F'_v	allowable shear stress, adjusted (psi, ksi)
f_v	horizontal shear stress (psi)
F_V, F_H	vertical or horizontal force (lb, k)
F_x, F_y	force along the x or y axis (lb, k)
F_x	axial force (lb, k)
F_y	steel yield strength (psi, ksi)
$F_{v,e}$	edgewise (in-plane) reference shear stress of CLT (psi)
G_a	apparent diaphragm or shear wall shear stiffness from nail slip and panel shear deformation
$GA_{eff,f}$	flatwise effective shear stiffness of CLT (lb/ft)
G_t	panel rigidity through the thickness, in lb per inch of panel width
H	horizontal force (lb, k)
h	height of shear wall (ft)
h_1	height of 1st story (ft)
h_{sx}	the story height below level x
I	moment of inertia (in⁴)
I_E, I_e	importance factor for seismic
I_o	moment of inertia of individual element about itself (in⁴)
I_T	the total moment of inertia (in⁴)
I_w	importance factor for wind
J	polar moment of inertia (in⁴)
k	kips, 1000 lb
K	rigidity, stiffness
L	length of diaphragm or shear wall (in, ft)
L'	length of cantilever diaphragm (ft)
LL	live load (lb, k, plf, klf)
L_1	length of section 1 (ft)

L_{1-3}	length of section from grid line 1 to 3 (ft)
l_{brg}	length of bearing (in)
L_{embed}	length of embedment (ft)
L_{hdr}	length of header (ft)
L_r	roof live load (psf)
L_{sw}	length of shear wall (ft)
L_{TD}	length of transfer diaphragm (ft)
l_u	unbraced length of bending member (in, ft)
L_{wall}	length of wall (ft)
$L/W, L/d, L/b$	length to width (or depth) ratio
M	bending moment (in-lb, in-k, ft-lb, ft-k)
M_{max}	maximum bending moment (in-lb, in-k, ft-lb, ft-k)
M_{net}	net bending moment (in-lb, in-k, ft-lb, ft-k)
$M_o, M_{o/t}$	overturning moment (ft-lb, ft-k)
M_R	resisting moment (ft-lb, ft-k)
M_x	bending moment at distance x (in-lb, in-k, ft-lb, ft-k)
M_1	bending moment at grid line 1, or moment 1 (in-lb, in-k, ft-lb, ft-k)
n	number of fasteners in the same plane
n	number of connectors per panel at base of CLT shear wall
$n_{\parallel}$	number of slip planes at a CLT diaphragm connection parallel to the applied loads
$n_{\perp}$	number of slip planes at a CLT diaphragm connection perpendicular to the applied loads
o/t	overturning
P	concentrated load (lb, k)
$P_{\parallel}$	panel length parallel to the applied load (ft)
$P_{\perp}$	panel length perpendicular to the applied load (ft)
p, q	wind pressure (psf)
R	reaction (lb, k)
R	generic reference design value calculated following the NDS
R'	generic adjusted design capacity calculated following the NDS
R_{2L}	reaction on the left side of grid line 2 (lb, k)
R_A	reaction at grid line A (lb, k)
R_L, R_R	left or right reaction (lb, k)
S	regular spacing of fasteners in a CLT diaphragm (in)
S, SL	snow load (lb, k, plf, klf)
S, S_x	section modulus (in^3)
SBP	soil bearing pressure
SDC	seismic design category
SW1	shear wall 1

T	tension force (lb, k)
T	fundamental period of vibration of structure (sec)
TA	transfer area
TD1	transfer diaphragm 1
TD	transfer diaphragm
t_e	effective shear thickness of plywood
Typ	typical
V	shear force (lb, k)
V	vertical force (lb, k)
V_V, V_H	vertical or horizontal shear force (lb, k)
V_{max}	maximum shear force (lb, k)
V_n	the average uniform load per nail (lb)
V_n	the nominal shear capacity of a fastener or connector (lb)
V_n'	the average non-uniform load per nail (lb)
V_{sw2}	shear force applied to shear wall 2 (lb, k)
V_s	flatwise reference shear capacity of CLT (lb/ft)
V_{TL}, V_{total}	total shear force (lb, k)
V_u, V_{uom}	ultimate (nominal) shear (lb, k)
V_{wall}	shear force applied to a wall (lb, k)
V_x	total shear force at distance x (lb, k)
V_{2L}	shear force on the left side of grid line 2 (lb, k)
v_{3AB}	uniform unit shear at grid line 3, from A to B (plf, klf)
v	uniform unit shear (plf, klf)
v_{diaph}	uniform unit shear in the diaphragm (plf, klf)
$v_{d,ASD}$	ASD unit shear demands in the diaphragm (plf, klf)
$v_{d,LRFD}$	LRFD unit shear demands in the diaphragm (plf, klf)
v_{max}	maximum uniform unit shear (plf, klf)
v_{net}, v_n	net uniform unit shear (plf, klf)
v_n	nominal diaphragm or shear wall shear capacity (plf)
V_r'	adjusted design shear based on effective depth (lb, k)
v_{sw2}	uniform unit shear in shear wall 2 (plf, klf)
v_x	uniform unit shear at distance x (plf, klf)
v_{2L}	uniform unit shear on the left side of grid line 2 (plf, klf)
W	width of diaphragm, opening (ft)
W'	width of cantilever diaphragm (ft)
w	lateral uniform load, wind or seismic (plf, klf)
w_{lw}	lateral uniform load due to wind, leeward pressures (plf, klf)
W_{TD}, w_{TD}	Width of transfer diaphragm (ft)
w_{ww}	lateral uniform load due to wind, windward pressures (plf, klf)
w_{3-5}	uniform load from grid line 3 to 5 (plf, klf)

w_E	lateral uniform load due to seismic (plf, klf)
w_{strip}	uniform load applied to a 1 ft wide strip across the structure (plf, klf)
w_x	uniform load applied along a distance x (plf, klf)
x	distance x (ft)
$\bar{\bar{x}}$	distance to the neutral axis from base line (in, ft)
Z	reference shear capacity of a single fastener per NDS (lb)
Z'	adjusted allowable shear capacity of a single nail per NDS (lb)
Z^*	adjusted allowable short-term shear capacity of a single fastener per NDS (lb)
γ_D	force amplification factor for CLT diaphragm components
Δ	deflection (in)
Δ_{ADVE}	average displacement of vertical force-resisting elements
Δ_a	total vertical elongation at wall anchorage
Δ_{aeff}	total effective vertical elongation at wall anchorage
Δ_B	deflection at grid line B (in)
Δ_b, Δ_B	deflection due to bending (in)
Δ_c, Δ_{cs}	deflection due to chord slip (in)
Δ_e	deflection due to elongation of steel strap (in)
Δ_{max}	maximum deflection (in)
$\Delta_{rot.}$	deflection due to rotation (in)
Δ_{ns}	deflection due to nail slip (in)
Δ_s	deflection due to shear (in)
Δ_{strap}	deflection due to strap elongation and nail slip (in)
Δ_T, Δ_{TL}	total deflection (in)
δ_{diaph}	diaphragm displacement (in)
δ_{MDD}	maximum diaphragm displacement (in.)
δ_{RH}	horizontal rotational displacement
δ_{RV}	vertical rotational displacement
δ_{slip}	diaphragm and shear wall deflection component resulting from fastener slip (in)
δ_x	story drift at level x (in)
δ_{xe}	deflection at the location required determined by an elastic analysis (in)
ρ	redundancy factor
θ	stability coefficient for P-delta effects
ϕ_D	LRFD resistance factor for diaphragms
Ω_o	overstrength factor
Ω_D	ASD reduction factor for diaphragms

CHAPTER 1

Code Sections and Analysis

1.1 Introduction

For centuries, building codes have been developed to define the standards for the design and construction of structures. Opinions are often expressed that code requirements have become too complex; however, from the earliest of codes to our current standards, codes have changed in response to our increased understanding of materials and methods as well as our knowledge of the forces that are imposed on structures, particularly wind and seismic forces. This understanding has been greatly increased by past structural failures and from current state-of-the-art testing, research, and a better understanding of how buildings respond in an extreme loading event. In addition, changes to the code have been brought about by the reality that structures have become increasingly more complex as compared to structures previously built.

The most widely used and accepted code for building design standards in the United States is the International Building Code (IBC) published by the International Code Council (ICC).[1] The document references a compilation of design standards that have been developed through an open and transparent consensus process that represents all interested parties and stakeholders. ASCE/SEI 7-2016, *Minimum Design Loads for Buildings and Other Structures,* is published by the American Society of Civil Engineers and the Structural Engineering Institute[2] and is referenced from the 2021 IBC. Wood lateral-force-resisting systems are addressed in *National Design Specification for Wood Construction* (NDS-2018) and *Special Design Provisions for Wind and Seismic* (SDPWS-2021), which are both published by the American Wood Council.[3] The IBC-21, ASCE 7-16, NDS-18, and SDPWS-21 are codes and standards that will be discussed in the chapters that follow. Relative sections and definitions from these codes and standards are provided for quick reference and comparisons. The following code sections and definitions are not direct quotes and can contain additional clarifications and authors' comments.

1.2 IBC 2021 Code Sections Referencing Wind and Seismic[1]

Chapter 2

202.1 Definitions

Diaphragm: A horizontal or sloped system acting to transmit lateral forces to vertical elements of the lateral force-resisting system. When the term "diaphragm" is used, it shall include horizontal bracing systems.

Collector: A horizontal diaphragm element parallel and in line with the applied force that collects and transfers diaphragm shear forces to the vertical elements of the lateral force-resisting system or distributes forces within the diaphragm, or both. [Authors' note: Collectors are also used at areas of discontinuity in diaphragms and shear walls and can be oriented in the direction within the diaphragm or shear wall.]

Seismic Design Category: A classification assigned to a structure based on its risk category and the severity of the design earthquake ground motion at the site.

Seismic Force-resisting System: That part of the structural system that has been considered in the design to provide the required resistance to the prescribed seismic forces. [Authors' note: This term is synonymous with "lateral-force-resisting system," under wind or seismic forces.]

Chapter 16

1604.4   Analysis

This section requires that load effects on structural members and their connections shall be determined by and take into account equilibrium, general stability, geometric compatibility and both short- and long-term material properties; and that any system or method of construction used shall be based on a rational analysis in accordance with well-established principles of mechanics. Such analysis shall result in a system that provides a complete load path capable of transferring loads from their point of origin to the load-resisting elements.

Lateral forces shall be distributed to the various vertical elements of the lateral-force-resting system in proportion to their rigidities, considering the rigidity of the horizontal bracing system or diaphragm.

Chapter 23

2305   General design requirements for lateral force resisting systems.

2305.1   General:

Structures using wood-framed shear walls or wood-framed diaphragms to resist wind, seismic or other lateral loads shall be designed and constructed in accordance with AWC SDPWS and the applicable provisions of Sections 2305, 2306 and 2307.

2305.1.1

Openings in shear panels that materially affect their strength shall be fully detailed on the plans and shall have their edges adequately reinforced to transfer all shearing stresses.

2306.2 and 2306.3

Wood frame diaphragms and shear walls shall be designed and constructed in accordance with AWC SDPWS and the provisions of IBC Sections 2305, 2306 and 2307.

Also see Section 2308.4.4.1—openings in diaphragms in SDC B-F, and Section 2308.4.4.2—vertical offsets in diaphragms in SDC D and E.

1.3   ASCE 7-16 Sections Referencing Seismic[2]

Chapter 11

11.2   Definitions

The following definitions are provided for comparison to other code or standards definitions.

Boundary Elements: Portions along wall and diaphragm edges and openings for transferring or resisting lateral forces. Boundary elements include chords and collectors at diaphragms and shear wall perimeters, edges of openings, discontinuities, and re-entrant corners.

Diaphragm Boundary: A location where shear is transferred into or out of the diaphragm element. Transfer is either to a boundary element or to another lateral force-resisting element.

Diaphragm Chord: A diaphragm boundary element perpendicular to the applied load that is assumed to take axial stresses caused by the diaphragm moment.

Collector (Drag strut, tie, diaphragm strut): A diaphragm or shear wall boundary element parallel to the applied load that collects and transfers diaphragm shear forces to the vertical elements of the seismic force-resisting system or distributes forces within the diaphragm or shear walls. [Authors' note: A collector can also resist wind or other lateral forces.]

Chapter 12

12.1.3 Continuous Load Path and Interconnection (partial quote)

A continuous load path, or paths, with adequate strength and stiffness shall be provided to transfer all forces from the point of application to the final point of resistance. [Authors' note: Connections are considered as part of the complete load path.]

12.3 Diaphragm Flexibility, Configuration Irregularities, and Redundancy.

12.3.1 Diaphragm Flexibility.

The structural analysis shall consider the relative stiffnesses of diaphragms and the vertical elements of the lateral force-resisting system. The structural analysis shall explicitly include consideration of the stiffness of the diaphragm (i.e., semi-rigid modeling assumption).

12.3.1.1 Flexible diaphragm condition

12.3.1.2 Rigid diaphragm condition

12.3.1.3 Calculated flexible diaphragm condition

12.10 Diaphragm Chords and Collectors

12.10.1 Diaphragm design:

Diaphragms shall be designed for both shear and bending stresses resulting from design forces. At diaphragm discontinuities, such as openings or reentrant corners, the design shall assure that the dissipation or transfer of edge (chord) forces combined with other forces in the diaphragm is within the shear and tension capacity of the diaphragm.

12.10.2 Collector elements.

Collector elements shall be provided that are capable of transferring the seismic or wind forces originating in other portions of the structure to the elements providing resistance to those forces.

1.4 Important AWC-SDPWS-2021 Sections[3]

2.2 Terminology

Boundary Element: Diaphragm and shear wall boundary members to which sheathing shear forces are transferred. Boundary elements include chords and collectors at diaphragm and shear wall perimeters, interior openings, discontinuities, and reentrant corners.

Diaphragm Boundary: A location where shear is transferred into or out of the diaphragm sheathing. Transfer is either to a boundary element or to another lateral force-resisting element.

Collector: A diaphragm or shear wall boundary element parallel to the applied force that collects and transfers diaphragm shear forces to the vertical lateral force-resisting elements or distributes forces within the diaphragm or shear wall.

Chord: A diaphragm boundary element perpendicular to the applied load that resists axial stress due to the induced moment.

Diaphragm: A roof, floor or other membrane bracing system acting to transmit lateral forces to the vertical resisting elements. When the term "diaphragm" is used, it shall include horizontal bracing systems.

4.1.1   Design requirements.

The proportioning, design and detailing of engineered wood systems members, and connections in lateral force-resisting systems shall be in accordance with

- Reference documents in Section 2.1.2 and the provisions of this chapter and standard.
- Applicable building code, and ASCE 7.
- The seismic shear capacity shall be determined in accordance with Sections 4.1.4.1 and 4.1.4.2 for wind.

Structures resisting wind and seismic loads shall meet all applicable drift, deflections, and deformation requirements of this standard. A continuous load path, or paths, with adequate strength and stiffness shall be provided to transfer all forces from the point of application to the final point of resistance.

4.1.9   Boundary elements.

Shear wall and diaphragm boundary elements shall be provided to transfer the design tension and compression forces. Diaphragm and shear wall sheathing shall not be used to splice boundary elements. Diaphragm chords and collectors shall be placed in, or tangent to, the plane of the diaphragm framing unless it can be demonstrated the moments, shears, and deformations, considering eccentricities resulting from other configurations, can be tolerated without exceeding the framing capacity and drifts limits.

4.2   Sheathed wood frame diaphragms

4.2.1   Application Requirements

Wood-framed diaphragms shall be permitted to be used to resist lateral forces provided the in-plane deflection of the diaphragm, as determined by calculations, tests, or analogies drawn therefrom, does not exceed the maximum permissible deflection limit of attached load distributing or resisting elements. Framing members, blocking, and connections shall extend into the diaphragm a sufficient distance to develop the force transferred into the diaphragm. [Authors' opinion: The development length should be verified by calculation as demonstrated in this book or by other equivalent method.]

4.2.2   Diaphragm Aspect Ratios.

Size and shape of diaphragms shall be limited to the aspect ratios in Table 4.2.2.

4.2.3   Deflections

Alternatively, for wood structural panel diaphragms, deflection shall be permitted to be calculated using a rational analysis where apparent shear stiffness accounts for panel deformation and non-linear nail slip in the sheathing-to-framing connection.

4.3   Sheathed wood framed shear walls

4.3.3.1   Shear Wall Aspect Ratios.

The size and shape of shear walls shall be limited to the aspect ratios in Table 4.3.3 and Figure 4C for segmented shear walls, Figure 4D for FTAO shear walls and Figure 4E for perforated shear walls. [See Chap. 10 for suggested shear wall header, sill, and transfer diaphragm aspect ratio limits.]

4.5   CLT diaphragms (new in SDPWS 2021)

4.6   CLT shear walls (new in SDPWS 2021)

1.5   Sections Specifically Referencing Structural Irregularities

It is important to recognize and understand structural irregularities. A large portion of this book provides guidance on how to identify and solve force transfer across areas of discontinuities in irregular structures. The following sections are presented to show agreement between the codes and standards with regard to lateral-force-resisting systems that resist wind and seismic forces. These sections have been selected for their relevance to this book. These sections should be reviewed in their entirety when reading each chapter of the book.

1.5.1   ASCE 7-16

12.3.2.1 and Table 12.3-1 Horizontal structural irregularities
12.3.2.2 and Table 12.3-2 Vertical structural irregularities
12.3.3 Limitations and additional requirements for systems with structural irregularities
12.3.3.3 Elements supporting discontinuous walls or frames
12.3.3.4 Increase in forces caused by irregularities for seismic design Categories D through F
12.8.4.1 Inherent Torsion
12.8.4.2 Accidental Torsion

1.5.2   SDPWS-21

4.1.7 Horizontal distribution of shear
4.1.8 Vertical distribution of seismic force resisting systems strength
4.2.5.1 Torsional Irregularity
4.2.6 Open-front Structures

1.5.3   2018 IRC[4]

R301.2.2.6 Irregular Buildings

- Shear wall or braced wall offsets out-of-plane
- Lateral support of roof and floors. Edges not supported by shear walls or braced wall lines (cantilevers)
- Shear walls or braced wall offsets in plane
- Floor or roof opening
- Floor level offset—vertically
- Perpendicular shear wall and bracing—do not occur in two perpendicular directions.
- Wall bracing in stories containing masonry or concrete construction.

1.6   Complete Load Paths

Most of the texts and publications available today only address simple rectangular diaphragms, the analysis of which does not easily adapt to complex diaphragm and shear wall layouts. The layout of the lateral-force-resisting system shown in Figs. 1.1 and 1.2 demonstrate these types of problems. The vertical and horizontal offsets shown in the figures create discontinuities in the diaphragm, which require special collector and drag strut elements to establish complete load paths. Collectors and drag strut elements in diaphragms and in shear walls are a critical part of complex lateral-force-resisting systems. The analysis and design requirements for diaphragms under wind or seismic loading is a complicated topic that is prone to being misunderstood. Some of the confusion has been brought about by the location of lateral-force-resisting systems requirements within ASCE 7-16. Chapters 11 and 12 of that standard, which address seismic design, provide a complete and organized coverage of lateral-force-resisting systems, components, and requirements under seismic loading conditions. Chapters 26 through 31 address the analysis and application of wind loads and pressures on structures and on components and cladding. It does not, however, cover lateral resisting elements or systems or their design requirements in as much detail as seismic design section does. Some designers may interpret the lack of discussion of structural systems or elements in the wind chapters to imply that drag struts and collectors are not required for wind design; and that, diaphragm discontinuities do not have to be addressed if wind controls. Section 1604.9 of the 2021 IBC addressing wind and seismic detailing says, "Lateral-force-resisting systems shall meet seismic detailing requirements and limitations prescribed in this code and ASCE 7, excluding Chapter 14 and Appendix 11A, even when wind load effects are greater than seismic load effects." Diaphragms, drag struts, collectors, and shear walls function the same way regardless of if loads applied to the diaphragm are from wind, seismic, soil, or other pressures. All irregularities and/or discontinuities within a system of diaphragms and shear walls should be addressed. It is easy to overlook the definitions section when thumbing through the codes and standards, believing that the contents therein are already understood. A quick review will show that the definitions actually set the criteria and requirements for diaphragms, chords, collectors, and their design. In practical terms, all diaphragms must have boundary members consisting of drag struts, chords, collectors, or other vertical lateral-force-resisting elements. Collectors are required at all offsets and areas of discontinuity within the diaphragm, including at openings. These requirements also apply to shear walls. Forces at all discontinuities and openings must be dissipated or transferred into the diaphragm or shear wall without exceeding its design capacity. The codes and standards specify that the sheathing shall not be used to splice boundary elements or collectors. Furthermore, all diaphragms and shear walls shall contain continuous load paths along all boundaries and lines of lateral-force-resistance and across all discontinuities.

Irregular shaped structures similar to the one shown in Fig. 1.1 are commonly designed without properly addressing the irregularities contained therein. The structure exhibits multiple vertical and horizontal offsets in the diaphragm, cantilever diaphragms, few opportunities for shear walls at the exterior wall line and multiple vertical and horizontal discontinuities in the lateral load paths of the lateral-force-resisting system. Some designers may intuitively place tie straps with blocking throughout the structure without explicit purpose, in an ambiguous attempt to address discontinuities with no rationalization or supporting calculations. Such a judgment-based approach will easily miss connections that are required to develop a complete load path, even along straight lines of lateral force resistance. ATC-7 noted that failures have occurred because of the following[5]:

- Connection failures caused by incomplete load paths, incomplete designs, inadequate detailing, and inadequate installation (construction). Often, the size of wood chords for tension and compression forces is also ignored in the design, which can lead to failures.

- Designs included diaphragm shears and chord forces only, connection designs were not addressed.

- Designs did not include load paths that continued down to the foundation and into the soil.

- Designing to the maximum diaphragm and shear wall capacity (close nailing), while limiting the number of shear walls to a minimum (no redundancy) provides no room for substandard workmanship. This puts a high demand on diaphragms, shear walls, and connections.

Figure 1.1 Irregular shaped structure.

- Splitting, using smaller nails than specified, using different species of wood than specified, over-driving nails, and slack in light-gage metal straps.

Edward F. Diekmann provided an interesting note in his engineering module "Design of Wood Diaphragms"[6] regarding a misconception for the requirement of wood diaphragms and shear walls. He noted that it was unfortunate that interest in diaphragms and shear walls was developed primarily on the West Coast, which gave rise to an impression that they were required only because of the earthquakes that occur in that region. Because of this misconception, it appears that a large number of wood framed structures in many other regions of the country are apparently erected without thought as to how they are to be braced against wind forces. Another problem noted in ATC-7-1[7] was the lack of complete load paths and detailing. Engineering has become a highly competitive business. ATC-7-1 noted "Nothing is more discouraging to the conscientious engineer endeavoring to deal with lateral forces with all the detailing requirements on diaphragms and shear walls than to contemplate the absence of attention paid by some of his fellow engineers to the most basic shear transfer problems. It is a sobering experience to see structural plans for a wood framed apartment complex without a single wood-framing detail and to realize that you were not given the job because your proposed fee was too high." It is hoped that the information provided in this book will provide clarity to the importance of complete load paths and designs.

The diaphragm and shear wall layout shown in Fig. 1.2 is a good example of structures currently being designed and built. The code and standards definitions should be

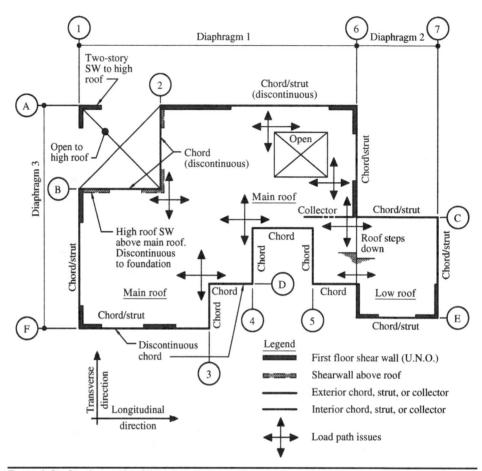

Figure 1.2 Continuous load path issues.

carefully reviewed for applicability to each irregularity discussed for this structure. In the transverse direction, two diaphragms exist. The main diaphragm is supported by the first-floor shear walls along grid lines 1 and 6. The low roof diaphragm is supported by the shear walls located at grid lines 6 and 7. The main diaphragm has multiple discontinuities and irregularities within the span which must be resolved. Starting at grid line 1A, it can be seen that a two-story entry condition exists, which is typical in many offices or shopping center complexes. The upper level is usually an architectural feature commonly referred to as a pop-up. The shear walls at grid line 1A are two stories in height and support the pop-up roof. The walls at grid line 2 and grid line B also support the pop-up roof but are discontinuous shear walls because they are supported by the main roof and do not continue down to the foundation. The pop-up section should be designed as a second story that transfers its forces as a concentrated load into the main diaphragm. The diaphragm sheathing and framing is often omitted below the pop-up section at the main roof level. Diaphragm boundary members are not allowed at the main diaphragm level at grid line A from 1 to 2 or at line 1 from A to B, due to architectural constraints. This condition creates a horizontal offset in the roof diaphragm in the transverse and longitudinal directions. The offset disrupts the diaphragm chords, creating an offset diaphragm. Because of the offset, a question arises on how to provide continuity in the chord members and transfer its disrupted force across the offset. It also raises a question on how to dissipate the disrupted chord force into the main diaphragm, at grid line 2B. Creating complete load paths to transfer all the discontinuous forces into the main diaphragm can be very complicated and challenging. There are multiple offsets at grid lines C, D, and F between 3 and 6. These offsets also cause a disruption in the diaphragm chords and struts and must also have their disrupted chord forces transferred into the main diaphragm by special means. The large opening in the diaphragm in-line with grid line 5 causes a disruption in the diaphragm web and requires the transfer of concentrated forces into the main diaphragm at each corner of the opening. The opening as well as the multiple offsets reduce the stiffness of the diaphragm. Diaphragm shears will increase at all areas of discontinuities because of the additional shears that are created by the transfer of the disrupted chord forces into the main diaphragm. The low roof diaphragm which is located between grid lines 6 and 7 from C to E is offset vertically from the main diaphragm. The low roof is supported on three sides by shear walls. The diaphragm boundary along grid line C is unsupported unless the boundary element along that line is transferred into the main diaphragm by a vertical collector in bending that extends into the main diaphragm. This example may appear to be extreme to some; however, such structures are becoming more commonplace in current practice and design.

The establishment of a complete load path does not end by providing boundary element along the entire length of the lateral-force-resisting line. It must also include all the connections necessary to make members in the line of lateral-force-resistance act as a unit and transfer the shears and forces from the diaphragm sheathing into the boundary elements, then into the vertical force-resisting elements and finally down into the foundation. The lateral forces must then be transferred safely into the soil without exceeding the soil capacity. The drawings and calculations must be complete and clear so that the engineer can assure that the load paths are complete from the point of application of the loads to the foundation. In addition, to a clearly defined load path, supporting calculations and drawings should be developed to assist the plans examiner in an efficient and accurate review of the documents. The drawings provide the contractor with the details necessary to construct the structure per the design, and the building inspector to verify compliance with the construction documents. Clear and thorough documentation of load paths can save countless hours of misunderstanding, construction errors, and revisions. In some cases, those errors may not even be realized because of the lack of clarity and the final product may not meet the intended design.

Figures 1.3 through 1.6 provide examples of maintaining complete load paths through various framing configurations. Figure 1.3 shows sloped roof trusses connected to exterior wood bearing walls. In configuration A, the diaphragm shears are transferred into full depth solid blocking installed between the trusses, then into the double top plate of the wall by shear clips and/or toenailing, then from the double top plate into the wall sheathing. As can be seen in Fig. 1.4, a common complaint for this configuration is the difficulty of providing for ventilation through the solid blocking. Figure 1.4 is a photograph of framing that is similar to configuration A. The photo shows that the blocking between the trusses is not the full depth of the trusses at the point of bearing. This is often done to provide roof ventilation. However, this prevents the installation of the boundary nailing for the diaphragm because the nails at this location cannot transfer shear across the air gap. Boundary nailing is required by code for all engineered diaphragms, as verified by diaphragm testing and the principles of mechanics and must

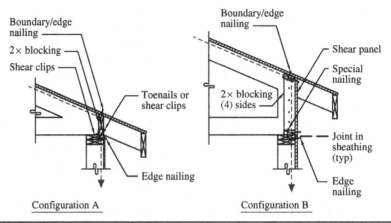

FIGURE 1.3   Example complete load paths—roof sections.

FIGURE 1.4   Photo of incomplete load path—blocking issue.

be installed. In addition, the roof sheathing in Fig. 1.4 is not supported by the blocking at the wall location to prevent its buckling under lateral shear forces (in-plane axial force applied to the sheathing). It should also be pointed out that code does not allow the diaphragm sheathing to act as the boundary element or to act as the splice for a boundary element. Under the blocking configuration shown in the figure, boundary nailing cannot be installed to transfer the shear forces into the wall top plate so the diaphragm shears would have to be transferred into the truss top chord at the first nail located back from the blocking. The effective nail spacing at that location would be one nail at 24″ o.c. parallel to the shear wall. It should be obvious that this excessive nail spacing would not be capable of resisting the applied shears. Therefore, a complete load path does not exist and the transfer of diaphragm shears into a boundary element cannot be obtained. The structural detailing for the shear transfer at this location was correctly shown on the drawings but was ignored or overlooked during construction.

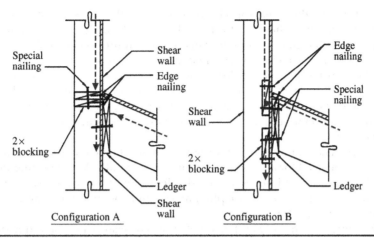

FIGURE 1.5 Example complete load paths—low roof sections.

There have been many debates on the necessity of full depth blocking at exterior wall lines, especially in areas of low to moderate seismicity. Some stakeholders are pushing for partial height blocking or to eliminate the blocking entirely. These efforts are not consistent with the goal of providing a complete lateral load path and should be scrutinized for rationality and substantiation. Attempting to transfer diaphragm

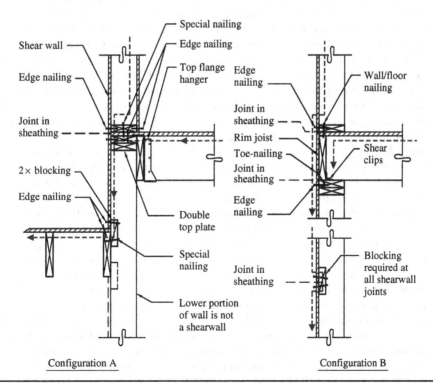

FIGURE 1.6 Example complete load paths—floor/roof sections.

shears through the truss top chord would put the truss top chord in cross-grain bending at the truss heel joint if partial-height blocking was used and could potentially pull off the gang-nail plate. This type of failure has been observed in the field. In the case where blocking is eliminated, the truss would have to transfer the shears into the wall top plate by roll-over action. Substantial testing for gravity plus roll-over forces or gravity plus cross-grain bending forces should be evaluated on gang-nail trusses before serious consideration can be given to reducing the full depth blocking requirements. Also, trusses would have to include these rotational forces in their design.

The APA has conducted tests on sloped mobile home roof diaphragms, which demonstrated the need for a complete diaphragm load path. An interesting mode of failure was that the gang-nail plates at the ridge line joint of the trusses were pulled apart by opposing shear forces in the diaphragm sheathing because blocking was not provided for the sheathing at the ridge joint.

In configuration B of Fig. 1.3, a prefabricated shear panel consisting of 2× members with plywood sheathing replaces the solid blocking. The load path is the same regardless of the blocking material installed, transferring the diaphragm shears down to the top plate. Larger vent holes can be cut in the shear panel sheathing to allow for ventilation. This condition accommodates the use of deep heel trusses that have become popular for creating roof overhangs and for energy purposes allowing deeper insulation. Recent editions of the IBC and IRC now show a deep heel truss condition in Figures 2308.6.7.2(2) and R602.10.8.2(3), respectively.

Figure 1.5 shows the condition at walls where low roofs frame into the walls at mid-height of the studs. The lateral load path is defined by the dashed arrows. Configuration A shows the condition where the ledger is attached directly to the wall sheathing. The exterior wall sheathing can be terminated at the low roof elevation if the wall is not acting as a shear wall. If the wall is acting as a shear wall, the sheathing should be installed full height of the wall. The lower roof shears are transferred from the diaphragm sheathing into the ledger and double wall blocking, into the sheathing and then down to the foundation. Condition B shows the condition where the wall sheathing is disrupted at the interface of the wall and low roof ledger. The shear from the upper wall sheathing is transferred into the blocking, into the ledger, back into the lower blocking and then back into the wall sheathing. The low roof shears are transferred into the ledger, then into the lower blocking and wall sheathing.

Figure 1.6 shows two floor framing sections. Joints in the wall sheathing can occur at many locations in the floor framing area. There are no guarantees where these joints will occur unless the joint locations are specifically detailed in the drawings. The nailing required to establish a complete load path should be based on the worst-case scenario assuming that the joints will fall at the locations shown in configuration B; and that, the sheathing will not be lapped onto the rim joist and blocking, unless specifically detailed on the drawings. Configuration A shows a condition where the upper floor joists are hangered off the wall in a semi-balloon framing condition instead of platform framing as shown in configuration B. The wall shear is transferred from the upper wall into the wall double top plate below and then back into the outer sheathing. The floor shears are transferred from the floor sheathing into the double top plate of the wall below, then back out into the wall sheathing. The low floor or roof sheathing is nailed to the edge joist. These shears are then transferred from the lower floor or roof through the blocking that is nailed to the edge joist, then down into the wall sheathing below as required. Configuration B represents the common method of platform framing a floor onto a bearing wall. Since sheathing joints usually occur at the upper wall bottom plate and lower wall top plate locations, the upper wall and floor shears must be transferred by nailing into the rim joist, then down into the lower wall top plate by toe nailing or shear clips. All of these figures should callout the complete nailing, clip, splice straps, and blocking necessary to provide a continuous load path. Calculations should be completed to verify the adequate transfer of all forces and shears. A mistake commonly made occurs when a detail is taken from a "Typical Detail Book" and is applied to a set of drawings without verifying that the capacity of the connections will actually meet or exceed the applied shears. It is sometimes assumed that the detail will work for any load that is applied. When applying a typical detail developed in-house or by others, it is the responsibility of the engineer to understand the load capacity of the detail and to be able to recognize when the capacity is exceeded.

Figure 1.7 is a typical interior or exterior shear wall elevation along a line of lateral-force resistance. In this case, the load path under discussion is the transfer of shears and lateral forces from the roof diaphragm down to the soil. The continuous rim joist or wall top plates and beams can be used as the diaphragm boundary elements (drag struts). If blocking occurs between the joists in lieu of a continuous rim joist, the diaphragm shears are transferred through the blocking into the drag members and shear wall at the wall top plate level. All the nailing, clips, splice straps, and blocking necessary to provide a complete continuous load path along the drag line must be detailed and installed correctly. The wall shears are transferred through the wall, into the bottom plate, and then into the foundation by anchor bolts. The wall overturning forces are resisted by dead loads and/or hold downs that are embedded into the foundation. The foundation

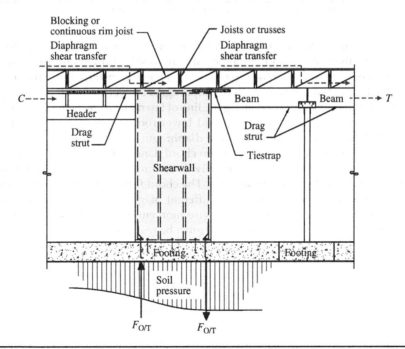

FIGURE 1.7 Complete load path to foundation—roof at same elevation.

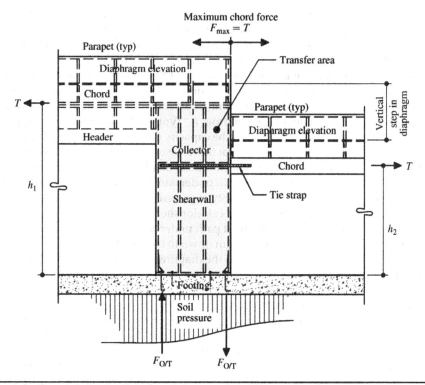

Figure 1.8 Complete load path to foundation—roof at different elevations.

must be designed to have the strength necessary to transfer all these forces plus gravity loads into the soil without exceeding the allowable soil bearing pressures. The load path is not complete until the forces are completely transferred into the soil. Figure 1.8 shows the condition where the roof diaphragms are vertically offset. If drag forces are applied in the same direction along the line of lateral resistance, the shear and overturning moments caused by the upper and lower roofs are additive to the transfer shear wall. When loads are applied to the diaphragm perpendicular to the wall line, the boundary members act as a diaphragm chord. Under this loading condition, a transfer wall is required to connect the vertically discontinuous chords. Figure 1.8 shows chord forces applied to the wall at the offset. The chord forces are equal in magnitude but act in opposite directions, occurring at different heights. This causes a counterclockwise moment that is larger than the clockwise moment. A net moment will result acting in the counterclockwise direction, which must be resisted by a hold down anchor.

Assuming no dead load (for simplicity):

$$M_1 = T(h_1)$$
$$M_2 = T(h_2)$$
$$F_{O/T} = \frac{M_1 - M_2}{L_{wall}}$$

The actual force transfer through this wall is somewhat complicated and will be addressed in detail in Chap. 7.

It is important to provide documents that can verify that a complete load path has been provided. Experience has shown that lateral load paths are occasionally framed incomplete in the field because details have not been completely or clearly defined in the drawings, assuring that a complete lateral system can be provided. Structural drawings can vary widely from region to region and from firm to firm depending on the prevalent lateral force in the area and individual office practices. Although not specifically required, the lateral drawings should include a simple key plan to show the diaphragm boundaries and required nailing, all drag struts/collector locations, special nailing requirements, shear walls and/or frame locations, and necessary structural sections. Defining collectors on the drawings assures that these members are highlighted as important lateral elements requiring special load path transfer of diaphragm shears down into the collector beam or truss. Whenever a preengineered truss is used as a strut or collector, it is also important to add the truss elevation with the applied forces and its location on the truss on the plans, so that the truss manufacture can properly design the truss for those forces. Grid lines are often convenient for ease of communication over the phone or in written forms to identify specific locations. Wall elevations should be provided when walls contain openings that require special force transfer connections, anchoring, or special nailing requirements.

1.7 Methods of Analysis

The examples in this book provide methods of analyzing complex diaphragms and shear walls. Each chapter contains one or two examples that demonstrate the method or methods being discussed in the chapter. Problems are located at the end of the chapter that are variations of the examples, each of which have a special lesson or point of interest. As shown in those examples, the relocation of a single shear wall can significantly change the distribution of forces through a structure. Unless noted otherwise, the lateral loads used in the examples are generalized and can represent wind, seismic, or soil loads at either an allowable stress (ASD) or strength (LRFD) design level. The applied loads are assumed to be the results of the individual's generation of forces to the structure, which are appropriately factored up or down to fit the load combination and design method being used. Some of the examples are carried out using more decimal places than would normally be used in common practice. The intent is to provide better closure of the diaphragm chord and collector force diagrams.

The typical sign convention used in this book is shown in Fig. 1.9. One-foot by one-foot square sheathing elements are used to show the direction of the shears acting on the sheathing elements or collectors and chords. The figure shows typical positive and negative sheathing elements when loaded in the transverse and longitudinal directions. The figure also shows representative portions of force diagrams for collectors, struts, and chords. For transverse loading, a positive force is drawn above the line representing tension. A negative force is drawn below the line representing compression. In reality, it does not make a difference which side of the line the forces are drawn as long as the construction of these diagrams is consistent throughout the analysis. This is because the force in a member will change from tension to compression upon the reversal of the direction of the loads.

As a prerequisite, the reader should have a working knowledge of the analysis and design of simple rectangular diaphragms, simple shear walls, and should know how to calculate wind and seismic forces to structures. The methods for calculating wind and

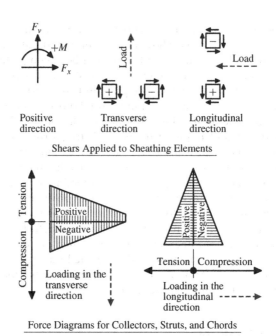

FIGURE 1.9 Standard sign convention.

seismic forces are not included in this book. A cursory review on the analysis of simple diaphragms and shear walls is presented as an introduction or refresher before reviewing the more advanced examples contained in this book.

The examples and methodology presented should be verified by the reader for its accuracy and applicability prior to use on a project.

1.8   References Containing Analysis Methods for Complex Diaphragms and Shear Walls

1. *Design of Wood Diaphragms*[6] by Edward F. Diekmann, *Journal of Materials Education*, University of Wisconsin, Madison, August 1982

This paper is one of a set of modules on Wood: Engineering Design Concepts that was prepared for the Fourth Clark C. Heritage Memorial Workshop at the University of Wisconsin, Madison, August 1982, published seriatim in the *Journal of Materials Education*. The paper is a very important document, which provides a fairly comprehensive coverage of basic diaphragm and shear wall analysis and connection design. Of greater importance and relevance to this book are the presentations on the following:

- Diaphragm continuity issues

- Diaphragms with openings

- Diaphragms with horizontal offsets (notches)

- Diaphragms with vertical offsets
- Collector analysis at diaphragm discontinuities
- Transfer of disrupted chord forces within the diaphragm
- Shear walls with openings

   The examples are clear and easy to follow. It is surprising, and at the same time unfortunate, that this paper was not provided in a major publication where it could have been more readily accessed by the engineering community.

2. *Diaphragms and Shear Walls*[8] by Edward F. Diekmann, S.E., *Wood Engineering and Construction Handbook*, Chapter 8, 3rd ed.

   Chapter 8 of the book is devoted to simple diaphragms and shear walls. Most of the material on basic diaphragms and shear walls that was included in reference 1 has been repeated here. Diaphragms with openings covered in reference 1 have also been included in the chapter. The information provides a comprehensive coverage of simple systems but is limited with regard to complex systems.

3. *ATC-7-1 Proceedings of a Workshop on Design of Horizontal Diaphragms*[6] 1980

   The objective of the workshop was to evaluate current knowledge and practice in the design and construction of horizontal wood diaphragms, examine the needs and priorities for immediate and long-range research required to minimize gaps in current knowledge, to improve current practice, and to provide state-of-the-art practice papers for the development of a guideline for the design of horizontal wood diaphragms. The document included several case studies on (1) the field performance of wood diaphragms subjected to wind and seismic loading conditions, (2) the performance of mechanical fasteners in wood diaphragms, (3) analysis methods for horizontal diaphragms, (4) a very basic discussion of irregular shaped diaphragms, and (5) details for the transfer of forces from the diaphragm to the vertical force-resisting elements. A final list of some of the recommendations developed from the workshop included the following:

- Develop mathematical models and analysis methods to predict the inelastic response of diaphragms
- Develop a simplified analytical model to predict deflections of diaphragms
- Perform additional dynamic tests using either cyclic loads or input from realistic earthquake motions
- Determine what, if any, size effects exist in the performance of diaphragm tests
- Determine, by tests, distances required for ties and collectors to spread loads into the diaphragm
- Evaluate, by tests, current assumptions associated with sub-diaphragms
- Determine the effects of the size and location of openings on the force distribution and deformation of diaphragms
- Determine the necessity of code enforced aspect ratios

4. *ATC-7 Guidelines for the Design of Wood Sheathed Diaphragms*[5] by the Applied Technology Council, Berkeley, California, September 1981

The guideline was prepared by H.J. Brunnier Associates and guided by an advisory panel comprised of Noel R. Adams, Edward F. Diekmann, Byrne Eggenberger, Ronald L. Meyes, Roland L Sharpe, and Edward J. Teal.

The document was considered to be the state-of-the-art at the time of publication. Most of the information contained in the guideline continues to be of value and relevance today. Twelve design examples are included in the guide, in addition to discussions on the topics listed below. Some of the concepts are not fully developed; however, there is a considerable amount of information contained therein for the novice and the experienced professional.

- Basic diaphragm discussions
  a. Basic components and stresses
  b. Girder analogy
  c. Truss analogy
  d. Moment couple series
- Diaphragms with openings
- Continuous diaphragms
- Sub-diaphragms
- Irregular diaphragms
- Diagonal and straight sheathed diaphragms
- Diaphragm deflections
- Load transfer through the diaphragm

5. *APA Research Report 138*[9] by John R. Tissell, P.E. and James R. Elliott, P.E., Technical Services Division

The report included eleven tests of diaphragms on panelized systems, high load diaphragms, diaphragms with openings, field glued diaphragms, and diaphragms with framing spaced at 5 ft. Of particular interest is Appendix E, which provided an example on the analysis of a diaphragm with openings. The example was based on the design method described in ATC-7, which was developed by Edward F. Diekmann, S.E.

6. *Design of Wood Structures*[10] by D.E. Breyer, J.F. Fridley, D.G. Pollock, and K.E. Cobeen

This book is perhaps the best known and most widely used book on the design of wood members, connections, diaphragms, and shear walls. It provides a very comprehensive coverage on the design of simple wood structures.

7. *2018 IBC Structural/Seismic Design Manual, Volume 2*[11] by SEAOC

Volume 2 provides one of the most comprehensive and state-of-the-art coverages on the analysis and design of light framed structures, and masonry and concrete tilt-up walls with flexible diaphragms. All of the design examples are located in high seismic zones.

8. *Diaphragms and Diaphragm Chords*[12] by Uno Kula, P.E., C.E., S.E., SEAOC 2001 70th Annual Convention Proceedings

The paper was a short abstract on the analysis of irregular shaped diaphragms containing openings and horizontal offsets of the diaphragm chords. Although the number of examples was limited, the method described for analyzing chord and collector forces was fairly clear and complete. The analytical method was consistent with the method developed by Edward F. Diekmann, S.E., as presented in ATC-7.

9. *Guide to the Design of Diaphragms, Chords, and Collectors*[13]

The guide provides examples for the analysis and design of multistory rectangular diaphragms with cantilever sections and interior openings.

- Four-story concrete diaphragms with concrete collectors
- Three-story wood diaphragms with wood collectors
- Four-story flexible steel deck diaphragm with steel beam collectors
- Four-story concrete filled steel deck diaphragm with steel beam collectors

Clear, thorough examples are provided for the design of the chords and collectors. The guide is based on the 2006 IBC and ASCE/SEI 7-05.

10. NEHRP Seismic Design Technical Brief No. 10[14]—*Seismic Design of Wood Light-Frame Structural Diaphragm Systems: A Guide for Practicing Engineers* by Kelly Cobeen, J. Dan Dolan, Douglas Thompson, John W. van de Lindt

1.9 References

1. International Code Council (ICC), *International Building Code, 2021 with commentary*, ICC, Whittier, CA, 2021.
2. American Society of Civil Engineers (ASCE), *ASCE/SEI 7-16 Minimum Design Loads for Buildings and Other Structures*, ASCE, New York, 2016.
3. American Wood Council (AWC), *Special Design Provisions for Wind and Seismic with Commentary (SDPWS-21)*, Leesburg, VA, 2021.
4. International Code Council (IRC), *International Residential Code, 2018 with commentary*, ICC, Whittier, CA, 2018.
5. Applied Technology Council (ATC), *Guidelines for Design of Horizontal Wood Diaphragms*, ATC-7, Applied Technology Council, Redwood, CA, 1981.
6. Diekmann, E. F., "Design of Wood Diaphragms," *Journal of Materials Education*. Fourth Clark C. Heritage Memorial Workshop, Wood Engineering Design Concepts, University of Wisconsin, WI, 1982.
7. Applied Technology Council (ATC), *Proceedings of a Workshop on Design of Horizontal Wood Diaphragms*, ATC-7-1. Applied Technology Council, Redwood, CA, 1980.
8. Diekmann, E. F., "Diaphragms and Shear Walls," Chap. 8, *Wood Engineering and Construction Handbook*, 3rd ed., K.F. Faherty and T.G. Williamson (eds.), McGraw-Hill, New York, 1999.
9. APA Research Report 138, Plywood Diaphragms. APA Form E315H, APA—The Engineered Wood Association, Engineering Wood Systems, Tacoma, WA, 2000.

10. Breyer, D. E., Martin, Z., and Cobeen, K. E., *Design of Wood Structures ASD*, 8th ed., McGraw-Hill, New York, 2000.

11. Structural Engineers Association of California (SEAOC), IBC Structural/Seismic Design Manual, Volume 2, SEAOC, CA, 2018.

12. Structural Engineers Association of California (SEAOC), SEAOC 2001 70th Annual Convention Proceedings, Diaphragms and Diaphragm Chords, Uno Kula, SEAOC, CA, 2001.

13. *Guide to the Design of Diaphragms, Chords and Collectors Based on the 2006 IBC and ASCE/SEI 7-05*, Dr. Timothy Mays, National Council of Structural Engineers Association (NCSEA), 2009.

14. National Institute of Standards and Technology (NIST), NEHRP Seismic Design Technical Brief No. 10, *Seismic Design of Wood Light-Frame Structural Diaphragm Systems, A Guide for Practicing Engineers*, U.S. Department of Commerce, 2014.

CHAPTER 2

Diaphragm Basics

2.1 Introduction

The methods for analyzing and designing simple rectangular box systems have been common knowledge for decades. A number of publications and textbooks that have been written to date provide a fairly complete coverage on the topic. However, a natural progression of architectural creativity has taken structures from simple rectangular floor plans to ones that contain horizontal and vertical offsets and other irregularities that complicate load paths. Due to time constraints, the length of undergraduate classes on wood design often limits the coverage of diaphragms to basic simple rectangular systems. The extent of mentoring after graduation varies greatly, and because there are very few books or examples that explain how to analyze complex diaphragms and structures, some individuals may not have an in-depth understanding how complex diaphragms and their components really work. This may cause some engineers to approach the analysis and design of these irregular shaped structures as though they are still rectangular diaphragms. A cursory review of simple diaphragms and their components will be provided here as a base from which to extend the discussion into irregular shaped diaphragms.

2.2 The Basic Lateral-Force-Resisting System

The structure shown in Fig. 2.1 represents a typical bearing wall system, also known as the box system. Lateral forces are resisted by the flexible wood roof diaphragm and light-framed shear walls at the upper level and by the flexible wood floor diaphragm and concrete or masonry shear walls at the lower level. Wind loads are typically transmitted to the roof diaphragm as a uniform load by wind pressure imposed on the vertical studs. Seismic forces are transmitted through the roof diaphragm by inertial forces on the mass of the diaphragm and tributary height of the walls. These loads are resisted by the roof diaphragm which acts like a horizontal beam, or deep girder. Girder analogy, as described in ATC-7,[1] assumes that the flanges of the diaphragm only take tension and compression forces due to bending and the web takes the entire shear. Plywood diaphragm testing has indicated that the girder analogy is an acceptable predictor of the performance of wood sheathed diaphragms; and that, the use of the analogy for the determination of the chord forces is conservative. Figure 2.2 shows the resulting shear and moment diagrams for a uniformly loaded rectangular diaphragm. Plywood diaphragms behave slightly different from shallow beams due to the width of the diaphragm and the jointed construction. The APA noted that it has been shown that shear

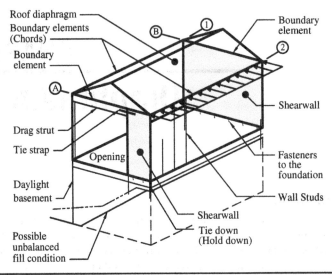

Figure 2.1 Typical box system loaded in the transverse direction.

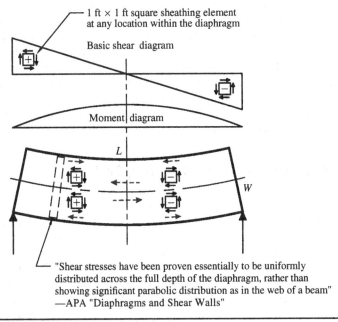

Figure 2.2 Shear distribution into a simple diaphragm.

stresses have been proven to be essentially uniform across the full width of the diaphragm, rather than showing significant parabolic distribution as in the web of a shallow solid beam. Other publications have also noted the same observation, citing the tests were not set up in a manner to accurately record the exact distribution. The assumption of uniform shear distribution is widely accepted in the engineering community as a useful assumption for design. The basic shear diagram for a diaphragm under uniform loads is plotted as shown in the figure, positive shears occurring on the left side of mid-span of the diaphragm and negative shear occurring on the right side. A symbol representing 1×1 ft^2 pieces of sheathing, referred to as "sheathing elements," has been added to the basic shear diagram to show the direction of the unit shear forces (plf) acting on the edges of the sheathing elements. The maximum moment occurs at the point on the shear diagram where the unit shear is zero. A simple rectangular diaphragm uniformly loaded is analyzed as follows:

$$w = \text{uniform distributed load}$$

$$L = \text{length of diaphragm}$$

$$W = \text{width of diaphragm (often referred to as "b or d")}$$

$$x = \text{distance from support under consideration}$$

$$R = \frac{wL}{2} = V = \text{diaphragm reaction at the support}$$

$$v = \frac{V}{W} = \text{diaphragm unit shear at the support}$$

$$V_x = R - wx = \text{shear at any distance } x \text{ from the support}$$

$$M_{\max} = \frac{wL^2}{8} = \text{maximum moment}$$

$$M_x = Rx - \frac{wx^2}{2} = \text{moment at any distance } x \text{ from the support}$$

$$F_{\text{chord}} = \frac{M}{W} = \text{chord force}$$

Code limits the maximum length to width aspect ratios of flexible wood and untopped steel deck diaphragms. Two tables regarding allowable aspect ratios for wood diaphragms and steel deck diaphragms have been included in Fig. 2.3 for the convenience of the reader. The limitations were established based on tests conducted on diaphragms by the APA and by the Steel Deck Institute. The limitations on aspect ratios apply to simple rectangular diaphragms as a whole. Currently, there are no guidelines or limitations for complex, irregular shaped diaphragms. Similar limitations could be applied to the diaphragm as a whole and to each individual section of the diaphragm that has been broken down into simple rectangular sections for the ease of analysis (e.g., segments around an opening in the diaphragm, sections near offsets, transfer diaphragms). The schematic plan in the upper right of Fig. 2.3 demonstrates a condition where all individual sections of the diaphragm and the diaphragm as a whole complies with the allowable aspect ratios listed in the tables, but by observation, this configuration would make the diaphragm far too flexible. The diaphragm must be stiff enough to prevent stability problems in the members supporting the diaphragm, prevent excessive shear and forces within the diaphragm, and limit drift in accordance with code allowable. Sound engineering judgment should be used when checking allowable aspect ratios.

Code requires that all boundaries of a diaphragm be supported by struts, chords, shear walls, frames, or other vertical lateral-force-resisting elements. Although vertical

4.2.2 Maximum Diaphragm Aspect Ratios

Diaphragm Sheathing Type	Maximum L/W Ratio
Wood structural panel, unblocked	3:1
Wood structural panel, blocked	4:1
Single-layer straight lumber sheathing	2:1
Single-layer diagonal lumber sheathing	3:1
Double-layer diagonal lumber sheathing	4:1

Courtesy, American Wood council, Leesburg, VA.

NDS-SDPWS Table 4.2.2
Maximum Diaphragm Aspect Ratios
(Horizontal or sloped diaphragm)

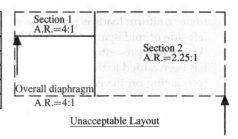

Steel Deck Diaphragms

Flexibility category	F Flexibility factor	Maximum span in feet for masonry or concrete walls	Allowable Aspect Ratio			
			Rotation not considered in diaphragm		Rotation considered in diaphragm	
			Masonry or concrete walls	Flexible walls	Masonry or concrete walls	Flexible walls
Very flexible	More than 150	Not used	Not used	2:1	Not used	1.5:1
Flexible	70 – 150	200	2:1 or as required for deflection	3:1	Not used	2:1
Semi-flexible	10 – 70	400	2.5:1 or as required for deflection	4:1	As required for deflection	2.5:1
Semi-rigid	1 – 10	No limitation	3:1 or as required for deflection	5:1	As required for deflection	3:1
Rigid	Less than 1	No limitation	As required for deflection	No limit	As required for deflection	3.5:1

Courtesy, ICC Evaluation Services, LLC, Whittier, CA.

Table 3 - Diaphragm Flexibility Limitation
ER-3056

See latest Evaluation Report for applicable footnotes and limitations.

Figure 2.3 Allowable diaphragm aspect ratios for flexible diaphragms.

lateral-force-resisting elements are not present at the end of cantilever diaphragms, a boundary member is required to act as a diaphragm chord when loads are applied parallel to the cantilever length. When the structures shown in Figs. 2.1 and 2.4 are loaded in the transverse direction by wind, or by seismic loads as shown in Fig. 2.5, both longitudinal walls at grid lines 1 and 2 act as diaphragm chords that serve as boundary elements. In accordance with the definitions included in Section 2.2 of SDPWS[2] and Section 11.2 of ASCE 7,[3] boundary members and boundary elements are elements that are parallel to the applied load that collects and transfers diaphragm shear forces to the vertical elements of the lateral-force-resisting system. Collectors and struts function in the same manner and are basically interpreted as being the same thing. It is the authors' preference to designate a drag strut as a perimeter boundary element that collects shears from one side only. A collector is an interior boundary element that collects shear from both sides. The full-length shear wall at grid line B is a boundary element which serves to support one end of the diaphragm. The partial length shear wall at grid line A also serves as boundary members and provides the support for the other end of the diaphragm. The wall at grid line A receives uniformly distributed diaphragm shears along the full width of the diaphragm. Since the wall

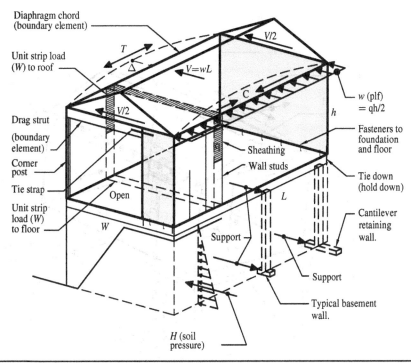

Figure 2.4 Wind load distribution into a diaphragm.

extends only a short distance across the width of the diaphragm, a boundary member known as a drag strut must be installed to support the remainder of the diaphragm and transfer the collected shears back into the top of the shear wall. Both shear walls act like vertical cantilever beams or diaphragms that resist the reactions of the diaphragm and keep the diaphragm from sliding off the structure. The diaphragm forces at the top of the walls cause overturning and sliding forces in the walls, which are transferred into the foundation for stability.

Figure 2.6 shows the elevations of the walls located at grid lines A and B. Boundary members are also required for shear walls, as shown in the figure. Overturning can be resisted by the dead load that is applied to the wall. Additionally, hold downs are required at each end of the shear wall if the dead load is not large enough to resist the overturning moment. Sliding is resisted by nailing or other method of connection of the wall to the floor and/or foundation. The floor diaphragm and its connection to the foundation must be designed to resist the tributary wind or seismic loads from the walls above plus soil loads in accordance with IBC[4] Section 1610 and ASCE 7 Sections 11.7 and 11.8. The foundation must be designed to safely transfer the lateral-forces into the soil without exceeding the allowable soil bearing pressure.

2.2.1 Shear Capacity of Nailed Sheathing

Once the diaphragm and shear wall forces and shears have been calculated, the sheathing and nailing requirements must be determined. Diaphragm and shear wall unit shear capacities are calculated from tabulated values using SDPWS Section 4.1.4. The

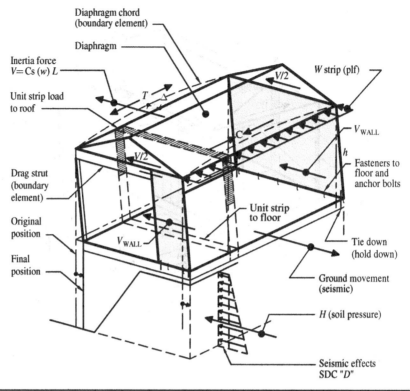

FIGURE 2.5   Seismic load distribution into a diaphragm.

shear capacities provided in SDPWS Tables 4.2A through 4.2D for diaphragms and Tables 4.3A through 4.3D for shear walls are nominal unit shear capacities, which must be reduced for ASD or LRFD unit shear capacities. Prior editions of the tables broke the nominal shear values down into two columns, one for seismic (column A) and one for wind (column B). The 2021 SDPWS has combined seismic or wind design into a single nominal shear value, typically equal to the older wind nominal value. ASD and LRFD capacities are determined by reduction factors as noted in Section 4.1.4.1 for seismic and Section 4.1.2 for wind.

For seismic design of diaphragms and shear walls, the ASD allowable shear capacity shall be determined by dividing the nominal shear capacity by the ASD reduction factor of 2.8 and the LRFD factored shear resistance shall be determined by multiplying the nominal shear capacity by a resistance factor of 0.50. For wind design of diaphragms and shear walls, the ASD allowable shear capacity shall be determined by dividing the nominal shear capacity by the ASD reduction factor of 2.0 and the LRFD factored shear resistance shall be determined by multiplying the nominal shear capacity by a resistance factor of 0.80. No further increases are permitted. Further decreases may be required per the footnotes of the tables, particularly when using framing other than Doug-Fir Larch or Southern Pine with a moisture content equal or less than 19 percent at the time of construction.

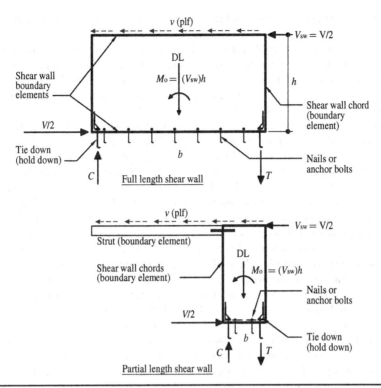

v (plf)

$V_{sw} = V/2$

DL

$M_0 = (V_{sw})h$

Shear wall boundary elements

h

Shear wall chord (boundary element)

$V/2$

Tie down (hold down)

b

Nails or anchor bolts

C

Full length shear wall

T

v (plf)

$V_{sw} = V/2$

Strut (boundary element)

DL

Shear wall chords (boundary element)

$M_0 = (V_{sw})h$

Nails or anchor bolts

$V/2$

b

Tie down (hold down)

C

T

Partial length shear wall

FIGURE 2.6 Shear wall boundary elements.

Occasionally, the question is asked if a higher shear can be obtained by using a thicker sheathing. SDPWS commentary notes that the use of "minimum nominal panel thickness" in the tables is to accommodate use of the tabulated nominal unit shear capacities. Structural panels of greater thickness can be used provided nails with the prescribed dimensions are used and that bearing length of the nail into framing exceeds the prescribed minimum bearing length. This allowance recognizes that greater structural panel thickness will also develop the strength of the prescribed nailing upon which the nominal capacities are based. However, a greater shear capacity will not be developed because the nails control the shear capacity by bending and nail slip.

2.3 Load Distribution into a Diaphragm

Wind pressure and seismic forces are the primary lateral forces that are applied to roof and floor diaphragms and their supporting shear walls. IBC Section 1609.1.1, with some exceptions, requires the determination of wind loads to be in accordance with ASCE 7 Chapters 26 through 30. Any method can be used provided the method is applicable to irregular shaped structures and separates the wind loads onto the windward, leeward, and side walls of the building and properly accesses the internal forces in the main wind-force-resisting system (MWFRS). The wind pressure on the sheathing is transferred uniformly into the supporting studs as shown in Fig. 2.4. The combined internal and external pressures multiplied by half of the stud height (qh/2) plus the wind pressures acting on the roof surface apply a uniform strip load (plf) to the roof and floor diaphragm.

Seismic forces are applied to the diaphragms of a structure when the ground moves laterally under the structure, causing inertia forces proportional to the mass of the structure to resist the resulting acceleration, as shown in Fig. 2.5. For a single story, the seismic force applied to the roof diaphragm and shear walls is usually based on 1-ft wide strips, in which the seismic weight consists of the dead load of the upper half of each wall that occurs across the section plus the roof dead load of the unit strip. The seismic weight of the unit strip is multiplied by the seismic response coefficient, C_s, and is applied as a uniform lateral load to the roof diaphragm. For multistory buildings, the diaphragm inertial design forces developed in the roof or floor diaphragms are calculated by ASCE 7 Section 12.10.1.1, Eq. 12.10-1. The dynamic response of multistory buildings is a complex multi-modal response. The diaphragm design force for a floor, F_{px}, can be higher than the vertical distribution of seismic forces F_x, because the peak diaphragm acceleration does not necessarily coincide with the peak shear in the story below. The in-plane seismic shear resulting from the self-weight of end walls should be included when calculating the end wall overturning. This in-plane load is often ignored for light-framed walls; however, exterior walls can have significant self-weight. The seismic weight tributary to the floor diaphragm is also broken down into 1-ft wide strips, which consists of the dead load of the lower half of the upper walls across the section plus the floor dead load of the unit strip. If half of the weight of the walls above the floor are included in the seismic weight of the level above that under consideration, then half, not all, of the weight of these walls above need to be included in the seismic weight. This seismic diaphragm design forces calculated per ASCE Section 12.10.1.1 are then applied as a uniform lateral load to the floor diaphragm.

The distribution of wind and seismic forces into and out of the diaphragm have identical load paths. Soil loads applied to the floor diaphragm are commonly ignored at the first floor if the backfill acts at full height on opposing walls of the foundation, creating a balanced condition, with the logic that the diaphragm is restrained from movement. However, for conditions of unbalanced backfill or first floors with daylight basements loaded on one side only, the soil pressures must also be applied to the floor diaphragm in addition to the wind and seismic forces.

Typically, two types of foundation walls are used for below grade walls, cantilever retaining walls and typical (standard) basement walls as shown in Fig. 2.4. The stability of cantilevered retaining walls is provided by the weight of the soil over the extended footing (heel) and increased soil pressure on the interior (toe) side of the footing. The weight of the soil pushing down on the heel of the footing resists overturning of the wall. Out-of-plane lateral support at the top of a cantilever wall is not required and transmission of soil pressures into the diaphragm does not occur. In contrast, the standard basement wall has a narrow continuous footing at the base of the wall that cannot resist overturning due to the small projection of the footing edges beyond the face of the wall. Stability of this wall is provided by the lateral support of the floor diaphragm at the top of the wall and by the concrete slab on grade at the bottom of the wall. This type of wall is a restrained wall that is designed for at-rest soil pressures, except as noted below in high seismic areas. The force applied to the floor diaphragm in this case would be equal to the reaction of the upper floor wall studs caused by wind pressure or seismic forces plus the reaction of the basement wall from the soil pressure, or the seismic strip load plus the reaction of the basement wall from the soil pressure. Additionally, for seismic design categories (SDCs) D through F, dynamic lateral soil pressures because of earthquake motion must also be applied in accordance with ASCE 7 Sections 11.8.3 and 12.1.5. As an example, assume the following:

Given: Assume wind controls the design.

The addition of dynamic soil pressures is not required.

$h_1 = 10$ ft upper story plus 1 ft of projected area for the width of the floor
$h_2 = 9$ ft basement
$p = 18$ psf, wind pressure
$q = 65$ psf, soil equivalent fluid pressure (at-rest)

Uniform load to the diaphragm from the soil and wind pressure:

$$w = p\left(\frac{h_1}{2} + 1\right) + \frac{q(h_2)^3}{6h_2} = p\left(\frac{h_1}{2} + 1\right) + \frac{q(h_2)^2}{6}$$

$$= 18\left(\frac{10}{2} + 1\right) + \frac{65(9)^2}{6} = 985.5 \text{ plf}$$

This is a significant load which would make it difficult to provide a reasonable diaphragm design. The ASD of the connections for shear transfer from the diaphragm to the foundation walls must include the load duration factor of $C_D = 0.9$, because soil pressure is a permanent load, in accordance with the NDS.[5,6] The shear per linear foot in the diaphragm and connections will be very high given the calculated load in the example unless special detailing and framing is considered.

Many structures have interior shear walls which break the diaphragm into separate sections or multiple spans. Interior shear walls are usually added when the allowable aspect ratio of the diaphragm is exceeded, when the demand on a wall line is too high, or when additional redundancy is desired. Traditionally, the approach used to analyze this condition has been to treat each span as a simply supported diaphragm, ignoring any continuity of the flange at the interior supports. History has shown that diaphragms have performed satisfactorily using the simple span beam analogy. Figure 2.7 shows a single and a multiple span diaphragm. Based on the simple span diaphragm analogy where the diaphragm is idealized as flexible, the force to each wall line will be distributed in accordance to its tributary width as shown in the figure.

Single-span diaphragm:

$$R = \frac{wL_1}{2} = V_{SW1} = V_{SW2}$$

Two-span diaphragms:

$$V_{SW1} = \frac{wL_1}{2}$$

$$V_{SW2} = \frac{w(L_1 + L_2)}{2}$$

$$V_{SW3} = \frac{wL_2}{2}$$

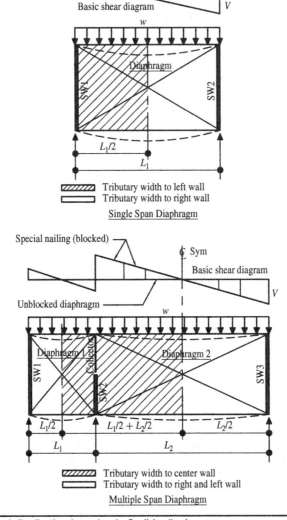

FIGURE 2.7 Load distribution into simple flexible diaphragms.

Although the simple span analogy for multiple-span diaphragms is widely used and accepted, there have been a significant number of discussions regarding the need to consider the possibility of continuity across the interior support locations. Diekmann[7] noted that the assumption that flexible diaphragms spanning between multiple supports can be adequately analyzed by ignoring any continuity and treating each segment as a simple beam. This is a reasonable assumption provided that no chord continuity exists at the interior support. In reality, continuity in the chord must exist to provide continuity in the boundary member under longitudinal loading, which would be designed to resist axial forces associated with drag elements. References 6, 7, 8, and

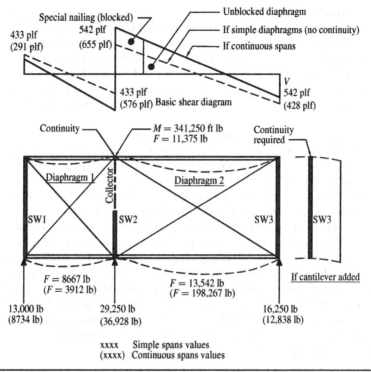

FIGURE 2.8 Load distribution into flexible diaphragms assuming chord continuity.

9 all briefly discuss the issue of continuity; however, nothing has been definitively presented as a guide for applying this to an analysis. A reasonable question to ask would be, if continuity must be considered at cantilevers as shown in Fig. 2.8, why shouldn't some degree of continuity be considered at interior supports, especially if a semi-rigid diaphragm exists? Chord continuity at the interior support would increase the diaphragm reaction to that wall. Conservatively, the analysis of multiple-span diaphragms could be enveloped with the maximum governing values used to assure adequacy of the design. ATC-7 notes that the distribution of shears to the various vertical lateral resisting elements and the degree of continuity at the interior supports are subject to various interpretations; and that, there is no logical reason for not considering the diaphragm to have some degree of continuity. Presently, the issue of chord continuity at interior supports is left up to the discretion of the design professional. Figure 2.8 assumes full continuity of the diaphragm at the interior support of the two-span diaphragm shown in Fig. 2.7. The two-span diaphragm is analyzed first as two simple spans, and then as a continuous span as shown in the figure. Assumed chord continuity at the interior support causes the force at the interior wall to increase by 26 percent for this example and reduces the force to the outer walls. The inclusion of chord continuity can be useful when there is a need to reduce the load to an end wall that has a marginal shear wall capacity. However, careful consideration of diaphragm flexibility/stiffness should be given to assure the distributions of forces within the diaphragm can be realized as assumed. Because diaphragms deflections are significantly from shear

deformations, which are ignored in classic Euler beam theory, more accurate representations of multiple-span diaphragms use Timoshenko beam modeling or detail finite element modeling. If a cantilever section is added to the diaphragm to the right of SW3, full continuity must exist to maintain the stability of the cantilever section and semi-rigid or rigid diaphragm stiffness is required. Chord splices in areas of continuity must be detailed or called out on the drawings; otherwise, there is no guarantee that continuity can be achieved. The absence of detailing could lead to a failure in the diaphragm chord.

Several of the examples presented in this book analyze complicated diaphragm layouts by breaking them down into smaller more manageable sections, and then by analyzing each section and its effects on the adjacent sections individually. To do so, the uniform load distributed into each section is calculated in proportion to the width of the section. This concept can be seen in Sec. 4.7 and in Chap. 8. For wind loading, the internal pressures are added to the windward and leeward pressures and are then combined as a unit uniform load which can be applied to the windward side of the diaphragm. At areas where diaphragms contain openings, it can be advantageous to apply the pressures on both sides of the diaphragm opening in accordance with their respective windward and leeward pressure distribution. Examples showing this type of distribution can be seen in Chaps. 5 and 8. Applying the separate wind pressures to both sides of the opening will reduce the shears and nailing requirement in both sections. Calculation of the seismic unit strip loads in the individual sections is in accordance with the width of each section as previously described for wind. This method of applying forces to a diaphragm with large openings can be seen in Chap. 5.

2.4 Diaphragm Deflection

The method of calculating the deflection of a simple rectangular diaphragm has been established and refined over the past few decades by extensive testing and research conducted by the American Plywood Association (APA).[8] The building code sections and SDPWS[6] sections addressing diaphragm deflection require that deflection be checked to verify that the diaphragm and any attached element will maintain their structural integrity under the design loads. It is interesting to note that this requirement has been in the code since the 1940s. ATC-7 noted that in this regard, there is an inconsistency between the code requirement and the standard of practice. It has become common to assume that diaphragms, regardless of their shape, still act like simple rectangular diaphragms. It is often believed that if the allowable aspect ratios are maintained, deflections will be within acceptable limits and calculations are not necessary. For regular diaphragms this may be a reasonable assumption; however, it would be prudent to verify the deflection of irregular shaped diaphragms due to their non-uniform shape, areas of discontinuities and changes in stiffness, and the demand placed on their performance. It is also prudent to look at the deflections of very long spanning diaphragms where the mid-span deflection may be much larger than the deflections at the supporting lines. It is important to know that ASCE 7, Section 12.3.1 requires that the structural analysis shall consider the relative stiffness of the diaphragm and vertical elements of the seismic-force-resisting system. While many sheathed wood-frame diaphragms can be assumed flexible in ACSE Section 12.3.1.1, other diaphragms may require a more careful comparison of the relative stiffnesses of lateral-force-resisting elements. This cannot be done without determining the deflections.

2.4.1   Simple Rectangular Diaphragm Deflection

The method of calculating deflection for simple rectangular diaphragms is covered in SDPWS and nearly every publication regarding diaphragm design. The four-term equation found in the SDPWS commentary and several APA publications is based on tests of simple span rectangular diaphragm of constant width, which are uniformly loaded and have uniform nailing with all edges blocked throughout its entire length:

$$\Delta = \frac{5vL^3}{8EAW} + \frac{vL}{4Gt} + 0.188Le_n + \frac{\Sigma(\Delta_c X)}{2W} \qquad \text{SDPWS Eq. C4.3.2-1}$$

where
A = Area of chord (in²)
W = Diaphragm width (ft)
E = Elastic modulus of chords (psi)
e_n = Nail deformation (in)
Gt = panel rigidity through the thickness (pounds per lineal inch, pli, of panel width)
L = Length of diaphragm (ft)
V = Maximum shear due to design loads (plf)
Δ = Calculated deflection (in)
$\Sigma(\Delta c X)$ = Sum of individual chord splice slip values on both sides of the diaphragm, each multiplied by its distance to the nearest support

The first term of the equation represents the bending deflection, the second term the shear deflection, the third term is for the contribution of nail slip, and the last term is for chord splice slip. The deflection calculated assumes that the maximum deflection occurs at mid-span. The deflection of an unblocked diaphragm is approximately 2.5 times the deflection calculated above if the spacing of the framing is 24″ o.c. or less, and 3.0 times if the framing spacing is more than 24″ o.c., as noted in APA "Diaphragms and Shear Walls, Design and Construction Guide."[8] The deflection of the diaphragm will significantly increase if the nail spacing increases the closer it gets to the centerline of the diaphragm, as is a common design optimization. The constant of 0.188 in the nail slip term of the equation applies to diaphragms with uniform nailing only. ATC-7 and APA Diaphragm and Shear Wall Construction Guide, Appendix C recommends that whenever a variation in nail spacing occurs, the constant is increased in proportion to the average load on each nail with non-uniform nailing compared to the average load that would be present if uniform nail spacing had been maintained (see Fig. 2.9).

$$0.188\left(\frac{V'_n}{V_n}\right)$$

where
V'_n = the average non-uniform load per nail
V_n = the average uniform load per nail.

For the values shown in the Fig. 2.9:

$$0.188\left(\frac{\text{Area}_2 + \text{Area}_3}{\text{Area}_1}\right) = 0.188\left(\frac{0.5(100 + 75)20 + 0.5(125)50}{0.5(100)70}\right) = 0.262''$$

In recent editions of the SDPWS, a simplified three-term version of the four-term deflection equation has been created by combining the second and third terms of the equation.

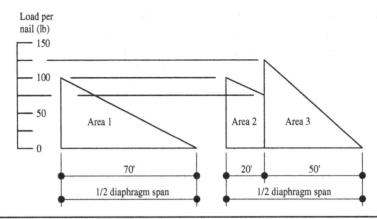

Figure 2.9 Deflection non-uniform nailing. (*Courtesy, APA—The Engineered Wood Association.*)

The intent of this modification is to provide a simpler linear calculation of the deflection in lieu of the non-linear calculation of the four-term equation. A graph comparing the differences between the two equations is shown in SDPWS commentary, Figure C4.3.2. It will also be discussed in greater detail in Sec. 9.6 and Fig. 9.11 of this book.

$$\Delta = \frac{5vL^3}{8EAW} + \frac{0.25vL}{1000G_a} + \frac{\sum(x\Delta_C)}{2W} \qquad \text{SDPWS Eq. 4.2-1}$$

where $W = b$ from the older four-term equation, and

$$G_a = \frac{1.4v_{S(ASD)}}{\dfrac{1.4v_{S(ASD)}}{G_v t_v} + 0.75e_n}$$

      G_a = Apparent diaphragm shear stiffness from nail slip and panel shear deformation.

$1.4v_{S(ASD)}$ = 1.4 times the ASD unit shear capacity for seismic. The value of 1.4 converts ASD level forces to strength level forces for seismic.

2.4.2 Complex Diaphragm Deflection

The diaphragms shown in the Fig. 2.10 are not rectangular in shape and therefore do not directly relate to the standard requirements for code allowable aspect ratios. It is suggested that as a starting point, the allowable aspect ratios should serve as guidance for the size of the main diaphragm as a whole or diaphragm sections individually (e.g., reduced diaphragm sections, transfer diaphragms). Once the basic shape has been reviewed and there are no sections that appear to be too flexible, have lack of strength or stiffness, then deflections design checks can be made. The deflection of offset diaphragms will be greater than that of a rectangular diaphragm, due to the change in stiffness of the diaphragm at the offsets. Deflection should be checked for irregular shaped diaphragms with questionable flexibility or have a Type 3 horizontal irregularity.

     Complex diaphragms require a more involved process to calculate deflections because of several variables. Examine the diaphragms shown in Fig. 2.10. Each of the diaphragm configurations shown have varying degrees of stiffness caused by the different widths of the diaphragm, represented by thickened lines shown below the

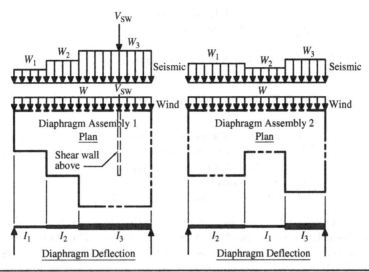

Figure 2.10 Diaphragm deflection irregular shapes.

diaphragms and designated as I_1, I_2, and I_3. Under wind loading, the applied loads are typically uniform along the entire length of the diaphragm, with the exception of a buildup of end zone wind pressure. However, changes in the area exposed to wind, such as changes in the roof height can lead to varying wind load along the length of a diaphragm. It may be common to assume a uniform load, but sometimes when "sharpening the pencil" a non-uniform wind load may be considered. When seismic controls, the uniform loads vary in accordance with the width of the diaphragm. Concentrated loads can also occur in the loading pattern if vertically discontinuous interior shear walls are located above the diaphragm and require the transfer of their shear through the diaphragm into adjacent shear walls below. The procedure for calculating the deflection under non-uniform loads and varying moments of inertia can be accomplished by virtual work to derive the bending and shear deflection, then modifying the other terms of the basic deflection equation in accordance with the configuration of the model as presented in ATC-7.[1] The maximum deflection will not always occur at midspan of the diaphragm. Deflections can now be determined by using available software or by customizing electronic spreadsheets, thereby eliminating the need to resort to lengthy hand calculations. The equations below follow the method used in ATC-7 and U.S.D.A. Forest Service Research Note-FPL-0210 for a single offset diaphragm.[9]

$$\Delta_{TL} = \Delta_B + \Delta_S + 0.188 L e_n + \frac{\sum(\Delta_c X)}{2b},$$

where $\quad \Delta_B = \int_a^b \frac{mM}{EI_1} dx + \int_b^c \frac{mM}{EI_2} dx$, and

$$\Delta_S = \frac{bt}{2GA^2}\left[\int_a^b \frac{wx}{\left(\dfrac{b'}{b}\right)^2} dx + \int_b^c wx\, dx\right] = \int_0^L \frac{vV}{Gt} dx$$

where Δ_B = Bending deflection term
Δ_S = Shear deflection term

$\int_a^b \dfrac{mM}{EI_1}\,dx$ = Bending deflection integrated over the interval from a to b.

$\dfrac{b'}{b}$ = shallower width (depth) of diaphragm divided by full width of diaphragm

$\dfrac{vV}{Gt}$ = multiplication of the area of the shear diagram (due to a general loading) and the ordinate of the shear diagram due to a unit load applied at the desired point of shear deflection.

G = Shear modulus
t = effective thickness of the sheathing

U.S.D.A. Forest Service Research Note-FPL-0210 provides a simplified energy method for calculation of the shear deflection of beams. The overall process is ideal for irregular shaped diaphragms because it addresses the different moments of inertia and varying loads and eliminates the need to derivate complex equations.[10]

2.5   Diaphragm Boundary Elements

Diaphragm boundary elements, chords, drag struts, and collectors are perhaps the most misunderstood elements of a diaphragm. A diaphragm boundary is defined as a location where shear is transferred into or out of the diaphragm sheathing. A boundary element can consist of a drag strut, chords, collector, or another lateral-force-resisting element such as a shear wall, braced frame, or moment resisting frame. All edges of a diaphragm are defined as the boundaries of the diaphragm, which must be supported by boundary elements as described above, except in the case of a cantilever diaphragm. Whenever an interior shear wall line is installed, it breaks the diaphragm down into two separate diaphragms and therefore defines an interior boundary for both diaphragms. If a partial length shear wall is installed, a drag strut or collector connecting to the shear wall is required to extend across the full width of the diaphragm to provide complete support for both diaphragm edges. Figure 2.11 shows two diaphragm configurations, a single-span diaphragm and a two-span diaphragm. In both cases, the diaphragms are loaded in the transverse direction. The top and bottom edges of both diaphragm configurations consist of partial length shear walls and/or a continuous boundary member, which act jointly as the diaphragm chords. The left edge of the single span diaphragm is supported by a full-length shear wall. The right edge is supported by a partial length shear wall and a drag strut. A continuous load path is provided at all boundaries of these diaphragms and are in compliance with the code. The two-span diaphragm is assumed to be two separate simple spans in accordance with accepted engineering practice, designated as diaphragm 1 and diaphragm 2. Diaphragm 1 is supported on the left and right edges by partial length shear walls SW1 and SW2, respectively. Since neither of the walls supporting diaphragm 1 extend the full width of the diaphragm, a collector must therefore be installed to support the remaining edges of the diaphragm. Diaphragm 2 is supported on the left edge by SW2 and its connected collector that also supports the right edge of diaphragm 1. The right edge is

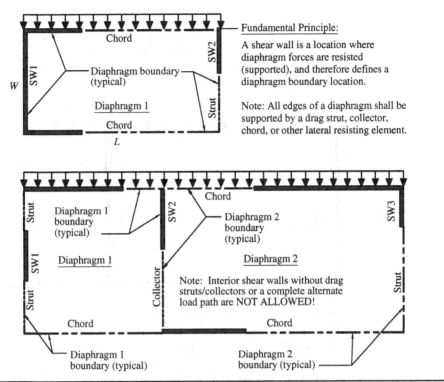

Fundamental Principle:

A shear wall is a location where diaphragm forces are resisted (supported), and therefore defines a diaphragm boundary location.

Note: All edges of a diaphragm shall be supported by a drag strut, collector, chord, or other lateral resisting element.

Note: Interior shear walls without drag struts/collectors or a complete alternate load path are NOT ALLOWED!

FIGURE 2.11 Diaphragm boundary elements.

supported by a partial length shear wall and a drag strut. Both diaphragm sections of the two-span diaphragm are in compliance with code requirements. In the longitudinal direction, both wall lines that support the diaphragm consist of shear walls and drag struts along their entire length. Both diaphragms are in compliance with code in both directions.

Figure 2.12 demonstrates how the collectors coming off shear walls SW1 and SW2 function, showing that they collect the unit transfer shear from the diaphragm sheathing and transfer the accumulated shear forces into the shear walls. The drag strut or collector must be tied to the shear wall to provide a complete continuous load path; otherwise, tearing can occur at that location and the edge of the diaphragm would become unsupported. Unfortunately, partial length interior shear walls are often installed without the addition of a collector. Some have argued that if the total shear from the diaphragm can be developed along the shear wall length, then why is a collector required. This neither complies with code requiring a continuous boundary element nor provides complete support for the entire edge of the diaphragm. Sheathing element symbols are placed at the positive and negative shear areas of the diaphragm, adjacent to shear walls SW1 and SW2, as shown in Fig. 2.12. The shear that is transferred into the shear walls and collectors is equal to and opposite in direction to the shear acting on the edge of the sheathing elements. The shears that are transferred from each side of the collector at the interior support line are acting in the same direction and do not cancel

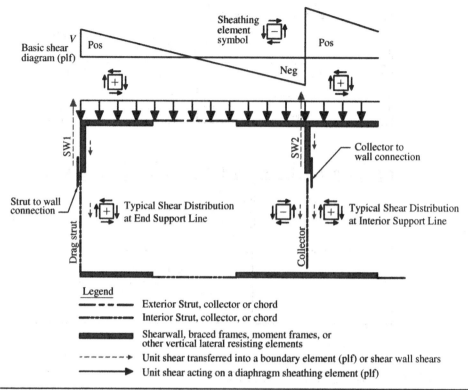

Basic shear diagram (plf)

Sheathing element symbol

Pos   Pos   Neg

SW1   SW2

Strut to wall connection

Drag strut

Collector to wall connection

Typical Shear Distribution at End Support Line

Collector

Typical Shear Distribution at Interior Support Line

Legend

— — — — Exterior Strut, collector or chord

———————— Interior Strut, collector, or chord

▬▬▬▬▬ Shearwall, braced frames, moment frames, or other vertical lateral resisting elements

– – – – – ➤ Unit shear transferred into a boundary element (plf) or shear wall shears

———————➤ Unit shear acting on a diaphragm sheathing element (plf)

Figure 2.12 Shear transfer into boundary elements.

each other out, creating a net uniform shear force. If a collector is not installed, a significant force will develop that cannot be transferred into the wall. A complete review of struts and collectors is presented in Sec. 2.6 and Chap. 3.

The layout for an "L" shaped diaphragm is shown in Fig. 2.13. The diaphragm is loaded in the transverse direction. Without the installation of a collector along grid line 2 from A to B, the horizontal section of the diaphragm would have to span between grid lines 1 and 3, which would cause it to deflect as shown. Under this condition, tearing will occur at the inside corner of the building unless a collector is installed as shown. A truss, beam, or girder commonly occurs at this location, which can conveniently serve as the required collector if properly designed. Installing the collector and making a connection to SW3 eliminates the tearing problem and provides a complete load path along grid line 2. Diaphragms 1 and 2 are then created as shown, which function the same as the two-span diaphragm previously discussed. The issues and solutions are the same for loading in the longitudinal direction. A collector is required to be installed along grid line B from 1 to 2 to prevent tearing at the inside re-entrant corner. All boundary edges of each diaphragm are supported as shown and as required by code.

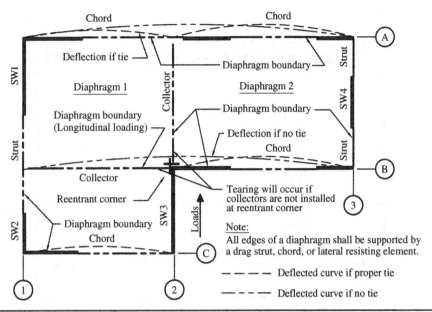

Figure 2.13 Boundary elements "L" shaped buildings—transverse loading.

The diaphragm shown in Fig. 2.14 is a single simple span diaphragm when loaded in the transverse direction. It is considered a cantilever diaphragm with a back-span when loaded in the longitudinal direction, which is shown in Fig. 2.15. For loading in the transverse direction, Fig. 2.14 shows a collector placed along grid line B and diaphragm chords located at grid lines A and C. The end walls along grid lines 1 and 2 consist of partial length shear walls and drag struts. The boundary elements are located along grid lines A, C, 1, and 2. Occasionally, engineers try to use the collector along grid line B as the diaphragm chord. This cannot happen because the diaphragm shear is considered to be uniform in magnitude across the full width of the diaphragm. The directions of the shears that are transferred into each side of the collector at grid line B are equal in magnitude but act in opposite directions and therefore cancel out to a net zero shear value, resulting in no force being applied to the collector at that location.

As a rule, the unit shears acting on adjacent sheathing elements will continue to cancel each other out until they reach an area where the magnitude of the unit shears vary, or they come to a boundary member where the shear is finally transferred.

Figure 2.15 shows the same diaphragm loaded in the longitudinal direction. In this case, the diaphragm support lines occur at grid lines A and B. The collector and shear walls at grid line B act as an interior support for the diaphragm and is therefore considered to be an interior boundary line. In summary, the boundary member at grid line C acts as a chord member for transverse loading and the boundary member at grid line B acts as a collector that completes the line of interior support for longitudinal loading.

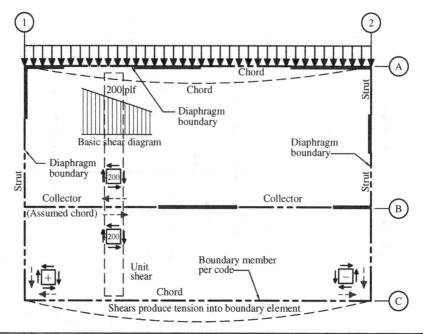

FIGURE 2.14 Cantilever diaphragm—transverse loading.

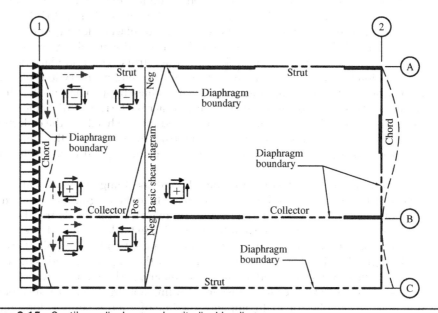

FIGURE 2.15 Cantilever diaphragm—longitudinal loading.

2.6   Drag Struts and Collectors

Code defines a drag strut as a diaphragm boundary element parallel to the applied load that collects and transfers diaphragm shear forces to the vertical elements of the lateral-force-resisting system or distributes forces within the diaphragm. For a collector, code refers to the definition of a drag strut. The main design requirements for collectors are provided in ASCE 7 Sections 12.10.2 and 12.10.2.1:

12.10.2   Collector Elements.
Collector elements shall be provided that are capable of transferring the seismic forces originating in other portions of the structure to the element providing the resistance to those forces.

> *12.10.2.1   Collector Elements Requiring Load Combinations Including Overstrength for Seismic Design Categories C through F.* This section structures assigned to Seismic Design Category C, D, E, or F. Collector elements and their connections, including connections to vertical elements, shall be designed to resist the maximum of the following:
>
> 1. Forces calculated using the seismic load effects including overstrength of Section 12.4.3 with seismic forces determined by the equivalent lateral force procedure of Section 12.8, F_x, or the modal response spectrum analysis procedure of Section 12.9.1.
>
> 2. Forces calculated using the seismic load effects including overstrength of Section 12.4.3 with seismic forces determined by Eq. (12.10-1), F_{px}, and
>
> 3. Forces calculated using the load combinations of Section 2.3.6 with seismic forces determined by Eq. (12.10-2), $F_{px\,min}$. Transfer forces as described in Section 12.10.1.1 shall be considered.

Exception: In structures or portions thereof braced entirely by wood light-framed shear walls, collector elements and their connections, including connections to vertical elements, need only be designed to resist forces using the load combinations of Section 2.3.6 with diaphragm seismic inertial design forces determined in accordance with Section 12.10.1.1.

Figure 2.16 shows a typical arrangement of struts, collectors, and chords. The figure has been included to provide clarity and consistency in the terminology used in the examples of this book. The author's preference is to label a strut as a collector element that collects shears from one side only and is located around the perimeter of the diaphragm. A collector is an interior element that collects diaphragm shears from both sides. Regardless of what they are called, they both work the same way and serve the same function. The role these members play under transverse and longitudinal loading is shown in the figure. For transverse loading, the members along grid lines 2 and 3 receive shears on both sides of the members and therefore can be classified as collector, or in this case, transfer diaphragm chords. The disrupted chord at grid line 2B is extended into transfer diaphragm TD1 by a collector, which transfers the disrupted force out to the main chords located at grid lines A and C. A transfer diaphragm is nothing more than a sub-diaphragm. Shears are applied on both sides of this member, which makes it a collector. The boundary member coming off the shear wall on grid line 1 receives shears from one side only which classifies it as a drag strut. The members along grid lines A, B, and C receive shears from one side only, but are classified as chords for the direction of loading, by definition. For longitudinal loading, the transverse member along grid line 1 acts as a diaphragm chord. The transverse members along grid lines 2

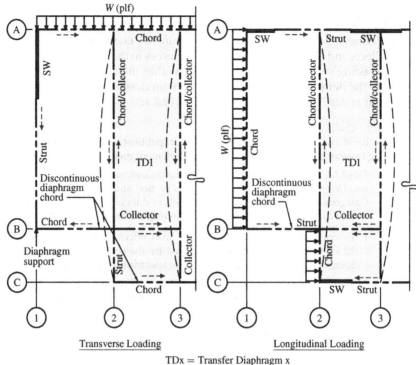

Figure 2.16 Strut, collector, and chord terminology.

and 3 act as chords for the transfer diaphragm, which receives shears on both sides of the members. These members can be classified as chords and/or collectors. The diaphragm section bounded by grid lines 1 to 2 from A to B is supported by a boundary member at grid line B that receives shears from one side only and can be classified as a drag strut.

Drag struts, collectors, and diaphragm chords are shown as solid or dashed lines on the framing plans. Figure 2.17 shows a framed wall elevation which could represent the exterior wall along grid line A of Fig. 2.16. The elevation shows two types of shear walls, one with an opening and one without. The diaphragm boundary drag element transferring the collected diaphragm shear force into the top of these shear walls can consist of two different types of framing members. The first possibility would be a continuous rim joist located directly below the diaphragm sheathing. The shear from the sheathing is transferred directly into the rim joist by nailing. All joints in this member must be spliced by a steel strap or lapped and nailed to blocking. The drag forces are transferred into the shear walls by toe nailing the rim joist to the wall top plate or by using light gauge metal shear clips. If a continuous rim joist is used as the strut or collector, the shear clips typically need to be installed only over the shear wall. At an interior wall line, the continuous rim joist is normally replaced by 2× blocking or prefabricated shear panels. It becomes impractical to use the blocking as the drag element because of the closely spaced joints. Each joint would have to be spliced to transfer the tension and compression forces. In this case, the diaphragm shears are transferred

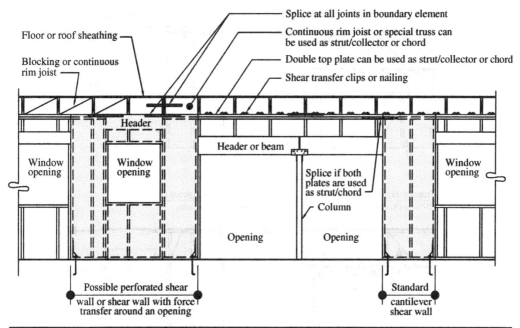

Splice at all joints in boundary element

Continuous rim joist or special truss can be used as strut/collector or chord

Floor or roof sheathing

Double top plate can be used as strut/collector or chord

Blocking or continuous rim joist

Shear transfer clips or nailing

Header

Header or beam

Window opening

Window opening

Window opening

Splice if both plates are used as strut/chord

Column

Opening

Opening

Possible perforated shear wall or shear wall with force transfer around an opening

Standard cantilever shear wall

FIGURE 2.17 Examples of struts/collectors/chords at exterior boundaries.

into the blocking and then down into the wall double top plate, which will serve as the drag element. The shear transfer connection from the blocking to the plates must occur at each 2× block or blocking panel along the entire length of the wall line. All the joints in the wall plate must be spliced by a steel strap or lapped and nailed plate splices as required by design. The transfer of the drag forces into the wall is accomplished by the fastening of the shear wall sheathing to the wall plates. This will complete the required continuous load path to the top of the wall. The farther away the drag element is from the diaphragm sheathing, the harder it becomes to transfer the shears to the drag member. The transfer of forces from the diaphragm sheathing into the boundary elements is an important part of the lateral-force-resisting system and should be completely designed and detailed in the construction documents.

Interior shear walls pose a special set of problems for the connection of collectors and drag struts to the wall. Figures 2.18 through 2.21 address collector or drag strut problems when the roof framing is oriented parallel to the shear wall. Figures 2.22 and 2.23 illustrate collector or drag strut problems when the roof framing is oriented perpendicular to the wall. Whenever an interior shear wall is required, common practice is to utilize an adjacent roof or floor truss as the collector when the framing is oriented parallel to the wall. Figure 2.18 shows several truss to shear wall connection configurations where the truss or collector falls directly over the wall. Whenever a shear wall is framed to the underside of the truss, configuration C of Fig. 2.19 is typically used. This detail allows the diaphragm shear to be transferred directly into the wall top plate through the bottom chord of the truss. High shears normally associated with interior shear walls often require light gauge shear clips for the transfer of the shear into the shear wall. These clips usually require several "common" type nails confined within a small space. The clips are typically 4 to 5 in in length. Care must be taken to avoid

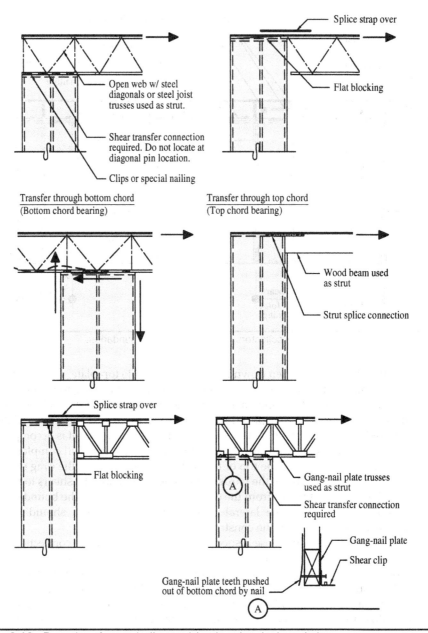

FIGURE 2.18 Examples of struts/collectors/chords at interior boundaries.

splitting the top or bottom chord of the truss when the clip spacing is close or popping off the gang-nail plate of the truss when the clip is nailed to the truss at a gang-nail plate location, as shown in the lower right detail of Fig. 2.18, detail "A." On one specific project, the design required a very narrow, high load interior shear wall to be designed, similar to the wall shown at the lower right of Fig. 2.18. The truss fell directly over the

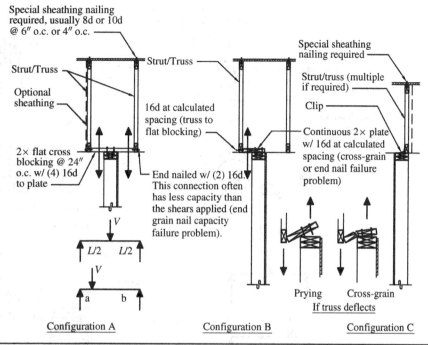

Special sheathing nailing required, usually 8d or 10d @ 6″ o.c. or 4″ o.c.

Strut/Truss

Optional sheathing

2× flat cross blocking @ 24″ o.c. w/ (4) 16d to plate

Strut/Truss

16d at calculated spacing (truss to flat blocking)

End nailed w/ (2) 16d. This connection often has less capacity than the shears applied (end grain nail capacity failure problem).

V

$L/2$ $L/2$

V

a   b

Configuration A

Special sheathing nailing required

Strut/truss (multiple if required)

Clip

Continuous 2× plate w/ 16d at calculated spacing (cross-grain or end nail failure problem)

Prying   Cross-grain
If truss deflects

Configuration B   Configuration C

FIGURE 2.19 Typical collector sections—parallel to shear wall.

wall, with the wall located in the center third of the truss span, where the highest bottom chord tensile stress occurred. The ends of the wall fell halfway between the bottom chord joints. The connection for the transfer of the collector force to the wall called out for "Simpson A35 clips" at 8″ o.c. on each side of the truss chord, staggered (4″ o.c. net spacing). The collector force was not called out on the framing plan or in the framing key notes, for the truss manufacturer. A review of the engineering calculations for the trusses revealed that the transfer of the collector forces through the truss and the effects of the high concentration of connection fasteners on the chord member were not addressed. Secondary bending in the bottom chord due to wall rotation and the effects of an interior support was also not addressed. Whenever this occurs in platform framed trusses, vertical web members should be installed at each end of the shear wall to prevent secondary bending in the bottom chord of the truss. The truss was designed as a simple span, ignoring the fact that the truss went into bearing at the shear wall. The integrity of the bottom chord of the truss and functionality of the design was highly questionable. In addition, no inspections were made on the installation or the connections. In consideration of the importance of the lateral-force-resisting system, it is necessary to ensure coordination of all the components including those designed and provided by others such as the pre-engineered truss manufacturer. These example conditions illustrate that there were several different members and connections that are susceptible to failure which could prevent the loads being transferred into the wall. Other details in Fig. 2.18 show conditions where shear transfer must be made through the top chord of the truss. In these cases, care must be taken in selecting the proper tie

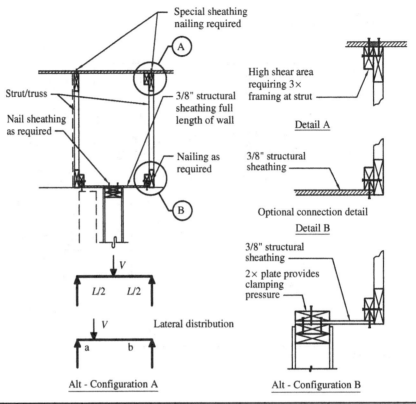

FIGURE 2.20 Collector/shear wall transfer issues.

strap that has a reasonable nail spacing that will avoid splitting the truss top chord member. For the gang-nail truss, the strap nailing and gang-nail plate occur at the same location which causes high-density nailing, splitting, and disturbance of the gang-nail connection. For high force collectors, double trusses could be considered to reduce connection difficulties.

There are no guarantees where the trusses will fall with respect to the wall location unless a specific truss layout pattern is shown on the drawings, or an additional truss is called out to be installed directly over the shear wall. Figure 2.19 shows three widely used typical details that cover most of the likely truss locations placed adjacent to the wall. Configuration A shows the condition where the shear wall falls at any location between two trusses. Shear is transferred through special nailing of the diaphragm sheathing into the collector truss. Occasionally plywood sheathing is applied to the face of the truss so that the collector forces are transferred through the sheathing similar to a shear panel or shear wall and not as a truss. Consideration needs to be given to attic ventilation and other code requirements for this condition. In some cases, it may be possible to incorporate a sheathed truss with a code required draft stop. The bottom chord of the truss is typically nailed into the end grain of the flat blocking over the wall with two 16d common nails. The flat blocking is typically spaced at 24″ o.c. or 48″ o.c. depending on demand and is nailed to the top plate of the shear wall with four 16d

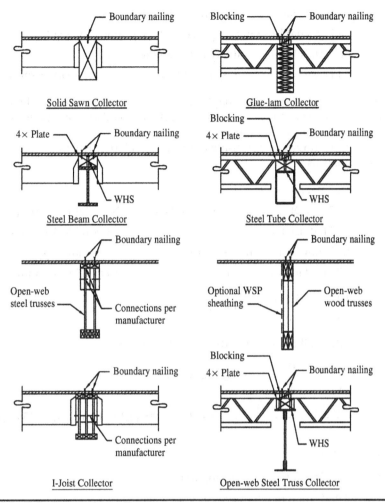

FIGURE 2.21 Collector member options.

common nails. Calculations verifying the capacity of the connections and members that are a part of the complete load path should be submitted as part of the calculation package. It would not be safe to assume that the detail works for all load conditions because they came out of a book of typical details. Since the main intent of the detail is for lateral loading, the effects of gravity loading are often ignored. To demonstrate this problem, configuration A was analyzed for gravity loads assuming the following:

1. Truss span, $L = 40$ ft

2. Truss spacing $= 24''$ o.c.

3. Dead load of the roof has already been applied to the trusses before the installation of the horizontal blocking.

4. Roof snow load (S) $= 25$ psf-uniform load to each truss $= 50$ plf.

5. The shear wall is located at mid-span of the truss.

6. The 2 × 6 flat blocking is spaced at 24″ o.c. along the wall.

7. The wall is located halfway between the trusses.

8. Load case = $D + (Lr$ or $S)$, Dead load plus roof live load or snow

Assume that the nailing of the diaphragm sheathing to the truss is adequate to transfer the required collector force into the truss and that all other nailing is in accordance with the detail shown in configuration A.

The resulting gravity force applied to each end of each piece of flat blocking is 366 lb by analysis.

Check the flat blocking

Assume 2 × 6 DF-Larch no. 2, $F_b = 900$ psi   NDS Table 4A Supplement

$C_{fu} = 1.15$ flat use factor   NDS Table 4A adjustment factors

$C_D = 1.15$ snow load duration factor   NDS Table 2.3.2

$C_F = 1.3$ Size factor for 2 × 6   NDS Table 4A adjustment factors

$$S = \frac{bd^2}{6} = \frac{5.5(1.5)^2}{6} = 2.06 \text{ in}^3$$

Assuming the wall is located mid-span of the blocking

$$M = \frac{PL}{4} = \frac{0.731(2)}{4} = 0.366 \text{ ft-k, where } L = 2'$$

$$f_b = \frac{M(12)}{S} = \frac{0.366(12)}{2.06} = 2.132 \text{ ksi} > 900(1.15)(1.15)(1.3) = 1548 \text{ psi allow.}$$

Therefore, N.G.

Check nailing of the flat blocking to the truss.

(2) 16d end-grain nailing used for the connection

$C_{eg} = 0.67$ end grain factor, NDS Section 12.5.2

Capacity = $2(141)(0.67)(1.15) = 217$ lb < 366 lb   ∴N.G.

The shear on the nailing is even greater if the wall is offset closer toward one of the trusses.

It is often assumed there is sufficient capacity in the blocking and connections. However, reality shows that under gravity loads the flat blocking can fail in bending and the nailing connection of the blocking to the bottom chord of the truss will fail in shear. The detail was created for the purpose of transferring the lateral collector forces into the shear wall. Gravity loads plus lateral load combinations are usually not considered. The framing shown cannot transfer lateral forces to the wall if it has already been weakened or has failed under gravity loading. The point of this exercise is to show that connections that are required to establish "complete load paths," as required by code, means that they shall be capable of resisting all gravity loads, lateral loads, and all relevant load combinations in order to maintain those load paths. An additional area of concern is the interior support condition of the trusses caused by the shear wall.

Usually, standard trusses are used for the collector on each side of the wall, which are designed as simple span trusses. If full or partial support occurs at the interior wall, a reversal of stress in the truss members is likely to occur. The total lateral force that can be applied to the wall for the detail shown is equal to the lesser of

Sheathing nailing
$10d @ 6''$ o.c. $= 88(2)(1.6)(1.1)(40') = 12{,}390$ lb, where $C_{di} = 1.1$ NDS 12.5.3 and Table 12Q

End grain nailing
$V = 2(141)(0.67)(1.6)6 = 1814$ lb, where $C_D = 1.6$ seismic/wind and six connections of $(2)16d$ nails each are used.

Blocking to wall
$V = 4(141)(1.6)(3) = 2707$ lb, where three connections of $(4)16d$ nails are used.

The maximum force is 1814 lb, which is controlled by the end-grain nail capacity. This is a much lower force than what would normally be expected to occur at an interior shear wall.

Configuration B of Fig. 2.19 is commonly used whenever a shear wall is located within a foot of the designated collector. Shear is transferred through special nailing of the diaphragm sheathing into the collector truss. The bottom chord of the truss is typically nailed into a flat 2×12 member that is installed over the entire length of the wall. The flat blocking is nailed to the shear wall with one or two rows of nailing. Gravity loading to this connection could cause a different problem if the wall is in the middle of the truss span where the truss deflection is greatest, and the wall is narrow. Once the truss deflects, the flat 2×12 member could fail either in cross-grain bending or by prying on the nails. A second concern can occur if the shear wall is offset from the collector, which causes an eccentrically applied force at the top of the wall that must be resisted by an opposing force at the ends of the wall. It should be remembered that the blocking is connected to the bottom chord of the truss, which could cause out-of-plane bending in the truss bottom chord if rotation occurs at the ends of the wall. The eccentricity should be resisted by lateral bracing at the ends of the wall. Configuration C of Fig. 2.19 is the most direct method of transferring collector forces into the shear wall but must be properly designed. This might require an additional truss to be installed directly over the shear wall if the collector forces are large. It is obvious that the details shown in Fig. 2.19 have not been fully developed by calculations. The completed load path is only as good as its weakest link. In some cases, there is not a complete inspection of the potential problem areas after the structure has been built.

Perhaps a more practical approach could be to modify the details as shown in Fig. 2.20. The 3/8" or 15/32" plywood sheathing would allow the truss to deflect and still transfer the lateral force to the wall. As a final note regarding configuration A of Figs. 2.19 and 2.20, the offset wall condition shown also increase the shear applied to the closest truss similar to gravity loads. Whenever the shear in the diaphragm sheathing is high, requiring $3\times$ framing, an additional $2\times$ member can be added to the truss chords as shown in detail A of Fig. 2.20 to allow staggering of the nails in the sheathing. Special instructions, such as those listed below, should be provided to the truss manufacturer, and should be noted on the drawings:

- Identify the designated collectors and the collector forces on the drawings.
- Provide an elevation of the collector truss on the drawings with all forces that are applied as well as all support conditions for the manufacturer.

- Callout the sheathing nailing to the collector on the drawings.
- Identify any special support conditions in a schematic diagram.
- Provide the controlling section cut showing the transfer of the shear forces into the wall.

It is important to provide structural observations on the lateral-force-resisting systems even if special inspections or observations are not required by code, especially at interior shear walls and at diaphragm and shear wall discontinuities. For many designers it is common practice to include the recommendation or requirement of such inspections on the construction documents. While such inspections may be outside of the scope of the typical code-required framing inspection, the building official can require the structural observation based on the recommendation of the designer.

Figure 2.21 shows several options for framing members that can serve as a collector element. When using pre-engineered trusses or joists, it is important to coordinate the feasibility of adding axial forces to their components.

The transfer of collector forces into shear walls is further complicated when the trusses are oriented perpendicular to the wall as shown in Figs. 2.22 and 2.23. Figure 2.22 shows an all-too-common condition where interior shear walls are placed within a structure without the installation of a drag strut or collector, which is a violation of the code. The assumption is often made that the ceiling board, roof sheathing, or truss system will transfer the diaphragm forces into the wall without the use of a collector. However, the transfer of lateral forces through the truss system is outside of the scope of the truss design unless specifically called for on the construction documents by

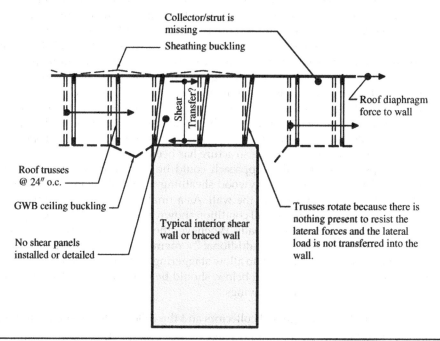

Figure 2.22 Interior shear wall without shear panels.

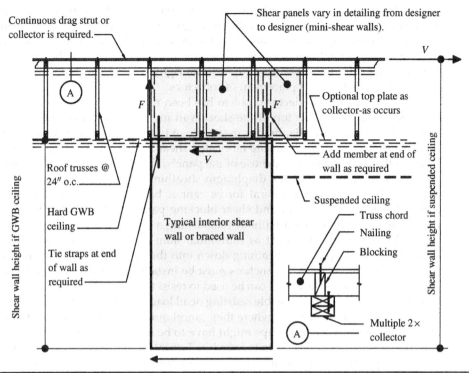

Continuous drag strut or collector is required.

Shear panels vary in detailing from designer to designer (mini-shear walls).

Optional top plate as collector-as occurs

Add member at end of wall as required

Roof trusses @ 24" o.c.

Suspended ceiling

Truss chord

Hard GWB ceiling

Nailing

Typical interior shear wall or braced wall

Blocking

Tie straps at end of wall as required

Multiple 2× collector

Shear wall height if GWB ceiling

Shear wall height if suspended ceiling

Figure 2.23   Interior shear wall with shear panels and collector added.

the design engineer. In addition, the code does not allow the diaphragm sheathing to be used as a boundary member or to splice a boundary member. While IBC Sections 2306.2.1 and 2508.6 allow gypsum wall board ceiling diaphragms, the practicality and performance of this type of diaphragm is questionable due to the brittle nature of the gypsum board and the reduced allowable aspect ratios of the gypsum wall board diaphragms. The IBC contains provisions for the use of gypsum board diaphragm ceilings, but limits their use based on a maximum diaphragm aspect ratio of 1½:1 and requires blocking of the perimeter edges among other requirements. In addition, the allowable tabular shear value in Table 2508.6 is required to be reduced by 50 percent for SDC's D, E. and F. The use of gypsum diaphragms require inspection per IBC Section 110.3.6 and would definitely warrant special detailing.

Installation of drag struts or collectors can easily end up being partially addressed at an isolated interior shear wall located beneath perpendicular trusses. Two conditions typically exist. Either the isolated individual wall section is located within a continuous wall line and the wall double top plate can be used as the collector, or the wall line does not continue on either side of the shear wall and it is truly isolated, in which case, an alternate method should be used. One of several possible approaches is to install a long continuous drag strut or collector consisting of a continuous light gauge steel strap placed over the diaphragm sheathing nailed into flat blocking placed between the truss top chords below. The functionality of this type of collector depends on a quality installation and should be given careful attention and inspection. Misalignment of the strap and the blocking below and/or large gaps at the ends of the blocking may prevent the detail from functioning as intended by the engineer. In the event the wall is truly isolated and double top plates are not available to act as collectors, on edge blocking can be installed between the trusses with continuous 2× members below the top chord that can act as the collector as shown at the bottom right of Fig. 2.23. See Chap. 9 for additional options and vertical placement of collectors.

Assuming that the collector problem has been resolved, a method of transferring the collector force into the top of the shear wall must be developed. The most direct method transferring the diaphragm shears down to the shear wall or wall top plate is by using pre-engineered or on-site fabricated shear blocking panels that are installed between the trusses. Without the use of the panels, the trusses would rotate and could possibly allow buckling of the diaphragm sheathing and ceiling board as shown in Fig. 2.22. In this case, the lateral forces cannot be transferred down to the wall. Figure 2.23 shows a collector and shear blocking panels added. Unfortunately, shear panels are sometimes added without establishing a complete load path into the wall below. The blocking panels act as individual mini-shear walls, which transfers the shears from the diaphragm sheathing down into the main shear wall or collector. As with any shear wall, tie down anchors must be installed at each end of each blocking panel unless the truss dead load can be used to resist the overturning forces. If the overturning forces exceed the available resisting dead loads of the truss, the shear forces can be transferred across the truss, where they cancel out to zero until they get to the ends of the shear wall, where tie straps might have to be installed. These anchors are often overlooked. Typical generic details found in Typical Detail Books usually show shear panels of equal height and width. These square panels do not always reflect actual field conditions. The closer a shear wall gets to the ridge line of a sloped roof, the taller the shear panel will be which increases the overturning forces in the individual panels. The depth of the truss at that location can, in some cases, exceed 12 ft in height. For this case, the wall could be treated as a two-story wall with the blocking panels acting as a second-floor shear wall. The shear and overturning forces of the blocking panels can be transposed into the top of the wall below for easier calculation.

2.6.1   Calculating Strut and Collector Forces

A brief discussion on two methods of calculating strut or collector forces is presented in Fig. 2.24. The "traditional method" is straight-forward and is more suited for simple rectangular diaphragms and shear walls. A comprehensive coverage of the traditional method can be found in *Design of Wood Structures*[10] and SEAOC's *Structural/Seismic Design Manual*, Vol. 2.[11] The traditional method assigns the shear wall shears as positive shears and the diaphragm shears as negative shears. The diaphragm and shear wall shears are diagrammatically placed side by side as shown at the lower left of Fig. 2.24. These values are combined resulting in a net shear at the shear wall. The strut or collector force is equal to the area of the net shear diagram. The "visual shear transfer method" is shown at the lower right of the figure. Diekmann[7] and Kula[12] both demonstrated in their presentations and examples that a visual representation of how shears are transferred from the sheathing elements into boundary members and collectors is critical to solving and understanding complex diaphragms force diagrams. Briefly, the basic shear diagram (plf or klf) is developed for the two-span diaphragm. The diaphragms are assumed to act as two simple spans. The left half of each shear diagram is positive shear, and the right half of each diagram is negative shear. Sheathing elements are

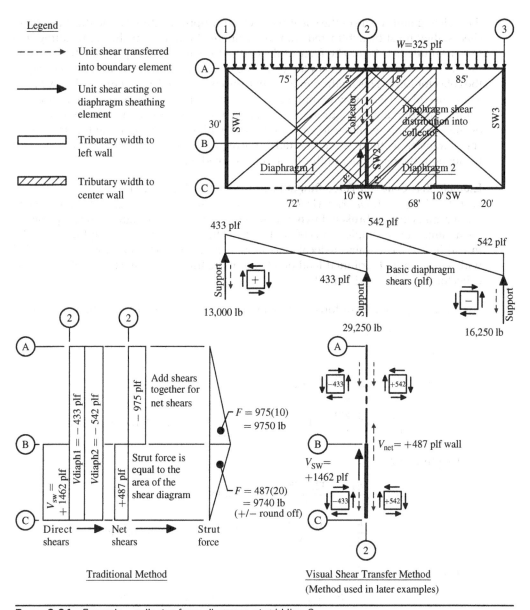

FIGURE 2.24 Example—collector force diagrams at grid line 2.

placed at the positive and negative shear locations on each side of the collector showing the direction of the shears acting on each edge of the sheathing elements. These shears are applied to the drawing as a visual reference. The direction of the shears being transferred into the collector from the positive and negative sheathing elements are called transfer shears. Both of the transfer shears act in the same direction collectively causing the collector to be in compression. The collector force is the sum of the diaphragm unit shears acting on each side of the collector multiplied by the length of the collector. Since the resisting unit shears of the shear wall act in the opposite direction to the diaphragm shears, they must be subtracted from the combined diaphragm shears, resulting in a final resulting net shear. The force at the end of the shear wall at grid line B is equal to the net shear times the length of the wall. The force at the end of the wall must be equal the collector force. This method may seem to be complicated at first but will prove to be a very useful and easy way to understand the overall flow of shears and forces through complex diaphragm configurations. The advantage of the method is that the direction of the shears and resulting forces can be seen without resorting to tables. The visual shear transfer method is well suited for irregular shaped diaphragms and is used throughout the remainder of this book unless noted otherwise.

Example 2.1: Collector Force Diagrams at Grid Line 2 (Fig. 2.24)

Part A—Using the Traditional Method: The diaphragm shown in Fig. 2.24 is a two-span diaphragm, which is broken down into diaphragm 1 and diaphragm 2. The diaphragms are assumed to be simple spans of 80 ft and 100 ft, respectively. The diaphragm is 30 ft in width. The shear walls are located as shown on the plan. The diaphragms are loaded with a uniformly distributed load of 325 plf in the transverse direction. Determine the collector force along grid line 2.

R_{2L}, V_{2L}, v_{2L} = shears or forces acting on the left side of grid line 2 (typ. notation)
$$w = 325 \text{ plf}$$

Diaphragm 1:

$$R_1 = \frac{wL}{2} = \frac{325(80)}{2} = 13,000 \text{ lb} = -R_{2L}$$

$$v_1 = \frac{R_1}{W} = \frac{13,000}{30} = 433 \text{ plf}, \ v_{2L} = \frac{-R_{2L}}{W} = -433 \text{ plf negative shear}$$

Diaphragm 2:

$$R_{2R} = \frac{wL}{2} = \frac{325(100)}{2} = 16,250 \text{ lb}$$

$$v_{2R} = \frac{R_{2R}}{W} = \frac{16,250}{30} = 542 \text{ plf}, \ v_3 = -542 \text{ plf negative shear}$$

Forces at grid line 2:

$$R_2 = R_{2L} + R_{2R} = 13,000 + 16,250 = 29,250 \text{ lb}$$

$$v_{SW2} = \frac{R_2}{L_{SW2}} = \frac{29,250}{20} = 1462 \text{ plf}$$

The magnitude of the shear would suggest an increase in the wall length is warranted.

Determination of net transfer shears acting on the collector:

$$v_{net} = -433 - 542 = -975 \text{ plf}$$

Note: Since both diaphragms are trying to slide off their support at grid line 2 in the direction of the applied loads, the diaphragm shears transferred into the collector are acting in the negative direction or toward the shear wall. They are additive and therefore are both negative in value. The shear wall shears act in the opposite direction and are therefore positive. The net shears at the shear wall are

$$v_{net} = v_{SW} - v_{2L} - v_{2R} = +1462 - 975 = +487 \text{ plf}$$

Determination of the collector force:

Summing from grid line A, collector force at grid line B:

$$F_B = -975(10) = -9750 \text{ lb, compression}$$

The direction of the shears transferred into the collector is pushing on the wall as expected.

Force at grid line C:

$$F_C = -9750 + 487(20) = 10 \text{ lb}$$

This is close enough (round off error); therefore, the force diagram close to zero.

Note: The force diagrams are just a means of verifying $\Sigma F = 0$. All force diagrams must close to zero, or an error exists. The determination of the direction of the shears that are applied from a simple diaphragm and distributed into the shear walls is based on a good understanding of how simple diaphragms work. Making a determination of the direction of the shears distributed in a complex diaphragm would not be easy. For complex diaphragm layouts, using the shear transfer method eliminates the need to assume the direction of the shears, because they are shown.

Part B—Using the Visual Shear Transfer Method: The sheathing elements for the left and right diaphragms are placed adjacent to the collector and shear wall on grid line 2. The shears that are transferred into the boundary elements along the grid line are both headed in the same direction and therefore should be added together. Since they are pushing against the wall, the collector will be in compression. The shear wall shears, and diaphragm transfer shears oppose each other and should be subtracted from each other to get a net shear. The net shear is acting in the opposite direction to the collector shears. The force diagrams are generated by multiplying the net shears by the length of the boundary elements under consideration and summing the forces along the line. ▲

This quick comparison of the two methods shows that the major difference between the two methods is basically a visual one. The main steps are the same. Irregular shaped diaphragms can get very complicated because of the number of discontinuities and complicated load paths required to transfer the discontinuous forces. The visual shear transfer method will prove to be very advantageous in solving these problems.

Example 2.2: Strut Force Diagrams at Grid Line C (Fig. 2.25)

Using the same diaphragm in the previous example, determine the strut forces along grid line C. The loads are applied to the diaphragm in the longitudinal direction. It is assumed that the forces to each wall are in proportion to their length in accordance with SDPWS Section 4.3.5.5.1, exception 1. As an alternate, the equal deflection method of that section can be used. Note that all shears at shear walls must be in terms of net shears (plf) before force diagrams can be calculated.

$$w = 325 \text{ plf}$$

$$R_C = \frac{325(30)}{2} = 4875 \text{ lb}, \quad v_{\text{diaph}} = \frac{4875}{180} = -27.08 \text{ plf}$$

Note: The direction of the diaphragm shear transferred into the strut is negative. The direction of the shear at the shear wall is positive.

$$v_{\text{SW}} = \frac{R_C}{\Sigma L_{\text{SW}}} = \frac{4875}{2(10)} = +243.75 \text{ plf}$$

Determine the net shears at the shear wall:

$$v_{\text{net}} = v_{\text{SW}} - v_{\text{diaph}} = +243.75 - 27.08 = +216.67 \text{ plf}$$

Determination of the strut force:

Summing from grid line 1, the force at the start of the first shear wall is equal to

$$F_{\text{start}} = -27.08(72) = -1949.76 \text{ lb compression}$$

The force at the end of the first shear wall is equal to

$$F_{\text{end}} = -1949.76 + 216.67(10) = +216.94 \text{ lb tension}$$

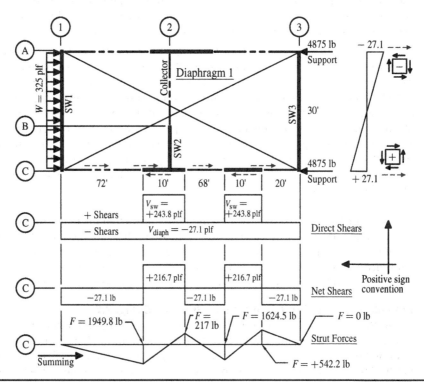

Figure 2.25 Example—strut force diagrams at grid line C.

The force at the start of the second shear wall is equal to

$$F_{\text{start}} = +216.94 - 27.08(68) = -1624.5 \text{ lb compression}$$

The force at the end of the second shear wall is equal to

$$F_{\text{end}} = -1624.5 + 216.67(10) = +542.2 \text{ lb tension}$$

The force at grid line 3 is equal to

$$F_3 = +542.2 - 27.08(20) = 0 \text{ lb}$$

Therefore, the force diagram closes to zero. ▲

   Even though, in this example, all the strut forces are less than the calculated chord forces, this will not always be the case. It is important to verify the chord and strut forces in both directions to determine the controlling forces in a chord or drag element. Unfortunately, this is not always done.

   Partial length struts or collectors are frequently used to transfer discontinuous chord and strut forces into the diaphragm as shown in the Fig. 2.26. Whenever the framing is perpendicular to the collector, the installation of the collector is usually accomplished by using a continuous light gauge steel strap applied over the sheathing and nailed into flat blocking below. A common problem that occurs in the field happens when the flat blocking is not installed directly under the steel strap location, and the nails from the strap only penetrate the sheathing. The full capacity of the strap cannot be developed due to an insufficient nail penetration. The strap and blocking are intended to act as a collector but is typically designed for tension forces only. Compression forces are often ignored. Collectors and struts must be designed to take both tension and compression forces because of load reversal. Proper detailing is required to assure that the strap and blocking assembly will be capable of resisting compression forces. To achieve this, the blocking must be installed in a level position and must be in full bearing against the truss chords or other framing members. The use of Z-clips will help maintain a level position. Option 1 (section) shown in Fig. 2.26 shows common installation problems that can occur during the installation of the strap and flat blocking. The blocking is often installed by toe nailing the blocking to the truss chords, to reduce the costs by eliminating the Z-clips. There is no guarantee that the blocking will be installed in a level position using this method. A lack of quality control in cutting the blocking could also cause gaps between the blocking and the truss chord. The tension/compression forces would then be transferred through the diaphragm sheathing, which is not allowed by code. The successful performance of this assembly depends on quality control and field inspection. To eliminate the problems associated with a multiple jointed member effectively transferring compression forces, continuous 2× plates could be installed to the bottom of the truss chords to act as the collector, if open web trusses are used. Blocking panels and the continuous 2× plates could be used if solid web trusses or joists are used as shown in option 2 (section). The transfer of shear and compression forces through the strap and blocking assembly can be seen in Fig. 2.27. Each block acts as a mini collector. The individual blocks collect the unit shears from the diaphragm sheathing and transfer that accumulated force into the truss top chord by bearing perpendicular to the grain. The force is then transferred across the truss chord and into the next block. The total force on the right side of the

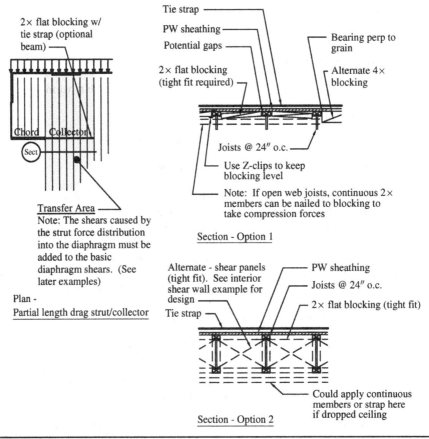

FIGURE 2.26   Example of a partial strut/collector.

detail is equal to $\Sigma F = F_1 + F_2 + F_3 + F_4$. Bearing stress perpendicular to the grain on the truss chords should be checked. The wide face of the blocking should be increased as necessary to reduce the bearing stress to acceptable levels. The location of bearing of the block on the truss chord is assumed to act at mid-depth of the flat blocking, which causes an eccentrically applied force into the blocking and produces vertical forces at the ends of the block, which is usually negligible.

In current practice, the length of the strap embedment into the main diaphragm section is usually an arbitrary distance that is based on the shear capacity of the basic diaphragm nailing at that location. This assumption is based on ASCE 7 Section 12.10.1, which states that "the design of the collector shall ensure that the dissipation or transfer of edge (chord) forces combined with other forces in the diaphragm is within shear and tension capacity of the diaphragm." This approach is also reinforced by the commentary which states that "the required development length is determined by dividing the axial force in the sub-chord by the shear capacity (in force/unit length) of the main diaphragm." As an example, assume that the existing diaphragm shear capacity is based on nailing that will allow a unit shear of 320 plf. The embedment length would then be calculated by dividing the collector force by 320 to get the required length.

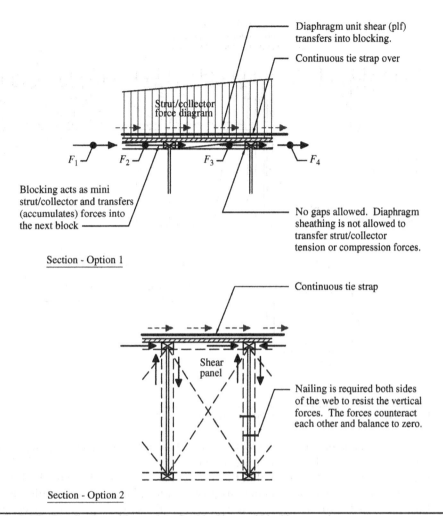

Diaphragm unit shear (plf) transfers into blocking.

Continuous tie strap over

Strut/collector force diagram

Blocking acts as mini strut/collector and transfers (accumulates) forces into the next block

F_1 F_2 F_3 F_4

No gaps allowed. Diaphragm sheathing is not allowed to transfer strut/collector tension or compression forces.

Section - Option 1

Continuous tie strap

Shear panel

Nailing is required both sides of the web to resist the vertical forces. The forces counteract each other and balance to zero.

Section - Option 2

FIGURE 2.27   Distribution of forces into flat blocking and blocking panels.

Although that approach is in the code, it is only a partial solution and can result in an inadequate collector length because it does not take into account the effects of the increase in diaphragm shears at that location caused by the discontinuous force being transferred into the main diaphragm. The basic diaphragm shear is already applied at that location and the shear from the transferred force is in addition to the basic diaphragm shear. Both shears must be combined as the code states, which could result in a longer collector length or tighter nailing spacing. It is commonly assumed that simply lapping the strap collector with the offset diaphragm chord will automatically and safely transfer the chord force across the offset. This assumption is also a partial solution because it does not take into account the rotation forces caused by the eccentric overlap of the chords, as will be discussed in detail in Chap. 3. Another issue associated with steel tie strap is strap elongation, which must be controlled to limit minor displacement between the framing members.

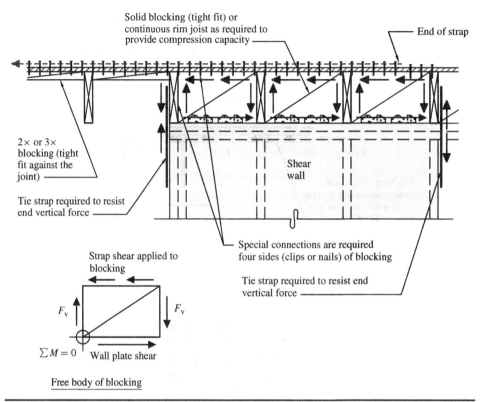

Solid blocking (tight fit) or
continuous rim joist as required to
provide compression capacity

End of strap

2× or 3×
blocking (tight
fit against the
joint)

Tie strap required to resist
end vertical force

Shear
wall

Special connections are required
four sides (clips or nails) of blocking

Tie strap required to resist end
vertical force

Strap shear applied to
blocking

F_v

F_v

$\sum M = 0$ Wall plate shear

Free body of blocking

Figure 2.28   Strut/collector with blocking or shear panels.

Occasionally, tie strap and blocking are called out on the roof framing plan with a total strap length and a minimum lap of the strap to the wall. However, calculations are usually not provided to verify that the transfer has effectively been made. They should be part of the construction documents provided for plan review and fully detailed on the drawings. Figure 2.28 shows a drag strut or collector connection to a shear wall. The collector, in this case, consists of flat blocking and a tie strap. The connection must resist tension and compression forces as previously discussed. The details showing the required connections are not always shown on the drawings. It is assumed that the contractor will install it correctly. The key issues for this connection are as follows:

- All blocking must be in full bearing against the joist/trusses to transfer the compression forces and must be within acceptable limits of the truss manufacturer.
- Solid blocking or blocking panels must be installed the full length of the strap and overlap with the wall.
- The solid blocking or blocking panels and its connections must be designed to take the entire strut or collector force.
- The solid blocking or blocking panels and their connections over the shear wall must be designed as a mini-shear wall, which resist sliding and overturning forces at each end of the panels.

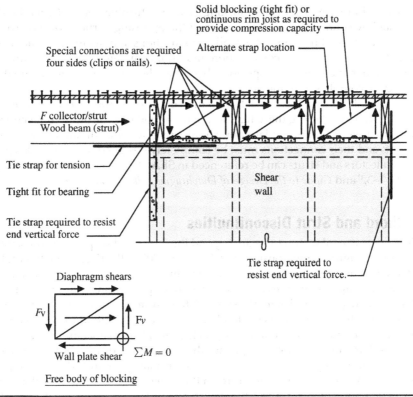

Solid blocking (tight fit) or
continuous rim joist as required to
provide compression capacity

Special connections are required
four sides (clips or nails).

Alternate strap location

F collector/strut
Wood beam (strut)

Tie strap for tension

Tight fit for bearing

Tie strap required to resist
end vertical force

Shear
wall

Tie strap required to
resist end vertical force.

Diaphragm shears

F_V

F_V

Wall plate shear $\Sigma M = 0$

Free body of blocking

FIGURE 2.29 Strut/collector with beam.

- Tie straps, shear clips or nailing should be installed at each end of the blocking panels to resist overturning and transfer these forces across the joists or trusses, if required by calculation.

- The diaphragm sheathing cannot be used to transfer the tension or compression forces, in accordance with code.

Figure 2.29 shows another method of connecting drag strut or collector forces to a shear wall that is often used. The collector, in this case, consists of a solid beam. The collector shown bears against the joists and blocking to resist the compression force. The blocking and its connections are extended back onto the wall far enough to develop the required compression force. A tie strap can be lapped and nailed onto the bottom of the beam to take the tension force and lap directly onto the wall double top plate to transfer the tension force into the wall. The wall plate then serves as a continuation of the strut or collector across the wall. As an alternate, a steel strap can be installed over the top of the beam and onto the shear panels over the full length of the wall. The tension force is then transferred down through the shear panels and into the wall. The connection of the blocking to the other framing members must be developed on all four sides of the blocking as shown in the figure and the free body diagram below the detail. The total shear applied to the blocking connectors to the wall is the sum of the collector force to each block plus the diaphragm unit shear. The resisting forces occur at the wall plate and at each end of the blocking. The overturning force of each block must be resisted by connections which counteract the opposing vertical forces across each joist or by tie straps at the end of the wall. If connections and detailing are not called out on the drawings, improper construction and an incomplete load path is almost assured. The builder and building inspector rely on the plans and details to ensure that the construction complies with the design.

Some designs stop at the determination of the collector or strut force and the size of the tie strap, neglecting the actual design of the collector, strut, or connections. This can lead to a weak link in the lateral-force-resisting system. All lateral-force-resisting elements and their connections must be completely designed. Complete design examples for collectors and struts can be referenced in SEAOC's *Structural/Seismic Design Manual*, Vols. 1–3,[11] and *Guide to the Design of Diaphragms, Chords and Collectors*.[13]

2.7   Chord and Strut Discontinuities

Most new irregular structures are beyond the intent and scope of the prescriptive "Conventional Light-Frame Construction," Section 2308 of the IBC, or the International Residential Code (IRC).[14] A lack of understanding of how forces are transferred through irregular shaped structures has led to very poor performing buildings with incomplete lateral load paths and lateral-force-resisting systems, as discussed in ATC-7. Compounding these problems is the tendency to reduce the lateral-force-resisting system to its absolute minimum while taking each element to its maximum capacity (capacity equals demand). This not only pushes the limits of the design, but also leaves no room for design oversights, construction errors, or quality control. Part of the problem has been a lack of engineering literature on the subject of analyzing irregular shaped structures. Complex structures like the one shown in Fig. 2.30 are a good example of buildings containing multiple irregularities and complicated load paths. The photo shows that the roof diaphragms are cantilevered, are offset vertically and horizontally, and contain challenging vertical and lateral load paths. Structure like these should be studied in depth to get a feel of how the structure will response to the loads being

Figure 2.30   Photo of diaphragm discontinuities.

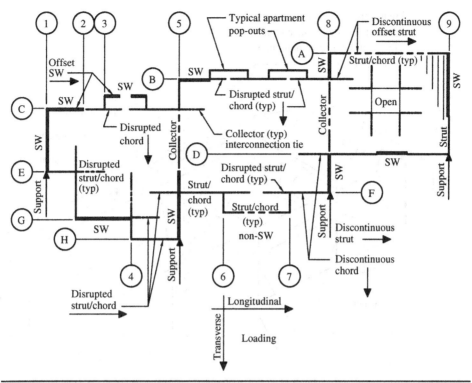

FIGURE 2.31 Typical diaphragm discontinuities.

distributed through the structure. Figure 2.31 shows a layout that is representative of many single-family residences, apartments, and commercial structures. The plan has multiple offsets, reentrant corners, and openings in the diaphragm, which cause a disruption in the chords and struts. The design issues for this layout are as follows:

Grid Line Location	Design Issues
Transverse loading	
3C	Disrupted chord
2E	Disrupted chord, offset diaphragm
4G	Disrupted chord, offset diaphragm
6F and 7F	Disrupted chord
6B and 7B	Disrupted chord
8.5C	Opening in diaphragm, discontinuity in the diaphragm web
Longitudinal loading	
2C and 3C	Offset shear walls, disrupted strut
2E and 4H	Offset struts, disrupted chords
6B, 7B, 6F, and 7F	Disrupted and offset struts
8B	Disrupted offset strut
8.5C	Opening in diaphragm, discontinuity in the web
5C, 8B, 8D, and 5F	Reentrant corners

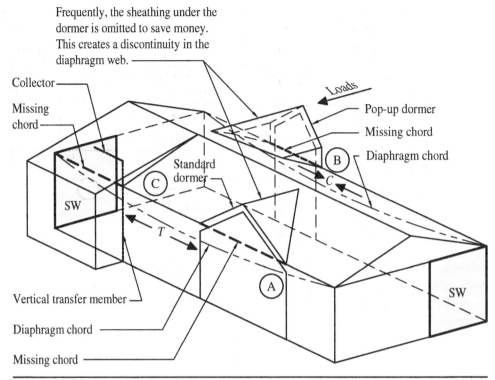

Frequently, the sheathing under the
dormer is omitted to save money.
This creates a discontinuity in the
diaphragm web.

Collector

Missing chord

Loads

Pop-up dormer

Missing chord

Diaphragm chord

Standard dormer

B

C

C

SW

T

A

Vertical transfer member

Diaphragm chord

Missing chord

SW

FIGURE 2.32   Examples of diaphragm chord/strut discontinuities.

Figure 2.32 shows a typical residential structure that is a simple rectangular diaphragm with supporting shear walls at each end of the structure. Entry areas with dormers have been placed at locations A and B. For architectural reasons, the diaphragm chord is not allowed to continue through the openings. Frequently, one or both entry walls are set back from the exterior wall or the diaphragm sheathing under the dormers is omitted, creating a disruption in the diaphragm chord and web. Because of the lack of horizontal members at these diaphragm chord locations, hinges are created at the ridge and at the bottom of the rafter to wall or main roof connections. Since the hinges cannot transfer the diaphragm chord forces across the discontinuity, the load path is transferred through the roof diaphragm, which is the stiffer element. The diaphragms under this framing condition would have to be designed as having an intermediate horizontal offset or notch. A pop-out section has been included at location C. The roof joists span across the main wall line and bear on the pop-out wall. This bearing condition causes a vertical offset in the diaphragm, preventing the main roof diaphragm chord from passing through the pop-out area. If the short transverse sidewall at this location cannot act as a shear wall or another shear wall does not exist along this line of lateral force resistance, then a discontinuous diaphragm chord is created. The transfer of the disrupted chord force across the vertical offset at C can be difficult. Although not ideal, one possible method of transferring the disrupted chord force across the vertical offset at location C is shown in Figs. 2.33 and 2.34. The vertical offset can vary in height. A possible method of transferring the force across the offset can be seen in Figs. 2.33 and 2.34. A vertical post or column could be cantilevered above the low roof to transfer the

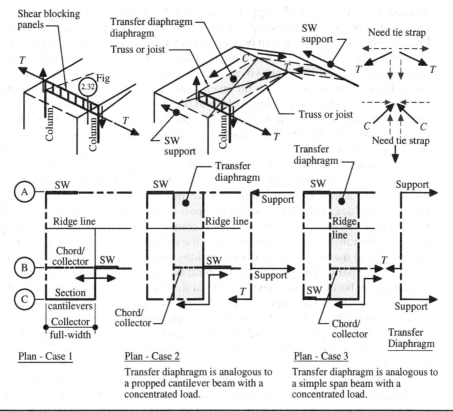

FIGURE 2.33 Transfer of chord/strut forces at C.

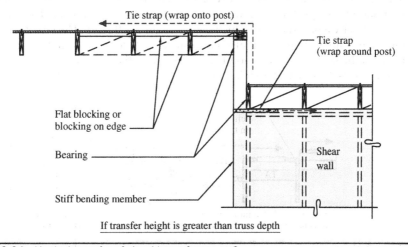

FIGURE 2.34 Vertical transfer of chord/strut forces at C.

disrupted chord force. In the upper connection, tie straps are installed up the post and over the top of the roof. The strap is nailed onto the blocking between the joists. For the lower connection, a tie strap can be wrapped around the post and is nailed into the wall top plate. Figure 2.33 shows three possible configurations for the lateral system layouts at the vertical offset. The first two cases occur when a shear wall is located in the main line of lateral force resistance at grid line B. Case 1 covers the condition where the width of the pop-out section is not very wide, and the chord can be extended across the entire width of the raised roof. Case 2 covers the condition where the width of the pop-out section is too wide to extend the chord across the entire width. In this case, the chord can be extended a short distance into a transfer diaphragm, where the disrupted force is distributed into the diaphragm similar to a propped cantilever beam. In case 3, the main line of resistance occurs at grid line C. The chord can be extended a short distance into a transfer diaphragm, where the disrupted force is distributed into a transfer diaphragm similar to a simple span beam with a concentrated load. If roof joists are used in a sloped roof, the vertical components of the tension and compression forces of the transfer diaphragm chords at the ridge must be addressed as shown at the upper right of the figure. These configurations create very flexible structures and questionable load paths. The vertical post should be stiff enough to adequately transfer the force without becoming too flexible. Lateral displacements should be checked.

The offsets shown in Figs. 2.33 and 2.35 demonstrate several other types of discontinuities caused by wall offsets. The main problems caused by these offsets are the disruptions of the diaphragm chord and drag struts, complicating the transfer of forces across the offset shear walls. A complete discussion and example will be presented in Sec. 12.3.

The photo shown in Fig. 2.36 shows an example of the architectural features of a modern home. The section above the garage door opening has a high sloped cantilevered roof with a cantilever length equal to the back-span length. The roof back span is supported off the end of a second-floor cantilever diaphragm which forms the ground level covered entry. The shear wall at the gable end above the garage door is triangular in shape and has a window opening at the left edge. The two vertical shear wall chords fall at the end of the second-floor cantilever and near mid-span of the portal frame (garage door opening) wall below. The shear wall is vertically discontinuous to the foundation. The portal frame was not tested for this type of load. The entire section is supported by a portal frame, which could create a soft story. The structure is located on the edge of an escarpment. It also has a steep roof slope that will produce a high wind force on the structure. The second story section to the right also exhibits a horizontally in-plane offset shear wall that is discontinuous to the foundation. The roof diaphragms

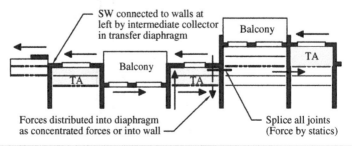

Figure 2.35 Offset exterior walls.

Figure 2.36 Photograph of complex roof.

have large openings in the webs to accommodate the dormers. The horizontal and vertical offsets and reentrant corners create multiple discontinuities and complicated load paths. The structure is in a high seismic zone as well as a high wind exposure category due to the hillside escarpment. The entire structure is required to be engineered.

Figure 2.37 shows another common framing condition that causes problems with transferring forces across a discontinuity in the diaphragm chord or strut. In this case, an open vaulted roof section is added at the center section of the building. The roof framing members at the vaulted area run in opposite direction from the roof framing to the left and right sections. For loading in the transverse direction, shear walls are placed at grid lines 1, 3, and 4, which create two separate diaphragms, diaphragm 1 and diaphragm 2. The logical locations of the chords for diaphragm 1 would occur at grid lines B and D, which would be required to extend across the vaulted roof section. The pop-out sections between lines A and B and between D and E can be assumed to be carried by diaphragm 1. For loading in the longitudinal direction, only one diaphragm exists, which spans between grid lines B and D and has a length that extends from 1 to 4. Shear walls along grid line B are assumed to act in the same line of lateral force resistance which must be connected by a strut that extends through the vaulted area of the roof. The strut along grid line D is discontinuous at grid line 3D and is offset from the supporting shear walls along grid line C. The strut forces must also be transferred through the vaulted roof area and across the offset to provide a complete load path. If a shear wall were to occur along grid line D, then an offset shear wall condition would exist, if the walls are assumed to act in the same line of resistance. The collector at grid line C from 2.5 to 3 is also required to prevent tearing at the reentrant corner. Section "A" shows a cross section of the diagram of the chord and strut force across the vaulted area. The ridge beam must be stiff enough and have adequate strength to support gravity loads plus the vertical component forces from the diaphragm chords or struts in accordance with all relevant code load combinations. A ridge board is not stiff enough to be used in lieu of the ridge beam because it will deflect under the vertical force components and create a three hinged arch condition which cannot transfer the lateral force across the pinned joints. Additional framing members are required to support these vertical forces at each side of the vaulted area and transfer them to the foundation.

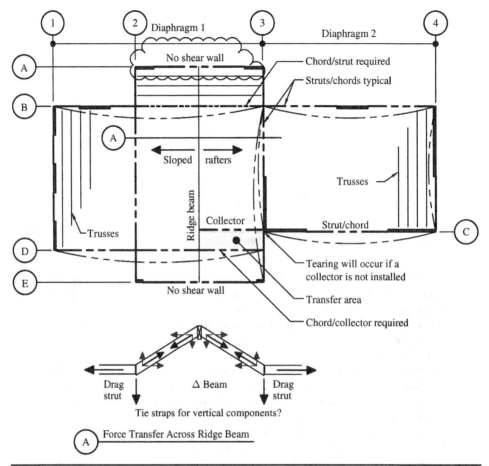

Figure 2.37 Other types of discontinuities.

Tie straps or other appropriate connections must be installed to resist the horizontal and vertical components shown in Section "A." The rafters or beams that are being used to act as the struts or collectors must also be designed for combined tension, compression, and gravity forces and loads. Applicable over-strength factor must be applied to all struts, collectors, and their connections in accordance with ASCE 7-16 Section 12.10.2.1. This framing configuration and load path can create significant displacements along the line of lateral force resistance. Although this faming scheme is a possible solution, it is not recommended.

2.8   Sub-diaphragms

The topic of sub-diaphragms is comprehensively covered in *Design of Wood Structures* and in SEAOC's *Seismic Design Manual*, Vol. 2. A brief review is provided here for the reader's convenience. A common mode of failure during major seismic events in high seismic areas has been the loss of support of a roof or floor due to an inadequate connection between concrete and/or masonry walls and flexible wood or untopped steel deck roof or floor diaphragms. The lessons learned from these failures have prompted significant changes in the code. The failures occurred partly due to underestimating the magnitude of the forces that are applied to the connection from the concrete or masonry wall, and partly due to poor detailing of the connection itself. Detailing of the connection prior to the code changes were typically done in a way that caused cross-grain bending in a wood ledger that was attached to the wall. The concept of continuous cross ties and sub-diaphragms was developed to safely distribute the out-of-plane wall forces of masonry or concrete walls into the diaphragm without exceeding the capacity of the diaphragm. Early discussions on this topic can be found in ATC-7 and ATC7-1.[15] Currently, ASCE 7 Sections 12.11 and 12.11.2.2 specifically require continuous cross ties and sub-diaphragms are required for structures with masonry or concrete walls tied to flexible diaphragm in SDCs C through F. Code requires the following:

1. Where anchor spacing exceeds 4 ft, the wall shall be designed to span horizontally between the anchors.

2. Continuous ties shall extend the full distance between the diaphragm chords. Additional chords may be installed to form sub-diaphragms spanning between and transferring their force into the continuous cross ties. The maximum aspect ratio of a sub-diaphragm shall not exceed 2.5:1.

3. In wood diaphragms, the continuous cross ties are in addition to the diaphragm sheathing. The diaphragm sheathing shall not be considered effective as providing the ties or struts required by ASCE 7 Section 12.11.2.2.3.

4. In steel decking diaphragms, the decking can be used for the cross tie when the loads are applied parallel to the ribs but not perpendicular to the ribs.

5. Anchorage shall not be accomplished by the use of toenails, nails in withdrawal, or by cross-grain bending of a ledger.

6. Connections (anchors) shall extend into the diaphragm a sufficient distance to develop the force.

The intent of item 6 is not always understood. Anchors that are typically used for the connection of the walls to the diaphragm or sub-diaphragms usually consist of light gauge straps nailed to flat blocking or framing member. It is often interpreted from item 6 that the anchors and blocking only need to extend into the sub-diaphragm a short distance. However, in accordance with item 2 and ATC-7, the sub-diaphragm collectors between the continuous cross ties (the anchor and its collector) are also required to extend to the interior chord of the sub-diaphragm. This would assure that the force is uniformly distributed the full depth of the sub-diaphragm, preventing localized tearing.

Figure 2.38 shows a typical layout for the cross tie and sub-diaphragms where the structure is loaded in the transverse direction. The sub-diaphragms are placed along grid lines A and C as shown. The sub-diaphragms are supported by the end walls located at grid lines 1 and 5 and by the continuous cross ties located at grid lines 2, 3, and 4. These cross ties usually consist of beams or girders connected with straps or other connectors. Out-of-plane anchors are connected to collectors that extend the full depth of the sub-diaphragm and are spaced a uniform distance across the sub-diaphragm. The anchors apply a concentrated load to the sub-diaphragm as shown at the bottom left of the figure. However, standard accepted practice is to apply the

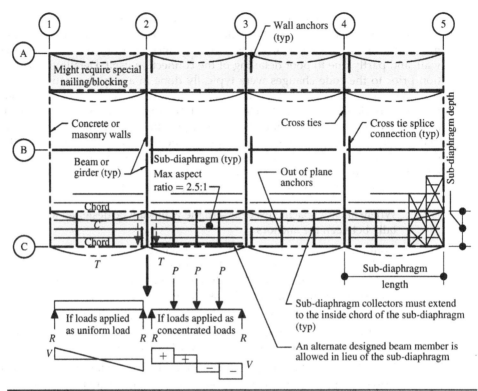

Figure 2.38 Sub-diaphragms for loading in the transverse direction.

out-of-plane forces as a uniform load, which is conservative. The depth of the sub-diaphragm can be increased to reduce the nailing requirements of the sub-diaphragm if necessary. The continuous cross ties and their connections are to be designed for tension and compression forces plus gravity loads as required by code. For loading in the longitudinal direction, sub-diaphragms are placed along grid lines 1 and 5 as shown in Fig. 2.39. The continuous cross tie occurs at grid line B. Special consideration should be given to the framing layout within the sub-diaphragms. The sub-diaphragms near grid line 1 show a panelized roof system with two layout configurations. The panels in the sub-diaphragm between grid lines A and B are laid out in a staggered pattern. In that configuration, framing members acting as the collectors will not always line up from panel to panel, which would prevent a continuous cross tie to the inside chord of the sub-diaphragm. An installation of an additional framing member would be required. A note regarding special nailing along the collector should be included on the drawings. The edges of the panels in the sub-diaphragm between grid lines B and C line up. In that case, the members acting as the collector line up. Tie straps are required to connect the collectors across the continuous panel joint in both configurations, otherwise a continuous load path across the depth of the sub-diaphragm is not provided. Other girder lines can also be used as continuous cross ties, as occurs in larger box type buildings to help reduce the sub-diaphragm depths and nailing. Continuous cross ties and sub diaphragms apply to concrete or masonry walls supported by a flexible diaphragm and are typically found in large box warehouse buildings. Using light-framed or

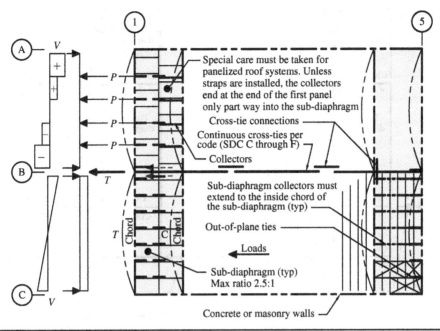

Special care must be taken for panelized roof systems. Unless straps are installed, the collectors end at the end of the first panel only part way into the sub-diaphragm

Cross-tie connections

Continuous cross-ties per code (SDC C through F)

Collectors

Sub-diaphragm collectors must extend to the inside chord of the sub-diaphragm (typ)

Out-of-plane ties

Loads

Sub-diaphragm (typ)
Max ratio 2.5:1

Concrete or masonry walls

Figure 2.39 Sub-diaphragms for loading in the longitudinal direction.

cross-laminated timber walls would eliminate the need for continuous cross ties, sub diaphragms, and significantly reduce the forces to the roof diaphragm.

2.9 Introduction to Transfer Diaphragms

The successful transfer of discontinuous chord or strut forces through a diaphragm or shear wall largely depends upon the use of a transfer diaphragm, commonly referred to as a sub-diaphragm, or by the use of a transfer area, which are discussed in detail in Sec. 2.10. Transfer diaphragms provide the following functions:

- A portion of a diaphragm used to transfer discontinuous chord or strut forces at horizontal and vertical offsets. See Figs. 2.40, 2.44, and 2.45.

- A diaphragm or portion of a diaphragm used to transfer shear forces from discontinuous shear walls. See Figs. 2.41 and 2.42.

- Provide lateral support for building sections that do not have other means of lateral support. See Fig. 2.43.

- A portion of a diaphragm used to dissipate member forces at the corners of an opening in a diaphragm or shear wall.

- A transfer diaphragm should have a similar stiffness (aspect ratio) as the main diaphragm.

The simplest forms of transfer diaphragms are shown in Fig. 2.40. The transfer diaphragm is a portion of a larger diaphragm that is used to transfer a disrupted chord or strut force into the body of the main diaphragm. An analogy of this transfer diaphragm

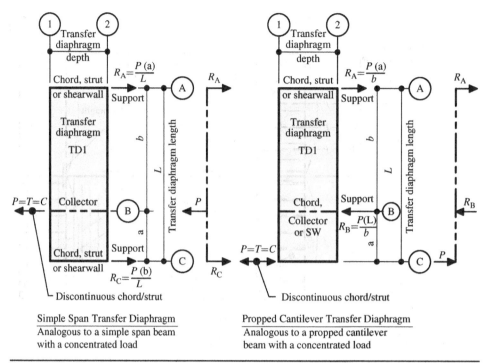

FIGURE 2.40 Simple span and propped cantilever transfer diaphragms.

shows that it acts like a simple span beam with a concentrated force, which transfers the disrupted member force out to the main chords or boundary elements by simple beam action, thereby providing a complete load path. The supports for the simple span transfer diaphragm occur at grid lines A and C. The propped cantilever transfer diaphragm has the disrupted chord or strut force applied at the end of the cantilever at grid line C, which causes the transfer diaphragm to act like a propped cantilever beam. Cantilevers can occur at both ends of a transfer diaphragm with interior supports. The supports for this transfer diaphragm occur at grid lines A and B. Transfer diaphragms are thoroughly covered in Chap. 3 by comprehensive examples and show that they can be very powerful analytical tools for transferring forces across areas of discontinuities.

Figure 2.41 shows the condition where a discontinuous interior shear wall is supported by the floor diaphragm below. The main diaphragm in this case becomes a transfer diaphragm which transfers the lateral force of the discontinuous wall into the shear walls below. This condition creates a Type 4 horizontal irregularity as noted in ASCE 7 Table 12.3-1 that triggers special overstrength and increase factors of Sections 12.10.1.1, 12.10.3.3, 12.3.3.3, and 12.3.3.4. The beam supporting the discontinuous wall is also used as the collector and must extend across the full width of the diaphragm to uniformly distribute the force into the diaphragm.

It is important to callout and fully detail all struts, collectors, and their nailing requirements on the drawings; otherwise, the system may not be installed in accordance with the intent of the design. An example of why this is important can be seen in Fig. 2.42. There is no guarantee where the sheathing joints will occur with respect to the

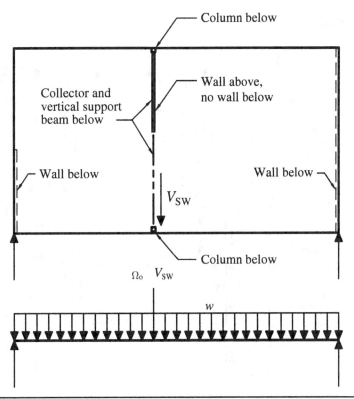

Figure 2.41 Discontinuous shear wall—transfer diaphragm.

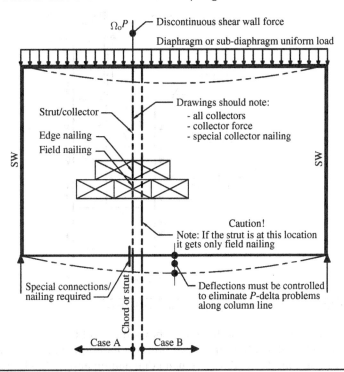

Figure 2.42 Special collector nailing issues.

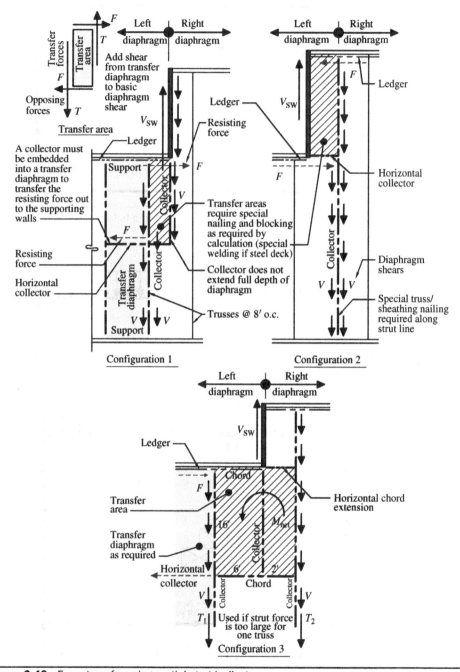

Figure 2.43 Force transfer using partial strut/collectors.

collector location. Case A of the figure shows the condition where a collector is placed at a plywood joint location. If special uniform nailing is not clearly called out on the drawings, the collector will receive half edge nailing and half field nailing because the staggered sheathing joints. Case B shows the collector placed at a location between

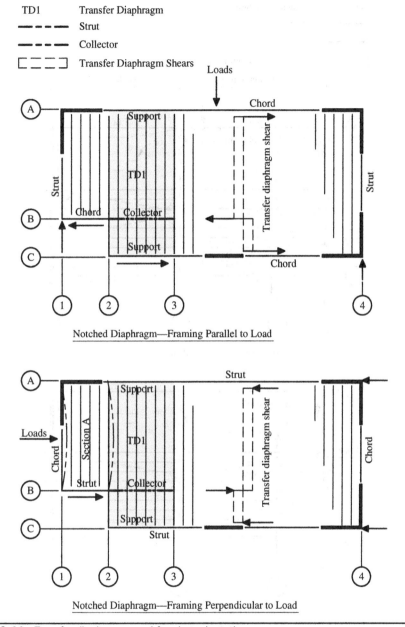

FIGURE 2.44 Transfer diaphragms and framing orientation.

panel joints. In this case, the collector could receive only field nailing (12″ o.c.). The result could lead to a redistribution of the lateral forces that are not anticipated.

Occasionally, the strut or collector coming off a shear wall or frame cannot be fully extended across the full width of the diaphragm. In that event, an alternate load path must be established to provide a complete load path across the diaphragm in

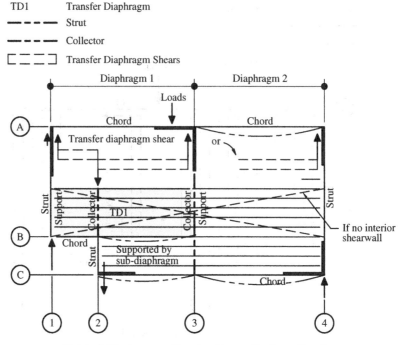

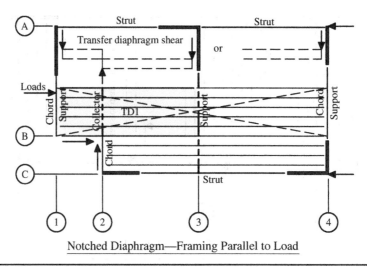

Figure 2.45 Transfer diaphragms and framing orientation.

accordance with the code. Figure 2.43 shows three partial strut or collector configurations. In configuration 1, the trusses do not line up with the shear wall. Therefore, an alternate load path is required to provide a complete load path across the diaphragm. This can be accomplished by using an adjacent truss that is offset to the left or right of the shear wall. A partial length strut or collector can be extended off a shear wall and lapped onto one of the adjacent trusses, which can be used as the main collector. A horizontal collector at the end of the short collector coming off the shear wall projects to the left penetrating the full depth of the transfer diaphragm to the left, completing the load path. The unit diaphragm shears, v, that are transferred into the collectors from the left and right diaphragms are shown in the figure. The collective shear force in the left collector acts toward the bottom of the page. That force is transferred across the transfer area into the shear wall and partial length in-line collector, which reacts in the opposite direction. The offset opposing forces cause a force couple that causes the transfer area to rotate in a counterclockwise direction. The rotation is resisted by the wall ledger that is connected to the wall at the top of the transfer area, and by the horizontal collector at the bottom of the transfer area that is embedded into the transfer diaphragm at the left. The horizontal collector force is transferred out to the exterior walls through the transfer diaphragm. This completes the load paths. Configuration 2 occurs when a collector cannot be extended from the end of a shear wall. In this case, the truss that is offset to the right of the shear wall that extends across the full width of the diaphragm can be used as the collector. The diaphragm shears transferred into the collector from the left and right diaphragms are shown in the figure. The collector force acts toward the bottom of the page and is transferred across a transfer area into the shear wall. The resisting force at the shear wall acts in the opposite direction to the collector force. These opposing offset forces cause the transfer area to rotate in a clockwise direction. The rotation is resisted by the wall ledgers at the top and bottom of the transfer area. Occasionally, drag forces are so large that one truss cannot be used as the collector. Configuration 3 shows the condition where a partial length collector extends off the shear wall. A collector truss is placed on each side of the partial length collector to reduce the forces applied to each collector truss. If the collector trusses are offset equal distances from the partial length collector, then the force in each collector truss will be equal. If the trusses are offset unequal distances from the partial collector, an unbalanced moment condition will occur that will cause the transfer area to rotate in a counterclockwise direction. The rotation is resisted by the wall ledger that is connected to the wall at the top of the transfer area, and by the horizontal collector at the bottom of the transfer area that is embedded into the transfer diaphragm, similar to configuration 1.

It is important to note that the transfer diaphragms and transfer areas already receive shear from the basic diaphragm shear. The additional shear caused by the transfer of the disrupted collector force must be added to the basic diaphragm shear.

The biggest challenge of transferring forces across discontinuities is the direction of the framing and the location of the shear walls. Several framing and wall layouts are shown in Figs. 2.44 through 2.47 which show how possible alternate load paths can be created across the discontinuities. The upper plan in Fig. 2.44 is loaded in the transverse direction with supporting shear walls located at grid lines 1 and 4. The framing is oriented parallel to the applied load. The horizontal offset at grid line B creates a discontinuity in the diaphragm chord at grid line 2B. The disrupted chord is extended into the transfer diaphragm located between grid lines 2 and 3 by a collector, where its force is then transferred out to the main diaphragm chords located at grid lines A and C. When that same diaphragm is loaded in the longitudinal direction as shown in the lower plan of Figure 2.44, the load path through the discontinuity is the same as the transverse loading. For this case, the diaphragm can be visualized as two separate diaphragms, which consist of the section between grid lines 1 and 2 from A to B (Section A) and the section between grid lines 2 and 4 from A to C. Section A is supported by the

TD1 Transfer Diaphragm
— - — - — Strut
— - — - — Collector
$\mathsf{C} \ \mathsf{\bar{\ }} \ \mathsf{\bar{\ }} \ \mathsf{\bar{\ }} \ \mathsf{J}$ Transfer Diaphragm Shears

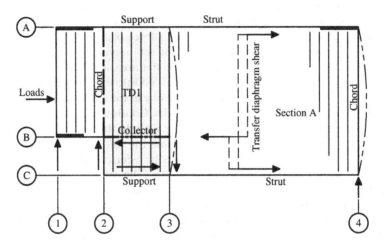

Propped Cantilever Diaphragm—Framing Perpendicular to Load

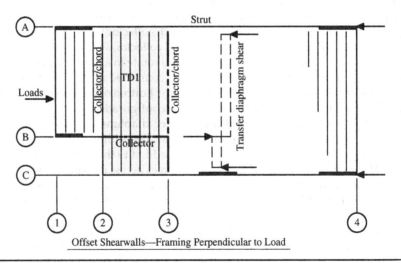

Offset Shearwalls—Framing Perpendicular to Load

FIGURE 2.46 Transfer diaphragms and framing orientation.

shear wall at grid line A and by the strut located at grid line B. Since there is no vertical lateral resisting element along grid line B, the support of the strut must be provided by the transfer diaphragm. The strut force is transferred out of the transfer diaphragm into the supports located at grid lines A and C.

The upper diaphragm shown in Fig. 2.45 has the framing oriented 90 degrees from the previous examples. Assume that the interior shear wall shown has not yet been

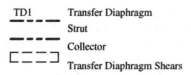

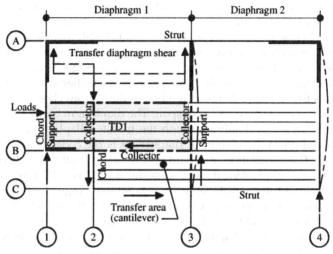

Propped Cantilever Diaphragm—Framing Parallel to Load

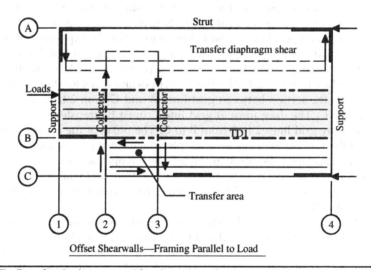

Offset Shearwalls—Framing Parallel to Load

FIGURE **2.47** Transfer diaphragms and framing orientation.

installed. The supporting walls for the diaphragm would then be located at grid lines 1 and 4. The offset of the chord from C to B causes a disruption of the chord at grid line 2B, which creates an offset diaphragm. Since the framing is oriented parallel to the chord, the transfer diaphragm must also be oriented in that direction. Without the interior shear wall, the transfer diaphragm must extend between grid lines 1 and 4. It should be remembered that all transfer diaphragms or sections being used as individual diaphragms must meet code allowable aspect ratios and have a similar stiffness to the main diaphragm; and that, collectors that are embedded into transfer diaphragms must extend the entire depth of the transfer diaphragm. Maintaining the required aspect ratio for the transfer diaphragm could cause the depth of the transfer diaphragm to extend from grid line A to B, making it impractical because of the collector length. An alternate approach would be to analyze the diaphragm as a simple span diaphragm that occurs between grid line 1 to 4 from A and B. This assumes that that the main section complies with the allowable aspect ratio. The section between grid line 2 and 4 from B to C would be supported off the main diaphragm. If an optional interior shear wall is installed as shown in the figure, two separate diaphragms are created. The diaphragm on the left has a horizontal offset in the bottom chord at grid line 2B and the diaphragm on the right is a simple span. The diaphragm between grid lines 1 and 3 is analyzed as an offset diaphragm with the transfer diaphragm oriented horizontally. The right support for the transfer diaphragm is the collector coming off the interior shear wall. The installation of the interior shear wall allows a shorter and shallower transfer diaphragm, making it more economical and easier to construct. The diaphragm at the bottom of the Fig. 2.45 shows the same diaphragm loaded in the longitudinal direction. The diaphragm support walls are located along grid lines A and C. The top chord occurs along grid line 1 which is offset to grid line 2, requiring the diaphragm to be designed as an offset diaphragm. The transfer diaphragm is the same configuration as the previous example.

All the diaphragms in Figs. 2.46 and 2.47 are loaded in the longitudinal direction. The upper diaphragm shown in Fig. 2.46 has supporting walls along grid lines A and B only. Since there are no vertical supporting elements along grid line C, the diaphragm section between grid lines A and C from 3 to 4 must be supported by a strut along grid line C, which is embedded into and supported off the end of the transfer diaphragm. The transfer diaphragm functions like a propped cantilever beam. The upper diaphragm shown in Fig. 2.47 also has its supporting walls located along grid lines A and B and has the section between grid lines A and C from 2 to 4 supported by the strut along line C. The drag force at grid line C must be transferred across a transfer area bounded by grid lines 2 to 3 from B to C. The offset of the strut and collector on each side of the transfer area causes a couple which must be resisted by the offset chord at grid line 2 and the collector at grid line 3, which oppose the couple forces. The collector at grid line 3 transfers the opposing force into the shear wall at that line. The disrupted chord at grid line 2 is embedded into a transfer diaphragm that has its supports located at grid lines 2 and 3.

The lower diaphragm shown in Fig. 2.47 has supporting walls along grid lines A, B, and C. The wall lines at lines B and C are assumed to act in the same line of resistance. For that to happen the same transfer area can be used to connect the two shear walls. The shared shear wall forces that are transferred across the offset produce a couple that is resisted by collectors embedded into a full-length transfer diaphragm oriented parallel to the applied load, which might have to extend in depth from grid line B to A due to the width of the diaphragm. While the framing examples presented cover most of the conditions encountered in today's structures, other layouts and load paths are possible. The ability to visualize how forces are applied and transferred through areas of discontinuity is key to solving the load path problems. These examples will be fully developed in later chapters.

2.10 Introduction to Transfer Areas

Transfer areas are similar to transfer diaphragms but are generally smaller in scale and use adjacent nonstructural walls or transfer diaphragms to resist the rotational forces from the offset. Transfer areas can be useful when trying to connect lateral resisting elements across offsets when it is not practical to install a full transfer diaphragm. As demonstrated in the previous sections, the opposing forces across the offset cause a force couple, which must be resisted by intersecting walls or distributed into a transfer diaphragm as shown in Fig. 2.48. The objective is to utilize a nonstructural wall to resist the rotational forces if it is possible not to exceed the wall sheathing capacity or require a hold down. Higher forces would require the resisting wall to be designed and detailed as a shear wall. In some cases, the return wall does not have sufficient strength or length to resist the force. In that event, the resisting force must be distributed into a transfer diaphragm or by other means.

Take for example in the plan, transfer area 1, TA1, distributes its rotational forces into the perpendicular walls at each end of the transfer area. Transfer area 2, TA2, distributes its right rotational force into the transfer diaphragm because the wall has no strength, and on the left side, into the perpendicular wall.

The plan section on the right side of the figure shows the framing around a typical transfer area with the direction of the forces acting around the perimeter. The transfer area shears are equal to the basic diaphragm shears at that location plus the shears that the transfer forces cause. It is important to note on the drawings the boundary nailing

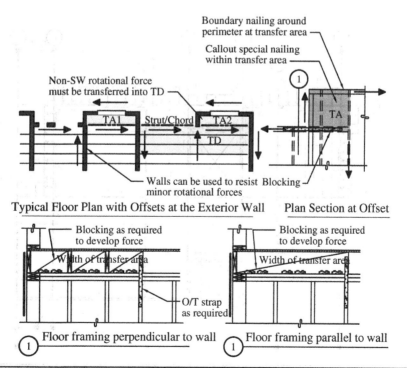

FIGURE 2.48 Transfer areas.

requirements around the transfer area and the tie strap requirements. Section 1 shows possible details that can be used to transfer the rotational forces into the top of the wall plate for floor framing that occurs in either direction. The blocking doesn't necessarily need to extend the full length of the wall, only far enough to transfer the force without causing a net overturning force. Once the transfer has been made, the wall top plate will distribute the force into the sheathing the full length of the wall.

2.11 Problem

Problem P2.1: Complex Diaphragm, Load Distribution

The section of the diaphragm between grid lines 1 and 6 is similar to the same section of the diaphragm shown in Example 8.1 of this book. Therefore, the analysis is the same with the exception of the cantilever section to the right of grid line 6. The cantilever section's top chord is tied to a collector that distributes the chord force into transfer diaphragm TD2.

Given: The structure shown in Fig. P2.1 is a one-story structure with shear walls placed at the exterior walls only. Wind loads are shown acting in the transverse and longitudinal directions.

Find:

1. Break the wind loads shown acting in the longitudinal into strip loads w_1 through w_9. Determine their magnitude.

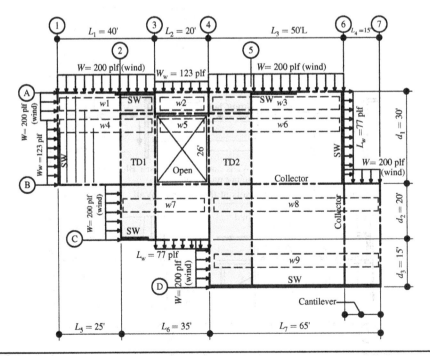

FIGURE P2.1

2. Determine seismic strip loads w_1 through w_9.

$DL_{roof} = 20$ psf

$DL_{walls} = 10$ psf

Wall height $= 15$ ft

Wind load $= 200$ plf as shown

Occupancy category II, $I_e = 1$

$C_s = 0.112$ (ASD)   ▲

2.12   References

1. Applied Technology Council (ATC), *Guidelines for Design of Horizontal Wood Diaphragms*, ATC-7, ATC, Redwood, CA, 1981.
2. American Wood Council (AWC), *Special Design Provisions for Wind and Seismic with Commentary*, AWC, Leesburg, VA, 2021.
3. *American Society of Civil Engineers (ASCE). 20222016. ASCE/SEI 7-22 16 Minimum Design Loads for Buildings and Other Structures*, ASCE/SEI 7-16, ASCE, New York, 2016.
4. International Code Council (ICC), *International Building Code, 2021 with commentary*, ICC, Whittier, CA, 2021.
5. American Wood Council (AWC), *National Design Specification for Wood Construction and supplement*, AWC, Leesburg, VA, 2018.
6. American Wood Council (AWC), *Commentary on the National Design Specification for Wood Construction*, AWC, Leesburg, VA, 2018.
7. Diekmann, E. F., "Design of Wood Diaphragms," *Journal of Materials Education*, 1982, Fourth Clark C. Heritage Memorial Workshop, Wood Engineering Design Concepts University of Wisconsin, WI.
8. APA—The Engineered Wood Association, *Design/Construction Guide-Diaphragms and Shear Walls*, APA Form L350, APA—The Engineered Wood Association, Engineering Wood Systems, Tacoma, WA, 2007.
9. U.S.D.A. Forest Service, *Research Note-FPL-0210*, USDA, Madison, WI, 1970.
10. Breyer, D. E., Martin, Z., and Cobeen, K. E., *Design of Wood Structures ASD*, 8th ed., McGraw-Hill, New York, 2020.
11. Structural Engineers Association of California (SEAOC), *IBC Structural/Seismic Design Manual*, Volumes 1–3, SEAOC, CA, 2018.
12. Kula, U., "Diaphragms and Diaphragm Chords," *SEAOC 2001 70th Annual Convention Proceedings*, Structural Engineers Association of California (SEAOC), Sacramento, CA, 2001.
13. Mays, T., *Guide to the Design of Diaphragms, Chords and Collectors*, National Council of Structural Engineers Association, ICC publications, 4051 West Flossmoor Road, Country Club Hills, IL, 2009.
14. International Code Council (ICC), *International Residential Code, 2021 with commentary*, ICC, Whittier, CA, 2021.
15. Applied Technology Council (ATC), *Proceedings of a Workshop on Design of Horizontal Wood Diaphragms*, ATC-7-1, ATC, Redwood, CA, 1980.

CHAPTER **3**

Diaphragms with Horizontal End Offsets

3.1 Introduction

Up to this point, regular diaphragms have been the primary discussion. The distribution of shear through these assemblies is direct and typically occurs along straight lines into continuous boundary elements. Challenging architectural designs, which incorporate vertical and horizontal offsets, angular wall lines, large open spaces and fewer solid walls, prevent the establishment of direct load paths. These complex irregular shaped structures can require complicated and costly solutions. This chapter focuses on diaphragms with horizontal offsets located at the end supports of the diaphragm, commonly referred to as a "notched diaphragm." By definition, offsets in the diaphragm can create a Type 3 horizontal irregularity in accordance with ASCE 7 Table 12.3-1,[1] a Diaphragm Discontinuity Irregularity, where there is an abrupt discontinuity or variation in stiffness of the diaphragm. This type of irregularity applies to Seismic Design Category (SDC) D-F. The offsets are created when a portion of an exterior wall line is offset from the main wall line, which causes a disruption in the diaphragm chord or strut. By code, whenever this occurs, the disrupted chord or strut force must be transferred across the discontinuity through an alternate load path. At diaphragm discontinuities such as openings and reentrant corners, ASCE 7 Section 12.10 requires the following:

- The design shall ensure that the dissipation or transfer of edge (chord) forces combined with other forces in the diaphragm is within the shear and tension capacity of the diaphragm.

- For structures that have horizontal or vertical structural irregularities of the types indicated in Section 12.3.3.4, the requirements of that section shall apply.

- Collector elements shall be provided that are capable of transferring the seismic forces originating in other portions of the structure to the element providing resistance to those forces.

In addition, SDPWS[2] Section 4.1.9 states that the diaphragm (and shear wall) sheathing shall not be used to splice boundary elements.

Discontinuous chord or strut forces are often dissipated into the main body of the diaphragm using a continuous light gauge steel strap applied over the sheathing and

85

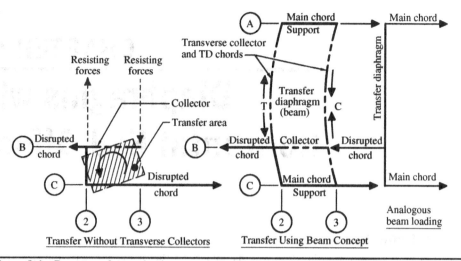

Figure 3.1 Force transfer mechanism.

nailed into flat blocking below. The strap and blocking are often designed incorrectly as previously discussed in Sec. 2.6. The approach of designing the steel strap and blocking for tension forces only and extending the strap an arbitrary distance into the diaphragm does not ensure that a complete load path has been established. Obviously, all chords, struts, and collectors receive both tension and compression forces due to the reversal of loads. The transverse collectors shown at grid lines 2 and 3 on the right side of Fig. 3.1, acting as the transfer diaphragm chords, are rarely included in the design. The free-body diagram shown on the left side of the figure clearly shows that the omission of those collectors allows rotation of the transfer area to take place because there is nothing present to oppose the rotating couple forces. Localized tearing and displacements due to increased shears and nail slip in the transfer area can occur if these members are omitted. Installing the missing transverse collectors for an arbitrary distance into the diaphragm can also create an incomplete load path. The possibility that localized distress within the diaphragm at the ends of the collectors could still exist. The free body at the right of the figure provides a method that not only eliminates rotational problems but also provides a complete load path as required by code. The method utilizes a portion of the diaphragm to the right of the discontinuity as a sub-diaphragm, or transfer diaphragm (TD), which receives the disrupted chord force as a concentrated force and distributes it out to the main diaphragm chords by simple beam action, thereby providing a complete load path.

3.2   Method of Analysis

Edward F. Diekmann, of GFDS Engineers in San Francisco,[3] developed a method of analyzing diaphragms with offsets (notches) and openings in the early 1980s. Although his method of analysis was published in journals and in brief presentations in various publications, the method does not appear to be widely known or used. Several of the examples in this book are based upon Mr. Diekmann's design method. Figure 3.2 describes the basic components of a horizontally offset (notched) diaphragm and

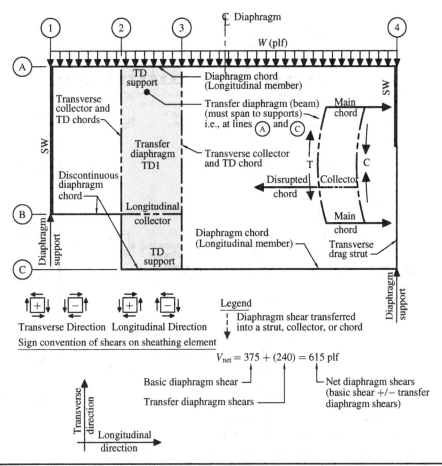

FIGURE 3.2   Offset diaphragm terminology.

illustrates the concept of transferring the disrupted chord force across the offset. Briefly, the disrupted chord is embedded into the transfer diaphragm by a collector. The transfer diaphragm acts like a beam with a concentrated load applied as depicted by the inset diagram. The transfer diaphragm transfers the disrupted chord force into its supports which are located at the top and bottom chords of the main diaphragm. The typical sign conventions for shears acting on diaphragm sheathing elements in the transverse and longitudinal loading directions, and a legend addressing the determination of the net shear within the transfer diaphragm have been included for reference. These sheathing elements play a key role in showing how the shears are distributed through the diaphragm. A complete description of this method is presented below.

To successfully solve the problem of transferring forces through discontinuous load paths, it is important to understand how shears are distributed into and out of a simple diaphragm. Diekmann[3] and Kula[4] both demonstrated in their presentations and examples a visual representation showing how shears are transferred from the sheathing into boundary members and collectors is critical to solving and understanding complex diaphragms. As previously discussed in Chap. 2, this is accomplished by using sheathing

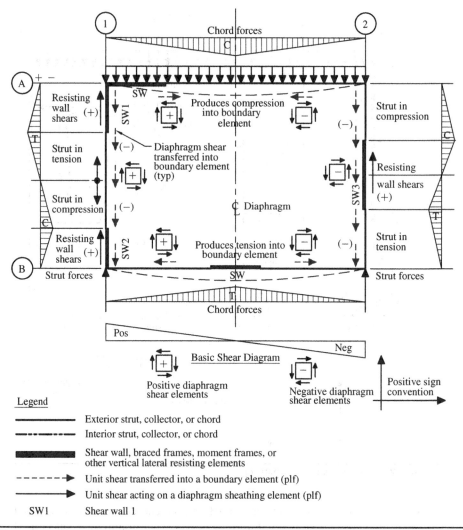

Figure 3.3   Shear distribution into a simple diaphragm.

element symbols and transfer shears. Sheathing element symbols are represented by a 1-foot by 1-foot piece of sheathing with the direction of the shears shown acting on each edge of the sheathing element. Transfer shears are the shears that are transferred from the edge of the sheathing element into a boundary element (e.g., chord, collector, strut, shear wall). Figure 3.3 represents a simple span diaphragm loaded in the transverse direction. The basic shear diagram is drawn below the diaphragm plan. The left half of the diagram is in positive shear and the right half is in negative shear. Sheathing element symbols are placed below the basic shear diagram to show the direction of shears acting on the edges of the sheathing elements in the positive and negative areas of the shear diagram. The positive and negative sheathing elements are placed on the plan at the top and bottom chords, and at the diaphragm support lines for reference. The shears transferred from these sheathing elements into the boundary elements are also shown on the plan. The shears transferred into the boundary elements are equal in magnitude but act in opposite directions to the shear acting on the sheathing elements. At grid line 1, the diaphragm shears that are transferred into the boundary elements are acting in the negative direction. The wall shears that resist these transfer shears act in the opposite direction and are positive. The direction of the shears transferred into the strut produces tension forces in the upper half of the strut because it is pulling away from the upper wall and compression forces in the lower half because it is pushing against the lower wall. The strut force diagram is drawn at the left of grid line 1 and is constructed by summing the net shears along the line of lateral-force-resisting system. At grid line 2, the wall is located near the center of the wall line, with struts placed on the plan above and below the wall. The direction of the shears transferred into the strut above the wall shows that it is acting in compression. The strut below the wall is acting in tension. The strut force diagram is drawn at the right of grid line 2. The direction of the shears transferred into the top and bottom chord shows that they are in compression and tension, respectively, in accordance with typical diaphragm behavior. From this example, the sheathing elements provide a valuable visual reference on how the shears are distributed through the diaphragm and into the boundary elements and whether the boundary element forces are in tension or compression. This method would typically not be utilized for simple diaphragms, but the value of its use will become even more apparent when analyzing a complex diaphragm.

Initially, load paths across discontinuities are typically assumed based on engineering judgment and then verified by analysis. Throughout the analysis, construction of the collector/chord force diagrams will prove or disprove the assumed load paths and load distributions. These force diagrams are a visual representation that the sum of all forces is equal to zero. In some cases, the force diagrams will not close to zero, which indicates that an error exists in the design assumptions, calculations that have been made, or round off issues have occurred. In that event, the engineer must evaluate the initial results and adjust the analysis accordingly. Solving for only a few key members will leave the design incomplete and could possibly leave a major error undiscovered.

The method of analyzing a diaphragm with a horizontal offset (notch) using a transfer diaphragm is summarized below, reference Fig. 3.4.

Procedure:

1. Solve for the diaphragm reactions.
2. Construct the basic shear diagram (unit shears-plf or klf), with special emphasis of the area within the transfer diaphragm as shown in the figure.
3. Find the shear values at critical locations on the basic shear diagram (e.g., all collector locations and areas of discontinuities).
4. Find the chord force at the discontinuity (offset at 2B), by cutting a free-body diagram at grid line 2 and solving for the bottom chord force at the discontinuity.
5. Determine the additional shear applied to the transfer diaphragm from the transfer of the disrupted chord force (transfer diaphragm shears) and determine whether they are positive or negative.
6. Determine the net diaphragm shears in the transfer diaphragm area by adding or subtracting the transfer diaphragm shears from the basic diaphragm shears.

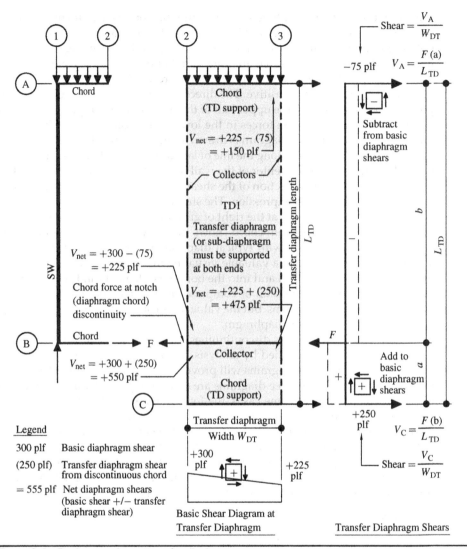

FIGURE 3.4 Basic procedure—transfer diaphragm.

7. Place all the sheathing elements and the direction of the shears being transferred into all the chords, struts, and collectors on the plan.

8. Determine ALL the strut, collector, and chord forces. The force is equal to the area of the shear diagram acting on the element being examined. Sum the forces along the length of the collectors or line of lateral-force-resistance. The force diagrams must close to zero or an error exists.

9. Determine the diaphragm nailing requirements.

10. Design all struts, collectors, chords, and connections, applying the over-strength factor as required by code and/or by engineering judgment.

The section of the diaphragms to the left of the discontinuity, bounded by grid lines 1 and 2 from A to B shown in Fig. 3.4, represents the section of the diaphragm with a reduced diaphragm width. The reduced width causes a disruption in the diaphragm chord and increases the diaphragm shears. To the right grid line 3 is the main body of the diaphragm. The section bounded between grid lines 2 and 3 from A to C has been designated as a transfer diaphragm, which is used to transfer the disrupted chord force out to the main diaphragm chords. The discontinuous chord at grid line 2B must be extended the full width of the transfer diaphragm by means of a collector. The disrupted chord at that location imposes its force on the transfer diaphragm like a simple span beam with a concentrated load. If the chord force is acting to the left, the transfer diaphragm reactions (shear) must act to the right. The shear diagram to the right of the transfer diaphragm shows the distribution of the discontinuous chord force and the additional resulting shears acting on the transfer diaphragm. The transfer diaphragm unit shears are equal to the transfer diaphragm reactions divided by the width of the transfer diaphragm, W_{TD}.

$$F = \text{disrupted chord force}$$

$$L_{TD} = a + b$$

The total shear and unit shears acting on the transfer diaphragm at grid lines A and C are

$$V_A = \frac{F(a)}{L_{TD}}, \qquad v_A = \frac{V_A}{W_{TD}} = \text{transfer diaphragm unit shears at grid line A}$$

$$V_C = \frac{F(b)}{L_{TD}}, \qquad v_C = \frac{V_C}{W_{TD}} = \text{transfer diaphragm unit shears at grid line C}$$

The direction of the transfer diaphragm shear (reactions) acting at the supports should be carefully noted. It represents the direction of the shear acting on the edge of the sheathing element closest to the chord member, not the shear that is transferred into the chord. By placing a sheathing element symbol adjacent to the transfer diaphragm reactions and completing the direction of the shear forces on the remaining sides of the sheathing element, determination of whether the transfer diaphragm shears are positive or negative can be made in accordance with the typical sign convention shown in Fig. 3.3. The main diaphragm is already stressed in shear from the uniform load, as calculated in the basic shear diagram. The additional transfer diaphragm shears created by the disrupted chord force must be added to or subtracted from the basic diaphragm shears to accurately account for the combined localized effects within the transfer diaphragm. This complies with the code requirements noted in Sec. 3.1 of this text and ASCE7-16 Section 12.10.1.

The portion of the basic shear diagram that is acting at the transfer diaphragm area is located below the transfer diaphragm for reference. The basic diaphragm shear is uniformly distributed across the entire length of the transfer diaphragm (L_{TD}). An example of a similar complete basic shear diagram can be referenced in Fig. 3.8. In Fig. 3.4, assume that the basic shear to the right of grid line 2 is +300 plf and +225 plf on the left and right side of grid line 3. The basic shears are equal on the left and right side of grid line 3 because the diaphragm width is the same. Also assume that the transfer diaphragm shear is a negative 75 plf from grid line A to B, and a positive 250 plf from grid line B to C. By summing the basic shear and transfer diaphragm shear, the net shears within the transfer diaphragm become

From grid line 2C to line 2B (below the collector)

$$v_{net} = +300 \text{ plf basic shear} + (250) \text{ plf transfer diaphragm shear} = +550 \text{ plf}$$

From grid line 2B to line 2A (above the collector)

$$v_{net} = +300 \text{ plf basic shear} - (75) \text{ plf transfer diaphragm shear} = +225 \text{ plf}$$

From grid line 3B to line 3A (above the collector)

$$v_{net} = +225 \text{ plf basic shear} - (75) \text{ plf transfer diaphragm shear} = +150 \text{ plf}$$

From grid line 3B to line 3C (below the collector)

$$v_{net} = +225 \text{ plf basic shear} + (250) \text{ plf transfer diaphragm shear} = +475 \text{ plf}$$

Only the shear in the transfer diaphragm area is affected by the transfer of the disrupted chord force. The shears in the diaphragm sections to the left and right of the transfer diaphragm remain unaffected and are equal to the basic diaphragm shear. It is important to note, in this case, that the shears in the transfer area bounded by grid lines B and C from 2 to 3 are twice the magnitude of the basic diaphragm shears. Had the nailing patterns on the structural drawings been based on the basic shear diagram, the nails in the transfer area would be overstressed and deformation within the transfer area could occur due to increased nail slip. Calculations can show that the transfer area shears can be up to three times greater than the basic shears.

Upon determination of the net diaphragm shears, the net shear sheathing elements and the shears that are transferred from the elements into the struts and collectors should be added to the diaphragm plan to show the direction of the shear forces acting on the struts and collectors. Continuing the example, the applications of net shears and transfer shears are applied to the collector as shown in Fig. 3.5. Observing the direction of the shears transferred into the collector at grid line B, a net shear in the sheathing of +225 plf occurs at the upper left of the collector and a net shear in the sheathing of +550 plf occurs below the collector at the same location. The directions of the shears transferred into the collector oppose each other with the largest shear value acting to the right. Visually, the direction of the shears indicates that the 225 plf shear must be subtracted from the 550 plf shear, with a resulting shear of 325 plf acting to the right. At the opposite end of the collector, the +150 plf shear must also be subtracted from the +475 plf shear, with a resulting shear value of 325 plf also acting to the right. These will not always be the same shear value.

$$\text{Shear left} = 550 - (225) = 325 \text{ plf, acting to the right (tension)}$$

$$\text{Shear right} = 475 - (150) = 325 \text{ plf, acting to the right (tension)}$$

The resulting shear diagram is shown at the bottom of the figure. The force on the collector is the average shear multiplied by the length of the collector, or simply put, the area of the shear diagram. The resulting force should equal the calculated chord force at the discontinuity. If it does not, an error exists.

$$F = \frac{(325 + 325)(L_{collector})}{2}, \text{ tension, the direction of force is acting to the right}$$

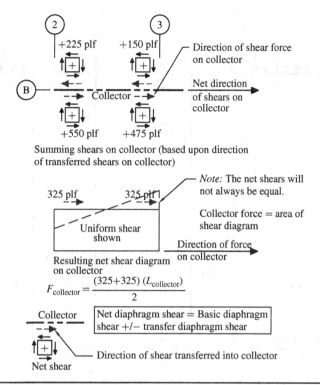

FIGURE 3.5 Shear distribution into a collector.

Tables could be generated to compile the diaphragm shear, transfer shear, and net shear data. However, as with tables, it is often difficult to mentally visualize the overall picture of how the shears are flowing through the diaphragm. The method shown above provides a clear visual representation of the direction of the shears that are transferred into the elements and visually verifies if the chords, struts, or collectors are in tension or compression.

Diekmann[3] and ATC-7[5] recommended that the aspect ratio of the transfer diaphragm should be similar to the main diaphragm, with a maximum aspect ratio of 4:1. Code limits the aspect ratio for blocked wood diaphragms to 4:1 and 3:1 for unblocked diaphragms. This should be carefully considered when determining the initial size of the transfer diaphragm. Whenever tie straps are installed, the tendency is to make the extension of the collector as short as possible to minimize construction costs, especially when the framing is oriented perpendicular to the collector. This can create two problems: first, transfer diaphragm shears will be high, which will increase the nailing requirements, cause splitting, and potentially create localized failures; second, if the collector length is too short, the required aspect ratio of the transfer diaphragm cannot be maintained, and the transfer diaphragm will not be stiff enough to distribute the shears as anticipated. To prove the first point, a comparison of transfer diaphragms with aspect ratios of 2:1 and 4:1 will be examined as shown in Fig. 3.6. A chord force of 6000 lb is applied to each transfer diaphragm. The basic diaphragm shear diagrams acting at the transfer

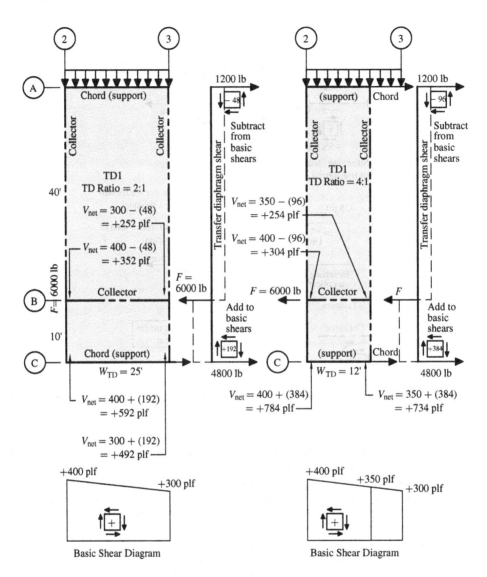

Maximum aspect ratios for transfer diaphragms	
Type	Maximum L/W Ratio
Wood structural panel, nailed all edges (blocked)	4:1
Wood structural panel, blocking omitted at intermediate joints	3:1

Figure 3.6 Transfer diaphragm aspect ratio.

diaphragm area are as shown at the bottom of the figure. The transfer diaphragm reaction and unit shears at grid lines A and C are

Transfer Diaphragm Reactions	A/R = 2:1	A/R = 4:1
$R_A = \dfrac{6000(10)}{50} = 1200$ lb	$v_A = \dfrac{1200}{25} = -48$ plf	$v_A = \dfrac{1200}{12.5} = -96$ plf
$R_C = \dfrac{6000(40)}{50} = 4800$ lb	$v_C = \dfrac{4800}{25} = +192$ plf	$v_C = \dfrac{4800}{12.5} = +384$ plf

The direction of the transfer diaphragm reaction at grid line A produces a negative shear, while the direction of the reaction at grid line C produces a positive shear.

Net Shears	A/R = 2:1	A/R = 4:1
Grid line 2 from B to C	$v_{net} = +400 + (192) = +592$ plf	$v_{net} = +400 + (384) = +784$ plf
Grid line 3 from B to C	$v_{net} = +300 + (192) = +492$ plf	$v_{net} = +350 + (384) = +734$ plf
Grid line 2 from A to B	$v_{net} = +400 - (48) = +352$ plf	$v_{net} = +400 - (96) = +304$ plf
Grid line 3 from A to B	$v_{net} = +300 - (48) = +252$ plf	$v_{net} = +350 - (96) = +254$ plf

It may be obvious that the transfer diaphragm with the highest aspect ratio will have the highest shear. However, the difference between the two examples should be carefully examined to evaluate the full impact on the design. Each transfer diaphragm will have its advantages and disadvantages.

Aspect Ratio = 2:1

Advantages:

- Entire transfer diaphragm does not have to be blocked (due to aspect ratio); only the area between grid lines B and C (due to the higher shear value).
- Nail spacing is reasonable and 2× framing can be used.
- Transverse collector forces will be smaller.

Disadvantages:

- The collector at grid line B is longer.
- The transfer diaphragm is larger.

Aspect Ratio = 4:1

Advantages:

- The collector at grid line B is shorter.
- The transfer diaphragm is smaller.

Disadvantages:

- Nail spacing is closer and $3\times$ framing might have to be used to reduce splitting problems.
- Transverse collector forces will be much higher.
- Entire transfer diaphragm must be blocked, based on the aspect ratio.
- The design capacity can approach demand, no room for poor workmanship.

The net transfer diaphragm shear above the collector at grid line B is usually less than the basic diaphragm shear, depending on the location of the discontinuous force. Consideration should be given for the final nailing to be based on the largest shear at that area (i.e., basic diaphragm shear vs. net shear).

Optimizing the transfer diaphragm and collector system requires careful consideration. Issues critical to the layout and design of a transfer diaphragm are

1. Aspect ratios
2. The magnitude of the resulting shears—capacity versus demand
3. Direction of the framing
4. Framing sizes to avoid splitting issues
5. Collector length
6. Diaphragm blocking requirements

Special attention should also be given to ASCE 7 Section 12.10.2.1 regarding the design of collectors and required use of the over-strength factor, omega, Ω_o, and its exception.

3.3 Development of Member Forces

Figure 3.7 can serve as a useful guide in understanding how a force can be developed in a collector. The basic rule of thumb is that a force can only be developed if there is a net shear acting on the element.

Partial plan at grid line 2 from A to B shows positive shears of varying magnitude acting on the sheathing elements on both sides of the vertical collector, meaning that the direction of the shears acting on each edge of the sheathing element act in the opposite direction. Since the shears that are transferred onto the collector are the same magnitude as the sheathing element but act in the opposite direction, the net shears acting on the collector, in all cases, will be the sum of the magnitudes of the shears transferred to the collector. In this case, they will be subtracted from each other. The net transfer shears acting on the horizontal collector at grid line B to the right of grid line 2 is determined in the same manner. The horizontal collector to the left of grid line 2 and vertical collector below grid line B have transfer shears acting on only one side and therefore are equal to the sheathing shears but act in the opposite direction.

Partial plan at grid line 3 from A to B shows the condition where a negative sheathing shear is located on the left side of the collector and a positive shear on the right side of grid line 2, both of varying magnitude. In this case, the shears acting on each edge of the sheathing element act in opposite directions, but the shears transferred into the collector act in the same direction, requiring them to be added together to get the net shear acting on the collector. For the horizontal collector to the right of grid line 3 and the

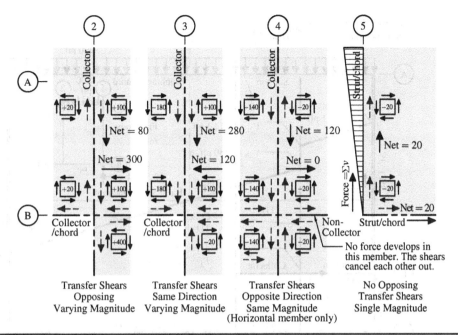

FIGURE 3.7   Development of member forces.

horizontal and vertical collectors with shears one side only, the determination of net shears is identical to the case at grid line 2.

Partial plan at grid line 4 along B shows negative shears having the same magnitude acting on the sheathing elements, on both sides of the horizontal collectors. Since the transfer shears are of equal magnitude but act in opposite directions, they will cancel out to zero and no force will be developed. However, the shears applied to the vertical collector above grid line B will result in a net shear as shown.

Partial plan at grid line 5 shows the condition where transfer shears act on only one side of the collector and therefore are equal to the sheathing shears but act in the opposite direction.

3.4   Single Offset Diaphragms

All the tools that are necessary to analyze a horizontally offset diaphragm have been reviewed and discussed in the preceding sections. The following examples show how those tools are applied to solve the transfer of the discontinuous chord or strut force across the offset in both the transverse and longitudinal directions. The direction of the roof or floor framing will be ignored until a familiarization with the techniques of solving these problems has been established.

Example 3.1: Single Offset Diaphragm, Analysis in the Transverse Direction

The diaphragm shown in Fig. 3.8 has a length of 125 ft and a width of 50 ft, with a 15-ft horizontal offset between grid lines 1 and 2. A 200-plf load is applied uniformly along the entire length of the diaphragm. The width of the diaphragm between grid lines A and B is 35 ft. The diaphragm chord is discontinuous at grid line 2B because of the

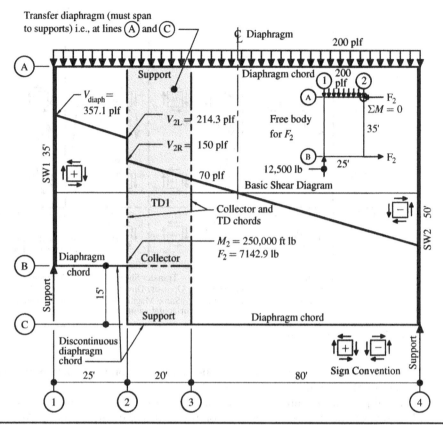

FIGURE 3.8 Example 3.1—Single offset diaphragm, analysis in the transverse direction.

offset. The chord is extended into the transfer diaphragm that lies between grid lines 2 and 3, by means of a longitudinal collector. The transfer diaphragm is supported at the main diaphragm chord locations at grid lines A and C, making the transfer diaphragm length 50 ft with a width of 20 ft. The aspect ratio of the transfer diaphragm is 2.5:1, which is the same as the main diaphragm. Transverse collectors are placed along the entire length of the transfer diaphragm at grid lines 2 and 3 to act as chord members for the transfer diaphragm. The sign convention for the diaphragm sheathing elements is shown at the lower right of the figure.

Construction of the basic shear diagram: The unit shear in the diaphragm must be calculated before adjusting the shears in the transfer diaphragm area for the influence of the disrupted chord force. Shear values are required at all collector locations. From Fig. 3.8:

$$R_1 = R_4 = \frac{wL}{2} = \frac{200(125)}{2} = 12,500 \text{ lb reaction at grid lines 1 and 4}$$

$$v_1 = \frac{R_1}{W_{\text{left}}} = \frac{12,500}{35} = +357.1 \text{ plf unit shear at grid line 1}$$

$$V_2 = R_1 - wx = 12,500 - 200(25) = 7500 \text{ lb shear at grid line 2}$$

$$v_{2L} = \frac{V_2}{W_{left}} = \frac{7500}{35} = +214.3 \text{ plf unit shear left side of grid line 2}$$

$$v_{2R} = \frac{V_2}{W_{right}} = \frac{7500}{50} = +150 \text{ plf unit shear right side of grid line 2}$$

$$V_3 = 12{,}500 - 200(45) = 3500 \text{ lb shear at grid line 3}$$

$$v_3 = \frac{3500}{50} = +70 \text{ plf unit shear at grid line 3}$$

$$v_4 = \frac{-12{,}500}{50} = -250 \text{ plf unit shear at grid line 4}$$

Determination of the chord force at grid line 2B by cutting free-body diagram at grid line 2 (see Fig. 3.8):

$$M_{2B} = R_1 x - \frac{wx^2}{2} = 12{,}500(25) - \frac{200(25)^2}{2} = 250{,}000 \text{ ft lb}$$

$$F_{2B} = \frac{M_{2B}}{W} = \frac{250{,}000}{35} = 7142.9 \text{ lb}$$

Note: The chord force at that location is also equal to the area of the basic shear diagram between lines 1 and 2.

Determination of transfer diaphragm shears (see Fig. 3.9): The disrupted chord force transferred into the transfer diaphragm affects the diaphragm unit shears in that area.

Disrupted chord force = 7142.9 lb

Summing moments about grid line C:

$$V_A = \frac{7142.9(15)}{50} = 2142.9 \text{ lb shear force at grid line A}$$

$$v_A = \frac{V_A}{W_{TD}} = \frac{2142.9}{20} = 107.1 \text{ plf unit shear in the transfer diaphragm at grid line A}$$

The direction of the reaction at grid line A is acting to the right. Applying the shear forces on the remaining sides of the sheathing element shows that the transfer diaphragm shear at grid line A is negative.

Summing moments about grid line A:

$$V_C = \frac{7142.9(35)}{50} = 5000 \text{ lb}, \qquad v_C = \frac{5000}{20} = +250 \text{ plf}$$

The direction of the reaction at C is also acting to the right. Applying the shears on the remaining sides of the sheathing element shows that the transfer diaphragm shear is positive.

Determination of diaphragm net shears (see Fig. 3.9): It is useful to overlay the basic shear diagram and transfer diaphragm shear diagram onto the diaphragm plan. The process of adding and subtracting shear values is greatly simplified because both shear diagrams are side by side for easy visual reference. It should be reemphasized that only the

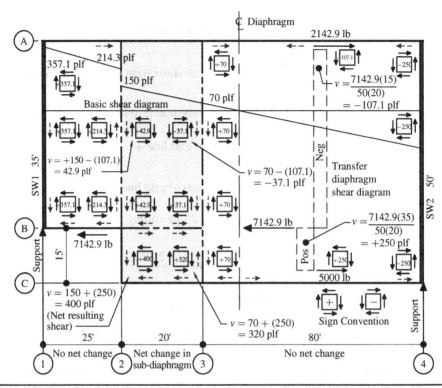

FIGURE 3.9 Transfer diaphragm and main diaphragm net shears.

shear within the transfer diaphragm is affected by the transfer of the disrupted chord force. The shears in the diaphragm sections to the left and right of the transfer diaphragm remain unaffected and are equal to the basic diaphragm shear. Once the net shears are determined, the representative sheathing elements and the shears that are transferred onto each chord or collector should be applied to the plan. The net shears in the transfer diaphragm are calculated as follows:

Location (Grid Line)	Net Shear
1 from A to B	$v = +357.1$ plf
2 from A to B (left side)	$v = +214.3$ plf
2 from A to B (right side)	$v = +150 - (107.1) = +42.9$ plf
2 from B to C (right side)	$v = +150 + (250) = +400$ plf
3 from A to B (left side)	$v = +70 - (107.1) = -37.1$ plf
3 from B to C (left side)	$v = +70 + (250) = +320$ plf
3 from A to C (right side)	$v = +70$ plf

The diaphragm nailing patterns can be determined from the calculated net diaphragm shears.

Determination of longitudinal collector/chord forces (see Fig. 3.10): The collector/chord forces are determined by averaging the shears along the member segment being reviewed and then multiplying them by the member segment length, which is the same

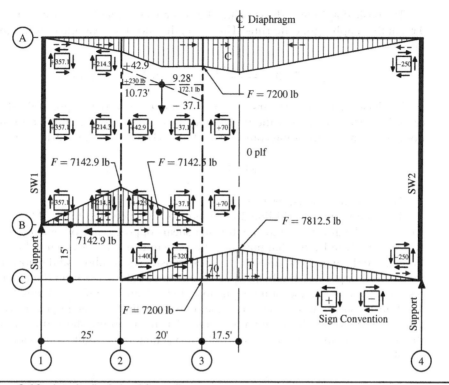

FIGURE 3.10   Longitudinal chord force diagrams.

thing as taking the area of the shear diagram along the length of the member being analyzed. It is important when learning the method of analysis that all the force diagrams are calculated and placed on the plan, as a visual verification of the analysis. Starting with the forces along grid line A, the chord force at grid line 2A is equal to

$$F_{2A} = \frac{(357.1 + 214.3)}{2}(25) = 7142.9 \text{ lb}$$

Note that this is equal to the chord force at grid line 2B that was previously calculated, as it should be. The direction of the shears transferred into the chord shows that the chord is in compression. The net shear at grid line 2 changes between A and B from a $+42.9$ plf to a -37.1 plf at grid line 3. From similar triangles, the shear becomes zero at 10.73' from grid line 2. The net shear diagram is drawn near grid line A for reference. The area of the positive shear represents a positive change in the chord force of $+230$ lb. The area of the negative shear decreases the chord force by -172.1 lb. The maximum chord force occurs at grid line 2 + 10.73', which is equal to

$$F_{2+10.73} = 7142.9 + 230 = 7373 \text{ lb compression}$$

The chord force at grid line 3A is equal to

$$F_{3A} = 7373 - 172.1 = 7200 \text{ lb compression}$$

The basic diaphragm shear value at grid line 3 is equal to +70 plf and changes to 0 plf at the diaphragm centerline. Therefore, the chord force at the diaphragm centerline is

$$F_{CL} = 7200 + \frac{(70 + 0)}{2}(17.5) = 7812.5 \text{ lb compression}$$

The basic diaphragm shears change to negative shear to the right of the diaphragm centerline. The direction of the transferred shears is now going in the opposite direction, indicating an opposing compression force. The chord force at grid line 4 equals

$$F_4 = 7812.5 + \frac{(0 - 250)}{2}(62.5') = 0 \text{ lb}$$

The force diagram closes to zero.

The chord force at grid line 2B was previously calculated to be 7142.9 lb tension. Notice that the direction of the net transfer shears opposes each other about grid line 2, which visually indicates that both members are in tension, as expected. Proceeding to the collector between grid lines 2 and 3, the transfer shears above and below the collector at line 2 oppose each other, requiring one to be subtracted from the other. The net resulting shear at that location is equal to $(+400 - 42.9) = +357.1$ plf acting to the right. The transfer shears above and below the collector at line 3 are acting in the same direction, requiring one to be added to the other. The net resulting shear at the right end of the collector is $(320 + 37.1) = +357.1$ plf, also acting to the right. The collector force is equal to

$$F_{collector} = \frac{(357.1 + 357.1)}{2}(20) = 7142.5 \text{ lb or } 357.1 \text{ plf} \times \text{collector length}$$

The chord and collector forces are equal and opposite to each other; therefore, the force diagram at grid line B closes to zero.

Next examine the chord forces along grid line C, starting at grid line 2. The shear at line 2 is +400 plf and +320 plf at line 3. Taking the average of the shears and multiplying by the chord length, the chord force at grid line 3 becomes

$$F_{3C} = \frac{(400 + 320)}{2}(20) = 7200 \text{ lb tension}$$

This is equal to the chord force calculated at grid line 3A. The shear at line 3 is +70 plf and 0 plf at the centerline of the diaphragm. The chord force at the centerline of the diaphragm is

$$F_{CL} = 7200 + \frac{(70 + 0)}{2}(17.5) = 7812.5 \text{ lb tension}$$

Since the diaphragm shears and transfer shears are the same from line 3 to line 4 as those calculated along grid line A, the remaining chord forces are the same as grid line A. The direction of the transfer shears into the chord indicate that the chord is in tension acting in an opposite direction about the centerline of the diaphragm. All the longitudinal force diagrams close to zero and follow basic diaphragm action.

Determination of transverse collector/chord forces (see Fig. 3.11): The net diaphragm and transfer shears for the transverse collectors are shown on the plan.

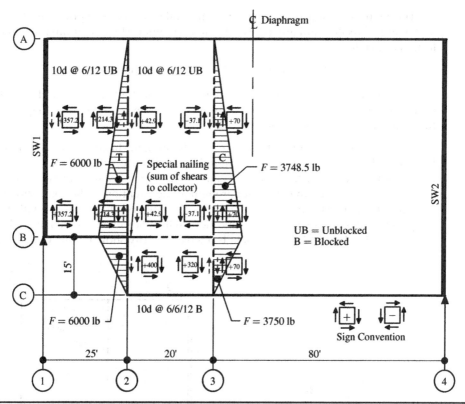

Figure 3.11   Transverse collector force diagrams.

Starting at grid line 2C, the collector along grid line 2 from B to C has a uniform net shear of +400 plf. The force at grid line 2B is equal to

$$F_{2B} = 400(15) = 6000 \text{ lb}$$

The direction of the transfer shear shows that the member is in tension. The net transfer shears on each side of grid line 2 between A and B are applied in opposite directions and are to be subtracted from each other. The resulting direction of the transfer shear is in the opposite direction to the drag member between B and C, which also indicates tension. Continuing from grid line 2B to grid line A, the force in the collector is equal to

$$F_A = 6000 - (214.3 - 42.9)(35) = 0 \text{ lb}$$

Therefore, the force diagram closes.

The collector along grid line 3 from B to C has a net shear of +320 plf on the left side of the member and a unit shear of +70 plf on the right side. The transfer shears act in opposite directions and are to be subtracted from each other. The force in the collector at grid line 3B is equal to

$$F_{3B} = (320 - 70)(15) = 3750 \text{ lb compression}$$

The net transfer shears on each side of the collector, between grid lines A and B, act in the same direction and are to be added to each other. The maximum force in the collector is equal to

$$F_{3B} = (37.1 + 70)(35) = 3748.5 \text{ lb compression}$$

This force is equal and opposite in direction to the collector between B and C, with some rounding error. Therefore, the force diagram closes. Since all the force diagrams close, the entire analysis has been verified. ▲

Special nailing is required to transfer the diaphragm shears into all the chords, collectors, and struts, and should be carefully called out on the framing plans and in related details. These critical members and their nailing are not always clearly called out on the plan and are therefore not brought to the attention of the contractor as being essential elements. The location of these critical members will not always fall at sheathing panel edges and can occur at any location on the plan, even at field nailing locations as referenced in Fig. 3.12. Typical details often arbitrarily show nailing of the sheathing

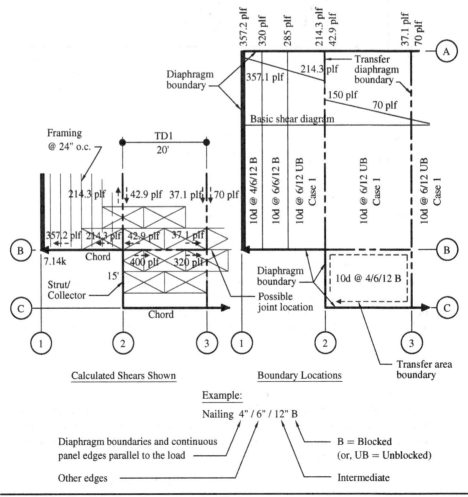

Figure 3.12   Sheathing layout versus strut nailing.

to trusses or beams that serve as collectors as edge or boundary nailing. Collector forces/shears can easily exceed typical diaphragm edge or boundary nailing capacities, which can create a deficiency in transferring the required diaphragm shears into the collector. This special nailing should be called out equal to or greater than the demand for the full length of the collector, and chord members of the transfer diaphragms. The entire perimeter of the transfer area should be treated as a boundary, as noted in the "Boundary locations plan" of the figure and receive boundary nailing. When high shear loads are transferred from each side, the nailing equivalent to two rows of boundary or edge nailing may be needed.

Engineering judgment should be used in the assignment of nailing at the transfer diaphragm area above the collector at grid line B. Even though the resultant net shears can be lower than the basic diaphragm shears, it is recommended that consideration should be given to providing enough shear capacity in that area to equal the greater of the net shear or the unadjusted basic diaphragm shears.

Frequently, the seismic and wind loads applied to the diaphragms are close in magnitude. When this occurs, there is sometimes a tendency to calculate the wind and seismic loads and then design the diaphragm for the controlling case without verifying that that load condition controls for all components of the diaphragm. For example, assume that the wind load to the diaphragm is 300 plf and the seismic load is 260 plf. If the designer decides that wind controls and the structure is in seismic category C or greater, the over-strength amplification factor of the collector force requirement, if applicable, would not be applied and the collector would be under-designed for seismic loading. The maximum shears and forces acting on the elements of the diaphragm due to wind and seismic loading conditions should be verified before completing the design. Neglecting to do so may result in a design not compliant with the code.

Example 3.2: Single Offset Diaphragm, Analysis in the Longitudinal Direction, Strut Offset In

The diaphragm shown in Fig. 3.13 can be designed by assuming that the smaller section of the diaphragm located between grid lines A and B from 1 to 2 acts as an independent diaphragm section. Both ends of this section are supported by struts that are tied into the main diaphragm that lies between grid lines A and C. The strut at line B supports one end of the diaphragm and extends into the transfer diaphragm that occurs between grid lines 2 and 3. The intent of placing a limited number of shear walls along grid lines A and C is to simplify the example and demonstrate the method of calculating the drag strut forces, not to suggest that this is a reasonable layout since the lack of redundancy would be a significant issue. The transfer diaphragm and the location of its members from the previous example have been maintained. The diaphragm is broken down into two sections, from grid lines 1 to 2 and from lines 2 to 4. The uniform load of 200 plf is distributed to these sections in proportion to their individual widths.

Load distribution within the diaphragm (see Fig. 3.13):

$$\text{Loads between A and B from 1 to 2} = \frac{200(25)}{125} = 40 \text{ plf}$$

$$\text{Loads between A and B from 2 to 4} = \frac{200(100)}{125} = 160 \text{ plf}$$

$$\text{Loads between B and C from 2 to 4} = 200 \text{ plf}$$

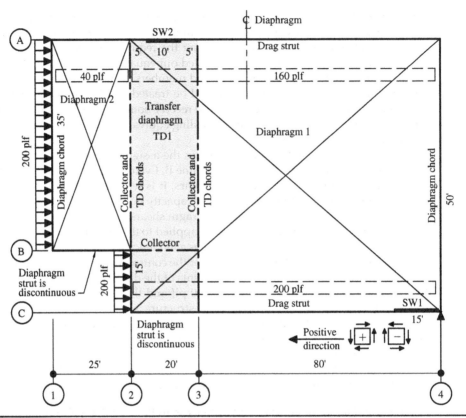

Figure 3.13 Example 3.2—Single offset diaphragm, analysis in the longitudinal direction.

Construction of the basic shear diaphragm (see Fig. 3.14):

Diaphragm 2 between grid lines A and B from 1 to 2:

$$R_A = R_B = \frac{wL}{2} = \frac{40(35)}{2} = 700 \text{ lb} \qquad v_{diaph} = \frac{R_A}{W} = \frac{700}{25} = 28 \text{ plf}$$

Place the positive and negative sheathing elements and the shears that are transferred into the boundary elements on the plan.

Diaphragm 1 between grid lines A and C from 2 to 4:

$$R_A = \frac{[200(15)(7.5) + 160(35)(32.5)]}{50} = 4090 \text{ lb negative shear}$$

$$v_A = \frac{R_A}{W} = \frac{-4090}{100} = -40.9 \text{ plf}$$

$$R_C = \frac{[160(35)(17.5) + 200(15)(42.5)]}{50} = 4510 \text{ lb}$$

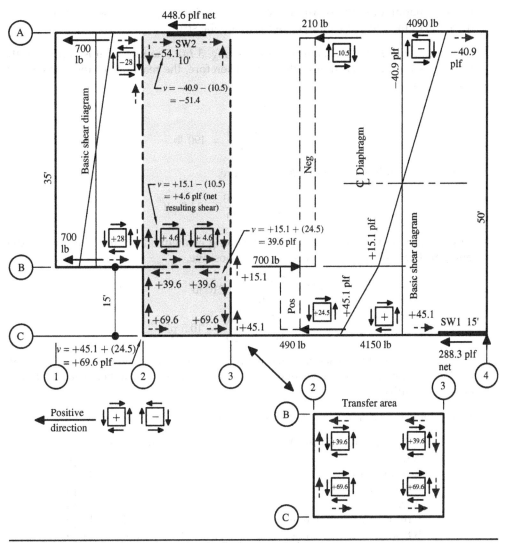

Figure 3.14 Transfer diaphragm and main diaphragm net shears.

$$v_C = \frac{R_C}{W} = \frac{4510}{100} = +45.1 \text{ plf}$$

$$V_B = 4500 - 200(15) = 1510 \text{ lb}$$

$$v_B = \frac{V_B}{W} = \frac{1510}{100} = +15.1 \text{ plf}$$

Place the positive and negative sheathing elements and the shears that are transferred into the boundary elements on the plan. The reaction of diaphragm 2 is transferred into the discontinuous strut at grid line B, which is then distributed into the transfer diaphragm. The disrupted strut force is in compression as indicated by the direction of the shears that are transferred into the strut.

Determination of transfer diaphragm shears (see Fig. 3.14): The force that is applied to the transfer diaphragm is acting to the right. Therefore, the reactions of the transfer diaphragm must act to the left.

$$R_B = 700 \text{ lb}$$

$$V_C = \frac{700(35)}{50} = 490 \text{ lb}$$

The transfer diaphragm depth is 20 ft; therefore,

$$v_C = \frac{V_C}{W_{TD}} = \frac{490}{20} = 24.5 \text{ plf}$$

Applying the remaining shear on the sheathing element shows that the transfer diaphragm shear between grid lines B and C is positive.

$$V_A = \frac{700(15)}{50} = 210 \text{ lb}$$

$$v_A = \frac{V_A}{W_{TD}} = \frac{210}{20} = 10.5 \text{ plf}$$

The direction of the reaction is also acting to the left. Applying the remaining shears on the sheathing element shows that the transfer diaphragm shear between grid lines A and B is negative.

Determination of diaphragm net shears (see Fig. 3.14):

Location (Grid Lines)	Net Shears
A from 1 to 2	−28 plf
A from 3 to 4	−40.9 plf
B from 1 to 2	+28 plf
A from 2 to 3	$v = -40.9 - 10.5 = -51.4$ plf
B from 2 to 3 (above line)	$v = +15.1 - 10.5 = +4.6$ plf
B from 2 to 3 (below line)	$v = +15.1 + 24.5 = +39.6$ plf
C from 2 to 3	$v = +45.1 + 24.5 = +69.6$ plf
C from 3 to 4	+45.1 plf

Determination of net shears at shear walls (see Fig. 3.14): Always observe the direction of the shear transferred into the boundary members.

$$v_{SW2} = \frac{wL}{2L_{SW2}} = \frac{200(50)}{2(10)} = +500 \text{ plf, where } L_{SW2} = 10 \text{ ft}$$

$$v_{diaph} = -51.4 \text{ plf from previous calculation}$$

$$v_{net} = v_{SW2} - v_{diaph} = 500 - 51.4 = +448.6 \text{ plf}$$

$$v_{SW1} = \frac{200(50)}{2(15)} = +333.3 \text{ plf, where } L_{SW1} = 15 \text{ ft}$$

$$v_{\text{diaph}} = -45.1 \text{ plf}$$

$$v_{\text{net}} = 333.3 - 45.1 = +288.3 \text{ plf}$$

An enlarged plan of the shears applied between grid lines B and C from 2 to 3 is shown at the lower right of the figure.

Determination of longitudinal and transverse strut/chord forces (see Fig. 3.15): Starting with the longitudinal strut along grid line A from 1 to 2, the strut force at line 2 is equal to

$$F = 28(25) = 700 \text{ lb compression, which is the same as at grid line 2B.}$$

The net shear in the transfer diaphragm between grid line 2 and the start of SW2 is −51.4 plf. Continuing 5 ft into the transfer diaphragm to the start of shear wall SW2, the strut force becomes

$$F = 700 + 51.4(5) = 957 \text{ lb compression}$$

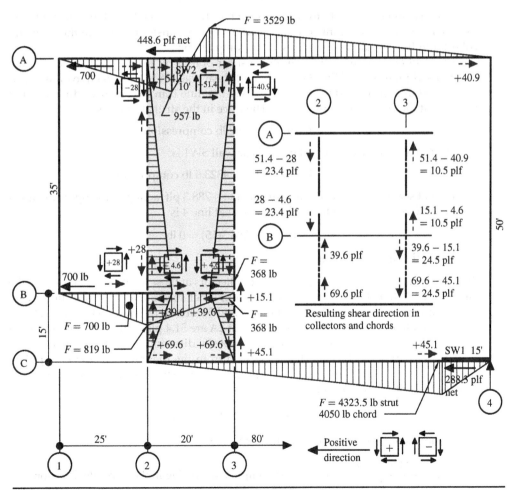

Figure 3.15   Longitudinal and transverse strut/collector force diagrams.

The net shear at the shear wall area is +448.6 plf as previously calculated, which acts in a direction opposite to the strut force and therefore must be subtracted from the strut force. The force at the end of the shear wall is equal to

$$F = 957 - 448.6(10') = -3529 \text{ lb tension}$$

Continuing to the end of the line at grid line 4, the force becomes

$$F = -3529 + 51.4(5) + 40.9(80) = 0 \text{ lb tension}$$

Therefore, the diagram closes.

For the longitudinal strut along grid line B from 1 to 2, the strut force at line 2 is equal to

$$F = 28(25) = 700 \text{ lb compression}$$

Observing the direction of the transferred shears on each side of the collector between grid lines 2 and 3, the shear value of 4.6 plf must be subtracted from the shear value of 39.6 plf on the opposite side of the collector. The force in this collector is equal to

$$F = (39.6 - 4.6)20 = 700 \text{ lb compression}$$

This is equal and opposite to the direction of the strut between lines 1 and 2, and therefore closes the force diagram. The direction of the net transfer forces into the strut and collector indicates that the members are in compression.

The strut at grid line C has a uniform shear of 69.6 plf from grid line 2 to 3 and a shear of 45.1 plf from grid line 3 to the start of shear wall SW1. The direction of the shear transferred into the boundary members along that line is in the negative direction and shows that the strut is in compression. The force in the strut at line 3 is

$$F = -69.6(20) = 1392 \text{ lb compression}$$

The force in the strut at the start of shear wall SW1 is

$$F = 1392 + 45.1(80 - 15) = 4323.5 \text{ lb compression}$$

The net shears at the shear wall are equal to +288.3 plf, acting in the opposite direction. Therefore, the force at the end of the wall at line 4 is

$$F = 4323.5 - 288.3(15) = 0 \text{ lb}$$

Therefore, the diagram closes.

Transverse collectors at grid lines 2 and 3. The collectors at these locations also serve as the chords for the transfer diaphragm. The resulting net shears and their direction transferred into the transverse collectors at lines 2 and 3 can be referenced in the figure. It can be observed that the transfer shears applied at 2A are 51.4 plf on the right of the collector and 28 plf on the left. The shears act in opposite directions, which requires them to be subtracted from each other. It can be seen from the diagram that the net applied transfer shear is a uniform 23.4 plf compression from grid line A to B. Therefore, the force in the collector at grid line 2B is

$$F_{2AB} = 23.4(35) = 819 \text{ lb compression}$$

On the opposite side of line B, the collector force is

$$F_{2BC} = \frac{(39.6 + 69.6)}{2}(15) = 819 \text{ lb compression, acting in the opposite direction}$$

At grid line 3, the net applied transfer shear above line B is a uniform 10.5 plf and below line B the shears are a uniform 24.5 plf.

$$F_{3AB} = 10.5(35) = 368 \text{ lb tension}$$

$$F_{3BC} = 24.5(15) = 368 \text{ lb tension acting in the opposite direction}$$

All the force diagrams close and receive tension and compression forces in accordance with the anticipated diaphragm movement. Therefore, the analysis has been verified. ▲

Example 3.3: Single Offset Diaphragm, Analysis in the Longitudinal Direction, Offset Shear Walls

The diaphragm shown in Fig. 3.16 is the same layout as the previous example, except that shear walls are placed along grid lines A, B, and C. Normally, interior shear walls act as supports for the diaphragm, which creates a diaphragm boundary. Code requires that drag struts/collectors, vertical lateral resisting elements, or any combination thereof are required to occur along the entire length of a diaphragm boundary. To comply with this requirement for the shear wall at grid line B, a collector would have to extend from the end of the shear wall to grid line 4. Assuming that this can be

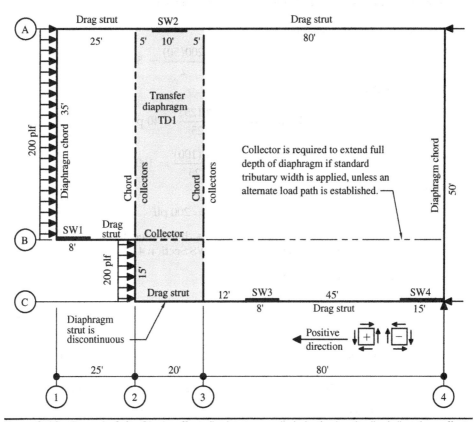

Figure 3.16   Example 3.3—Single offset diaphragm, analysis in the longitudinal direction, offset shear walls.

accomplished, the loads to the individual wall lines can be distributed to each shear wall on a tributary width basis. However, the installation of the collector is impractical because the collector would be over 100 ft in length and would be very difficult to construct if the joists or trusses are oriented in a direction perpendicular to the collector. A solution to the problem is to design the shear walls along grid lines B and C to act as though they are in the same line of lateral force resistance. Since the offset of the walls causes a discontinuity in the strut that would normally have provided a continuous load path connecting the shear walls, an alternate load path must be established across the discontinuity.

When considering lines B and C as acting in the same line of lateral force resistance, a mistake commonly made is to assume that lines A and B/C each resist half of the loads applied to the diaphragm based on a tributary width between grid lines A and C. Two uncertainties exist when shear walls are offset. The first is how much shear force is transferred into the area of discontinuity. The second is how much of the uniform load applied to the diaphragm is distributed into line B/C. Obviously, if shear wall 1 is located at grid line C, half of the load goes to grid line A and half goes to line C. The more SW1 is offset toward the center of the diaphragm, the more load it draws. Since the intent is to make the walls at lines B and C act as a unit, this causes a greater shift of loads to line B/C and a smaller portion to line A. To prove this point, we will initially assume that lines A and B/C receive equal tributary loads.

Load distribution into diaphragm (see Fig. 3.17):

$$R_A = R_{B,C} = \frac{wL}{2} = \frac{200(50)}{2} = 5000 \text{ lb}$$

Loads from A to B:

$$\text{From 1 to 2} = \frac{200(25)}{125} = 40 \text{ plf}$$

$$\text{From 2 to 4} = \frac{200(100)}{125} = 160 \text{ plf}$$

Loads from B to C:

$$\text{From 2 to 4} = 200 \text{ plf}$$

Determination of shear wall shears: The force is distributed to each shear wall in accordance with their length, which follows SDPWS Section 4.3.5.5.1, exception.

$$L_{SW1+SW3+SW4} = 8 + 8 + 15 = 31 \text{ ft}$$

$$V_{SW1} = \frac{L_{SW1}R_{B,C}}{\sum L_{SW}} = \frac{8(5000)}{31} = 1290.3 \text{ lb,} \qquad v_{SW1} = \frac{V_{SW1}}{L_{SW1}} = \frac{1290.3}{8} = +161.3 \text{ plf}$$

$$V_{SW2} = 5000 \text{ lb,} \qquad\qquad\qquad v_{SW2} = \frac{5000}{10} = +500 \text{ plf}$$

$$V_{SW3} = V_{SW1} = 1290.3 \text{ lb,} \qquad\qquad v_{SW3} = v_{SW1} = 161.3 \text{ plf}$$

$$V_{SW4} = \frac{15(5000)}{31} = 2419.4 \text{ lb,} \qquad\qquad v_{SW4} = \frac{2419.4}{15} = +161.3 \text{ plf}$$

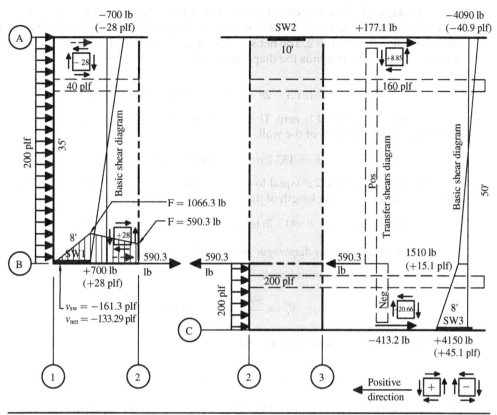

FIGURE 3.17 Basic diaphragm shear and transfer diaphragm shears.

Construction of the basic shear diagram (see Fig. 3.17):

Diaphragm section between grid lines A and B from 1 to 2

$$R_A = R_B = \frac{wL}{2} = \frac{40(35)}{2} = 700 \text{ lb}$$

$$v_{\text{diaph}} = \frac{R_A}{W_{1-2}} = \frac{700}{25} = 28 \text{ plf}$$

Diaphragm section between grid lines A and C from 2 to 4:

$$R_A = \frac{[200(15)(7.5) + 160(35)(42.5)]}{50} = 4090 \text{ lb}, \quad v_{\text{diaph}} = \frac{4090}{100} = -40.9 \text{ plf}$$

$$R_C = \frac{[160(35)(17.5) + 200(15)(42.5)]}{50} = 4510 \text{ lb}, \quad v_{\text{diaph}} = \frac{4510}{100} = +45.1 \text{ plf}$$

$$V_B = 4510 - 200(15) = 1510 \text{ lb}, \quad v_{\text{diaph}} = \frac{1510}{100} = +15.1 \text{ plf}$$

Determination of the force transferred into the transfer diaphragm (see Fig. 3.17): The force transferred into the transfer diaphragm can be determined by summing the shears along grid line B from 1 to 2. The net shear along the length of shear wall SW1 is equal to the shear wall shear minus the diaphragm shear.

$$v_{net} = +161.3 - 28 = +133.29 \text{ plf at the shear wall}$$

The force at grid line 1 is zero. The force at the end of the shear wall is equal to the net shear times the length of the wall.

$$F = 133.29(8) = +1066.3 \text{ lb tension}$$

The force at grid line 2 is equal to the force calculated above minus the diaphragm shear times the remaining length of the strut.

$$F_{2B} = 1066.3 - 28(17) = 590.3 \text{ lb tension, transferred to transfer diaphragm TD1}$$

Determination of the transfer diaphragm shears (see Fig. 3.17):

Summing moments about line A:

$$V_C = \frac{590.3(35)}{50} = 413.2 \text{ lb}$$

The width of the transfer diaphragm is 20 ft.

$$v_C = \frac{413.2}{20} = 20.66 \text{ plf}$$

The direction of the reaction at grid line C is acting to the right. Applying the remaining shears acting on the remaining sides of the sheathing element produces a negative shear.

Summing moments about line C:

$$V_A = \frac{590.3(15)}{50} = 177.1 \text{ lb}$$

$$v_A = \frac{177.1}{20} = 8.85 \text{ plf}$$

The direction of the reaction is acting to the right. Applying the remaining shear on the sheathing element produces a positive shear.

Determination of diaphragm net shears (see Fig. 3.18):

Location (Grid Lines)	Net Shears
A from 1 to 2	$= -28$ plf
B from 1 to 2	$= +28$ plf
A from 2 to 3	$v = -40.9 + (8.85) = -32.05$ plf
B from 2 to 3 (above line)	$v = +15.1 + (8.85) = +23.95$ plf
B from 2 to 3 (below line)	$v = +15.1 - (20.66) = -5.56$ plf
C from 2 to 3	$v = +45.1 - (20.66) = +24.4$ plf
C from 3 to 4	$= +45.1$ plf
A from 3 to 4	$= -40.9$ plf

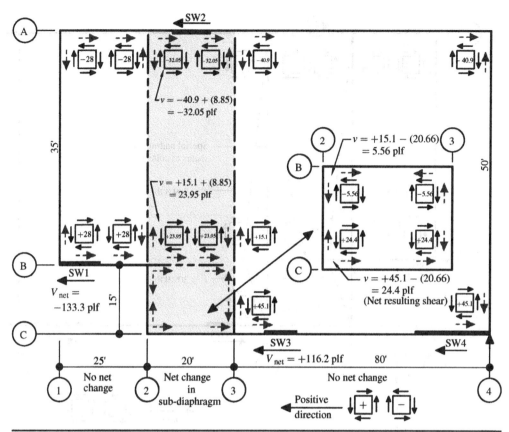

Figure 3.18 Net diaphragm shears.

The net shears in the area bounded by grid lines B and C from 2 to 3 are smaller than the previous examples. This is because the disrupted force transferred into the transfer diaphragm is in tension because of the presence of SW1 and pulling away from the loads in the larger diaphragm section, thereby reducing the net shears (i.e., SW1 supports part of the main diaphragm). Place all the net shear sheathing elements and the shears that are transferred into the boundary elements on the plan as shown in the figure.

Determination of transverse collector/chord forces (see Fig. 3.19): The net shear transferred into the collector at grid line 2B is $28 - 23.95 = +4.05$ plf and $32.05 - 28 = +4.05$ plf at grid line 2A. Both net shears act toward grid line B, indicating that the collector is in compression. Starting at grid line A, the force at grid line 2B is equal to

$$F_{2B} = 4.05(35) = 141.7 \text{ lb compression}$$

The net shears on the collector section between grid lines B and C changes from $+24.4$ plf to -5.56 plf as diagramed below the plan. From similar triangles, the total force on the collector at that section is the sum of the areas of the shear triangles.

$$F_{2B} = 149.33 - 7.73 = 141.6 \text{ lb}$$

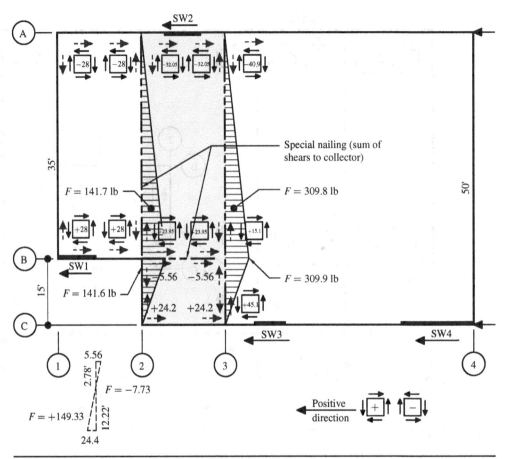

Figure 3.19   Transverse collector force diagrams.

This force is equal and opposite to the force on the opposite side of grid line B; therefore, the force diagram closes to zero and the collector is in compression. The net shears transferred into the collector at grid line 3A to 3B are $23.95 - 15.1 = +8.85$ plf and $40.9 - 32.05 = +8.85$ plf, respectively, indicating that there is a uniform net shear. Starting at grid line A, the force at grid line 3B is equal to

$$F_{3B} = 8.85(35) = 309.8 \text{ lb tension}$$

The net shears transferred into the collector at grid line 3B to 3C are $45.1 - 24.4 = +20.66$ plf and $15.1 + 5.56 = +20.66$ plf, respectively, again indicating a uniform shear. The force at grid line 3B is equal to

$$F_{3B} = 20.66(15) = 309.9 \text{ lb compression}$$

Determination of longitudinal collector/chord forces (see Fig. 3.20): The force in the strut at grid line B has already been determined up to grid line 2. The shears transferred above and below the collector are 23.95 plf and 5.56 plf, respectively. Since they are

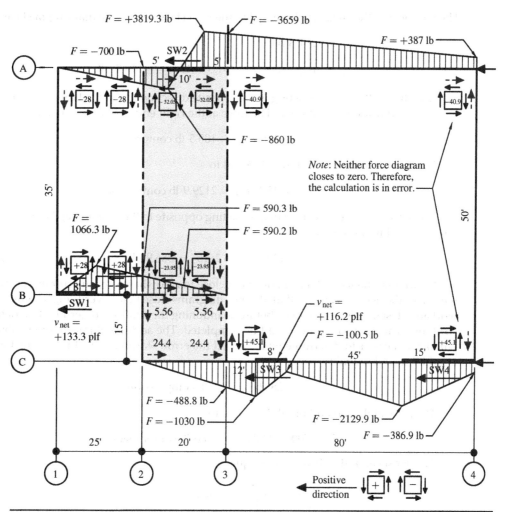

FIGURE 3.20   Longitudinal strut force diagrams.

going in the same direction, they must be added. Continuing to grid line 3, the force is equal to

$$F_{3B} = 590.3 - (23.95 + 5.56)(20) = 0 \text{ lb}$$

Therefore, the diagram closes.

Determine the forces along grid line C. Starting at grid line 2, the strut force at grid line 3C is equal to

$$F_{3C} = 24.4(20) = 488.8 \text{ lb compression}$$

The force at the left end of SW3 is equal to

$$F = 488.8 + 45.1(12) = 1030 \text{ lb compression}$$

The net shear at the shear wall is equal to the wall shear minus the diaphragm shear.

$$v_{SW} = 161.3 \text{ plf}$$

$$v_{net} = v_{SW} - v_{diaph} = 161.3 - 45.1 = +116.2 \text{ plf}$$

Since the wall shears are acting in an opposite direction to the strut force, they must be subtracted from the strut force. The force at the right end of SW3 is equal to

$$F = 1030 - 116.2(8) = 100.5 \text{ lb compression}$$

The force at the left end of SW4 is equal to

$$F = 100.5 + 45.1(45) = 2129.9 \text{ lb compression}$$

The net shear at SW4 is also 116.2 plf, acting opposite to the strut force; therefore, the force at grid line 4 is equal to

$$F_4 = 2129.9 - 116.2(15) = 386.95 \text{ lb}$$

Notice that the force diagram does not close to zero; therefore, an error exists. This reinforces the need to confirm that all force diagrams close before verifying a complete, workable design exists. However, before abandoning the analysis, determination of the strut forces along grid line A should be completed. The additional information gained will make it easier to determine what caused the error. Starting at grid line 1, the force at grid line 2A is equal to

$$F_{2A} = 28(25) = 700 \text{ lb compression}$$

The force at the left end of SW2 is equal to

$$F = 700 + 32.05(5) = 860 \text{ lb compression}$$

The net shear at the shear wall is equal to

$$v_{SW} = \frac{R_A}{L_{SW2}} = \frac{5000}{10} = 500 \text{ plf}$$

$$v_{net} = v_{SW} - v_{diaph} = 500 - 32.05 = 467.95 \text{ plf}$$

Acting in the opposite direction to the strut force. The force at the right end of SW2 is equal to

$$F = 860 - 467.95(10) = -3819.3 \text{ lb tension}$$

The force at grid line 3 is equal to

$$F_{3A} = -3819.3 + 32.05(5) = -3659 \text{ lb tension}$$

The force at grid line 4 is equal to

$$F_{4A} = -3659 + 40.9(80) = -387 \text{ lb tension}$$

This force diagram also does not close to zero by the same amount, indicating an error in the assumption of distribution by tributary width is incorrect. Figure 3.21 shows how to correct the problem. At grid line C, the shear wall shears are not high enough to reduce the strut forces to zero. At grid line A, the shear wall shears are too

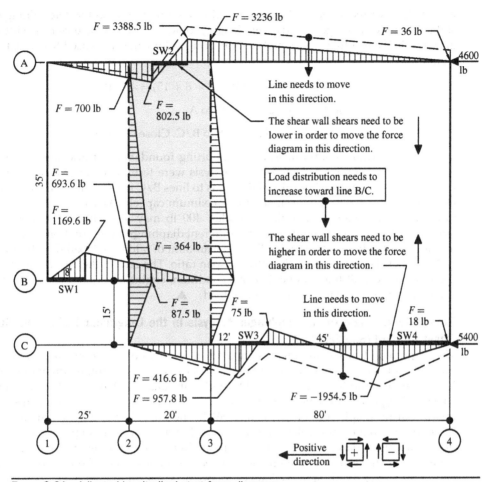

FIGURE 3.21   Adjusted longitudinal strut force diagrams.

high and leave a residual 387 lb strut force. It can be concluded that the loads applied to the diaphragm need to shift toward grid line B/C, which will increase the wall shears at lines B and C and reduce the wall shears at grid line A. Adjustment of the load to the lines of lateral force resistance by an amount approximately equal to the discrepancy should be close enough to close the force diagrams to zero. Until the designer gains experience by analyzing a significant number of problems, the process can be trial and error. Adjustment of the load to each wall line by 400 lb will revise the reactions at the wall lines to

$$R_{B/C} = 5400 \text{ lb}$$
$$R_A = 4600 \text{ lb}$$

Repeating the analysis, the force diagrams basically close as shown in Fig. 3.21. The small magnitude of the residual force at grid line 4 is negligible and can, by some, be within the spirit of the analysis, therefore be ignored. It can be concluded that the offset wall at grid line B draws a disproportionately larger load than the other walls in line B/C, which is contrary to the classical distribution based on tributary width, where

$R = wL/2$. As a recommended alternate and more direct solution for determining the shift, a weighted average (based on SW length) could be used for the tributary width to grid line A. If a residual force results, engineering judgment can be used to evaluate its significance.

$$[8 \times 35 + (8 + 15)50] / (8 + 8 + 15) = 46.1 \text{ ft}$$

$$46.1 \times 200 / 2 = 4612 \text{ lb to A}$$

$$50 \times 200 - 4612 = 5388 \text{ to B/C. Close enough.}$$

The shear walls, hold-downs, and supporting foundations at grid lines B and C might have been under-designed if the analysis were terminated after analyzing the transfer members at grid line 2B and the loads to lines B/C were assumed to be correct, especially if the walls were designed to a maximum capacity. When compared to significantly larger loads in real-life examples, 400 lb may seem like an insignificant amount to be concerned about; however, different diaphragm configurations will cause a different load distribution to line B/C. The amount of additional load distribution to wall line B/C is controlled by the offset to span ratio. The greater the offset, the greater the load to the wall line will be. For this condition, engineering judgment should be used regarding acceptance of the residual load. ▲

Example 3.4: Single Offset Diaphragm, Analysis in the Longitudinal Direction, Strut Offset Outward (See Fig. 3.22)

The diaphragm has a span of 60 ft with a 100 ft width and a 15-ft outward cantilever offset at grid line 2. A collector would normally be extended the full width of the cantilever section at grid line 2, which would simplify the analysis. However, because of the collector length, an alternate approach is to extend the collector only a short distance into a transfer diaphragm off the end of SW1. The offset in the diaphragm creates a discontinuity in the strut/collector at grid lines 1 and 2. The diaphragm can be broken down into two separate sections. The main diaphragm lies between grid lines 2 and 3 from C to D. The second diaphragm lies between grid lines 1 and 3 from A to C. The left end of the upper diaphragm section is supported by a strut that transfers its reaction, R_1, into the end of the propped cantilever transfer diaphragm that lies between grid lines B and C. The transfer diaphragm is tied to the shear wall at grid line 2 by the collector that is embedded into the transfer diaphragm.

A 200 plf uniform load is applied to the diaphragm. The load is distributed to each diaphragm section in accordance to its width. The basic shear diagrams, transfer diaphragm shears, and loading diagrams can be referenced in Fig. 3.23.

Load distribution into the diaphragm:

Loads between A and C

$$\text{from 2 to 3} = \frac{200(40)}{100} = 80 \text{ plf}$$

$$\text{from 1 to 2} = 200 \text{ plf}$$

Loads between C and D

$$\text{from 2 to 3} = \frac{200(60)}{100} = 120 \text{ plf}$$

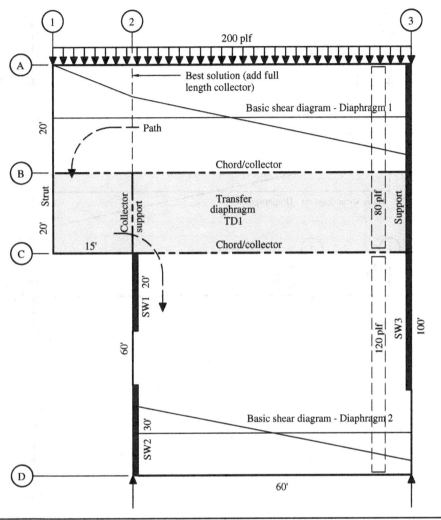

FIGURE 3.22   Example 3.4—Discontinuous strut, cantilever diaphragm.

Construction of the basic shear diagrams (see Fig. 3.23):

Upper diaphragm 1:

$$R_1 = \frac{[80(60)(30) + 200(15)(67.5)]}{75} = 4620 \text{ lb}, \qquad v_1 = \frac{4620}{40} = 115.5 \text{ plf}$$

$$R_3 = \frac{[200(15)(7.5) + 80(60)(45)]}{75} = 3180 \text{ lb}, \qquad v_3 = \frac{3180}{40} = -79.5 \text{ plf}$$

$$V_2 = R_1 - wx = 4620 - 200(15) = 1620 \text{ lb}, \qquad v_2 = \frac{1620}{40} = 40.5 \text{ plf}$$

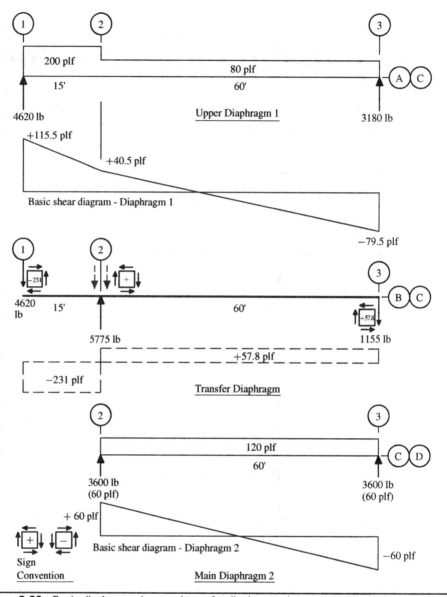

FIGURE 3.23 Basic diaphragm shear and transfer diaphragm shears.

Determination of transfer diaphragm shears (see Fig. 3.23): The left reaction of diaphragm 1, R_1, is supported off the end of the transfer diaphragm cantilever. Reference the direction of shears acting on the sheathing elements for the positive or negative shears.

$$R_2 = \frac{4620(75)}{60} = 5775 \text{ lb}, \quad v_2 = \frac{5775}{20} = -231 \text{ plf}$$

$$R_3 = \frac{4620(15)}{60} = 1155 \text{ lb}, \quad v_3 = \frac{1115}{20} = 57.75 \text{ plf}$$

Main diaphragm section from C to D, uniform load:

$$R_2 = R_3 = \frac{120(60)}{2} = 3600 \text{ lb}, \quad v_{2,3} = \frac{3600}{60} = 60 \text{ plf}$$

Determination of the diaphragm net shears (see Fig. 3.24):

Location (Grid Lines)	Net Shears
1 from A to B	+115.58 plf
3 from A to B	−79.5 plf
1 from B to C	$v = +115.5 - (231) = -115.5$ plf
2 from B to C (left)	$v = +40.5 - (231) = -190.5$ plf
2 from B to C (right)	$v = +40.5 + (57.75) = +98.25$ plf
3 from B to C	$v = -79.5 + (57.75) = -21.75$ plf

Determination of transverse collector/chord forces (see Fig. 3.25): The chord force along grid line A from 1 to 2, starting at grid line 1:

$$F_2 = \frac{(v_1 + v_2)x}{2} = \frac{(115.5 + 40.5)(15)}{2} = 1170 \text{ lb compression}$$

The diaphragm shear at grid lines 2 and 3 changes from +40.5 plf to −79.5 plf, respectively. The forces along that interval can be broken down into positive and negative forces by calculating the areas of the similar triangles shown in the figure. The maximum positive force in the chord is equal to F_2 plus the positive triangular area.

$$F_{max} = 1170 + 410 = 1580 \text{ lb compression}$$

This force is equal to the force of the negative triangular area. Therefore, the force diagram closes.

The transfer diaphragm chord force along grid line B, starting at grid line 1, is equal to

From 1 to 2,

$$F_{2B} = \frac{[(115.5 + 115.5) + (40.5 + 190.5)]}{2}(15) = 3465 \text{ lb tension}$$

From 2 to 3,

$$F_{2B} = \frac{[(98.25 - 40.5) + (79.5 - 21.75)]}{2}(60) = 3465 \text{ lb tension}$$

Therefore, the diagram closes.

The transfer diaphragm chord force along grid line C, starting at grid line 1, is equal to

From 1 to 2,

$$F_{2B} = \frac{(115.5 + 190.5)}{2}(15) = 2295 \text{ lb compression}$$

From 2 to 3,

$$F_{2B} = \frac{[(98.25 - 60) + (60 - 21.75)]}{2}(60) = 2295 \text{ lb compression}$$

The force diagram closes to zero.

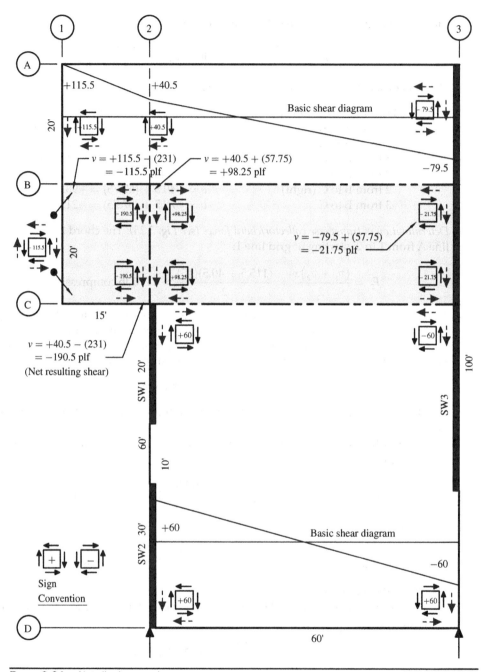

FIGURE 3.24   Net diaphragm shears.

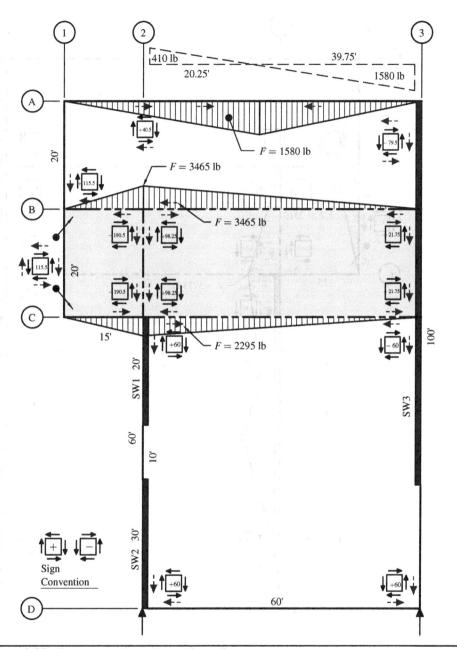

FIGURE 3.25 Transverse collector/chord force diagrams.

Determination of longitudinal strut/collector forces (see Fig. 3.26): The strut force along grid line 1 is equal to

From A to B,

$$F_{1B} = 115.5(20) = 2310 \text{ lb compression}$$

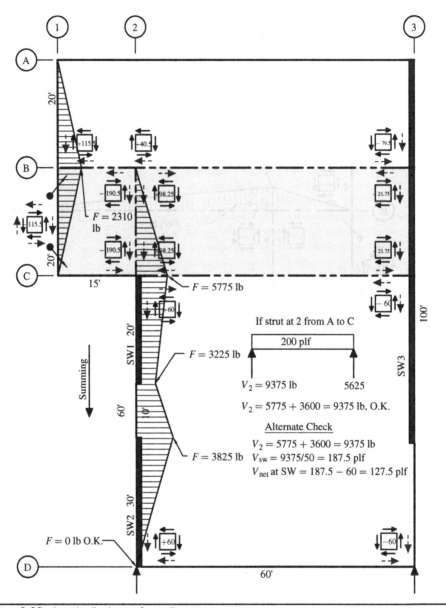

FIGURE 3.26   Longitudinal strut force diagrams.

From B to C,

$$F_{1B} = 115.5(20) = 2310 \text{ lb compression (opposite direction)}$$

The force diagram closes.

The strut force along grid line 2 is equal to

From B to C,

$$F_{2C} = -\frac{(98.25 + 190.5)}{2}(20) = -5775 \text{ lb compression}$$

The reaction of diaphragm 2 = 3600 lb; Therefore, the total shear to grid line 2 is

$$R_2 = 5775 + 3600 = 9375 \text{ lb}$$

Reference the alternate check in the figure.

$$v_{SW1} = v_{SW2} = \frac{9375}{(20 + 30)} = 187.5 \text{ plf}$$

$$v_{net} = v_{SW} - v_{diaph} = 187.5 - 60 = 127.5 \text{ plf at the shear walls}$$

At the end of SW1,

$$F = -5775 + 127.5(20) = -3225 \text{ lb compression}$$

At the end of SW1,

$$F = -3225 - 60(10) = -3825 \text{ lb compression}$$

At grid line D,

$$F_D = -3825 + 127.5(30) = 0 \text{ lb}$$

Therefore, the diagram closes.

Looking back at the direction of the transferred shears in Figs. 3.24 and 3.25, it can be seen that for the transverse members, the chord at grid line A is in compression because that section of the diaphragm was assumed to span from grid line 1 to 3. The transfer diaphragm was designed to act as a cantilever beam with a back span. The top chord at grid line B is in tension and the bottom chord at grid line C is in compression. For the longitudinal direction, all the struts are in compression. All the forces applied to the collectors, chords, and struts act as expected and have confirmed the analysis. ▲

3.5   Diaphragms Offset at Both Ends

Double offset diaphragms are created when two horizontal offsets occur at the exterior wall lines. This causes a disruption or discontinuity in the diaphragm chord at two locations as shown in Fig. 3.27. The analysis of double offset diaphragms is the same as a single offset diaphragm, with the exception that two transfer diaphragms are required to transfer the disrupted chord forces. A good example of this condition can be referenced in Fig. 3.27. This configuration is not an ideal lateral-force-resisting system layout due to a minimal number of vertical lateral-force-resisting elements and the reduced diaphragm widths at the offsets. Creating a system with the least number of shear walls possible is generally not good practice. Adding shear walls in locations similar to the optional layouts shown in Fig. 3.32 adds redundancy to the system and reduces the shears and forces within the diaphragm to acceptable limits. Keeping the system simple yet redundant can reduce the demand on the lateral resisting elements and in many cases reduce construction costs. These layouts are typically better performing systems.

Example 3.5: Diaphragm Offset at Both Ends, Analysis in the Transverse Direction, Multiple Chord Discontinuities (See Fig. 3.27)

The diaphragm in this example is 120 ft long by 50 ft in width. There is a 14-ft horizontal offset between grid lines 1 and 2 and a 30-ft horizontal offset between grid lines 5 and 6. Both of these offsets produce a notching effect and a discontinuity in the diaphragm

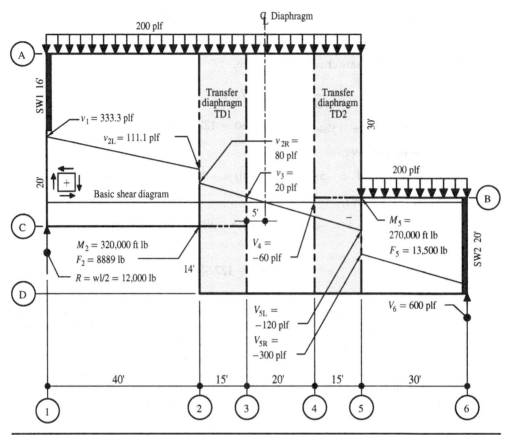

Figure 3.27 Example 3.5—Diaphragm offset at both ends, basic shear diagrams.

chords at grid lines 2C and 3B, as shown in Figs. 3.28 and 3.29. A collector at both locations extends into the transfer diaphragms marked TD1 and TD2, where the disrupted chord forces are transferred into the main diaphragm chords located at lines A and D. The diaphragm is loaded with a 200 plf uniform load. Note that the aspect ratio for the offset sections and the diaphragm as a whole is within allowable limits. However, it is also one of those questionable conditions as to the stiffness of the diaphragm as a whole, as previously mentioned in Fig. 2.3.

$$\text{Aspect ratio TD1 and TD2} = \frac{50}{15} = 3.33{:}1, \text{ which requires blocking.}$$

Construction of the basic shear diagram (see Fig. 3.27):

$$R_1 = R_6 = \frac{wL}{2} = \frac{200(120)}{2} = 12{,}000 \text{ lb}$$

$$v_1 = \frac{R_1}{W} = \frac{12{,}000}{36} = 333.3 \text{ plf}$$

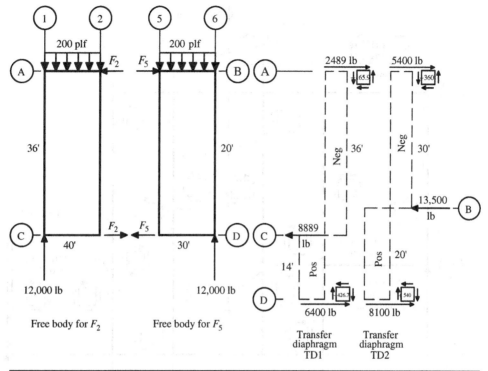

Figure 3.28   Free body of chord forces and transfer diaphragm shears.

$$v_6 = \frac{R_6}{W} = \frac{-12{,}000}{20} = -600 \text{ plf}$$

$$V_2 = R_1 - wx = 12{,}000 - 200(40) = 4000 \text{ lb}$$

$$v_{2L} = \frac{V_2}{W_{\text{left}}} = \frac{4000}{36} = 111.1 \text{ plf}$$

$$v_{2R} = \frac{V_2}{W_{\text{right}}} = \frac{4000}{50} = 80 \text{ plf}$$

$$V_3 = V_2 - w(15) = 4000 - 200(15) = 1000 \text{ lb}$$

$$v_3 = \frac{1000}{50} = 20 \text{ plf}$$

$$V_5 = R_1 - w(90) = 12{,}000 - 200(90) = -6000 \text{ lb}$$

$$v_{5R} = \frac{-6000}{20} = -300 \text{ plf, right side of grid line 5}$$

$$v_{5L} = \frac{-6000}{50} = -120 \text{ plf, left side of grid line 5}$$

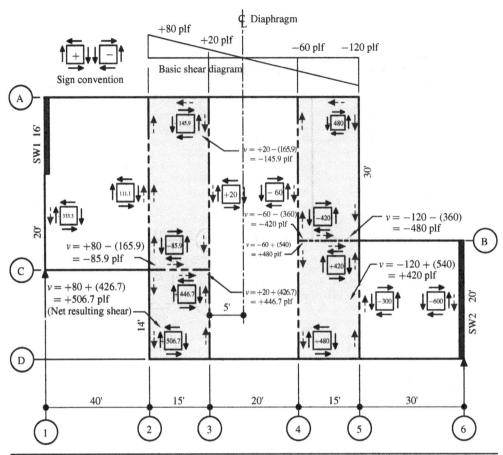

FIGURE 3.29 Net diaphragm shears.

$$V_4 = R_1 - w(75) = 12{,}000 - 200(75) = -3000 \text{ lb shear at grid line 4}$$

$$v_4 = \frac{-3000}{50} = -60 \text{ plf}$$

Calculation of the chord forces at the discontinuities (see Figs. 3.27 and 3.28):

$$M_2 = R_1 x - \frac{wx^2}{2} = 12{,}000(40) - \frac{200(40)^2}{2} = 320{,}000 \text{ ft lb at grid line 2}$$

$$F_2 = \frac{M_2}{W} = \frac{320{,}000}{36} = 8889 \text{ lb at grid line 2C}$$

$$M_5 = R_6 x - \frac{wx^2}{2} = 12{,}000(30) - \frac{200(30)^2}{2} = 270{,}000 \text{ ft lb at grid line 5}$$

$$F_5 = \frac{M_6}{W} = \frac{270{,}000}{20} = 13{,}500 \text{ lb at grid line 5D}$$

Determination of the transfer diaphragm shears (see Fig. 3.28):

Transfer diaphragm TD1:

$$V_D = \frac{8889(36)}{50} = 6400 \text{ lb}, \quad v_D = \frac{6400}{15} = 426.7 \text{ plf}$$

$$V_A = \frac{8889(14)}{50} = 2489 \text{ lb}, \quad v_A = \frac{2489}{15} = -165.9 \text{ plf}$$

Transfer diaphragm TD2:

$$V_D = \frac{13,500(30)}{50} = 8100 \text{ lb}, \quad v_D = \frac{8100}{15} = 540 \text{ plf}$$

$$V_A = \frac{13,500(20)}{50} = 5400 \text{ lb}, \quad v_A = \frac{5400}{15} = -360 \text{ plf}$$

Determination of diaphragm net shears (see Fig. 3.29):

Net shears between grid lines 1 and 2 (from basic shear diagram):

Location (Grid Line)	Net Shear
1	$v = +334$ plf
2	$v = +111$ plf

Net shears between grid lines 2 and 3:

Location (Grid Line)	Net Shear
2 (below C)	$v_{net} = +80 + (426.7) = +506.7$ plf
2 (above C)	$v_{net} = +80 - (165.9) = -85.9$ plf
3 (below C)	$v_{net} = +20 + (426.7) = +446.7$ plf
3 (above C)	$v_{net} = +20 - (165.9) = -145.9$ plf

Net shears between grid lines 4 and 5:

Location (Grid Line)	Net Shear
4 (below B)	$v_{net} = -60 + (540) = +480$ plf
4 (above B)	$v_{net} = -60 - (360) = -420$ plf
5 (below B)	$v_{net} = -120 + (540) = +420$ plf
5 (above B)	$v_{net} = -120 - (360) = -480$ plf

Net shears between grid lines 5 and 6 (from basic shear diagram):

Location (Grid Line)	Net Shear
5	$v = -300$ plf
6	$v = -600$ plf

Apply the net shear elements and their transfer shears onto the plan for a visual reference, see Fig. 3.29.

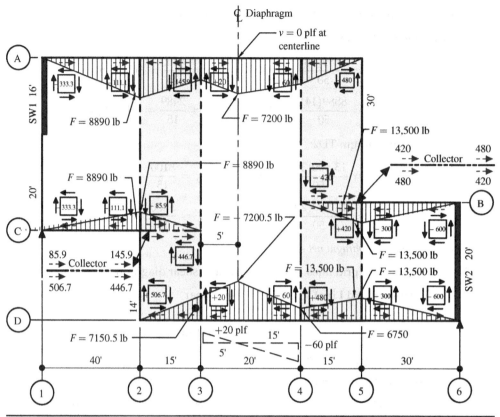

Figure 3.30 Longitudinal strut force diagrams.

Determination of longitudinal strut/chord forces (see Fig. 3.30): Strut forces along grid line A. The direction of transfer shears determines if the member is in tension or compression. Starting at grid line 1:

At grid line 2,

$$F_{2A} = \frac{(+333.3 + 111.1)}{2}(40) = 8890 \text{ lb compression}$$

At grid line 3,

$$F_{3A} = 8890 - \frac{(85.9 + 145.9)}{2}(15) = 7151.5 \text{ lb compression}$$

At diaph. c.l.,

$$F_{CL} = 7151.5 + \frac{(0 + 20)}{2}(5) = 7201.5 \text{ lb compression}$$

At grid line 4,

$$F_4 = 7201.5 - \frac{(0 + 60)}{2}(15) = 6751.5 \text{ lb compression}$$

At grid line 5,

$$F_5 = 6751.5 - \frac{(420 + 480)}{2}(15) = 1.5 \text{ lb compression}$$

Close enough, the diagram closes.

Strut force along grid line B, starting at grid line 4: The chord force on the right side of grid line 5 was calculated as 13,500 lb compression. The force on the left side of grid line 5 is equal to

$$F_{5B} = \frac{[(480 + 420) + (420 + 480)]}{2}(15) = 13,500 \text{ lb compression}$$

Also note that

$$F_{5-6} = \frac{600 + 300}{2}(30) = 13,500 \text{ lb compression}$$

Strut force along grid line C, starting at grid line 1: The chord force on the left side of grid line 2 was calculated as 8890 lb tension. The force on the right side of line 2 is equal to

$$F_{2C} = \frac{[(446.7 + 145.9) + (506.7 + 85.9)]}{2}(15) = 8890 \text{ lb tension}$$

Strut force along grid line D, starting at grid line 2:

At grid line 3,

$$F_{3D} = \frac{(-506.7 - 446.7)}{2}(15) = -7150.5 \text{ lb tension}$$

At diaph. c.l.,

$$F_{CL} = -7150.5 - \frac{(0 + 20)}{2}(5) = -7200.5 \text{ lb tension}$$

At grid line 4,

$$F_{4D} = -7200.5 + \frac{(0 + 60)}{2}(15) = -6750 \text{ lb tension}$$

At grid line 5,

$$F_{5D} = -6750 - \frac{(480 + 420)}{2}(15) = -13,500 \text{ lb tension}$$

At grid line 6,

$$F_{6D} = -13,500 + \frac{(300 + 600)}{2}(30) = 0 \text{ lb tension}$$

Therefore, the diagram closes to zero.

Determination of transverse strut/chord forces (see Fig. 3.31):

Forces at grid line 1, starting at line A:

$$V_{SW1} = 12,000 \text{ lb}$$

$$v_{SW1} = \frac{12,000}{16} = 750 \text{ plf}$$

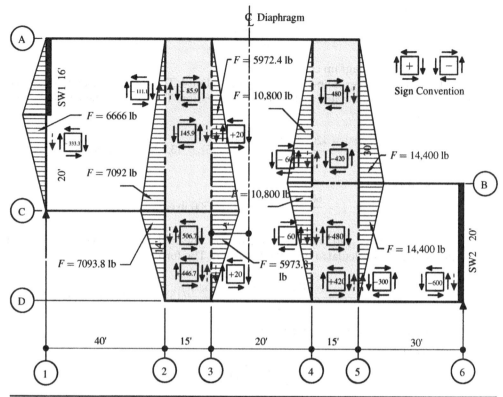

FIGURE 3.31 Transverse strut/chord force diagrams.

$$v_{\text{diaph}} = 333.3 \text{ plf}$$

$$v_{\text{net}} = 750 - 333.3 = +416.7 \text{ plf}$$

At the end of SW1, $F = 416.7(16) = 6667.2 \text{ lb}$

Strut force, $F = 333.3(20) = 6667.2 \text{ lb tension}$

The force diagram closes.

Forces along grid line 2, starting at line A:

At grid line 2, (above C), $F_{2C} = (111.1 + 85.9)(36) = 7092 \text{ lb tension}$

At grid line 2, (below C), $F_{2C} = 506.7(14) = 7093.8 \text{ lb tension}$

The force diagram closes.

Forces along grid line 3:

At grid line 3, (above C), $F_{3C} = (145.9 + 20)(36) = 5972.4 \text{ lb compression}$

At grid line 3, (below C), $F_{3C} = (446.7 - 20)(14) = 5973.8 \text{ lb compression}$

The force diagram closes.

Forces along grid line 4:

At grid line 4, (above B), $F_{4B} = (420 - 60)(30) = 10,800$ lb tension

At grid line 4, (below B), $F_{4B} = (480 + 60)(20) = 10,800$ lb tension

The force diagram closes.

Forces along grid line 5:

At grid line 5, (above B), $F_{5B} = 480(30) = 14,400$ lb compression

At grid line 5, (below B), $F_{5B} = (420 + 300)(20) = 14,400$ lb compression

The force diagram closes.

Diaphragm deflection (see Fig. 3.32): The double end offset diaphragm has three different widths due to the offsets at each end and should be analyzed as a single span beam with three different moments of inertia. It should be expected that the diaphragm just examined would have questionable performance, especially with respect to deflection. The

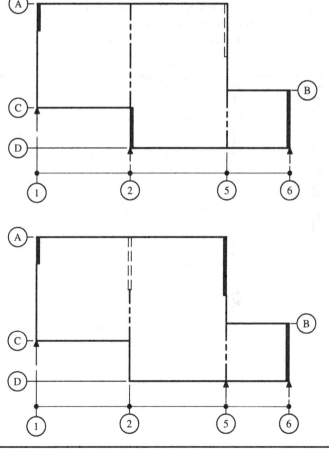

Figure 3.32 Optional layouts.

performance can be improved by adding interior shear walls which would increase redundancy, shorten the span, and reduce the demand on each of the individual shear walls. It would also resolve some of the deflection issues.

Optional Layouts (see Fig. 3.32): Optional layouts are available assuming there is flexibility in the architectural design. In the first option, an interior shear wall is installed at grid line 2. The resulting configuration produces a simple rectangular diaphragm on the left of grid line 2 and a single offset diaphragm on the right side of line 2. Adding an additional shear wall at grid line 5 would remove the offset condition and simplify the analysis. The second option is the reverse of option 1, where an interior shear wall is installed at grid line 5. The resulting configuration produces a simple rectangular diaphragm to the right of grid line 5 and a single offset diaphragm to the left of line 5. Adding an additional shear wall at grid line 2 would remove the offset condition. ▲

3.6 Problems

Problem 3.1: Single Offset Diaphragm, Analysis in the Longitudinal Direction, Discontinuous Strut Offset Outward (See Fig. P3.1)

Given: The diaphragm is 125 ft long with a 50 ft width. There is a 15-ft offset between grid lines B and C. The plan shows that shear walls are located at grid lines A and B only. The edge of the diaphragm at grid line C is supported by a strut that is tied into the end of the

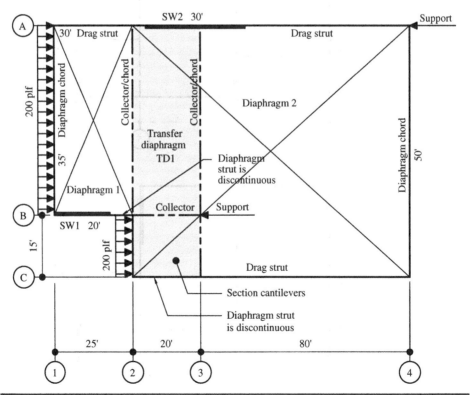

Figure P3.1

propped cantilever transfer diaphragm that occurs between grid lines 2 and 3. The strut at grid line C is offset outward from grid line B, which creates a discontinuity in the strut. The diaphragm is loaded in the longitudinal direction with a 200-plf uniform load.

Find:

1. Determine all the diaphragm shears.
2. Determine all chord, strut, and collector forces. ▲

Problem 3.2: Single Offset Diaphragm, Analysis in the Transverse Direction, Collector Past the Centerline of the Diaphragm (See Fig. P3.2)

Given: The structure shown in the isometric drawing and plan view drawings is 160 ft long with a 60 ft width. A pop-up roof section occurs between grid lines 1 and 2 from B to C. The roof sheathing has been omitted at the plane of the main roof diaphragm at this area, which creates a discontinuity in the diaphragm chord at grid line B. Grid line 2 is located at the centerline of the diaphragm. The isometric is provided to show that the additional wind loads applied to the pop-up must be transferred into the main diaphragm. A transfer diaphragm is placed between grid lines 2 and 3. The transfer diaphragm and collector coming off grid line B occurs in the negative shear area of the diaphragm. The diaphragm is loaded as shown in the isometric and plan. The pop-up should be designed as a second level.

Find:

1. Explain why the beam at grid line C from 1 to 2 cannot be used as part of the diaphragm chord.
2. Determine all the diaphragm shears. What is the maximum shear?
3. Determine all chord, strut, and collector forces. ▲

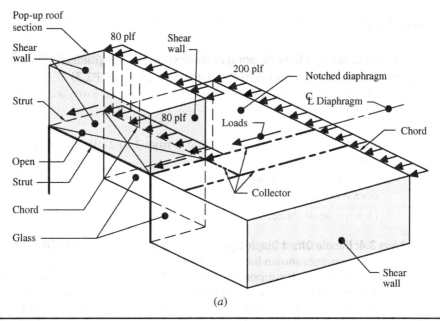

(a)

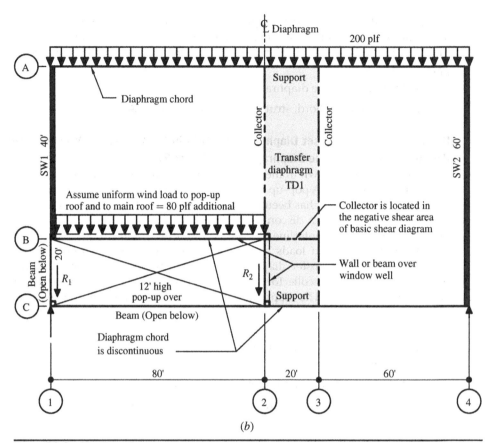

Figure P3.2 *(Continued)*

Problem 3.3: Single Offset Diaphragm, Analysis in the Longitudinal Direction, Transfer Diaphragm Past the Centerline of the Diaphragm (See Fig. P3.3)

Given: Using the same structure and loading presented in Prob. 3.2, analyze the diaphragm in the longitudinal direction.

Find:

1. Explain why the beam at grid line C from 1 to 2 plays a key part of the diaphragm.

2. Determine all the diaphragm shears. What is the maximum shear and where does it occur?

3. Determine all chord, strut, and collector forces. ▲

Problem 3.4: Double Offset Diaphragm, Analysis in the Transverse Direction (See Fig. P3.4)

Given: The diaphragm shown has a length of 140 ft and a width of 50 ft. The diaphragm has two offsets at the left support, a 15-ft offset between grid lines B and C, and a 10-ft offset between grid lines C and D. Both offsets cause discontinuities in the diaphragm chords. A uniform load of 200 plf is applied to the diaphragm in the transverse direction. Two transfer diaphragms, TD1 and TD2, are required to distribute the disrupted

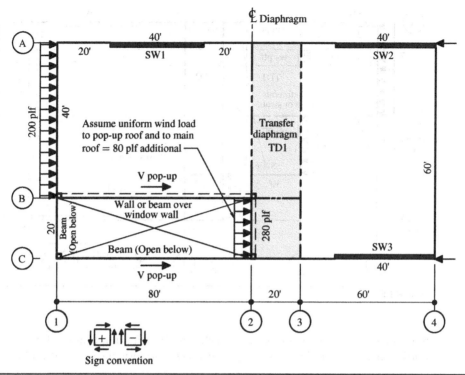

FIGURE P3.3

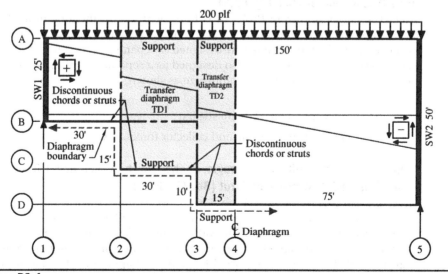

FIGURE P3.4

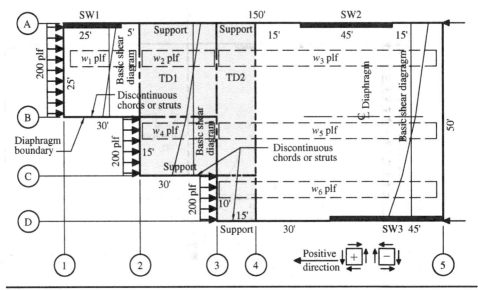

chord forces out to the main diaphragm chords. Transfer diaphragm TD1 supports the chord at grid line B and transfer diaphragm TD2 supports the chord at grid line C.

Find:

1. Determine all the diaphragm shears.

2. Determine all chord, strut, and collector forces. ▲

Problem 3.5: Double Offset Diaphragm, Analysis in the Longitudinal Direction, Multiple Struts Discontinuities (See Fig. P3.5)

Given: The diaphragm from Prob. 3.4 is to be analyzed in the longitudinal direction. Shear walls are located at grid lines A and D only. The diaphragm section between grid lines 1 and 2 and between 2 and 3 are designed as separate diaphragms. The diaphragm section between lines 3 and 5 is also designed as a separate diaphragm. A 200-plf uniform wind load is applied to the diaphragm as shown.

Find:

1. Determine all the diaphragm shears.

2. Determine all chord, strut, and collector forces. ▲

Problem 3.6: Double Offset Diaphragm, Analysis in the Longitudinal Direction, Offset Shear Walls, and Discontinuous Strut (See Fig. P3.6)

Given: The diaphragm shown in Prob. 3.5 is revised to have shear walls relocated at grid lines A, C, and D. The walls at lines C and D are assumed to act in the same line of resistance, which creates an offset shear wall condition. The diaphragm section between grid lines 1 and 2 is designed as a separate diaphragm which is supported by the strut at grid line B. The strut is supported by transfer diaphragm TD1 that lays between lines 2 and 3 from A to C. The diaphragm sections between grid lines 2 and 3 and

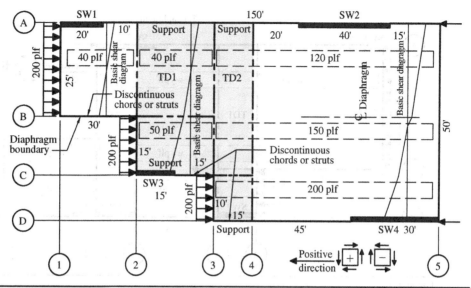

between grid lines 3 and 5 are interconnected by transfer diaphragm TD2. The uniformly distributed load of 200 plf is distributed to each of these sections in proportion to their individual widths.

Find:

1. Determine all the diaphragm shears.
2. Determine all chord, strut, and collector forces.   ▲

Problem 3.7: Diaphragm Offset at Both Ends, Analysis in the Longitudinal Direction, Discontinuous Struts (See Fig. P3.7)

Given: The diaphragm shown in Example 3.5 is revised to have the loads applied in the longitudinal direction. Shear walls are located along grid lines A and D only. The diaphragm section between grid lines 1 and 2 is designed as a separate diaphragm which is supported by shear walls at grid lines A and by a strut at grid line C. The strut at grid line C extends into transfer diaphragm TD1, connecting the smaller diaphragm section to the larger section between grid lines 2 and 5. The diaphragm section between lines 5 and 6 is also designed as a separate diaphragm, which is supported by a drag strut at grid line B that is connected to transfer diaphragm TD2. The uniformly distributed load of 200 plf is applied to the diaphragm as shown.

Find:

1. Determine all the diaphragm shears.
2. Determine all chord, strut, and collector forces.   ▲

Problem 3.8: Diaphragm Offset at Both Ends, Analysis in the Longitudinal Direction, Offset Shear Walls, and Discontinuous Strut (See Fig. P3.8)

Given: The diaphragm shown in Prob. 3.7 is revised to have loads applied in the longitudinal direction. Shear walls are now located at grid lines A, C, and D. The diaphragm

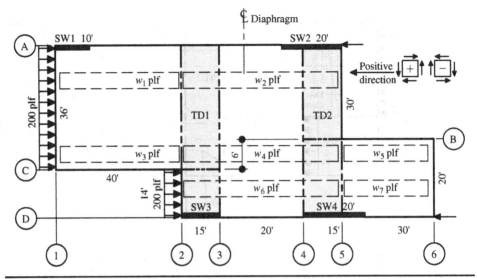

Figure P3.7

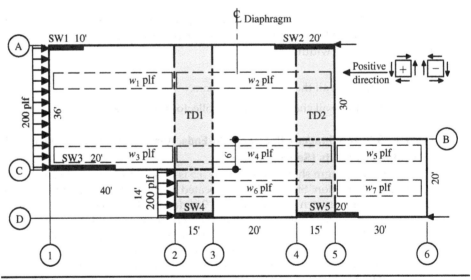

Figure P3.8

sections between grid lines 1 and 5 are interconnected by transfer diaphragm TD1. The diaphragm section between lines 5 and 6 is designed as a separate diaphragm, which is supported by a drag strut at grid line B that is connected to transfer diaphragm TD2. The uniformly distributed load of 200 plf is applied to the diaphragm as shown.

Find:

1. Determine all the diaphragm shears.

2. Determine all chord, strut, and collector forces. ▲

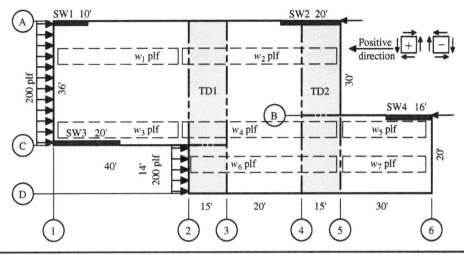

Problem 3.9: Diaphragm Offset at Both Ends, Analysis in the Longitudinal Direction, Offset Shear Walls, and Propped Cantilever Transfer Diaphragms (See Fig. P3.9)

Given: The diaphragm shown in Prob. 3.7 is revised to have loads applied in the longitudinal direction. Shear walls are now located at grid lines A, C, and D. The diaphragm sections between grid lines 1 and 5 are interconnected by transfer diaphragm TD1. The diaphragm section between lines 5 and 6 is designed as a separate diaphragm, which is supported by a drag strut at grid line B that is connected to transfer diaphragm TD2. The uniformly distributed load of 200 plf is applied to the diaphragm as shown.

Find:

1. Determine all the diaphragm shears.

2. Determine all chord, strut, and collector forces. ▲

3.7 References

1. American Society of Civil Engineers (ASCE), *ASCE/SEI 7-16 Minimum Design Loads for Buildings and Other Structures*, ASCE/SEI 7-16, ASCE, New York, 2016.

2. American Wood Council (AWC), *Special Design Provisions for Wind and Seismic with Commentary*, AWC, Leesburg, VA, 2021.

3. Diekmann, E. F., "Design of Wood Diaphragms," *Journal of Materials Education*, 1982, Fourth Clark C. Heritage Memorial Workshop, Wood Engineering Design Concepts University of Wisconsin, WI.

4. Kula, U., "Diaphragms and Diaphragm Chords," *SEAOC 2001 70th Annual Convention Proceedings*, Structural Engineers Association of California (SEAOC), Sacramento, CA, 2001.

5. Applied Technology Council (ATC), *Guidelines for Design of Horizontal Wood Diaphragms*, ATC-7, ATC, Redwood, CA, 1981.

CHAPTER 4

Diaphragms with Intermediate Offsets

4.1 Introduction

Structures with horizontal offsets (notches) located at the interior of the diaphragm span are common. Examples with a cutout or open area greater than 50 percent of the gross area of the diaphragm are classified as a Type 3 horizontal irregularity in accordance with ASCE 7[1] Table 12.3-1 and trigger special requirements in Seismic Design Categories (SDC) D, E, and F. Even in diaphragms not classified as having the Type 3 horizontal irregularity, the diaphragm load path around the offset should be evaluated. These offsets cause a reduction in the width of the diaphragm, creating a change in stiffness and discontinuity in the diaphragm chords or struts. These offsets sometimes occur at multiple locations along a chord or on opposite sides of the diaphragm. Unlike a single end offset, a transfer diaphragm must be installed on each side of the offset to receive the disrupted chord forces and dissipate them into the diaphragm, as shown in Fig. 4.1. This situation could also be used for the analysis of diaphragms with large openings, in which the upper or lower sections around the opening do not comply with a reasonable aspect ratio and become too flexible to transfer forces around the opening. The analysis methods used in Chap. 3 provide all the tools necessary to analyze these types of diaphragms. Placing shear walls at strategic locations can eliminate many of the difficulties caused by these offsets.

4.2 Intermediate Offset, Transverse Loading

The problem associated with diaphragms having an intermediate offset when loaded in the transverse direction is the discontinuity of the diaphragm chord caused by the offset. The reduction in the width of the diaphragm at the offset increases the chord force and diaphragm deflection at that location due to a decrease in the overall stiffness of the diaphragm. The disrupted chord force at the reduced section can be very large and can present some difficulty in transferring the force into the main body of the diaphragm.

Example 4.1: Intermediate Offset, Offset to the Left of Centerline, Analysis in the Transverse Direction

The diaphragm shown in Fig. 4.1 has a length of 150 ft and a width of 50 ft, with a 20-ft-long by 10-ft-wide offset which occurs between grid lines 2 and 3. The offset (notch) causes a disruption in the diaphragm chord at grid lines 2C and 3C. A 200-plf load is

145

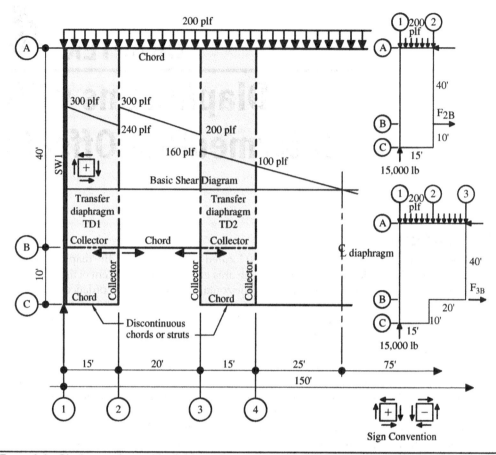

Figure 4.1 Diaphragm plan and basic shear diagrams.

applied uniformly along the entire length of the diaphragm in the transverse direction. The width of the diaphragm between grid lines A and B is 40 ft. The offset chord is extended into transfer diaphragms TD1 and TD2 by two longitudinal collectors which are located between grid lines 1 and 2, and between grid lines 3 and 4, respectively. The collectors extend the entire width of the transfer diaphragms. The transfer diaphragm is supported at the main diaphragm chord locations at grid lines A and C. Each transfer diaphragm has a length of 50 ft and a width of 15 ft, making the aspect ratios of the transfer diaphragms equal to 3.33:1 which requires it to be blocked in accordance with SDPWS[2] Table 4.2.2. The aspect ratio of the main diaphragm is equal to 3:1; therefore, they are of similar stiffness. Transverse collectors are placed along the entire width of the diaphragm at grid lines 1, 2, 3, and 4 to act as chord members for the transfer diaphragm. At grid line 1, the shear wall is full length of the diaphragm; therefore, the diaphragm shears and shear wall shears are equal but act in the opposite directions cancelling each other out to zero. However, there will be a residual transfer diaphragm chord force caused by the discontinuous chord force at grid line 2B. The sign convention for the diaphragm sheathing elements is shown at the lower right of the figure.

Construction of the basic shear diagram: Shear values are required at all collector locations, see Fig. 4.1.

$$R_1 = R_5 = \frac{wL}{2} = \frac{200(150)}{2} = 15{,}000 \text{ lb reaction at the supporting walls}$$

$$v_1 = \frac{R_1}{W} = \frac{15{,}000}{50} = 300 \text{ plf unit shear at grid line 1}$$

$$V_2 = R_1 - wx = 15{,}000 - 200(15) = 12{,}000 \text{ lb shear at grid line 2}$$

$$v_{2L} = \frac{V_2}{W_{left}} = \frac{12{,}000}{50} = 240 \text{ plf unit shear left side of grid line 2}$$

$$v_{2R} = \frac{V_2}{W_{right}} = \frac{12{,}000}{40} = 300 \text{ plf unit shear right side of grid line 2}$$

$$V_3 = R_1 - wx = 15{,}000 - 200(35) = 8000 \text{ lb shear at grid line 3}$$

$$v_{3L} = \frac{V_3}{W_{left}} = \frac{8000}{40} = 200 \text{ plf unit shear left side of grid line 3}$$

$$v_{3R} = \frac{V_3}{W_{right}} = \frac{8000}{50} = 160 \text{ plf unit shear right side of grid line 3}$$

$$V_4 = R_5 - wx = 8000 - 200(15) = 5000 \text{ lb shear at grid line 4}$$

$$v_4 = \frac{R_5}{W} = \frac{5000}{50} = 100 \text{ plf unit shear at grid line 4}$$

Determination of the chord force at grid line 2B:

$$M_{2B} = R_1 - \frac{wx^2}{2} = 15{,}000(15) - \frac{200(15)^2}{2} = 202{,}500 \text{ ft lb}$$

$$F_{2B} = \frac{M_{2B}}{DW} = \frac{202{,}500}{40} = 5062.5 \text{ lb}$$

Determination of the chord force at grid line 3B:

$$M_{3B} = R_1 - \frac{wx^2}{2} = 15{,}000(35) - \frac{200(35)^2}{2} = 402{,}500 \text{ ft lb}$$

$$F_{3B} = \frac{M_{3B}}{W} = \frac{402{,}500}{40} = 10{,}062.5 \text{ lb}$$

Determination of left transfer diaphragm shears, TD1 (see Fig. 4.2):

Disrupted chord force = 5062.5 lb at grid line 2B

$$V_A = \frac{5062.5(10)}{50} = 1012.5 \text{ lb,} \qquad v_A = \frac{V_A}{W_{TD}} = \frac{1012.5}{15} = 67.5 \text{ plf}$$

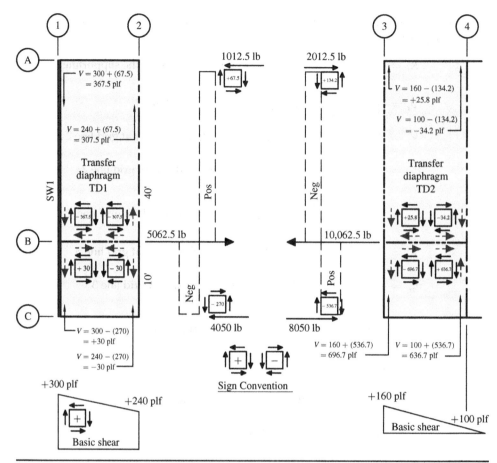

FIGURE 4.2   Transfer diaphragms and net diaphragm shears.

The direction of the reaction of the transfer diaphragm at A is acting to the left. Applying the remaining shears on the sheathing element shows that the transfer diaphragm shear is positive.

$$V_C = \frac{5062.5(40)}{50} = 4050 \text{ lb}, \quad v_C = v_A = \frac{V_C}{W_{TD}} = \frac{4050}{15} = -270 \text{ plf}$$

The direction of the transfer diaphragm reaction at C is also acting to the left. Applying the remaining shears on the sheathing element shows that the transfer diaphragm shear is negative.

Determination of transfer diaphragm net shears (see Fig. 4.2): Place the basic shear diagram and transfer diaphragm shear diagrams onto the diaphragm plan for easy reference. The process of adding and subtracting the shear values to find the net shears in the transfer diaphragm area is greatly simplified because both shear diagrams are side by side for easy visual reference. Once the net shears are determined, the representative shear elements and shears that are transferred into each chord or collector should also be placed on the plan. It should be reemphasized that only the shear within the transfer diaphragm is affected by the transfer of the disrupted chord force. The shears in the diaphragm sections to the left and right of the transfer diaphragm remain unaffected and are equal to the basic diaphragm shear. The net shears in the transfer diaphragm are calculated as follows:

Location (Grid Line)	Net Shear
1 from A to B	$v = +300 + (67.5) = +367.5$ plf
2 from A to B (left side)	$v = +240 + (67.5) = +307.5$ plf
2 from A to B (right side)	$v = +300$ plf
1 from B to C	$v = +300 - (270) = +30$ plf
2 from B to C	$v = +240 - (270) = -30$ plf

The diaphragm nailing patterns can be determined from the calculated net diaphragm shears.

Determination of right transfer diaphragm shears (see Fig. 4.2):

$$\text{Disrupted chord force} = 10{,}062.5 \text{ lb at grid line 3B}$$

$$V_A = \frac{10{,}062.5(10)}{50} = 2012.5 \text{ lb}, \quad v_A = \frac{V_A}{W_{TD}} = \frac{2012.5}{15} = -134.2 \text{ plf}$$

The direction of the reaction at A is acting to the right. Applying the remaining shears on the sheathing element shows that the transfer diaphragm shear is negative.

$$V_C = \frac{10{,}062.5(40)}{50} = 8050 \text{ lb}, \quad v_C = \frac{V_C}{W_{TD}} = \frac{8050}{15} = +536.7 \text{ plf}$$

The direction of the reaction at C is also acting to the right. Applying the remaining shears on the sheathing element shows that the transfer diaphragm shear is positive.

Determination of diaphragm net shears (see Fig. 4.2): Following the same procedure as for the left transfer diaphragm, the net shears in the transfer diaphragm are calculated as follows:

Location (Grid Line)	Net Shear
3 from A to B (right side)	$v = +160 - (134.2) = +25.8$ plf
3 from A to B (left side)	$v = +200$ plf
3 from B to C	$v = +160 + (536.7) = +696.7$ plf
4 from A to B (left side)	$v = +100 - (134.2) = -34.2$ plf
4 from B to C (left side)	$v = +100 + (536.7) = +636.7$ plf
4 from A to C (right side)	$v = +100$ plf

Determination of longitudinal collector/chord forces (see Fig. 4.3): The collector/chord forces are determined by adding or subtracting the transfer shears on each side of the collector or chord member based on their direction relative to each other. The resulting shears at each end of the member are averaged and multiplied by the member length, or simply taking the area of the resulting shear diagram along the length of the

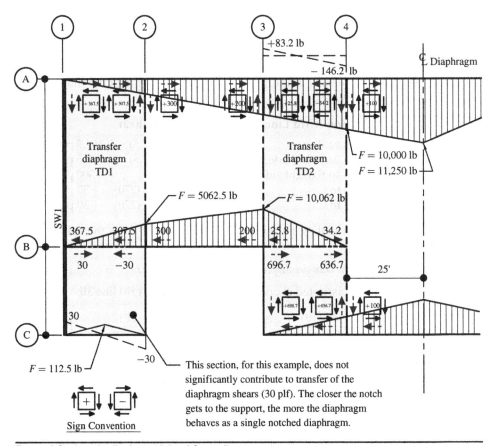

FIGURE 4.3 Longitudinal strut/chord force diagrams.

member being analyzed. It is important, but not necessary, that all the force diagrams are calculated and placed on the plan, as a visual verification of the analysis. Determine the forces along grid line B. Starting at grid line 1, the collector force at grid line 2B is equal to

$$F_{2B} = \left[\frac{(367.5 - 30) + (307.5 + 30)}{2}\right](15) = 5062.5 \text{ lb tension}$$

The direction of the net shears transferred into the collector indicates that the member is in tension and is equal to the chord force that was previously calculated. The chord force at grid line 3B is equal to

$$F_{3B} = 5062.5 + \frac{(300 + 200)}{2}(20) = 10,062.5 \text{ lb tension}$$

The collector force on the right side of grid line 3 is

$$F_{3B} = \left[\frac{(696.7 - 25.8) + (636.7 + 34.2)}{2}\right](15) = 10,063.5 \text{ lb}$$

This force is equal in magnitude but opposite in direction to the chord force on the left side of line 3. Therefore, the force diagram closes to zero.

The chord force along grid line A, starting at line 1, is as follows:

$$F_{2A} = \frac{(367.5 + 307.5)}{2}(15) = 5062.5 \text{ lb compression at grid line 2A}$$

This is equal to the chord force at grid line 2B. The chord force at 3A is equal to

$$F_{3A} = 5062.5 + \frac{(300 + 200)}{2}(20) = 10,062.5 \text{ lb compression}$$

This is equal to the chord force at grid line 3B. Note that the transfer shears are acting in the opposite direction. As a result, the forces along grid line A from 3 to 4 can be calculated using similar triangles to calculate the chord force at 4A, by adding the areas of the positive and negative shear triangles as shown in the figure.

$$F_{4A} = 10,062.5 + 83.2 - 146.2 = 10,000 \text{ lb compression}$$

$$F_{CL} = 10,000 + \frac{(100 + 0)}{2}(25) = 11,250 \text{ lb compression}$$

The chord force along grid line C, starting at grid line 3, is as follows:

$$F_{4C} = \frac{(696.7 + 636.7)}{2}(15) = 10,000 \text{ lb tension}$$

This force is equal to the chord force at grid line 4A.

$$F_{CL} = 10,000 + \frac{(100 + 0)}{2}(25) = 11,250 \text{ lb tension}$$

This force is equal to the chord force at grid line 4A.

Notice that the direction of all the transfer shears indicates that the members are in tension, as expected. The force diagram at grid line C closes to zero.

Determination of transverse collector/chord forces (see Fig. 4.4): The transfer shears for the transverse collectors are shown on the plan. The net shears on each side of grid line 2 between A and B are acting in opposite directions and the difference in the magnitudes is the net shear applied to the collectors.

The collector along grid line 2 from A to B has a uniform net shear of $307.5 - 300 = 7.5$ plf. The force at grid line 2B is equal to

$$F_{2B} = 7.5(40) = 300 \text{ lb tension}$$

The direction of the net transfer shear shows that the member is in tension. The direction of the transfer shears between B and C acts in the opposite direction, which also indicates the member is in tension. The force in the collector on the opposite side of grid line 2B is equal to

$$F_{2B} = 30(10) = 300 \text{ lb}$$

Therefore, the diagram closes.

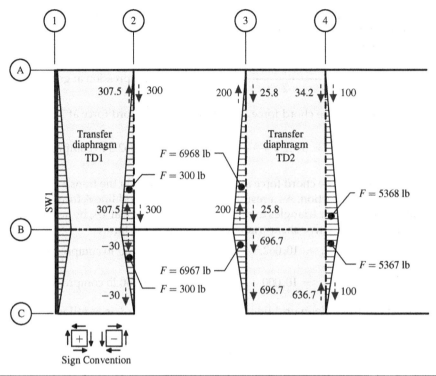

Figure 4.4   Transverse collector force diagrams.

The transfer shears applied to the collector along grid line 3 from A to B act in opposite directions and the difference in the magnitudes produces a net shear of $200 - 25.8 = 174.2$ plf. The force in the collector at grid line 3B is equal to

$$F_{3B} = 174.2(40) = 6968 \text{ lb tension}$$

The maximum force in the collector on the opposite side of grid line 3B is equal to

$$F_{3B} = 696.7(10) = 6967 \text{ lb tension}$$

This force is equal and opposite in direction to the collector between A and B. Therefore, the force diagram closes.

The force in the collector at grid line 4B between A and B is equal to

$$F_{4B} = (34.2 + 100)(40) = 5368 \text{ lb compression}$$

The maximum force in the collector on the opposite side of grid line 4B is equal to

$$F_{4B} = (636.7 - 100)(10) = 5367 \text{ lb compression}$$

This force is equal and opposite in direction to the collector between A and B. Therefore, the force diagram closes. Since all the force diagrams close, the entire analysis has been verified. ▲

4.3   Optional Layouts

Example 4.1 assumed that shear walls supporting the diaphragm were allowed at grid lines 1 and 5 only. Shear wall locations are often controlled by the floor plan layout. However, the engineer usually has some flexibility for the utilization and placement of those walls in the lateral-force-resisting system. The goal of the engineer should be to optimize the lateral-force-resisting system by strategically placing walls to eliminate discontinuities and irregularities, thereby creating some redundancy into the system. As an example, review Fig. 4.5. The diaphragm in Example 4.1 can be simplified by adding a shear wall at grid line 3. This creates two separate diaphragms, as shown at the bottom dimension line of the figure and eliminates the large chord force at grid line 3B. Diaphragm 1 is a single offset diaphragm, which requires a transfer diaphragm and remains to be a complex diaphragm, unless the collector at grid line B from 1 to 2 is used as a part of the diaphragm chord and the section below is considered as an independently supported diaphragm. The reduced span proportionately reduces the magnitude of the collector and chord forces. Diaphragm 2 becomes a simple diaphragm. A collector or interconnection tie is required at grid line 3B to provide some degree of continuity and to prevent separation at the reentrant corner, in accordance with IBC[3] Section 1604.4. The structural configuration becomes further simplified if an additional shear wall is added at grid line 2. The diaphragm is then broken down into three simple spans and all irregularities and discontinuities are removed. If a shear wall is installed at grid line 2 only, as shown in the figure, the

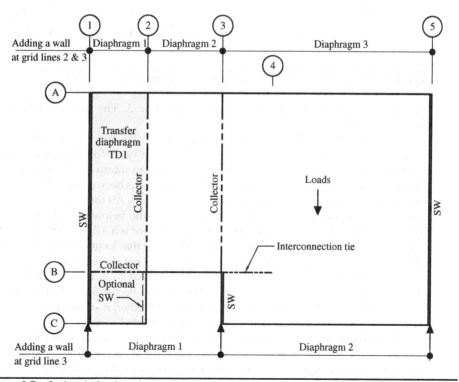

FIGURE 4.5   Optional plan layout.

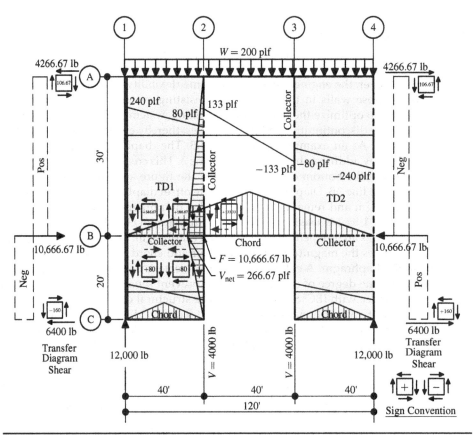

Figure 4.6 Analysis option 1—analyze as diaphragm with intermediate offset.

results are opposite to placing a wall at grid line 3. The offset diaphragm occurs between grid lines 2 and 5.

It has often been asked if transfer diaphragms must be used to solve this type of problem. The examples in the book represent only one way to do it, but not necessarily the only way. For example, Fig. 4.6 shows a plan with an intermediate offset with transfer diaphragms located on each side of the offset, which has been designed in the same manner as the previous example using transfer diaphragms. All the resulting calculations are shown in the figure, including the net shears above and below the collector at grid line B from 1 to 2. The shear above the left end of the collector is a +346.67 plf. Below the collector at that location is a +80 plf. The transfer shears at that location are acting in opposing directions. The difference in the magnitudes leaves a net shear applied to the collector of +266.67 plf acting in tension. On the right end of that collector, a +186.67 plf is acting above the collector and a −80 plf is acting below the collector. Since the transfer shears are acting in the same direction, they are added to each other, resulting in a +266.67 plf tension force. The sum of the forces is equal to the disrupted chord force.

As an alternate, it has also been asked if the lower sections can be hung off the main section, assuming the diaphragm chords are located at grid lines A and B only. The answer is yes you can. The results will be the same as using transfer diaphragms, with the same framing as shown in Fig. 4.7. The transfer shears for both the main diaphragm

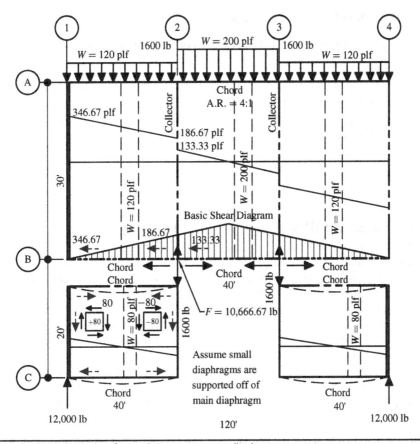

Figure 4.7 Analysis option 2—analyze as separate diaphragms.

and the chord of the lower section must be combined to get the final chord force at grid line B. If the results are not the same, then an error has been made. The difference in the analysis is that the lower supported diaphragm sections do not have shear walls at grid line 2 or 3, so the edges at those grid lines must be supported by collectors that are embedded the full width into the main diaphragm, which will create concentrated forces to the main diaphragm loading and the diaphragm section from grid lines 1 and 4 from A to B becomes a transfer diaphragm. The final loading for the main diaphragm is as shown in the figure. Using this technique can be very valuable when analyzing buildings with multiple minor offsets. It allows the creation of a continuous interior chord member and breaks the main diaphragm section down into a simple rectangular section that is easier to analyze.

4.4   Diaphragm Deflections

The diaphragm in Example 4.1 should be checked for deflection in accordance with the simplified method outlined in Chap. 2, or by using other rational methods. The diaphragm is analyzed as a single span beam with two different moments of inertia. The stiffer sections are located on each side of the offset.

Deflection Calculation Method

1. Determine the moments of inertia of the individual segments.

2. Construct the basic shear and moment diagram.

3. Calculate the bending deflection using a computer program, by hand calculations or other recognized methods.

4. Calculate the shear deflection using the simplified energy conservation method, or virtual work.

5. Calculate nail slip and chord slip. Include the effects of using a steel strap and blocking at the collectors.

4.5   Intermediate Offset, Longitudinal Loading

Analysis of diaphragms in the longitudinal direction is commonly ignored, thinking that the shears in that direction are less than in the transverse direction. In many cases, this assumption can be correct for simple diaphragms. However, even if the shears are lower, complete load paths are still required by code along each line of lateral force resistance. Whenever irregularities exist in the longitudinal direction, an analysis should be performed to verify the load paths and the transfer of forces across all discontinuities has been accomplished. Strut forces can also be larger than chord forces. As shown in Chap. 3, the results of the analysis can be surprising. Several shear wall configurations are presented in the examples and the problems located at the end of this chapter. These problems should be solved by the reader to help improve analytical skills and to learn the additional lessons caused by complicated load path issues.

Example 4.2: Intermediate Offset, Analysis in the Longitudinal Direction, Shear Walls in Line, Offset Strut

The diaphragm shown in Fig. 4.8 has a length of 50 ft and a width of 150 ft. Shear walls are located along grid lines A and C only. A 10-ft offset from grid line B to C creates a discontinuity in the strut. The offset strut is required to provide a continuous load path to connect SW3 and SW4 together and to prevent tearing at the inside corners of the offset. To accomplish this, collectors must be added to the ends of the offset strut that extend into the transfer diaphragms as shown. One way this analysis can be done is by breaking the diaphragm down into three sections, between grid lines 1 and 2, between 2 and 3, and between 3 and 5, designated as sections A, B, and C, respectively. The left edge of section B is assumed to be supported by transfer diaphragms TD1 and TD2 at grid line B. It is reasonable to assume that the left reaction of section B is equally distributed into both transfer diaphragms.

Load distribution into the diaphragm: The loads applied to the diaphragm, in this case, are caused by seismic forces. Assume that the total lateral seismic load (plf) to the diaphragm between grid lines A and B is 200 plf. The loads are distributed into each section in proportion to its width, as shown in Fig. 4.8.

$$w_{1-2} = \frac{wW_{\text{section}}}{W_{\text{total}}} = \frac{200(15)}{150} = 20 \text{ plf} \qquad \text{A to B}$$

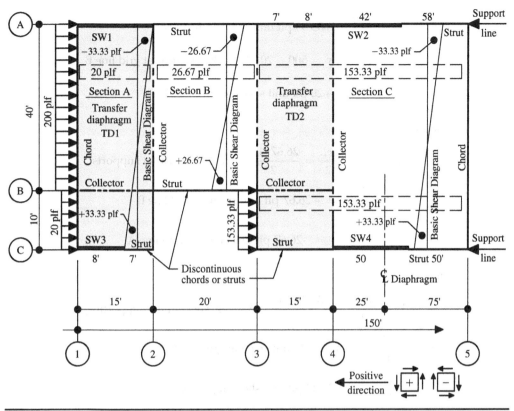

FIGURE 4.8   Diaphragm plan and basic shear diagrams.

$$w_{2-3} = \frac{200(20)}{150} = 26.67 \text{ plf} \qquad \text{A to B}$$

$$w_{3-5} = \frac{200(115)}{150} = 153.33 \text{ plf} \qquad \text{A to B}$$

$$w_{1-2} = 20 \text{ plf} \qquad \text{B to C by inspection}$$

$$w_{3-5} = 153.33 \text{ plf} \qquad \text{B to C}$$

Construction of the basic shear diagram:

Section A:

$$V_A = V_C = \frac{wL}{2} = \frac{20(50)}{2} = 500 \text{ lb shear at supporting walls}$$

$$v_A = \frac{V_A}{W_{\text{section}}} = \frac{-500}{15} = -33.33 \text{ plf unit shear at grid line A}$$

$$v_C = \frac{500}{15} = +33.33 \text{ plf unit shear at grid line C}$$

$$V_B = R_C - wx = 500 - 20(10) = 300 \text{ lb shear at grid line B}$$

$$v_B = \frac{300}{15} = +20 \text{ plf unit shear at grid line B}$$

Section B:

$$V_A = V_B = \frac{wL}{2} = \frac{26.67(40)}{2} = 533.4 \text{ lb shear at supports}$$

$$v_B = \frac{533.4}{20} = +26.67 \text{ plf unit shear at grid line B}$$

$$v_A = \frac{-533.4}{20} = -26.67 \text{ plf unit shear at grid line A}$$

Section C:

$$V_A = V_C = \frac{wL}{2} = \frac{153.33(50)}{2} = 3833.3 \text{ lb shear at the supporting walls}$$

$$v_C = \frac{3833.3}{115} = +33.33 \text{ plf unit shear at grid line C}$$

$$v_A = \frac{-3833.3}{115} = -33.33 \text{ plf unit shear at grid line A}$$

Determination of the strut force at grid line 2B and 3B (see Fig. 4.9):

$$R_B = 533.4 \text{ lb from 2 to 3}$$

$$F_{2B} = F_{3B} = \frac{533.4}{2} = 266.7 \text{ lb}$$

This is because the total shear at grid line B was assumed to be shared equally by transfer diaphragms TD1 and TD2.

Determination of transfer diaphragm shears, TD1 and TD2 (see Fig. 4.9): The strut force is pulling on transfer diaphragm TD1 and pushing on transfer diaphragm TD2. Since the length and depth of the transfer diaphragms are the same, and the strut force is acting in the same direction, the resulting shears are also the same.

$$F_{strut} = 266.7 \text{ lb}$$

$$V_A = \frac{266.7(10)}{50} = -53.34 \text{ lb}, \qquad v_A = \frac{-53.34}{15} = -3.56 \text{ plf}$$

The direction of the reaction at A is acting to the left. Applying the remaining shears on the sheathing element shows that the transfer diaphragm shear is negative.

$$V_C = \frac{266.7(40)}{50} = +213.36 \text{ lb}, \qquad v_C = \frac{213.36}{15} = +14.22 \text{ plf}$$

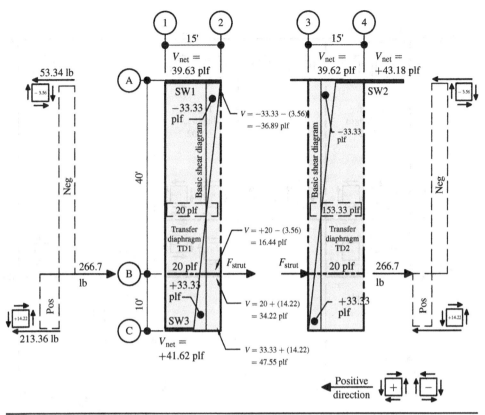

FIGURE 4.9 Transfer diaphragm and net diaphragm shears.

The direction of the reaction at C is also acting to the left. Applying the remaining shears on the sheathing element shows that the transfer diaphragm shear is positive.

Determination of diaphragm net shears (see Figs. 4.9 and 4.10): The net shears in the transfer diaphragm are calculated as follows:

Location (Grid Line)		Net Shear
TD1	A from 1 to 2	$v = -33.33 - (3.56) = -36.89$ plf
	B from 1 to 2 (above)	$v = +20 - (3.56) = +16.44$ plf
	B from 1 to 2 (below)	$v = +20 + (14.22) = +34.22$ plf
	C from 1 to 2	$v = +33.33 + (14.22) = +47.55$ plf
TD2	A from 3 to 4	$v = -33.33 - (3.56) = -36.89$ plf
	B from 3 to 4 (above)	$v = +20 - (3.56) = +16.44$ plf
	B from 3 to 4 (below)	$v = +20 + (14.22) = +34.22$ plf
	C from 3 to 4	$v = +33.33 + (14.22) = +47.55$ plf

The diaphragm nailing patterns can be determined from the calculated net diaphragm shears.

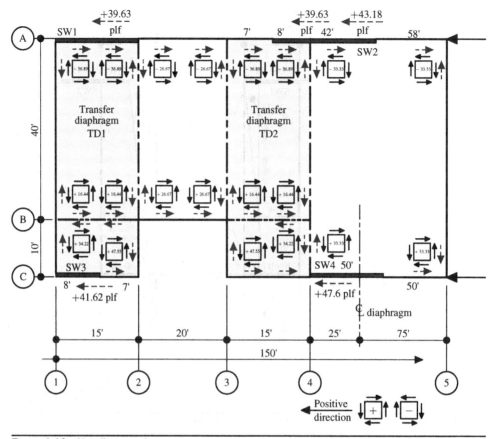

FIGURE 4.10   Net diagram shears.

Determination of net shears at shear walls (see Fig. 4.9): Note that the total load to grid line A is equal to the tributary of the uniform load of sections A, B, and C plus the proportional load of the strut force applied at grid line B from section B.

$$R_A = 500 + \frac{2(266.7)(10)}{50} + 533.4 + 153.33\left(\frac{50}{2}\right) = 4973.33 \text{ lb}$$

$$V_{SW1,2} = \frac{4973.33}{(15+50)} = 76.52 \text{ plf}$$

This assumes that the load distribution to each wall is in proportion to their length.

$$V_{SW3} = 500 + \frac{266.67(40)}{50} = 713.36 \text{ lb}, \qquad v_{SW3} = \frac{713.36}{8} = +89.17 \text{ plf}$$

$$V_{SW4} = 33.33(115) + \frac{266.67(40)}{50} = 4046.3 \text{ lb}, \qquad v_{SW4} = \frac{4046.3}{50} = +80.92 \text{ plf}$$

The net shears at the shear walls are equal to the wall shear minus the diaphragm shear.

$$v_{net} = v_{SW} - v_{diaph} = +76.52 - 36.89 = +39.63 \text{ plf at SW1}$$

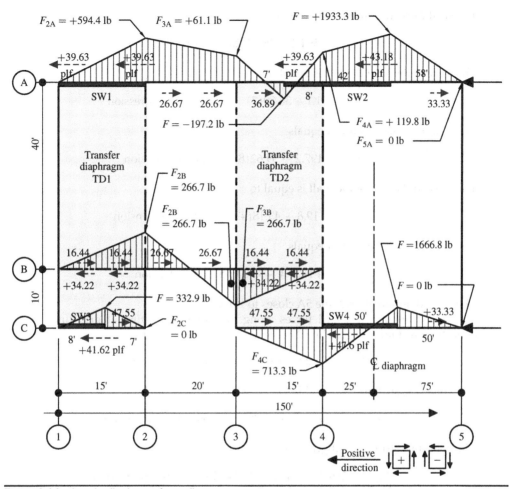

FIGURE 4.11 Longitudinal strut force diagrams.

$$v_{net} = +76.52 - 36.89 = +39.63 \text{ plf at SW2 (left 8')}$$

$$v_{net} = +76.52 - 33.33 = +43.18 \text{ plf at SW2 (right 42')}$$

$$v_{net} = +89.17 - 45.55 = +41.62 \text{ plf at SW3}$$

$$v_{net} = +80.92 - 33.33 = +47.6 \text{ plf at SW4}$$

The net diaphragm shears and shears transferred into the struts, collectors, and chords are placed on the plan in Fig. 4.10.

Determination of longitudinal strut forces (see Fig. 4.11): The direction of the resulting shears transferred into the struts, chords, and collectors will indicate whether the members are in tension or compression. Starting at grid line 1A, the strut force at grid line 2A is equal to

$$F_{2A} = 39.63(15) = +594.4 \text{ lb tension}$$

The strut force at grid line 3A is equal to

$$F_{3A} = +594.4 - 26.67(20) = +61.1 \text{ lb tension}$$

The force at the start of SW2 is equal to

$$F_{start} = +61.1 - 36.89(7) = -197.2 \text{ lb compression}$$

The strut force at grid line 4 equals

$$F_4 = -197.2 + 39.62(8) = +119.8 \text{ lb tension}$$

The force at the end of the wall is equal to

$$F_{end} = +119.8 + 43.18(42) = +1933.3 \text{ lb tension}$$

The strut force at grid line 4 equals

$$F = +1933.3 - 33.33(58) = 0 \text{ lb tension}$$

The force diagram at grid line 5A closes to zero.

Next examine the chord forces along grid line B. Starting at grid line 1B, the collector/strut force at grid line 2 becomes

$$F_{2B} = (34.22 - 16.44)(15) = +266.7 \text{ lb tension}$$

The collector/strut force at grid line 3 becomes

$$F_{3B} = +266.7 - 26.67(20) = -266.7 \text{ lb compression}$$

The collector/strut force at grid line 4 becomes

$$F_{4B} = -266.7 + (34.22 - 16.44)(15) = 0 \text{ lb compression}$$

The force diagram at grid line 4B closes to zero.

Starting at grid line 1C, the force at the end of SW3 is equal to

$$F = +41.62(8) = 332.9 \text{ lb tension}$$

The strut force at grid line 2 is equal to

$$F_{2C} = +322.9 - 47.55(7) = 0 \text{ lb tension}$$

Checking the line from 3C to 5C, the strut force at grid line 4C is equal to

$$F_{4C} = -47.55(15) = -713.3 \text{ lb compression}$$

The force at the end of SW4 is equal to

$$F = -713.3 + 47.6(50) = 1666.8 \text{ lb tension}$$

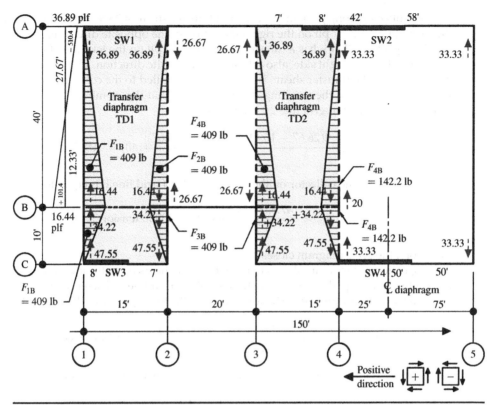

F_{1B} = 409 lb

FIGURE 4.12 Transverse collector force diagrams.

The force at grid line 5C is equal to

$$F_{5C} = 1666.8 - 33.33(50) = 0 \text{ lb tension}$$

All the longitudinal force diagrams close to zero and follow basic diaphragm action.

Determination of transverse collector/chord forces (see Fig. 4.12): The net diaphragm shears and transfer shears for the transverse collectors are shown on the plan. Note that the transfer shears at grid line 1 from grid line B to A change from +16.44 to −36.89 plf. Similar triangles are created to find the length of the positive and negative areas. The positive and negative forces are determined by taking the areas of these triangles. The chord force along grid line 1 from A to B at grid line 1B is equal to

$$F_{1B} = +101.4 - 510.4 = -409 \text{ lb compression}$$

The direction of the transfer shear shows that the member is in compression. On the opposite side of grid line 1B, the force is equal to

$$F_{1B} = \frac{(47.55 + 34.22)}{2}(10) = 409 \text{ lb compression}$$

Therefore, the force diagram closes.

The collector along grid line 2 at A has a transfer shear of 36.89 plf on the left side of the member and 26.67 plf on the right side acting in the opposite directions. The collector along grid line 2 at B has a transfer shear of 16.44 plf on the left side of the member and 26.67 plf on the right side, also acting in the opposite direction. The difference in the magnitudes of the transfer shears is the net shear applied to the collectors at each location. The direction of the resulting shears indicates that the member is in tension. The force in the collector at grid line 2B is equal to

$$F_{2B} = \frac{[(26.67 - 16.44) + (36.89 - 26.67)]}{2}(40) = 409 \text{ lb}$$

The strut force in the collector on the opposite side of grid line B is equal to

$$F_{2B} = \frac{(47.55 + 34.22)}{2}(10) = 409 \text{ lb tension}$$

Therefore, the force diagram closes.

Inspection of the figure shows that the transfer shears on both sides of the collector at grid line 3 are equal and opposite to the collector on line 2. Therefore, the forces are the same, but in compression. The force in the collector at grid line 4B is equal to

$$F_{4B} = \frac{[(36.89 - 33.33) + (20 - 16.44)]}{2}(40) = 142.4 \text{ lb tension}$$

The strut force in the collector on the opposite side of grid line B is equal to

$$F_{4B} = \frac{[(47.55 - 33.33) + (34.22 - 20)]}{2}(10) = 142.4 \text{ lb tension}$$

Therefore, the force diagram closes.

Since all the force diagrams close, the analysis has been verified. ▲

4.6   Diaphragms with Offset at the End Wall

Occasionally, an entrance occurs at an end wall. To demonstrate the problems transferring the diaphragm reactions across such an opening, an offset is created at the left end wall of the plan shown in Fig. 4.13. Example 4.3 analyzes the diaphragm with the sections above and below the opening connected by a strut at grid line 1. It would be a worthwhile exercise to analyze this offset end without the strut across the offset as shown in Prob. P4.3. The objective of the two exercises is to demonstrate the difference in the distribution of shears across the offset with and without the strut.

This is another example that can be used to analyze a diaphragm with a large opening where the section to the left of the opening does not meet reasonable aspect ratios and is too flexible to transfer forces across the opening.

Example 4.3: Intermediate Offset at End Wall with Strut, Analysis in the Transverse Direction (See Fig. 4.13)

The diaphragm has a length of 150 ft and a width of 56 ft. A 16-ft-long × 18-ft-wide offset is located at grid line 1, which supports a skylight that is incapable of transferring

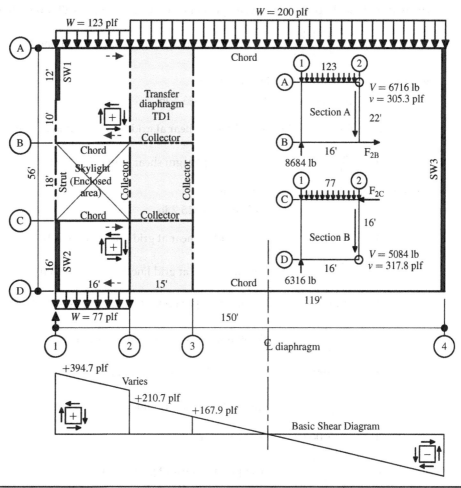

FIGURE 4.13   Basic shear diagrams and free-body diagrams.

diaphragm shears across the opening. A strut supports the left end of the skylight and connects the shear walls along grid line 1 together. The strut acts as part of the bound-ary member along the line of lateral force resistance. The sections above and below the opening are designated as sections A and B, respectively. Since the sections are fixed at grid line 2, the members at grid lines B and C from 1 to 2 are chord members for the sections, which are discontinuous. The disrupted chords are embedded into transfer diaphragm TD1 where their forces are then transferred to the exterior wall lines at A and C. The aspect ratio of TD1 is 3.73:1; therefore, the transfer diaphragm must be blocked. The 200-plf uniform load applied to the diaphragm is caused by wind. The offset area is enclosed by a wall at grid line 1 from B to C. The wind load is split into windward and leeward pressures between lines 1 and 2.

Construction of the basic shear diagram (see Fig. 4.13):

$$V_1 = V_4 = \frac{wL}{2} = \frac{200(150)}{2} = 15,000 \text{ lb shear at each support wall}$$

The width of the diaphragm between grid lines 1 and 2 is equal to 22 ft plus 16 ft.

$$v_1 = \frac{15,000}{22 + 16} = +394.7 \text{ plf unit diaphragm shear at grid line 1}$$

Section A: apply windward wind pressures.

$$R_1 = v_{\text{diaph}} W_{\text{sect A}} = 394.7\,(22) = 8683.4 \text{ lb}$$

$$V_2 = 8683.4 - 123(16) = 6716 \text{ lb, shear at grid line 2, section A (left)}$$

$$v_2 = \frac{6716}{22} = +305.3 \text{ plf unit diaphragm shear at grid line 2 (left)}$$

Section B: apply leeward wind pressures.

$$R_1 = v_{\text{diaph}} W_{\text{sect B}} = 394.7\,(16) = 6315.2 \text{ lb}$$

$$V_2 = 6315.2 - 77(16) = 5084 \text{ lb, shear at grid line 2, section B (left)}$$

$$v_2 = \frac{5084}{16} = +317.8 \text{ plf unit shear at grid line 2, section B (left)}$$

$$V_{2R} = 15,000 - (123 + 77)(16) = 11,800 \text{ lb shear at grid line 2 (right)}$$

$$v_{2R} = \frac{11,800}{56} = +210.7 \text{ plf unit shear at grid line 2 (right)}$$

$$V_3 = 11,800 - 200(15) = 8800 \text{ lb shear at grid line 3}$$

$$v_3 = \frac{8800}{56} = +167.9 \text{ plf unit shear at grid line 3}$$

$$V_4 = -15,000 \text{ lb shear at grid line 4}$$

$$v_{43} = \frac{-15,000}{56} = -267.9 \text{ plf unit shear at grid line 4}$$

Determination of the chord force at grid lines 2B and 2C (see Fig. 4.13):
Section A:

$$F_{2B} = \left[R_{1\,\text{sect A}}\,(x) - \frac{wx^2}{2} \right] \frac{1}{W_{\text{sect A}}} = \left[8684(16) - \frac{123(16)^2}{2} \right] \frac{1}{22} = 5600 \text{ lb}$$

Section B:

$$F_{2C} = \left[R_{1\,\text{sect B}}\,(x) - \frac{wx^2}{2} \right] \frac{1}{W_{\text{sect B}}} = \left[6316(16) - \frac{77(16)^2}{2} \right] \frac{1}{16} = 5700 \text{ lb}$$

As a point of interest, the reader should determine the shears and chord forces in sections A and B at grid line 2 if a 200-plf uniform load is applied to section A only instead of applying windward and leeward wind loads to sections A and B. This is commonly done but can cause section A to be over-designed and section B to be under-designed. Note that this example does not consider internal wind pressures which can

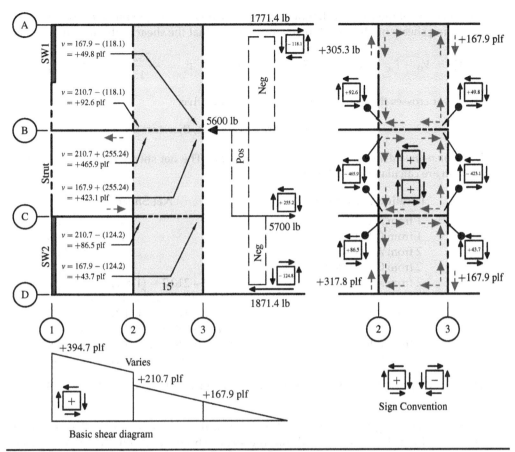

FIGURE 4.14 Transfer diaphragm and net diaphragm shears.

have a significant impact on the design around internal opening of diaphragms. Positive or negative pressure would apply to sections A and B simultaneously and cancel out when transferred to the shear walls; however, such internal pressures may increase the shear in the sheathing and chords and collectors around the entry/internal opening in the diaphragm.

Determination of transfer diaphragm shears, TD1 (see Fig. 4.14): The chord force at grid line B is in tension and is pulling on transfer diaphragm TD1. The chord force at grid line C is in compression and is pushing on the transfer diaphragm.

$$V_A = \frac{[5700(16) - 5600(34)]}{56} = -1771.4 \text{ lb}, \quad v_A = \frac{-1771.4}{15} = -118.1 \text{ plf}$$

The direction of the transfer diaphragm reaction at A is acting to the right. Applying the remaining shears on the sheathing element shows that the shear is negative.

$$V_D = \frac{[5600(22) - 5700(40)]}{56} = -1871.4 \text{ lb}, \quad v_D = \frac{-1871.4}{15} = -124.8 \text{ plf}$$

The direction of the transfer diaphragm reaction at D is acting to the left. Applying the remaining shears on the sheathing element shows that the shear is also negative.

$$V_C = V_D - F_{2C} = 1871.4 - 5700 = -3828.6 \text{ lb}, \qquad v_C = \frac{-3828.6}{15} = +255.24 \text{ plf}$$

The shear crosses the line; therefore, it becomes positive.

$$V_B = -3828.6 + 5600 = 1771.4 \text{ lb, therefore OK.}$$

Determination of diaphragm net shears (see Fig. 4.14): The net shears in the transfer diaphragm are calculated as follows:

Location (Grid Line)	Net Shear
1 from A to B	$v = +394.7$ plf
1 from C to D	$v = +394.7$ plf
2 from A to B (left)	$v = +305.3$ plf
2 from C to D (left)	$v = +317.8$ plf
2 from A to B	$v = +210.7 - (118.1) = +92.6$ plf
2 from B to C	$v = +210.7 + (255.24) = +465.9$ plf
2 from C to D	$v = +210.7 - (124.2) = +86.5$ plf
3 from A to B	$v = +167.9 - (118.1) = +49.8$ plf
3 from B to C	$v = +167.9 + (255.24) = +423.1$ plf
3 from C to D	$v = +167.9 - (124.2) = +43.7$ plf

The diaphragm nailing patterns can be determined from the calculated net diaphragm shears.

Determination of longitudinal collector/chord forces at B and C (see Fig. 4.15): The direction of the resulting shears transferred into the struts, chords, and collectors will indicate whether the members are in tension or compression. Starting at grid line 1B, the strut force at grid line 2B is equal to

$$F_{2B} = \frac{(394.7 + 305.3)}{2}(16) = +5600 \text{ lb tension}$$

The strut force at grid line 3B is equal to

$$F_{2B} = +5600 - \frac{(465.9 - 92.6) + (423.1 - 49.8)}{2}(15) = 0 \text{ lb tension}$$

The strut force at grid line 2C is equal to

$$F_{2C} = \frac{(394.7 + 317.8)}{2}(16) = +5700 \text{ lb compression}$$

The strut force at grid line 3C is equal to

$$F_{3C} = 5700 - \frac{(465.9 - 86.5) + (423.1 - 43.7)}{2}(15) = 0 \text{ lb tension}$$

All the longitudinal force diagrams close to zero and follow basic diaphragm action.

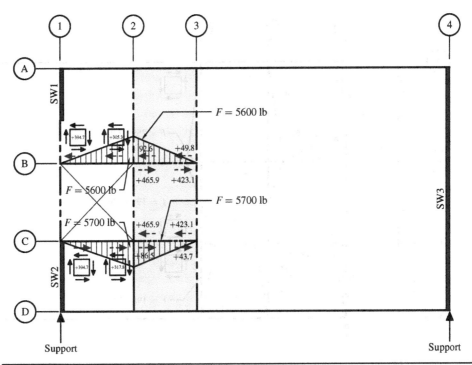

Figure 4.15 Longitudinal chord force diagrams.

Determination of net shears at shear walls (see Fig. 4.16): Note that since the line of lateral force resistance at grid line 1 is continuous, the total shear to each wall will be distributed in proportion to their lengths.

$$R_1 = 15,000 \text{ lb}$$

$$V_{SW1} = \frac{12(15,000)}{28} = 6428.6 \text{ lb}$$

$$V_{SW2} = \frac{16(15,000)}{28} = 8571.4 \text{ lb}$$

$$v_{SW1,2} = \frac{15,000}{(12+16)} = +535.7 \text{plf}$$

$$v_{diaph} = 394.7 \text{ plf}$$

$$v_{net} = 535.7 - 394.7 = +141 \text{ plf} \quad \text{net shear at the shear walls}$$

Determination of transverse strut/collector forces (see Fig. 4.16): Starting at grid line 1A, the strut force at the end of the wall is equal to

$$v_{net} = +141 \text{ plf net shear at the shear walls previously calculated}$$

$$F = v_{net}L_{SW1} = +141(12) = +1692 \text{ lb tension}$$

The strut force at grid line 1B is equal to

$$F_{1B} = +1692 - 394.7(10) = -2252 \text{ lb compression}$$

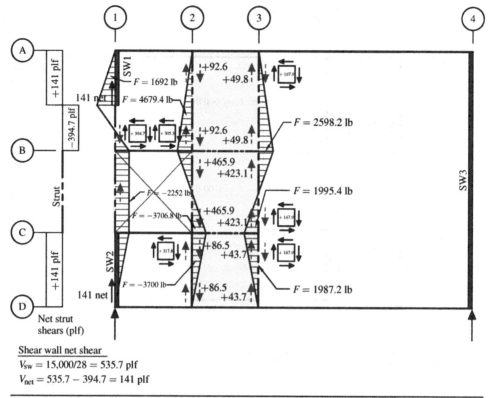

Figure 4.16 Transverse strut/collector force diagrams.

Since no additional shears are applied to the strut from B to C, the strut force remains constant to grid line C. The force diagram also shows that part of the strut force from the end of SW1 to grid line B is in compression and therefore transfers that force to SW2 through the strut.

The force at grid line 1D is equal to

$$F_{1D} = -2252 + 141(16) = 0 \text{ lb}$$

Therefore, the force diagram closes.

The collector along grid line 2 at A has a transfer shear of 305.3 plf on the left side of the member and 92.6 plf on the right side. The transfer shears are in opposite directions. The difference in the magnitudes is the net shear applied to the collectors. The force in the collector at grid line 2B is equal to

$$F_{2B} = (305.3 - 92.6)(22) = 4679.4 \text{ lb above line B}$$

The force in the collector on the opposite side of grid line B at line 2C is equal to

$$F_{2C} = 4679.4 - 465.9(18) = -3706.8 \text{ lb compression}$$

The force in the collector at grid line D is equal to

$$F_{2D} = -3706.8 + (317.8 - 86.5)(16) = -6 \text{ lb compression}$$

Therefore, the force diagram closes.
     The force in the collector at grid line 3B is equal to

$$F_{3B} = (167.9 - 49.8)(22) = 2598.2 \text{ lb compression}$$

The strut force in the collector at grid line 3C is equal to

$$F_{3C} = -2598.2 + (423.1 - 167.9)(18) = +1995.4 \text{ lb tension}$$

The force in the collector at grid line D is equal to

$$F_{3D} = +1995.4 - (167.9 - 43.7)(16) = 8.2 \text{ lb compression}$$

Therefore, the force diagram closes (round off).
     Since all the force diagrams close, the entire analysis has been verified. ▲

4.7 Problems

Problem 4.1: Intermediate Offset Diaphragm, Analysis in the Longitudinal Direction, Offset Shear Walls, Discontinuous Strut

Given: The diaphragm configuration shown in Fig. P4.1 has shear walls located at grid lines A, B, and C. Shear walls SW3 and SW4 are close enough to be considered to act in the same line of lateral force resistance and are connected by the strut/collector along grid line B from 1 to 3. The drag strut along line C from 3 to 5 supports the lower edge of the diaphragm section to the right of the offset and is supported off the end of transfer diaphragm TD2. The transfer diaphragm acts as a propped cantilever beam. The diaphragm loads are from seismic loads and are as shown on the plan.

Find:

1. All the diaphragm shears.

2. All the strut, collector, and chord forces. ▲

Problem 4.2: Intermediate Offset Diaphragm, Analysis in the Longitudinal Direction, Offset Shear Walls, Discontinuous Strut

Given: The diaphragm configuration is the same as the plan shown in Prob. 4.1, but now has an additional shear wall placed at grid line C between lines 4 and 5 as shown in Fig. P4.2. Shear walls SW3, SW4, and SW5 are assumed to act in the same line of resistance. The loads applied to the diaphragm are the same as Prob. 4.1.

Find:

1. All the diaphragm shears.

2. All the strut, collector, and chord forces. ▲

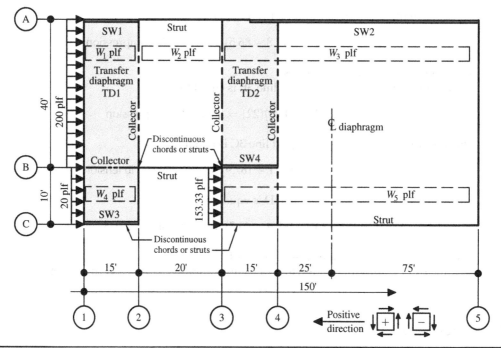

FIGURE P4.1

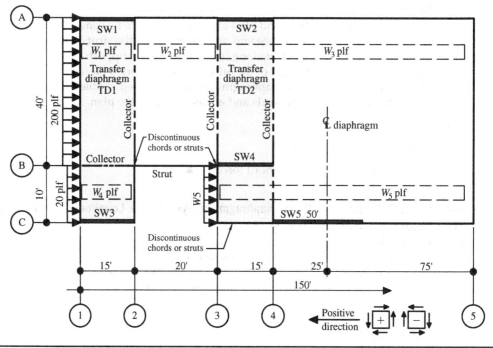

FIGURE P4.2

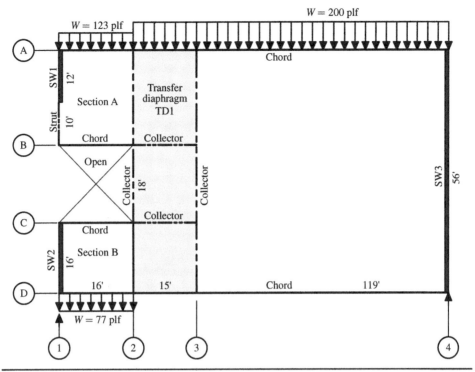

Problem 4.3: Intermediate Offset at End Wall Without Strut, Analysis in the Transverse Direction

Given: The diaphragm is identical to the diaphragm in Example 4.3 except that the strut at grid line 1 from B to C has been eliminated as shown in Fig. P4.3. The load to the diaphragm is the same as the previous example. The area at the offset is open; no wall occurs at grid line 1 from B to C. Code does not allow sheltering of wind loads. Normally, there will be windward and leeward pressures applied to each section above and below the offset.

Find:

    1. Explain what the difference is between this problem and Example 4.3.

    2. All the diaphragm shears.

    3. All the strut, collector, and chord forces.   ▲

4.8 References

1. American Society of Civil Engineers (ASCE), *ASCE/SEI 7-16 Minimum Design Loads for Buildings and Other Structures*, ASCE, New York, 2016.
2. American Wood Council (AWC), *Special Design Provisions for Wind and Seismic with Commentary* (SDPWS), AWC, Leesburg, VA, 2021.
3. International Code Council (ICC), *International Building Code, 2021 with Commentary*, ICC, Whittier, CA, 2021.

CHAPTER 5

Diaphragms with Openings

5.1 Introduction

Large openings in diaphragms can cause a Type 3 horizontal irregularity in accordance with ASCE 7 Table 12.3-1,[1] a Diaphragm Discontinuity Irregularity, where there is an abrupt discontinuity or variation in stiffness of the diaphragm. This type of irregularity triggers additional requirements in SDC D-F. Several articles and examples addressing the method of analyzing diaphragms with openings have been included in publications since the early 1980s. Most of those publications were based on a method of analysis developed by Edward F. Diekmann.[2] At the time these publications were developed, there had been only a few tests conducted on diaphragms with openings. The results of these tests are included in APA Research Report 138.[3] Openings in roof and floor diaphragms occur to accommodate stairwells, mechanical chases, skylights, atriums, etc. Until recently, there has been very little guidance on deciding when the size of these openings is large enough to require special reinforcing and a complete analysis, or small enough to simply ignore and provide minimal reinforcement. Diekmann suggested that if the opening is small enough to cause only minimal localized shear and bending effects within the diaphragm, then it can be considered as insignificant. ATC-7,[4] IBC,[5] and Diekmann[2] prescriptively recommended that at small openings minimal reinforcing at the corners of the opening should extend a minimum distance equal to the length or width of the opening in the direction under consideration, similar to 2021 IBC Figure 2308.4.4.1(1). In other words, the minimum distance left and right past each side of the opening would be equal to the length (L) of the opening and the minimum distance above and below the opening width would be equal to the opening width (b or W). For larger openings, a more rigorous review is required. ATC-7 and Diekmann recommended that all sections on all sides of the opening should comply with the allowable aspect ratio of SDPWS[6] Table 4.2.4. Any section not complying should be considered ineffective due to its lack of stiffness and thus be ignored in analyzing the distribution of forces across an opening. Figure 5.1 demonstrates this concept.

Diekmann provided an example in his paper on the "Design of Wood Diaphragms"[2] outlining a quick check on how to determine the significance of the opening size. In his review, it was assumed that half of the disrupted shear caused by the opening (i.e., the diaphragm unit shear to the right of the opening at grid line 2 times the opening width)

175

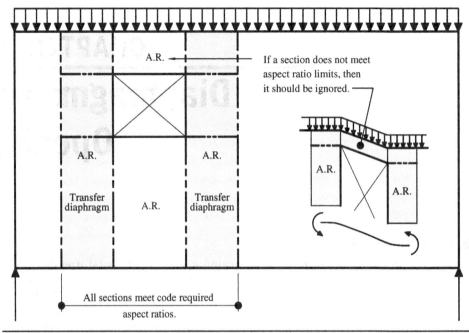

FIGURE 5.1 Individual section aspect ratios.

would be distributed into the diaphragm by a vertical collector installed above and below the opening width along grid line 2 between grid lines A to B and C to D, respectively, as shown in Fig. 5.2. The length of the collector would be determined by dividing the calculated force by the net shear resulting from the diaphragm nailing capacity minus the applied shear at that location (reserve capacity).

The forces in the horizontal collectors located at grid lines B and C were assumed to have an upper and lower bound occurring at the left and right side of the opening, respectively. The horizontal collector forces will be higher on the left side of the opening due in part by the higher diaphragm shears closest to the diaphragm support and the assumptions made. The lower bound assumes that half of the collected shears along the entire length of the opening is applied to right side of the opening. The upper bound assumes that the entire collected shear along the full opening length is applied to the left side of the opening. The length of the horizontal collector in each case is determined in the same manner as the collectors above and below the opening width. Roughly, the collected shear force in the left horizontal collectors at grid lines B and C was assumed to be equal to the average of the unit shear at the left and right end of the section above the opening multiplied by the opening length. Figure 5.2 shows two openings, a reasonably small opening, and a large opening. First, assume that a 12-ft-long × 10-ft-wide opening is located where the basic diaphragm unit shear is equal to 150 plf and the capacity of the nailing is equal to 290 plf. The total width of the diaphragm is 60 ft. The opening is located 10 ft below the diaphragm top chord as shown in configuration 1 of Fig. 5.2. The unit shear on the left side of grid line 2 would be equal to

$$v_L = \frac{\text{Total shear at grid line 2}}{\sum W_{\text{left}}} = \frac{150(60)}{(10+40)} = 180 \text{ plf}$$

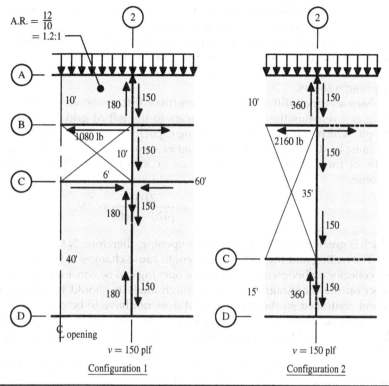

Figure 5.2 Effect of opening size on diaphragm.

The shear above and below the opening has increased to 180 plf, which is still less than the capacity of the diaphragm nailing. Therefore, changes to the nailing do not have to be made, and the required collector length between grid lines A to B and C to D along grid line 2 would be

$$F = 150(10) = 1500 \text{ lb, half going above and the other half going below the opening}$$

$$W_{\text{opening}} = 10', \text{opening width}$$

$$L_{\text{embed}} = \frac{F}{(\text{nailing capacity} - \text{applied shear})} = \frac{750}{(290 - 180)} = 6.82' < 10\text{-ft opening}$$

This length is less than the minimum recommended collector length.

Assume that the opening causes a lower bound horizontal collector force at grid line 2B equal to 180(6) = 1080 lb. Based on the reserve capacity of the nailing scheme, the required embedment of a collector into the diaphragm to the right of grid line 2 would be

$$W_{\text{opening}} = 12.0', \text{opening width}$$

$$L_{\text{embed}} = \frac{F}{(\text{nailing capacity} - \text{applied shear})} = \frac{1080}{(290 - 150)} = 7.71' < 12\text{-ft opening}$$

Since the embedment lengths of the vertical and horizontal collectors are less than the minimum suggested lengths, and the nail spacing does not have to be decreased, it can be assumed that the opening has no significant effect on the diaphragm at the right side of the opening. However, the upper bound effects at the left side of the opening must also be checked and, by inspection, should control because of the higher basic diaphragm shears.

Now assume a different opening width of 35 ft, as shown in configuration 2. Everything else remains the same. The shears to the left of grid line 2 would increase to 360 plf, which requires a higher nailing capacity, say 420 plf. Therefore, the nail spacing must be decreased to meet this higher shear capacity. Accordingly, the horizontal collector force at grid line 2B increases to 360(6) = 2160 lb, the embedment length becomes

$$L_{embed} = \frac{2160}{(420 - 360)} = 36',$$

which is greater than the length of the opening, therefore, N.G.

The nailing and the embedment length cause changes in the sheathing nailing and the collector embedment length so the opening can be considered to have a significant effect on the diaphragm, and a thorough analysis should be performed. The lower bound controlled so the upper bound does not have to be checked. If the opening is located in the diaphragm where the diaphragm unit shears are much higher, the size of the opening could have a greater effect on the diaphragm and would require a complete analysis.

Recent testing and corresponding paper by FPInnovations, "Design example: Designing for Openings in Wood Diaphragm"[7] provides a different and simple test for determining if an opening size can be ignored or must be designed. The paper noted:

It is strongly recommended that analysis for a diaphragm with an opening should be carried out except where all four of the following items are satisfied:

a. Width of the opening is no greater than 15 percent of diaphragm width.

b. Length of the opening is no greater than 15 percent of diaphragm length.

c. Distance from diaphragm edge to the nearest opening edge is a minimum of 3 times the larger opening dimension.

d. The diaphragm portion between opening and diaphragm edge satisfies the maximum aspect ratio requirement. (All sides of the opening.)

This is a much simpler approximate approach that can be used prescriptively. The required strap lengths are determined by calculations.

5.2   Method of Analysis

The application of wind and seismic loads to a diaphragm with openings was discussed in Sec. 2.3. For wind loading, it is advisable to apply the sum of the windward and leeward wind pressures, inclusive of both positive and negative interior pressures, at areas that do not contain an opening to the windward side of the diaphragm. At areas containing openings, windward, leeward, and internal pressures should be applied separately as shown in Fig. 5.3. Seismic loads are distributed in accordance with the dead loads that are tributary to the sections. A typical method of distributing uniform loads

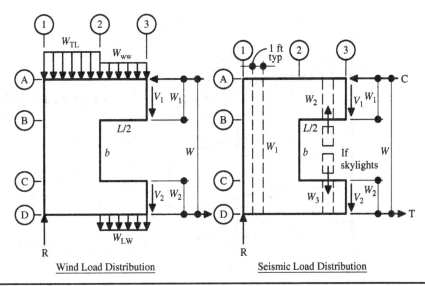

Figure 5.3 Diaphragm loading.

into a diaphragm utilizes a 1-ft-wide strip across the building, W_1, would be equal to the dead load of the upper half of all walls within the unit strip from grid lines A to D plus the dead load of the roof from A to D times the seismic response coefficient C_s. At the opening, the 1-ft strip, W_2, would be equal to the dead load of the upper half of all walls within the unit strip from grid line A to B, plus the roof dead load across the width W_1 plus half of the skylight dead load (if any), times the seismic response coefficient C_s. The 1-ft unit strip, W_3, would be equal to the dead load of the upper half of all walls within the unit strip from grid line C to D, plus the roof dead load across the width W_2 plus half of the skylight dead load, times C_s.

The method of analyzing diaphragms with openings has been described in ATC-7, APA Report 138, several papers by Diekmann and more recently by FPInnovations. The method of analysis assumes that a diaphragm with a large opening behaves like a Vierendeel truss (i.e., a parallel chord truss with only vertical web members; no diagonal web members at the opening). The vertical web-to-chord member joints are rigid connections, not pinned, which cause the truss to act like a moment resisting frame. The sections above and below the opening are assumed to have points of inflection at their mid-length, due to the shear acting on the section, as shown in Fig. 5.4. The local moment at the point of inflection due to the shear acting on the section is equal to zero; therefore, the local chord force in the members at the edge of the opening (grid lines 3B and 3C) is equal to zero. The chord forces at grid lines 3A and 3D are equal to the moment in the diaphragm at grid line 3 divided by the width of the diaphragm. The sections above and below the opening can be broken down into two separate elements, one on each side of the point of inflection as demonstrated in Fig. 5.5. The moments in each of these elements at grid lines 2 and 4 are equal to the shear at the centerline of the opening multiplied by the length of the element, plus or minus the moment due to the uniform load action on the element (reference Figs. 5.4 to 5.7). In the original method of analysis presented in previous publications,[2-3] the diaphragm was first analyzed without the opening to obtain flange force and web shear at the point of inflection then

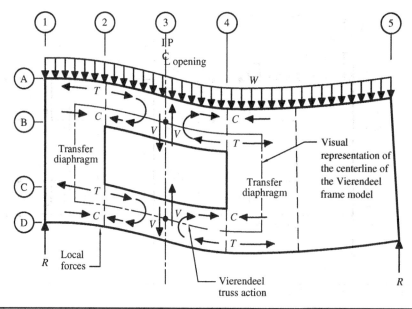

Figure 5.4   Basic procedure and forces.

added to or subtracted from the shears generated by a diaphragm with an opening using a table. However, the analysis can be simplified by constructing the basic shear diagram with an opening, then solving for the unknown forces by using free-body diagrams. The same resulting shears and chord forces will result. The shears and forces in the transfer diaphragm areas are then resolved by using the visual shear transfer method. Transfer diaphragms are placed on each side of the opening to receive the local horizontal forces produced by bending of the sections above and below the opening. Those forces are then transferred out to the main diaphragm chords at grid lines A and D. Stresses in the flange due to the local effects are then combined with the basic diaphragm action.

Analysis Procedures

1. Plot the basic shear diagram with the opening (Fig. 5.5).

2. Determine the vertical shears at all significant locations, grid lines 1 to 5 (see Fig. 5.5).

3. Break the sections above and below the opening into equal lengths each side of the inflection points as shown in Fig. 5.5.

4. Assuming $F_{3B} = F_{3C} = 0$ at the point of inflection, calculate the chord force F_3 at grid line 3.

5. Starting at grid line 4 and moving to the left (see Fig. 5.6)

   - Distribute the shear force, $V_{4'}$ into the upper and lower elements (I and III) in proportion to their widths, V_{4AB} and V_{4CD}.

   - Sum the shears acting on elements I and III from grid line 4 to grid line 3. The unit shears are equal to the calculated element shear divided by the element width.

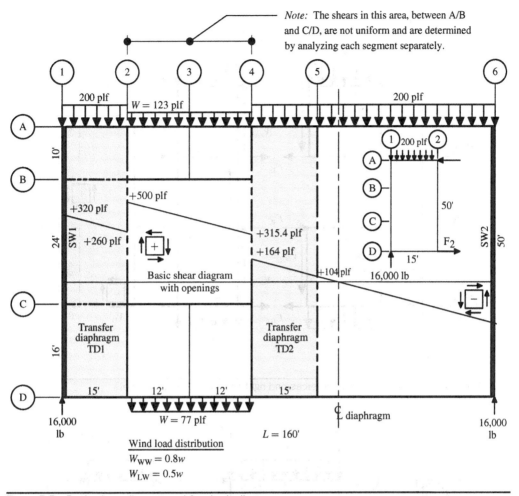

Note: The shears in this area, between A/B and C/D, are not uniform and are determined by analyzing each segment separately.

FIGURE 5.5   Basic shear diagram and free-body diagram.

- Sum the moments about the corners of each element to solve for the unknown chord forces F_{4A}, F_{4B}, F_{4C} and F_{4D}. Verify the calculated chord forces by summing horizontal forces acting on each element.

6. Continue at grid line 3 (Fig. 5.7).

   - Sum the shears acting on elements II and IV from grid line 3 to grid line 2. $V_{2AB} = V_{3AB} + w_1(L_{2-3})$ and $V_{2CD} = V_{3CD} + w_2(L_{2-3})$. The unit shears are equal to the calculated element shear divided by the element width.

   - Sum the moments about the corners of the element to solve for chord forces F_{2A}, F_{2B}, F_{2C} and F_{2D}, Figs. 5.6 and 5.7. For the loading in Fig. 5.5, the final forces are shown in Fig. 5.8.

   - Determine transfer diaphragm TD1 net shears caused by the local chord forces at grid lines B and C as shown in Fig. 5.9. The process is the same as outlined in Chaps. 3 and 4, using the visual shear transfer method.

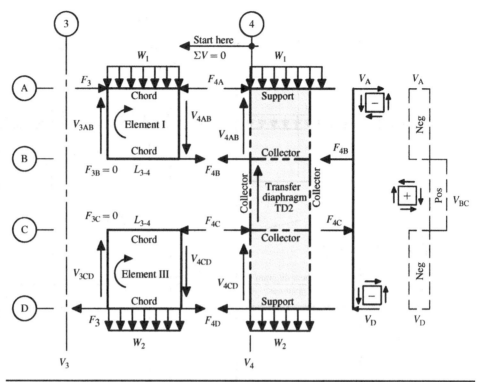

FIGURE 5.6   Free body of element forces and right transfer diaphragm shears.

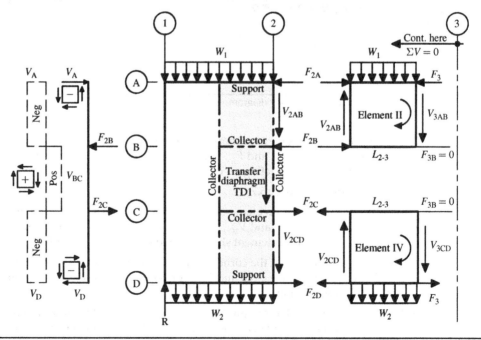

FIGURE 5.7   Free body of element forces and left transfer diaphragm shears.

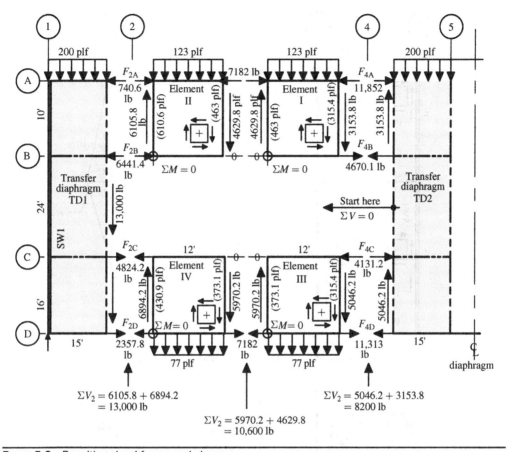

FIGURE **5.8**   Resulting chord forces and shears.

- Determine transfer diaphragm TD2 net shears caused by the local chord forces at grid lines B and C as shown in Fig. 5.10.
- Determine the transfer diaphragm vertical collector forces, Figs. 5.9 and 5.10.

7. Plot the force diagrams in the longitudinal directions, Fig. 5.11.

The longitudinal force diagrams at grid lines B and C confirm that the chord forces are equal to zero as assumed.

5.3   Single Opening in Diaphragm

APA Research Report 138 included tests on two diaphragms that were 48 ft long by 16 ft wide, one having two 8-ft by 8-ft openings symmetrically placed about the centerline of the diaphragm and another diaphragm with a 20-ft by 20-ft openings. The test diaphragms were identical to the diaphragm analyzed in Appendix E of the report. The report noted that the test results supported the assumption that the diaphragms behave similar to a Vierendeel truss. The mode of failure was by buckling of the plywood at

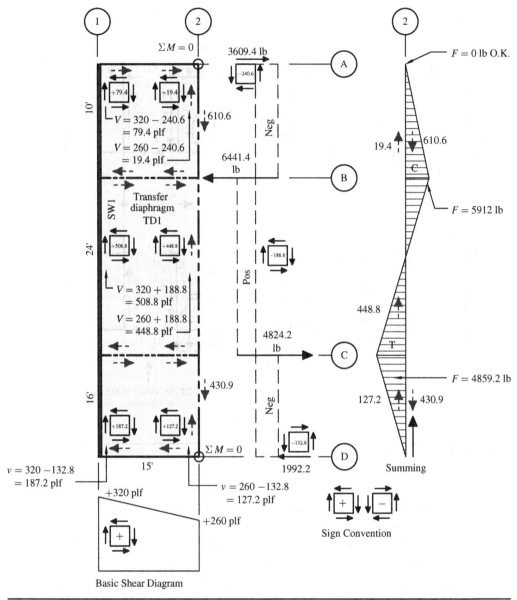

FIGURE 5.9   Left transfer diaphragm shears and transverse collector force.

the corners of the opening due to compressive forces. No attempt was made to modify or reinforce the boundary members around the openings in the test diaphragms. The method used to solve the problem in Appendix E followed the original Diekmann method as previously discussed. The use of the tables to determine the net shears and forces is somewhat abstract and can make it difficult to visualize the flow of shears and forces through the diaphragm. The analysis used in the example that follows can be

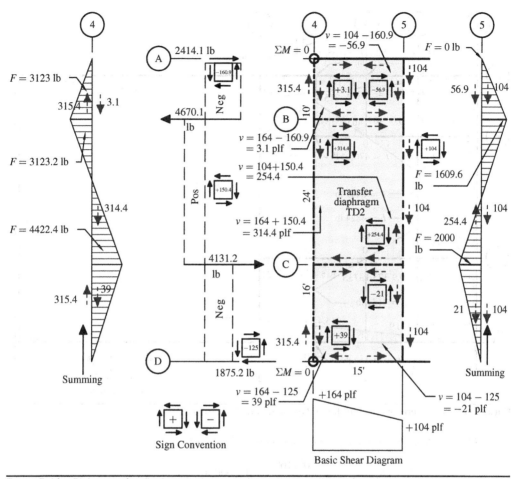

FIGURE 5.10 Right transfer diaphragm shears and transverse collector forces.

simplified by using the method described in Chap. 3 and as noted above, which eliminates the need to use tables.

Example 5.1: Single Opening in Diaphragm Horizontally and Vertically Offset (Fig. 5.5)

The diaphragm shown in Fig. 5.5 has a length of 160 ft and a width of 50 ft. A 24-ft by 24-ft opening from grid line 2 to 4 and B to C is offset horizontally and vertically from the center of the diaphragm as shown in the figure. A transfer diaphragm is located on each side of the opening, designated as TD1 and TD2. The aspect ratio of all the diaphragm sections and the transfer diaphragms are in compliance with code maximums. The aspect ratio of both transfer diaphragms is 3.33:1, which requires the transfer diaphragms to be blocked. A uniform load of 200 plf is applied to the diaphragm. At the opening, the wind load is split into windward and leeward wind loads above and below the opening as shown, 123 plf and 77 plf, respectively, which includes internal wind pressures.

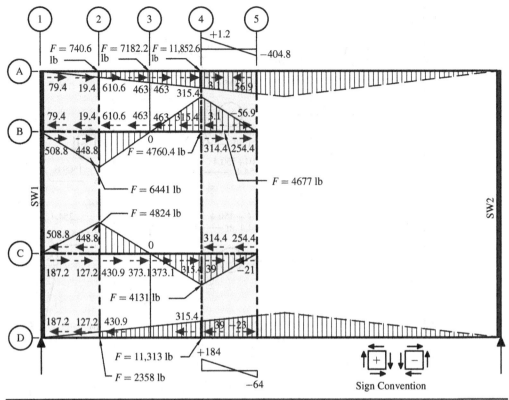

Figure 5.11 Longitudinal collector force diagrams and chord forces.

Develop the basic shear diagram (with opening) (Fig. 5.5):

$$R_1 = R_6 = \frac{wL}{2} = \frac{200(160)}{2} = 16,000 \text{ lb}$$

$$v_1 = \frac{R_1}{W} = \frac{16,000}{50} = +320 \text{ plf, unit shear at grid line 1}$$

$$V_2 = R_1 - wx = 16,000 - 200(15) = 13,000 \text{ lb}$$

$$v_{2L} = \frac{V_2}{W_{\text{left}}} = \frac{13,000}{50} = +260 \text{ plf, unit shear at grid line 2 (left)}$$

$$v_{2R} = \frac{V_2}{W_{\text{right}}} = \frac{13,000}{26} = +500 \text{ plf, unit shear at grid line 2 (right)}$$

$$V_3 = 13,000 - 200(12) = 10,600 \text{ lb, shear at grid line 3}$$

$$V_4 = 10,600 - 200(12) = 8200 \text{ lb}$$

$$v_{4L} = \frac{8200}{26} = +315.4 \text{ plf, unit shear at grid line 4 (left)}$$

$$v_{4R} = \frac{8200}{50} = +164 \text{ plf, unit shear at grid line 4 (right)}$$

The shear at grid line 5 is equal to +104 plf.

Find the chord forces (without opening) at grid line 3 (Fig. 5.8):

$$M_3 = R_1 x - \frac{wx^2}{2} = 16{,}000(27) - \frac{200(27)^2}{2} = 359{,}100 \text{ ft lb}$$

$$F_3 = \frac{M_3}{W} = \frac{359{,}100}{50} = 7182 \text{ lb at A and D}$$

Resolve shears and chord forces on elements I through IV (Figs. 5.6 and 5.8). Start at grid line 4 and sum shears to the left.

Shear distribution to elements I and III:

$$V_4 = 8200 \text{ lb.}$$

The shear at elements I and III are distributed to each element in accordance with their width.

$$V_{4AB} = \frac{W_{A-B}V_4}{\Sigma W_{\text{left}}} = \frac{10(8200)}{26} = 3153.8 \text{ lb,} \qquad V_{4CD} = \frac{16(8200)}{26} = 5046.2 \text{ lb}$$

$$v_{AB} = v_{CD} = \frac{8200}{26} = 315.4 \text{ plf, unit shear at grid line 4 in elements I and III}$$

Element I:

Shear at grid line 3:

$$V_{3AB} = 3153.8 + 123(12) = 4629.8 \text{ lb,} \qquad v_{3AB} = \frac{4629.8}{10} = +463 \text{ plf}$$

$$F_{3A} = 7182 \text{ lb, compression as previously calculated}$$

$$F_{3B} = 0 \text{ as previously assumed}$$

Summing moments about the lower left corner of element I:

$$F_{4A} = \left[7182(10) + 3153.8(12) + \frac{123(12)^2}{2} \right] \frac{1}{10} = 11{,}852 \text{ lb}$$

$$F_{4B} = 11{,}852 - 7182 - 0 = 4670.1 \text{ lb}$$

Note that this does not match the 500 plf shear calculated in the basic shear diagram. This is because the basic diaphragm shear was calculated using a uniform load of 200 plf. The actual load to elements I and II is 123 plf, demonstrating why the wind loads should be split to windward and leeward at the opening. Not doing so can cause the upper section to be over-designed and the lower section to be under-designed.

Element III:

$$V_{3CD} = 5046.2 + 77(12) = 5970.2 \text{ lb,} \qquad v_{3CD} = \frac{5970.2}{16} = 373.1 \text{ plf}$$

$$F_{4C} = \left[5046.2(12) + \frac{77(12)^2}{2} \right] \frac{1}{16} = 4131.2 \text{ lb}$$

$$F_{4D} = 7182 + 4031.2 = 11{,}313 \text{ lb}$$

Element II (Figs. 5.7 and 5.8): Continuing to the left from grid line 3. The shears and forces from element I and III at grid line 3 are transposed to elements II and IV, respectively.

$$V_{2AB} = 4629.8 + 123(12) = 6105.8 \text{ lb}, \qquad v_{2B} = \frac{6105.8}{10} = 610.6 \text{ plf}$$

$$F_{2A} = \left[4629.8(12) + \frac{123(12)^2}{2} - 7182(10) \right]\frac{1}{10} = -740.6 \text{ lb}$$

$$F_{2B} = 7182 - 740.6 = 6441.4 \text{ lb}$$

Element IV:

$$V_{2CD} = 5970.2 + 77(12) = 6894.2 \text{ lb}, \qquad v_{3CD} = \frac{6894.2}{16} = 430.9 \text{ plf}$$

$$F_{2C} = \left[5970.2(12) + \frac{77(12)^2}{2} \right]\frac{1}{16} = 4824.2 \text{ lb}$$

$$F_{2D} = 7182 - 4824.2 = 2357.8 \text{ lb}$$

Again, note that the unit shears do not match the shears calculated in the basic shear diagram at grid line 2. This is because the basic diaphragm shear was calculated using a uniform load of 200 plf. The actual load to elements III and IV is 77 plf.

Summing the shear on the elements above and below the opening at grid lines 4, 3, and 2 will equal the shears previously calculated on the basic shear diagram in Fig. 5.5. The sum of these shears is a check to verify if the analysis is accurate. The summary is shown at the bottom of Fig. 5.8.

Determine the left transfer diaphragm shears, net shears, and transverse collector forces (see Figs. 5.8 and 5.9):

Left transfer diaphragm shears:

$$V_A = [4824.2(16) - 6441.4(40)]\frac{1}{50} = -3609.4 \text{ lb}, \qquad v_A = \frac{-3609.4}{15} = -240.6 \text{ plf}$$

Completing the remaining shear forces on the sheathing element shows that the shear is negative.

$$V_D = [6441.4(10) - 4824.2(34)]\frac{1}{16} = -1992.2 \text{ lb}, \qquad v_D = \frac{-1992.2}{15} = -132.8 \text{ plf}$$

Completing the remaining shear forces on the sheathing element shows that the shear is negative.

$$V_{BC} = 1992.2 - 4824.2 = -2832 \text{ lb}, \qquad v_{BC} = \frac{2832}{15} = 188.8 \text{ plf}$$

Completing the remaining shear forces on the sheathing element shows that the shear is positive.

The basic diaphragm unit shears are plotted below transfer diaphragm TD1 for a visual reference.

Net diaphragm shears:

Location (Grid Line)	Net Shear
1 from A to B	$v = +320 - (240.6) = +79.4$ plf
1 from B to C	$v = +320 + (188.8) = +508.8$ plf
1 from C to D	$v = +320 - (132.8) = +187.2$ plf
2 from A to B (left side)	$v = +260 - (240.6) = +19.4$ plf
2 from B to C (left side)	$v = +260 + (188.8) = +448.8$ plf
2 from C to D (left side)	$v = +260 - (132.8) = +127.2$ plf

Collector forces at grid line 2, summing from grid line D:

Location (Grid Line)	Force
2 at C	$F_{2C} = (430.9 - 127.2)16 = 4859.2$ lb
2 at B	$F_{2B} = 4859.2 - (448.8)24 = -5912$ lb
2 at A	$F_{2A} = -5912 + (610 - 19.4)10 = 0$ lb closes

Determine the right transfer diaphragm shears, net shears, and transverse collector forces (see Figs. 5.8 and 5.10):

Right transfer diaphragm shears:

$$V_A = [4131.2(16) - 4670.1(40)]\frac{1}{50} = -2414.1 \text{ lb}, \quad v_A = \frac{2414.1}{15} = 160.9 \text{ plf}$$

Completing the remaining shear forces on the sheathing element shows that the shear is negative.

$$V_D = [4670.1(10) - 4131.2(34)]\frac{1}{50} = -1875.2 \text{ lb}, \quad v_A = \frac{1875.2}{15} = 125 \text{ plf}$$

Completing the remaining shear forces on the sheathing element shows that the shear is negative.

$$V_{BC} = 1875.2 - 4131.2 = -2256 \text{ lb}, \quad v_{BC} = \frac{2256}{15} = 150.4 \text{ plf}$$

Completing the remaining shear forces on the sheathing element shows that the shear is positive.

Net diaphragm shears:

Location (Grid Line)	Net Shear
4 from A to B	$v = +164 - (160.9) = +3.1$ plf
4 from B to C	$v = +164 + (150.4) = +314.4$ plf
4 from C to D	$v = +164 - (125) = +39$ plf
5 from A to B (left side)	$v = +104 - (160.9) = -56.9$ plf
5 from B to C (left side)	$v = +104 + (150.4) = +254.4$ plf
5 from C to D (left side)	$v = +104 - (125) = -21$ plf

Collector forces at grid line 4, summing from grid line D:

Location (Grid Line)	Force
4 at C	$F_{2C} = (315.4 - 39)16 = 4422.4$ lb
4 at B	$F_{2B} = 4422.4 - (314.4)24 = -3123.2$ lb
4 at A	$F_{2A} = -3123.2 + (315.4 - 3.1)10 = 0$ lb closes

Collector forces at grid line 5, summing from grid line D:

Location (Grid Line)	Force
5 at C	$F_{2C} = (104 + 21)16 = 2000$ lb
5 at B	$F_{2B} = 2000 - (254.4 - 104)24 = 1609.6$ lb
5 at A	$F_{2A} = 1609.6 - (104 + 56.9)10 = 0$ lb closes

Determine longitudinal chord/collector forces (Fig. 5.11):

Collector forces at grid line B, summing from grid line 1:

Location (Grid Line)	Force
B at 2	$F_{2B} = \left[\dfrac{(508.8 - 79.4) + (448.8 - 19.4)}{2}\right](15) = 6441$ lb
B at 3	$F_{3B} = -6441 + \dfrac{(610.6 + 463)}{2}(12) = 0$ lb
B at 4	$F_{4B} = 0 + \dfrac{(463 + 315.4)}{2}(12) = 4670.4$ lb
B at 5	$F_{5B} = 4670.4 - \left[\dfrac{(315.4 - 3.1) + (254.4 + 56.9)}{2}\right](15) = 0$ lb closes

Collector forces at grid line C, summing from grid line 1:

Location (Grid Line)	Force
C at 2	$F_{2C} = \left[\dfrac{(508.8 - 187.2) + (448.8 - 127.2)}{2}\right](15) = 4824$ lb
C at 3	$F_{3C} = 4824 - \dfrac{(430.9 + 373.1)}{2}(12) = 0$ lb
C at 4	$F_{4C} = 0 - \dfrac{(373.1 + 315.4)}{2}(12) = -4131$ lb
C at 5	$F_{5C} = 4131 - \left[\dfrac{(314.4 - 39) + (254.4 + 21)}{2}\right](15) = 0$ lb closes

Chord forces at grid line A, summing from grid line 1:

Location (Grid Line)	Force
A at 2	$F_{2A} = \dfrac{(79.4 + 19.4)}{2}(15) = 740.6$ lb
A at 3	$F_{3A} = 740.6 + \dfrac{(610.6 + 463)}{2}(12) = 7182.2$ lb
A at 4	$F_{4A} = 7182.2 + \dfrac{(463 + 315.4)}{2}(12) = 11,852.6$ lb
A at 5	$F_{5A} = 11,852.6 + 1.2 - 404.8 = 11,449$ lb

Chord forces at grid line D, summing from grid line 1:

Location (Grid Line)	Force
D at 2	$F_{2D} = \dfrac{(187.2 + 127.2)}{2}(15) = 2358 \text{ lb}$
D at 4	$F_{4D} = 2358 + \dfrac{(430.9 + 315.4)}{2}(24) = 11{,}313 \text{ lb}$
D at 5	$F_{5D} = 11{,}313 + 184 - 64 = 11{,}433 \cong 11{,}448 \text{ lb}$

The analysis shows that the net diaphragm shears in the areas located between grid lines 1 and 2 from A to B, between 1 and 2 from C to D, between 4 and 5 from A to B, and between 4 and 5 from C to D have been reduced below the basic unit shears for a diaphragm without an opening. The magnitude of the shear suggests that these areas can be unblocked with a 6″ o.c. edges and 12″ o.c. intermediate nailing pattern. However, the aspect ratio of the transfer diaphragm is equal to

$$\text{A.R.} = \frac{50}{15} = 3.33{:}1$$

which requires the transfer diaphragms to be blocked. All other areas have increased in shear and should be nailed accordingly. All the shears have been verified for this load case. Accordingly, all the chord, strut, and collector force diagrams have closed to zero which verifies that the analysis is correct. ▲

5.4 Diaphragm Deflection

ATC-7 noted that the introduction of an opening in a diaphragm has no effect on the deflection contributed by the flanges. There is a small contribution due to bending when web openings occur in actual diaphragms, but this is not reflected in the equation since all bending is assumed to be resisted by the flanges. The shear stresses in the remaining sections of the diaphragm on either side of the opening are increased since in normal construction practice there is less web material available. The effect of an opening on the deflection is determined from classical analysis used to derive the second term of the equation (shear deflection), using integrations over segments of the diaphragm. The procedure for designing diaphragms for non-uniform loads is the same as that for any beam subject to these loads. The determination must be done based on modifying the basic deflection equation by deriving the applicable coefficients from elastic beam theory. The bending deflection can be determined by performing a computer analysis. The remainders of the terms are derived using the method shown in Chap. 3.

5.5 Problem

Problem P5.1: Double Opening in Diaphragm, Horizontally and Vertically Offset (Fig. P5.1)

Given: The diaphragm shown in Fig. P5.1 has a length of 320 ft and a width of 80 ft. A 20-ft by 20-ft opening from grid line 2 to 3 and C to D is offset horizontally to the left of the center of the diaphragm as shown in the figure. A 40-ft by 40-ft opening from grid

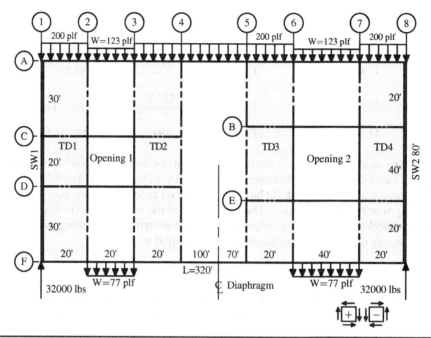

Figure P5.1

line 6 to 7 and B to E is offset horizontally to the right of the center of the diaphragm as shown in the figure.

Find: Determine the diaphragm shears and collector/chord/strut forces. ▲

5.6 References

1. American Society of Civil Engineers (ASCE), *ASCE/SEI 7-16 Minimum Design Loads for Buildings and Other Structures*, ASCE, New York, 2016.
2. Diekmann, E. F., "Design of Wood Diaphragms," *Journal of Materials Education*, 1982, Fourth Clark C. Heritage Memorial Workshop, Wood Engineering Design Concepts University of Wisconsin, WI.
3. APA—The Engineered Wood Association, APA Research Report 138, *Plywood Diaphragms*, APA Form E315H, APA—The Engineered Wood Association, Engineering Wood Systems. Tacoma, WA, 2000.
4. Applied Technology Council (ATC), *Guidelines for Design of Horizontal Wood Diaphragms, ATC-7*, Applied Technology Council, Redwood, CA, 1981.
5. International Code Council (ICC), *International Building Code, 2021 with commentary*, ICC, Whittier, CA, 2021.
6. American Wood Council (AWC), *Special Design Provisions for Wind and Seismic with Commentary*, AWC, Leesburg, VA, 2021.
7. *Design Example: Designing for Openings in Wood Diaphragm*, Canadian Wood Council, FPInnovations, 2665 East Mall, Vancouver, BC, 2013.

CHAPTER 6

Open-Front and Cantilever Diaphragms

6.1 Introduction

In the past, the demand or interest in utilizing cantilever/open-front diaphragms was limited to simple structures one or two stories in height. Complex building shapes and footprints are driving design procedures and code requirements to evolve for all lateral force-resisting systems and materials. The requirement for the structural analysis to consider the relative stiffnesses of diaphragms and the vertical elements of the seismic force-resisting system has been in the codes for a long time but this has often been ignored, in part, because of the thought that if allowable aspect ratios for diaphragms and shear walls were complied with that relative structural stiffness in wood structures would not be of concern. However, as buildings get taller, more complex, and more flexible, there is a greater need to understand the relative stiffness of diaphragms and shear walls, and multistory shear wall effects. It has become apparent that diaphragms are neither truly flexible nor truly rigid. Since open-front flexible diaphragms cannot transmit forces by torsion, they must be analyzed as semi-rigid or rigid. Semi-rigid diaphragms can be analyzed using the envelope method (i.e., analyzed as flexible first, then as rigid and taking the maximum forces). Full scale shake table testing and state-of-the-art research have caused increased concerns regarding how these diaphragms should be designed and significant changes were made to the 2021 edition of SDPWS[1] and its commentary.

Architecturally demanding exterior wall lines in modern structures do not always provide opportunities to use traditional design approaches. In mid-rise, multi-family buildings, corridor-only shear wall floor plans, similar to the plan shown in Fig. 6.1(d), are becoming a popular design approach due to the lack of available shear walls at the exterior wall lines. Low-rise retail buildings, such as the ubiquitous strip mall, are another building type where the open-front diaphragm is frequently employed.

All the examples and the problems included in this book are the type of layouts that would normally require a complete analysis. It should also be re-emphasized that most of the examples and problems covered could be considered to be aggressive designs and have been chosen to demonstrate the methods that can be used to analyze them.

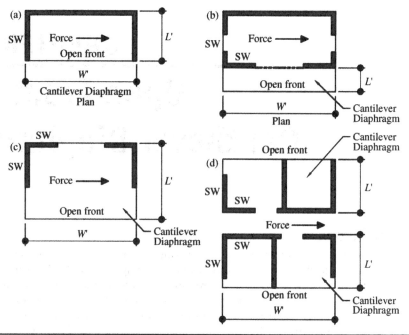

Figure 6.1   SDPWS Figure 4A.

The example plans should not be used in construction unless a complete, thorough analysis and engineering judgment has been exercised. Unfortunately, until recently there has been very little guidance or published examples available for the practicing engineer on how to analyze an open-front, cantilever diaphragm. To fill the gap, the authors developed a complete design example published by WoodWorks—Wood Products Council.[4] A brief overview will be presented here for reference and guidance. Only a partial set of calculations are included in the discussion that follows to demonstrate the design process being reviewed. The complete design example can be downloaded for free from www.woodworks.org.

6.2   Code Requirements

For seismic applications, many considerations during the design of cantilever diaphragms originate from the requirements of SDPWS Section 4.2.6.1 and in Fig. 6.1. There have been a significant number of changes from the 2008 edition of SDPWS to the 2021 edition regarding seismic loading of open-front diaphragms. The 2008 edition allowed a maximum cantilever length of 25 ft to be exceeded provided calculations could show that the drift could be tolerated. However, with the 2015 edition, the maximum length of the cantilever was changed from 25 ft to 35 ft with few exceptions. Current requirements for an open-front structure are as follows.

6.2.1   SDPWS 2021 Section 4.2.6.1: Open-Front Structures

For resistance to seismic loads, wood-frame diaphragms in open-front structures shall comply with all the following requirements:

1. The diaphragm conforms to Sections 4.2.8.1, 4.2.8.2, or 4.2.8.3.

2. The L'/W' ratio as shown in SDPWS Figure 4A (a) through (d) is not greater than 1.5:1 when sheathed in conformance with 4.2.8.1 or not greater than 1:1 when sheathed in conformance with 4.2.8.2 or 4.2.8.3. For open-front structures that are also torsionally irregular, as defined in 4.2.5.1, the L'/W' ratio shall not exceed 0.67:1 for structures over one story in height, and 1:1 for structures one story in height.

3. For loading parallel to the open side, diaphragms shall be modeled as semi-rigid or idealized as rigid, and the maximum story drift at each edge of the structure shall not exceed the ASCE 7 allowable story drift when subject to seismic design forces including torsion and accidental torsion and shall include shear and bending deformations of the diaphragm.[2]

4. The cantilevered diaphragm length, L', (normal to the open side) shall not exceed 35 ft.

SDPWS Section 4.2.5.1, which addresses torsionally irregular structures, requires a reduction in allowable aspect ratios. Compliance with the requirements can require balancing the design between the various elements of the lateral-force-resisting system to achieve the required structural stiffness.

6.3   Cantilever Diaphragm, Corridor-Only Shear Walls

The floor plan shown in Fig. 6.2 will be used for the purpose of discussion and guidance through the design process. The design example is in seismic design category (SDC) D. The plan is a one-story structure, symmetric about both axes for simplicity. Such a floor plan could be similar to a four-unit office complex or residential building and can be single or multiple stories. Short shear walls are placed along grid lines A and B in a symmetrical pattern as shown in the figure. Three shear walls are placed along each of corridor wall lines 2 and 3 to resist seismic forces in the North-South direction. The exterior walls that occur at grid lines 1 and 4 do not have enough stiffness to act as shear walls creating the open-front structure/cantilever diaphragm condition. A bearing wall is placed mid-width of the diaphragm to reduce the roof framing depth but will not be used as part of the lateral-force-resisting system. For north-south lateral loads, the diaphragm cantilevers in both directions from the corridor wall lines.

The roof height is at a 10-ft elevation producing the aspect ratios of the shear walls equal to 1.25:1 and 1:1, as shown in the plan. All shear walls comply with the allowable height to width, (h/b), aspect ratios of 2:1 or less in SDPWS Section 4.3.3 and Table 4.3.3. If aspect ratios are greater than 2:1, the shear capacity must be reduced for high-aspect ratio walls in accordance with SDPWS Section 4.3.3. This example follows the common, but not required, practice of using allowable stress design (ASD) for the force capacity design of the shear walls and diaphragms. Strength-level forces are used for shear wall and diaphragm deflections, story drift, and drift-based torsional irregularity checks.

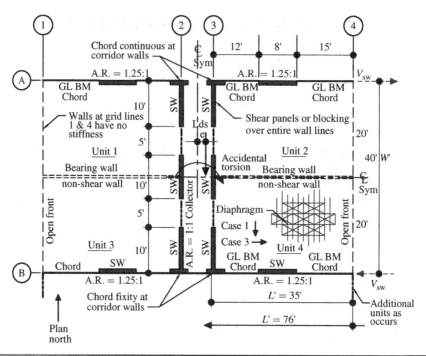

Figure 6.2 Example plan one story.

As with any design project, some assumptions typically need to be made before the design process can proceed (e.g., diaphragm flexibility, degree of torsional irregularity, redundancy). The SDPWS requires that cantilever diaphragms must be either semi-rigid or rigid. A rigid diaphragm can initially be assumed and verified by calculation later in design. In most cases, cantilever diaphragms are torsionally irregular. Based on the plan layout, making a best guess assumption at the start of the design as to its torsional irregularity can avoid iteration of the analysis later. Since the example plan is symmetric about both axes, the center of rigidity and center of mass occur at the same location; therefore, there is no inherent torsion. Inherent and accidental torsion are only considered for diaphragms that are not flexible. Accidental torsion, as defined by ASCE 7 Section 12.8.4.2, is an additional torsion force that is applied to the structure due to uncertainties inherent in the design, such as an unsymmetrical placement of mass in the real structure. To calculate the accidental torsion, the center of mass is assumed to be displaced from its calculated position by a distance equal to 5 percent of the dimension of the structure in the perpendicular to the direction of the applied force. In ASCE 7 the accidental torsion is applied in all buildings for determining whether a horizontal irregularity exists (e.g., torsional irregularity); but it need not be included in the structural design forces except when a torsional irregularity exists. In buildings with inherent torsion, the combined effect of accidental torsion and inherent torsion should be considered.

For open-front structures, the classification of the structure as having a torsional irregularity (Type 1a), or an extreme torsional irregularity (Type 1b), is especially important. Per ASCE 7 Section 12.3.3.1, structures assigned to SDC E and F are prohibited from an extreme horizontal torsional irregularity (Type 1b). Since the example structure is assigned to SDC D, this prohibition does not apply. A Type 1a horizontal irregularity was assumed for the design example.

ASCE 7 Section 12.8.4.2 was expanded from the prior edition to add the following:

Accidental torsion shall be applied to all structures for determination if a horizontal irregularity exists as specified in Table 12.3-1. Accidental torsion moments (M_{ta}) need not be included when determining the seismic forces E in the design of the structure and in the determination of the design story drift in Sections 12.8.6, 12.9.1.2, or Chapter 16, or limits of Section 12.12.1, except for the following structures:

1. Structures assigned to Seismic Category B with Type 1b horizontal structural irregularity.
2. Structures assigned to Seismic Category C, D, E, and F with Type 1a or Type 1b horizontal structural irregularity.

If the structure has a Type 1a or Type 1b horizontal irregularity, amplification of the accidental torsion is required per ASCE 7 Section 12.8.4.3, which can impact the design of components of the structure. Since the amplification factor must be applied early in the design process, the value cannot be calculated until the design has progressed significantly. A conservative value can be assumed early to avoid iteration of the analysis, which must be verified later.

Another decision that must be made is redundancy, which can be either verified at the start of the analysis or assumed, then verified at the end of the design.

The flow of the analysis for a cantilever diaphragm is as follows:

1. Establish building design criteria, design assumptions, and loading information.
2. Calculate main lateral-force-resisting system (MLFRS) seismic forces.
3. Perform an initial rigid diaphragm analysis (RDA) based on using shear wall stiffness values proportional to the wall lengths.
4. Determine the shear wall construction based on demand and adjust as required to account for anticipated drift limitation issues.
5. Calculate the nominal shear wall stiffnesses using shear wall design details (see Sec. 9.6).
6. Perform revised rigid diaphragm analysis using the nominal shear wall stiffnesses.
7. Verify shear wall designs with loads from revised RDA.
8. Calculate diaphragm design forces including torsional forces if required.
9. Design the diaphragm.
10. Verify assumption of semi-rigid or rigid diaphragm behavior for horizontal distribution of forces.
11. Check story drift limits.
12. Verify presence of torsional irregularities.
13. Verify accidental torsion amplification factor.
14. Verify redundancy factor.

6.3.1 Shear Wall Design Based on Rigid Diaphragm Analysis

In an RDA, distribution of loads to the walls depends on the location and stiffness of the walls. The stiffness of the walls depends on the construction details of the walls. Unlike in flexible diaphragm analysis, the loads to a wall line will vary with changes in the construction details of the shear wall. This creates a design process that is often inherently iterative.

For the initial RDA, using shear wall stiffness values proportional to the wall length or capacity (as per SDWPS 2021 Section 4.3.5.5.1, Exception 1) is often an expedient starting point to represent the initial wall stiffness. Only the relative stiffness between the walls is important to distribute lateral load between the walls. Since there are no walls that have an aspect ratio greater than 2:1, no reduction in shear capacities is required. In accordance with ASCE 7 Section 12.8.4.2, the accidental torsion moment needs to be applied to the design forces if the structure is in SDC D and has a Type 1a horizontal irregularity, as previously assumed. Based on the assumed stiffness values and assumptions for ρ and A_x, an initial RDA is performed using a spreadsheet tool to expedite the calculations.

Once the initial RDA design forces have been distributed to the walls, the shear wall construction details can be determined, after which, the nominal wall stiffness can be determined per Sec. 9.6.

Using the nominal stiffness values of the shear walls calculated above, the rigid diaphragm analysis calculations are updated to calculate the distribution of the final lateral loading to the shear walls. Because the initial design was based on a preliminary estimate of relative wall stiffness values, the selected shear wall design details need to be verified as adequate for the loads calculated with the nominal wall stiffnesses.

In the design example, the revised RDA based on nominal calculated stiffness value did not significantly change the load distribution among the shear walls. This is not always the case, especially where the lateral system includes a combination of relatively narrow and relatively long walls or hold downs with dramatically different anchorage slip values. If some components of the initial design were found to not have sufficient capacity at this step, then the design details and nominal shear wall stiffness values would need to be updated, the rigid diaphragm analysis recalculated, and the capacity of the shear walls checked again.

6.3.2   Diaphragm Design Forces

Roof and floor diaphragms are to be designed to resist the seismic design forces of the MLFRS but not less than the diaphragm inertial design forces based on estimated forces defined in ASCE 7 Section 12.10.1.1 and Eq. 12.10-1, including the minimum and maximum force limitations.

Cantilever Diaphragm Design

The diaphragm in the example plan shown in Fig. 6.2 acts like an open-front/cantilever on both sides of the corridor walls. The diaphragm chords are continuous across the full length of the diaphragm and therefore realtively self-restrained between grid lines 2 and 3. The WSP layout for loading in the longitudinal direction is Case 1 as shown in SDPWS Table 4.2A.

There are several triggers in ASCE 7 and in SDPWS that can require consideration of accidental torsion load to be applied to a structure. The drift requirements for open-front structures in SDPWS Sections 4.2.5.1 and 4.2.6.1 require the application of the accidental torsion and consideration of the deformations of the diaphragms. The method by which the accidental torsion is applied can impact the diaphragm deformation. In addition to directly applying a concentrated torsional moment in a 3D model,

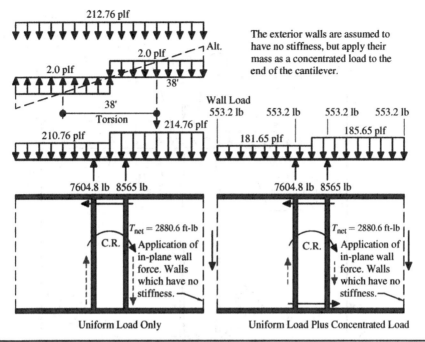

The exterior walls are assumed to have no stiffness, but apply their mass as a concentrated load to the end of the cantilever.

FIGURE 6.3 Torsional distribution method—not mandatory.

there are also several other possible methods of applying torsional forces into diaphragms and shear walls by hand calculations (see Fig. 6.3), or by using a spreadsheet. For the design example, the method on the right side of Fig. 6.3 was used to distribute the torsional moment into the diaphragm as an equivalent uniform load. The method shown in the figure is assumed to be a rational approach to determining diaphragm chord forces, shears, deflection, story drift, and torsional irregularities. The concentrated forces applied to the diaphragm represent the exterior facade walls, which can be heavier relative to other walls, or where walls become discontinuous to the foundation. As such, this method could result in slightly larger chord forces than the other methods.

6.3.3 Diaphragm Design

As noted in Sec. 6.3, the capacity design of the diaphragm will be done using allowable stress design and diaphragm deflection will be done using strength design. The governing diaphragm seismic design force is equal to the MLFRS seismic forces, F_x, used for the shear wall design or the diaphragm design forces, Fpx, calculated using ASCE 7 Eq. 12.10-1, whichever is larger. In many cases, the diaphragm design forces in multistory structures will be larger than the MLFRS forces. Each floor will have different acceleration histories and should be designed to resist an inertial force, F_{px}, which results from the peak response acceleration of that floor.

Shear walls placed along the diaphragm chord lines can affect the final chord forces. Unlike simple span diaphragms, rotational forces increase or decrease the chord forces depending on what side of the diaphragm is being analyzed. Narrow shear walls can

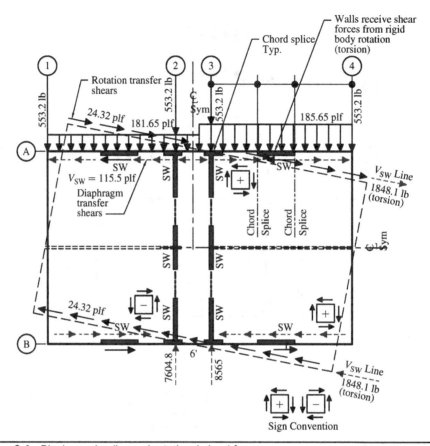

Figure 6.4 Diaphragm loading and rotational chord forces.

also influence the final chord forces, whereas longer walls will not. Because of this, the maximum chord forces must be determined before the diaphragm deflection can be calculated.

Figure 6.4 shows the direction of the diaphragm bending and rotational shears that are transferred into the chords. The maximum diaphragm shear occurs at the corridor support walls and is equal to the standard diaphragm shear plus the rotational shear. The shears transferred into the chords causing bending and uniform shears caused by rotation are shown in Fig. 6.5 at lines 1 and 2, respectively. By observation, the uniform rotational shears (dashed arrows) reduce the diaphragm shears causing chord bending forces on the left cantilever and increase the shears causing bending chord forces on the right cantilever. Once the diaphragm chord forces are determined from bending, line 3, they are added or subtracted from the rotational force, line 2, to determine the final chord forces. However, before that can be done, the rotational force diagram must be modified due to the presence of the shear walls located along the chord line. The shear wall at the left cantilever, located between grid lines 15 ft and 23 ft, resists rotation, and therefore increases the bending tension forces in the chord between 23 ft and 35 ft during rotation. The shear wall at the right cantilever, located between grid lines 53 ft and

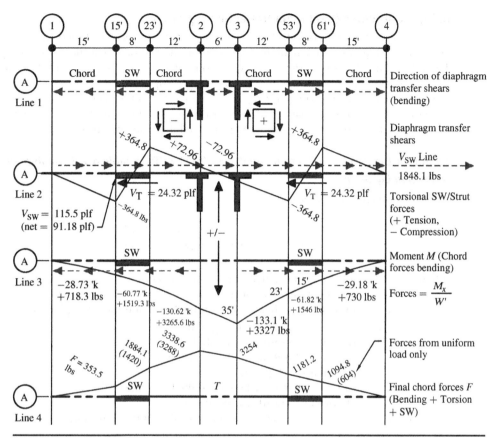

FIGURE 6.5 Maximum diaphragm chord forces.

61 ft, also resists rotation, thereby reducing the chord bending tension forces. The same process is used to calculate the compression chord forces at grid line B.

Line 1: Shows the direction of the diaphragm shears transferred into the top chord caused by bending.

Line 2: Shows the additional chord forces caused by diaphragm rotation.

Line 3: References the diaphragm moments and chord forces caused by bending $= \dfrac{M}{W'}$, where $W' = 40$ ft.

Line 4: The final chord forces are determined by observing the direction of applied shear transfer forces and combining the rotational forces determined on line 2 with the bending chord forces on line 3.

Overall, the presence of the short shear walls used in this example changes the chord forces by approximately 365 lb at the start of each shear wall. This suggests that the shear wall effects on the chord forces are minimal for this example but can be significantly larger if greater eccentricities occur with short shear walls located along the chord lines. Once the chord forces have been determined at the chord splice locations, the design and construction of the diaphragm can be completed.

6.3.4   Cantilever Diaphragm Deflection Equations

A task group of AWC's Wood Design Standards Committee developed the following approximate equations to calculate the maximum deflection of cantilever diaphragms. Three-term and four-term deflection equations were developed in a similar manner used to derive the deflection equations for simply supported diaphragms, one set for uniformly loaded cantilever diaphragms and the other set for point loaded cantilever diaphragms to represent heavy mass exterior walls. The three-term equation was used for the design example to determine diaphragm deflection and stiffness.

Cantilever Diaphragm Deflection Equations—2021 SDPWS Table 4.2.3 Diaphragm Deflection Equations

Three-term equation for uniform load:

$$\delta_{\text{Diaph Unif}} = \frac{3vL'^3}{EAW'} + \frac{0.5vL'}{1000G_a} + \frac{\sum x\Delta_C}{W'}$$

where
E = modulus of elasticity of diaphragm chords, psi
A = area of chord cross section, in^2
v_{max} = induced unit shear at the support from a uniform applied load, lb/ft
L' = cantilever diaphragm length, ft
W' = cantilever diaphragm width, ft
G_a = apparent diaphragm shear stiffness from nail slip and panel shear deformation, kips/in
x = distance from chord splice to the free edge of the diaphragm, ft
Δ_C = diaphragm chord splice slip, in
$\delta_{\text{Diaph Unif}}$ = calculated deflection at the free edge of the diaphragm, in

For a uniform load of w, the induced unit shear at the support $v = wL'/W'$.

Four-term equation for uniform load:

$$\delta_{\text{Diaph Unif}} = \frac{3vL'^3}{EAW'} + \frac{0.5vL'}{G_v t_v} + 0.376L'\, e_n + \frac{\sum x\Delta_C}{W'}$$

where
e_n = nail slip per SDPWS C4.2.2D for the load per fastener at v_{max}
$G_v t_v$ = panel rigidity through the thickness per SDPWS Commentary Table C4.2.3A

Similarly, the equations developed for a point load at the end of the cantilever are as follows:

Three-term equation for point load:

$$\delta_{\text{Diaph Conc}} = \frac{8vL'^3}{EAW'} + \frac{vL'}{1000G_a} + \frac{\sum x\Delta_C}{W'}$$

where
$\delta_{\text{Diaph Conc}}$ = calculated deflection at the free edge of the diaphragm, in
For the point load of P, the induced unit shear at the support, $v = P/W'$

Four-term equation for point load:

$$\delta_{\text{Diaph Conc}} = \frac{8vL'^3}{EAW'} + \frac{vL'}{G_v t_v} + 0.75\, L'\, e_n + \frac{\sum x \Delta_C}{W'}$$

6.3.5 Rigid Diaphragm Flexibility Check

The right cantilever will be used to verify the diaphragm flexibility because it has a higher load due to the distribution of the torsional moment and will produce the largest story drift. Diaphragm flexibility is covered in IBC[3] Section 1604.4, ASCE 7 Section 12.3.1, and SDPWS Section 4.1.7. These sections all refer to story drift for the determination of diaphragm flexibility but have slightly different requirements.

Per IBC Section 1604.4 and SDPWS Section 4.1.7, a simple span diaphragm can be idealized as rigid when:

$$\delta_{\text{MDD}} \le 2 * \Delta_{\text{ADVE}}$$

Where δ_{MDD} is the maximum in-plane deflection of the diaphragm and Δ_{ADVE} is the average drift of the adjoining vertical elements (e.g., the average story drift of a simple-span diaphragm structure). ASCE 7 does not provide a calculated rigid diaphragm condition. ASCE 7 Figure 12.3-1 relates the determination of the flexibility to a simple span diaphragm, which does not relate to a cantilever diaphragm.

While the cantilever diaphragm configuration is not specifically covered in ASCE 7, it is covered in SDPWS Section 4.1.7.2, which states:

> Cantilevered diaphragms shall be permitted to be idealized as rigid when the calculated maximum in-plane deflection of the diaphragm itself under lateral load is less than or equal to two times the deflection of vertical elements of the associated lateral force-resisting system of the story below used to determine the cantilever length, L' (see Figure 4A). Other diaphragms shall be permitted to be idealized as rigid when the calculated maximum in-plane deflection of the diaphragm itself under lateral load is less than or equal to two times the average deflection of adjoining vertical elements of the associated lateral force-resisting system of the story below. The deflections shall be determined under equivalent tributary lateral load.

Figure 6.6 (lower figure) shows a simple application of the conditions above that can apply in cantilever cases. For the cantilever diaphragm condition, the preferred method to check the calculated diaphragm flexibility is to use the drift of the closest shear wall line in the direction of the loading that defines the maximum cantilever length for Δ_{ADVE} as this is the only adjoining vertical element.

When calculating shear wall and diaphragm deflections for the determination of diaphragm flexibility and story drift, it is permitted to use $\rho = 1.0$ in accordance with ASCE 7 Section 12.3.4.1 (2). Given the assumption that the structure has a torsional irregularity and is assigned to SDC D, ASCE 7 Section 12.8.4.2 requires the inclusion of the accidental torsion in the determination of design story drift. Therefore, accidental torsion is applied and amplified by A_x in accordance with 12.8.4.3 (a previously assumed value that needs to be verified).

Under the wind provisions of ASCE 7, Chapter 27, Section 27.5.4 also requires that the structural analysis shall consider the stiffness of diaphragms and vertical elements of the main wind force-resisting system (MWFRS). Diaphragm flexibility requirements for wind conditions are embedded within the definitions of ASCE 7 Section 26.2, Definitions—Diaphragm, which states that diaphragms constructed of WSP are

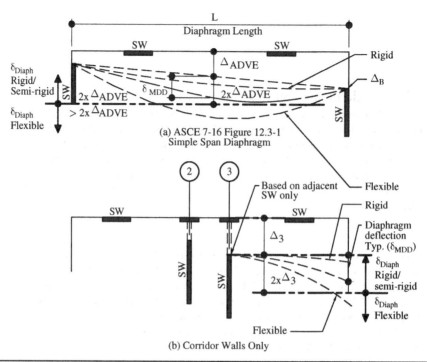

FIGURE 6.6 Calculated diaphragm flexibility methods for diaphragms.

permitted to be idealized as flexible. However, this option is not allowed for cantilever diaphragms and the diaphragm flexibility must be calculated.

Results from spreadsheet in example paper:

$$\delta_{Diaph\ 1} = 0.26 \text{ in, } \delta_{Diaph\ 4} = 0.265 \text{ in, Deflection at grid lines 1 and 4.}$$

$$\text{Deflection at grid line 3} = \frac{F}{1000k} = \frac{9412.2}{1000(43.54)} = 0.216 \text{ in, translation}$$

$$2 \times \Delta_3 = 0.432 \text{ in}$$

$$0.265 \text{ in} < 0.432 \text{ in}$$

Therefore, diaphragm can be idealized as rigid.

6.3.6 Story Drift Check

All structures with seismic loading need to meet the story drift limits of ASCE 7 Section 12.12.1. For structures designed using the equivalent lateral force procedure, the story drift values are determined from ASCE 7 Section 12.8.6:

The design story drift (Δ) shall be computed as the difference of the deflections at the centers of mass at the top and bottom of the story under consideration (Fig. 12.8-2). Where centers of mass do not align vertically, it is permitted to compute the deflection at the bottom of the story based on the vertical projection of the center of mass at the top of the story. Where allowable stress design is used, Δ shall be computed using the strength level seismic forces specified in Section 12.8 without reduction for allowable stress design.

For structures assigned to Seismic Design Category C, D, E, or F that have horizontal irregularity Type 1a or 1b of Table 12.3-1, the design story drift, Δ, shall be computed as the largest difference of the deflections of vertically aligned points at the top and bottom of the story under consideration along any of the edges of the structure. The deflection at level x (δ_x) (in or mm) used to compute the design story drift, Δ, shall be determined in accordance with the following equation:

$$\delta_x = \frac{C_d \delta_{xe}}{I_e} \qquad\qquad 12.8\text{-}15$$

where
C_d = deflection amplification factor in Table 12.2-1
δ_{xe} = deflection at the location determined by an elastic analysis
I_e = Importance factor determined in accordance with Section 11.5.1

For open-front structures, SDPWS Section 4.1.7.2 applies similar drift checks where the maximum story drift at each edge of the open-front structure shall not exceed the ASCE 7 allowable story drift when subject to seismic design forces including torsion and accidental torsion. This check shall include shear and bending deformations of the diaphragm, computed on a strength level basis. The SDPWS drift check applies at the edges of the structure for all open-front structures, with or without torsional irregularities. This example is symmetric, and the load case checked will create the maximum drift at the right cantilever. ASCE 7 Section 12.8.6 has a similar requirement for drift to be checked at the largest displacement along any edges of the structure if a Type 1 horizontal irregularity exists.

Drift consists of three components: diaphragm translation and diaphragm rotations from wall displacements and in-plane diaphragm deformations, as shown in SDPWS Figure C4.2.6B and Fig. 6.7.

$$\text{Drift } \Delta = \delta_{\text{Translation}} + \delta_{\text{Rotation}} + \delta_{\text{Diaph}}$$

The deflection from translation and rotation are based on the response of the shear walls under the rigid diaphragm assumption. The RDA distribution of loads is applied with $\rho = 1.0$ and A_x = assumed value. The displacement values shown in Fig. 6.8 are from the example paper. Using the nominal shear stiffness shear wall values to calculate the deflections of the shear walls:

$$\text{Drift } \Delta = \sqrt{\left(\delta_T + \delta_D \pm \delta_{RL}\right)^2 + \left(\delta_{RT}\right)^2}$$

Calculation results from example paper:

$$\delta_2 = 0.192 \text{ in}$$
$$\delta_3 = 0.216 \text{ in}$$
$$\delta_A = 0.081 \text{ in}$$
$$\delta_B = -0.081 \text{ in}$$

Calculating the longitudinal translation:

$$\delta_{\text{Translation}} = \frac{(\delta_2 + \delta_3)}{2} = \frac{(0.192 + 0.216)}{2} = 0.204 \text{ in (average)}$$

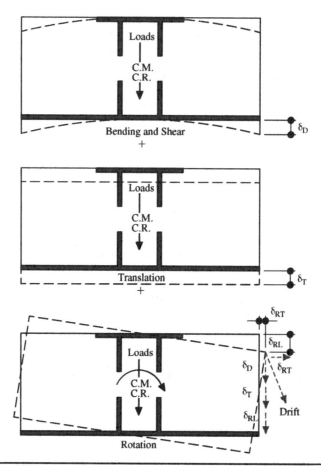

Figure 6.7 Story drift components.

Calculating the diaphragm displacement at grid 4:

Due to rigid diaphragm rotation, relative to the center of rigidity:

$$\delta_{RL} = \frac{\delta_A + \delta_B}{W'}(L' + 3) = (0.081 + 0.081)(38)/40 = 0.154 \text{ in, } \delta_{RT} = 0.081 \text{ in}$$

Example diaphragm deflection right cantilever:

$$\delta_D = \frac{3vL'^3}{EAW'} + \frac{0.5vL'}{1000G_a} + \frac{\sum A_C X_C}{W'} = \frac{3(233.2)(35)^3}{1,700,000(16.5)40} + \frac{0.5(233.2)35}{1000(25)}$$
$$+ 0.075 = 0.265 \text{ in}$$

The method for checking drift and torsional irregularities should include the diaphragm deflections as well as the rotational translation, δ_{RT}, in the drift values at the edges of the structure.

$$\text{Drift } \Delta = \sqrt{\left(\delta_T + \delta_D \pm \delta_{RL}\right)^2 + \left(\delta_{RT}\right)^2}$$

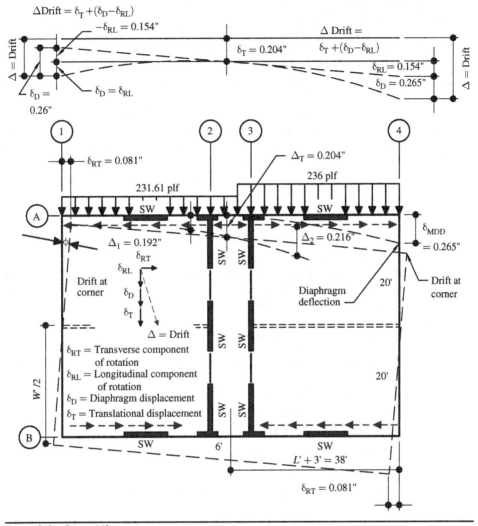

FIGURE 6.8 Story drift.

$$\text{Drift } \Delta_4 = \sqrt{(0.204 + 0.265 + 0.154)^2 + (0.081)^2} = 0.628 \text{ in}$$

$$\text{Drift } \Delta_1 = \sqrt{(0.204 + 0.26 - 0.154)^2 + (0.081)^2} = 0.320 \text{ in}$$

Combining all the terms of the diaphragm deflection at the edge of the structure:

$$C_d = 4, I_e = 1$$

$$\delta_x = \frac{C_d \delta_{xe}}{I_e}$$

$$\delta_x = \frac{C_d \delta_{xe}}{I_e} = \frac{4(0.628)}{1} = 2.51 \text{ in}$$

In ASCE 7 Table 12.12-1, the allowable story drift limit of $0.025\,h_{sx}$ or $0.02\,h_{sx}$ depends on the non-structural components and detailing. Under the first category, one of the requirements is that interior walls, partitions, ceilings, and exterior walls can accommodate the higher story drift limit. Most sheathed wood-framed walls can undergo the 2.5 percent drift level while providing life safety performance at the seismic design level; however, window systems or architectural finishes may not be able to perform without some damage. The selection of the higher 2.5 percent drift limit should be taken only with consideration of the non-structural wall and window performance, specifically designing and detailing the systems to prevent damage, and possibly under consultation with the architect and the authority having jurisdiction (AHJ) over the project. Otherwise, the 2 percent drift limit requirements should be used.

$$0.025\,h_{sx} = 0.025(10)(12) = 3.0 \text{ in} > 2.51 \text{ in}$$

Therefore, drift is OK for 2.5 percent limit, and

$$0.02\,h_{sx} = 0.02(10)(12) = 2.4 \text{ in} < 2.51 \text{ in}$$

Therefore, drift is not OK for 2 percent limit.

If drift values needed to be reduced, the portion of the drift coming from the deflection of the corridor walls and diaphragm deflection are roughly equal, which suggests that stiffening the walls or the diaphragms could help reduce the story drift at the building edge.

Options for providing additional stiffness:

    a. Increase the stiffness of the diaphragm and/or shear walls using the four-term deflection equations. Note both must use the four-term equation at the same time. Do not mix the three-term and four-term equations.

    b. Lengthen or add shear walls.

    c. Change the construction details:

        1. Increase the stiffness of the walls by adding sheathing on both sides.

        2. Increase the size (and stiffness) of hold downs.

        3. Increase the number of shear wall boundary studs.

        4. Increase the shear wall nailing.

6.3.7   Verification of Torsional Irregularity

An early assumption of this design was that the structure has a torsional irregularity, but not an extreme torsional irregularity.

> ASCE 7 Table 12.3-1 Type 1a Torsional Irregularity:
> Torsional irregularity is defined to exist where the maximum story drift, computed including accidental torsion with $A_x = 1.0$, at one end of the structure transverse to an axis is more than 1.2 times the average of the story drifts at the two ends of the structure. Torsional irregularity requirements in the reference sections apply only to structures in which the diaphragms are rigid or semi-rigid.

This is essentially the same definition of torsional irregularity found in SDPWS Section 4.2.5.1. ASCE 7 Table 12.3-1 defines a horizontal structural irregularity Type 1b—Extreme Torsional Irregularity as the same criteria with a limit of 1.4 instead of 1.2 for a

Type 1a irregularity. In Section 12.3.3.1, an extreme torsional irregularity, horizontal irregularity Type 1b, is allowed in structures assigned to Seismic Design Categories B, C, and D, but not in E, or F.

ASCE 7 Table 12.3-1 Irregularity Requirements: Type 1a and 1b irregularities both trigger Sections 12.3.3.4 and 12.8.4.3 (amplification of accidental torsion) for SDC C to F. The former section requires a 25 percent increase in the connections of the diaphragm to the vertical elements and collectors: and the collectors and their connection to the vertical force-resisting elements. ASCE 7 Table 12.3-1, Type 1a and 1b irregularities note that $A_x = 1.0$ when checking for torsional irregularities. Therefore, $\rho = 1.0$ and $A_x = 1.0$ are used in the torsional irregularity checks.

> *Exception:* Collector forces calculated using the seismic load effects, including over-strength of Section 12.4.3 need not be increased. The diaphragm shears do not have to be increased 25%. Section 12.8.4.3 requires an amplification of the accidental torsion.

Torsional Irregularity Check:

Calculated results from spreadsheet in example paper (similar to Fig. 6.8):

$\delta_{SW2} = 0.194$ in displacement at line 2

$\delta_{SW3} = 0.214$ in displacement at line 3

$$\delta_T = \frac{(\delta_{SW2} + \delta_{SW3})}{2} = 0.204 \text{ in displacement at center of mass (diaphragm translation)}$$

Rigid diaphragm rotation:

$$\delta_{SWA,B} = 0.065 \text{ in} = \delta_{RT} \text{ Transverse displacement at lines A and B from rigid}$$
$$\text{diaphragm rotation}$$

Using similar triangles, the longitudinal displacement at lines 1 and 4 from rigid diaphragm rotation is as follows:

$$\delta_{RL} = \frac{2\delta_{SWA,B}(L' + 3')}{W'} = 0.124 \text{ in}$$

Diaphragm deformations:

$$\delta_{D,1} = 0.256 \text{ in}$$
$$\delta_{D,4} = 0.260 \text{ in}$$

The method for checking a torsional irregularity can include the diaphragm deflections as well as the rotational translation, δ_{RT}, in the drift values at the edges of the structure.

$$\text{Drift } \Delta = \sqrt{\left(\delta_T + \delta_D \pm \delta_{RL}\right)^2 + \left(\delta_{RT}\right)^2}$$

$$\text{Drift } \Delta_4 = \sqrt{(0.204 + 0.260 + 0.124)^2 + (0.065)^2} = 0.592 \text{ in}$$

$$\text{Drift } \Delta_1 = \sqrt{(0.204 + 0.256 - 0.124)^2 + (0.065)^2} = 0.342 \text{ in}$$

$$\Delta_{avg} = \frac{0.592 + 0.342}{2} = 0.467 \text{ in}$$

Checking the irregularity criteria of ASCE 7 Table 12.3-1:

$$0.592 > 1.2(0.467) = 0.56 \text{ in}$$

Therefore, horizontal torsional irregularity Type 1a exists.

$$0.592 < 1.4(0.467) = 0.654 \text{ in}$$

Therefore, horizontal torsional irregularity Type 1b does not exist.

The building has a torsional irregularity (Type 1a) but not a (Type 1b), as originally assumed.

6.3.8   Amplification of Accidental Torsional Moment

Amplification of accidental torsion is intended to account for an increase in torsional moment caused by potential yielding of the perimeter SFRS (i.e., walls, shifting of center of rigidity) leading to dynamic torsional instability. When computing A_x for each level, absolute displacements, δ_x, of each level are used, not story drifts, Δ. The displacements used to find A_x are calculated using $\rho = 1.0$ and $A_x = 1.0$.

ASCE 7 Section 12.8.4.3 Amplification of Accidental Torsional Moment:
Structures assigned to Seismic Design Category C, D, E, or F, where Type 1a or 1b torsional irregularity exists as defined in Table 12.3-1 shall have the effects accounted for by multiplying M_{ta} at each level by a torsional amplification factor (A_x) as illustrated in Figure 12.8-1 and determined from the following equation:

$$A_x = \left(\frac{\delta_{max}}{1.2\delta_{avg}} \right)^2$$

where δ_{max} = maximum displacement at level x computed assuming $A_x = 1$
δ_{avg} = average of the displacements at the extreme points of the structure at level x computed assuming $A_x = 1$

The torsional amplification factor, (A_x), shall not be less than 1 and is not required to exceed 3.0. From the results of the example paper:

$$A_x = \left(\frac{\delta_{max}}{1.2\delta_{avg}} \right)^2 = \left(\frac{0.592}{1.2(0.467)} \right)^2 = 1.116 < A_x = 1.25 \text{ assumed}$$

As noted, when discussing the preliminary design assumption, an accurate determination of this value cannot be calculated until after deflections are calculated from a designed structural system.

6.3.9   Verification of Redundancy Factor

ASCE 7 requires the following for structures assigned to SDC D through F:

12.3.4.2   Redundancy Factor, ρ, for Seismic Design Categories D through F.
For structures assigned to Seismic Design Category D and having extreme torsional irregularity as defined in Table 12.3-1, Type 1b, ρ shall equal 1.3. For other structures assigned to Seismic Design Category D and for structures assigned to Seismic Design Categories E or F, ρ shall equal 1.3 unless **one** of the following two conditions is met, whereby ρ is permitted to be taken as 1.0. A reduction in the value of ρ from 1.3 is not permitted for structures assigned to Seismic

Design Category D that have an extreme torsional irregularity, Type 1b. Seismic Design Categories E and F are not specified because extreme torsional irregularities are prohibited (see Section 12.3.3.1).

a. Each story resisting more than 35 percent of the base shear in the direction of interest shall comply with Table 12.3-3.

b. Structures that are regular in plan at all levels provided that the seismic force-resisting systems consist of at least two bays of seismic force-resisting perimeter framing on each side of the structure in each orthogonal direction at each story resisting more than 35 percent of the base shear. The number of bays for a shear wall shall be calculated as the length of shear wall divided by the story height or two times the length of shear wall divided by the story height, h_{sx}, for light-frame construction.

The example structure does not have an extreme torsional irregularity or two bays of seismic force-resisting perimeter framing on each side of the structure in each orthogonal direction and is not considered "regular in plan" in the longitudinal direction, so condition "b" does not apply. The calculated condition "a" will be considered.

ASCE 7 Table 12.3-3 for shear walls or wall piers with a height-to-length ratio greater than 1.0 notes:

Removal of a shear wall or wall pier with a height-to-length ratio greater than 1.0 **within** any story, or collector connections thereto, would not result in more than a 33% reduction in story strength; nor does the resulting system have an extreme torsional irregularity (horizontal structural irregularity Type 1b). The shear wall and wall pier height-to-length ratios are determined as shown in Figure 12.3-2.

There are no walls that have an aspect ratio greater than 1:1 in the longitudinal direction. However, the removal of one shear wall at line A, which has an aspect ratio of 1.25:1, effects torsional resistance and is therefore removed for the extreme torsional irregularity check. If one shear wall symbolically fails (is removed), the center of rigidity would shift toward grid line B which would require re-calculation of the distribution of forces using RDA. Although this relocation of the center of rigidity creates inherent torsion in the transverse direction, symmetry between the center of rigidity and center of mass remains in the longitudinal direction. The elastic method shall be used to verify the redundancy.

For the complete design example, download the example paper from Woodworks, www.woodworks.org: Tabs-Learn, Light-frame.[4]

6.4   Open-Front One Side

The following problems are intended to clarify the importance of creating internal load paths within the diaphragms and to explain how to analyze them. These problems do not consider the relative differences in stiffnesses between the diaphragms and shear walls as required by code, ASCE 7 Section 12.3.1 and SPDWS Section 4.2.6. These omissions including rigid diaphragm analyses and non-enveloping procedures are done to simplify and reduce the size of the examples.

Example 6.1: Open-Front Diaphragm One Side—Boundary Member Forces

The code requirements for open-front structures were given in Sec. 6.2. For open-front diaphragms, the maximum length of the diaphragm normal to the open side shall not be more than 35 ft. The objective of Figs. 6.9 through 6.21 and Examples 6.1 through 6.3

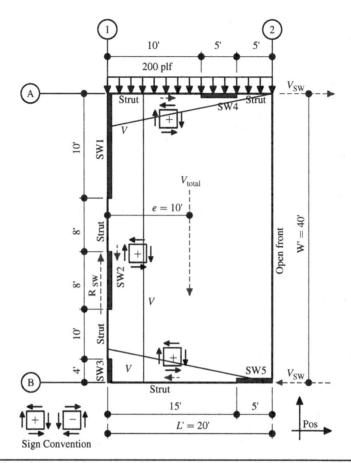

Figure 6.9 Open-front one side.

is to help understand how to calculate boundary member forces. For a complete analysis, the procedures outlined in Secs. 6.1 through 6.3 must also be done.

The open-front diaphragm shown in Fig. 6.9 has partial length wood framed walls located along grid lines 1, A and B. The wall line along grid line 2 is an open-front consisting of solid windows or door openings. The diaphragm is loaded in the longitudinal direction, parallel to grid lines 1 and 2. The only support for the diaphragm in the direction of the applied loads is the walls along grid line 1. Since the resultant force occurs at the center of mass at mid-length of the diaphragm, a large inherent eccentricity exists from the supporting walls to the location of the resultant load. The torsion created by the eccentric load is resisted by the shear walls located along grid lines A and B. Sheathing element shear forces have been placed on the plan to show the direction of the shear forces being transferred into the boundary members. The analysis of the chord forces for an open-front diaphragm (neglecting the effects of the short walls located along the chord line) is as follows:

L' = length of the diaphragm normal to the open side

W' = width or depth of the diaphragm, parallel to the applied load

w = uniform load

$V = wL'$ Total shear force and reaction at grid line 1

$e = \dfrac{L'}{2}$ Eccentricity

$M = Ve$ Torsional moment

$V_{sw} = \dfrac{M}{W'}$ Total shear force applied to the walls at grid lines A and B

The shear walls at grid lines A and B oppose the torsional moment.

If full length shear walls are used on all three sides of the diaphragm, there are no diaphragm chord forces along grid lines A and B because the chords are not restrained from bending and the wall shears are equal to the diaphragm shears, cancelling each other out. Because of this, the chord forces from bending are assumed to be zero and the first term of the deflection equation drops out. Although chord forces do not exist, strut forces do exist when partial length walls occur at the end walls due to rotation as shown below. Partial length shear walls are installed along all three lines of resistance in lieu of full-length walls, as shown in Fig. 6.9. This is done to demonstrate that the boundary members along grid lines A and B receive tension and compression forces by drag strut action, not chord action. The diaphragm is 20 ft long by 40 ft wide. The aspect ratio of the diaphragm is equal to 0.5, which is less than 1.5 allowed. Five-foot shear walls are installed at grid lines A and B, labeled SW4 and SW5, respectively. Three shear walls are placed at grid line 1, a 10-ft wall, an 8-ft wall, and a 4-ft wall, noted as SW1, SW2, and SW3, respectively. A 200-plf uniform load is applied to the diaphragm as shown in the figure.

$L' = 20$ ft < 35 ft, which is a prescriptive requirement.

$W' = 40$ ft

$w = 200$ plf

$V_2 = 0$ plf shear at grid line 2

$V_1 = V_{\text{Total}} = 200(20) = 4000$ lb, shear at grid line 1 which is equal
                     to the total shear force applied to the diaphragm.

$v_1 = \dfrac{V_1}{W'} = \dfrac{4000}{40} = 100$ plf unit diaphragm shear along grid line 1

$v_{\text{SW1, SW2, SW3}} = \dfrac{4000}{(10 + 8 + 4)} = +181.8$ plf unit shear wall shear at grid line 1

$v_{\text{net}} = 181.8 - 100 = +81.8$ plf net shear at shear walls SW1, SW2, and SW3
                (shear wall shear minus diaphragm shear)

$e = \dfrac{20}{2} = 10$ ft eccentricity

$V_{\text{SW4, SW5}} = \dfrac{V_1 e}{W'} = \dfrac{4000(10)}{40} = 1000$ lb, total shear force at grid lines A and B

$v_{\text{SW4, SW5}} = \dfrac{1000}{5} = 200$ plf unit shear wall shear at grid lines A and B

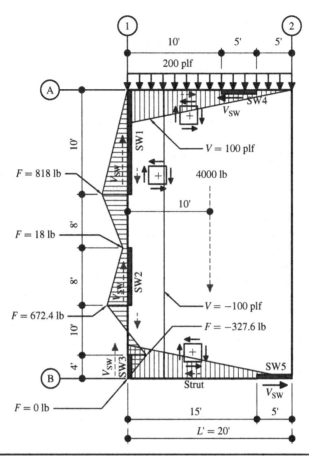

Figure 6.10 Shear diagrams and longitudinal strut force diagram.

Determination of strut forces along grid line 1, starting at grid line A (Fig. 6.10):

$F = +81.8(10) = +818$ lb tension at the end of shear wall SW1

$F = 818 - 100(8) = +18$ lb tension at the start of shear wall SW2

$F = 18 + 81.8(8) = +672.4$ lb tension at the end of shear wall SW2

$F = 672.4 - 100(10) = -327.6$ lb compression at the start of shear wall SW3

$F = -327.6 + 81.8(4) = 0$ lb at grid line B; therefore, the diagram closes.

Determination of strut forces along grid line A, starting at grid line 1 (Fig. 6.11): Since the direction of the shear transferred into the boundary element is in the positive direction, the diaphragm unit shear diagram will be plotted above the line. The direction of the shear wall unit shear is going in the negative direction; the shear wall unit shear diagram will be plotted below the line. It makes no difference which side of the line the shear diagrams are plotted as long as consistency is maintained in the entire analysis. The direction of the shears that are transferred into the struts show if the members are in tension or compression. The diaphragm unit shear varies from 100 plf at grid line 1 to zero at grid line 2. The

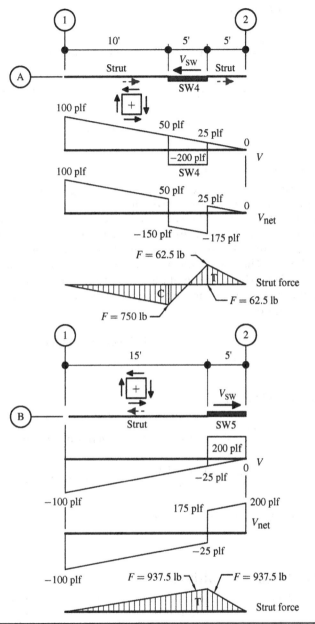

Figure 6.11 Strut force diagram at grid lines A and B.

magnitude of the diaphragm unit shears at the start and end of the shear wall are 50 plf and 25 plf, respectively. The net shear diagram is shown in the figure.

$$v_{net} = -200 + 50 = -150 \text{ plf net shear at the start of SW4}$$

$$v_{net} = -200 + 25 = -175 \text{ plf net shear at the end of SW4}$$

Starting at grid line 1:

$$F = \frac{(100 + 50)}{2}(10) = +750 \text{ lb compression at the start of SW4}$$

$$F = +750 - \frac{(150 + 175)}{2}(5) = -62.5 \text{ lb tension at the end of SW4}$$

$$F = -62.5 + \frac{(25 + 0)}{2}(5) = 0 \text{ lb at grid line 2}$$

Therefore, the force diagram closes.

Determination of chord force along grid line B, starting at grid line 1 (Fig. 6.11): The direction of the diaphragm unit shear transferred into the boundary element indicates a negative shear. The direction of the shear wall unit shear acts in the opposite direction indicating a positive shear. Positive shears will be plotted above the line, negative shears below the line. The diaphragm unit shear varies from 100 plf at grid line 1 to zero at grid line 2. The magnitude of the basic diaphragm unit shear at the start of the shear wall is 25 plf. The direction of the shears transferred into the boundary elements show whether the members are in tension or compression. The net shear diagram is shown in the figure.

$$v_{net} = +200 - 25 = +175 \text{ plf net shear at the start of SW5}$$

$$v_{net} = +200 - 0 = +200 \text{ plf net shear at the end of SW5}$$

Starting at grid line 1:

$$F = -\frac{(100 + 25)}{2}(15) = -937.5 \text{ lb tension at the start of SW5}$$

$$F = -937.5 + \frac{(175 + 200)}{2}(5) = 0 \text{ lb at grid line 2}$$

Therefore, the force diagram closes. ▲

6.5 Irregular Shaped Cantilevers

Cantilever diaphragms often contain multiple offsets similar to that shown in Fig. 6.12. This can cause some confusion in how to approach the analysis. By observation, the maximum drift will occur at the upper right corner of the diaphragm. One possible approach to determining the total drift would be to break the cantilever down into three rectangular sections, making the diaphragm deflection calculation easier. Starting with section 1, the diaphragm deflection can be calculated for that section alone. The deflection of section 2 can then be added to get the total diaphragm deflection at the outside corner. Since the diaphragm is semi-rigid or rigid, the rotational and translational displacements can be determined, and the total drift can be calculated in accordance with the method outlined in Example 6.1.

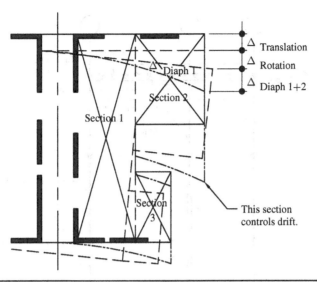

Δ Translation

Δ Rotation

Δ Diaph 1+2

This section
controls drift.

Figure 6.12   Irregular shaped cantilevers.

6.6 Open-Front Both Sides—Boundary Member Forces

Occasionally, open-front walls are desired on both sides of a diaphragm. At first glance, this layout may look questionable. However, the interior supporting wall is located close to the resultant load, which reduces the eccentricity. As a result, smaller forces are applied to the end walls than if the supporting walls are located at grid line 1 due to the reduced torsional moment. The diaphragm shown in Fig. 6.13 has collectors installed between the supporting interior shear walls to provide a complete load path the full-length along grid line 2. In the event a full-length collector cannot be installed, an alternate load path must be used as shown in Fig. 6.20. The diaphragm sections on each side of the interior shear walls must comply with the prescriptive requirements of SDPWS Section 4.2.6, Figure 4A. In this example, the cantilever lengths on each side of the interior walls are unequal, which produces an inherent torsion when loaded in the longitudinal direction. The analysis is similar to the open-front diaphragm in Example 6.1. End-zone wind pressures are applied to the end of the longer cantilever, for the worst case, which increases the torsional moment. If grid line 2 is in the middle of the diaphragm, creating equal cantilevers, a minimum eccentricity should be used in accordance with ASCE7-16 Section 12.8.4.2. The torsional moment is resisted by the end shear walls located at grid lines A and B.

Example 6.2: Open-Front Diaphragm Both Sides, Interior Shear Walls In-Line—Boundary Member Forces

The diaphragm shown in Fig. 6.14 is 80 ft wide by 36 ft in length, with interior shear walls located at grid line 2 and end walls placed at grid lines A and B. The maximum cantilever length is 20 ft. A full-length collector is installed along grid line 2 to complete the required diaphragm boundary. The location of grid line 2 creates a 16-ft cantilever on the left side of the wall line and a 20-ft cantilever on the right. The diaphragm will be analyzed with the wind loads shown in the figure. The end-zone wind pressure is

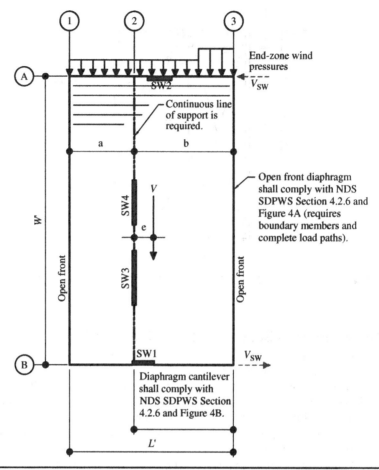

Figure 6.13 Open-front both sides, interior shear walls in-line.

placed at grid line 3 to produce the maximum eccentric load to the diaphragm. The strut force diagram at grid line A and the collector force diagram at grid line 2 will be determined.

Construction of basic shear diagram and wall shears (see Fig. 6.15):

$$R_2 = 200(32) + 250(4) = 7400 \text{ lb total shear at grid line 2}$$

Starting at grid line 1:

$$V_1 = 0 \text{ lb}$$

$$V_{2L} = -200(16) = -3200 \text{ lb, total shear on the left side of grid line 2}$$

$$v_{2L} = \frac{-3200}{80} = -40 \text{ plf diaphragm unit shear on the left side of grid line 2}$$

$$V_{2R} = R_2 - V_{2L} = 7400 - 3200 = +4200 \text{ lb right side of grid line 2}$$

$$v_{2R} = \frac{4200}{80} = +52.5 \text{ plf unit shear on the right side of grid line 2}$$

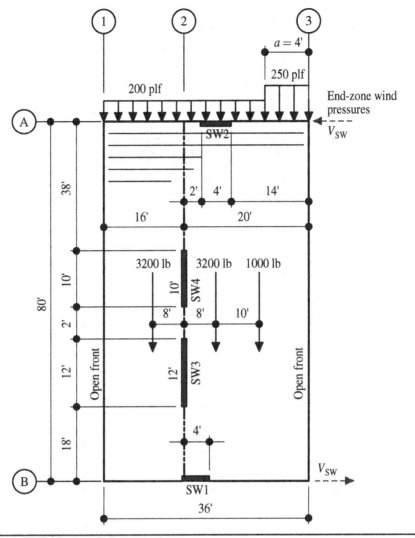

Figure 6.14 Application of diaphragm forces.

$$V_{2+2'} = V_{2R} - w(2') = 4200 - 200(2) = 3800 \text{ lb shear at the start of SW2}$$

$$v_{2+2'} = \frac{3800}{80} = +47.5 \text{ plf unit shear at the start of SW2}$$

$$V_{2+6'} = V_2 - w(6') = 4200 - 200(6) = 3000 \text{ lb shear at the end of SW2}$$

$$v_{2+6'} = \frac{3000}{80} = +37.5 \text{ plf unit shear at the end of SW2}$$

$$V_{2+10'} = 3000 - 200(10) = 1000 \text{ lb shear at the start of end-zone pressure}$$

$$v_{2+10'} = \frac{1000}{80} = +12.5 \text{ plf unit shear at the start of end-zone pressure}$$

$$V_3 = 1000 - 250(4) = 0 \text{ lb}$$

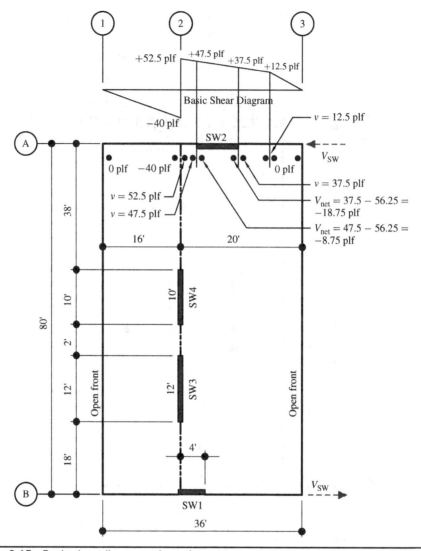

Figure 6.15   Basic shear diagram and net shear.

The resultant loads and distances from the wind loads are shown in Fig. 6.14.

$$\sum M_2 = +3200(8) + 1000(18) - 3200(8) = 18{,}000 \text{ ft-lb clockwise moment}$$

$$V_{\text{SW1, SW2}} = \frac{18{,}000}{80} = +225 \text{ lb shear force to shear walls SW1 and SW2}$$

$$v_{\text{SW}} = \frac{225}{4} = 56.25 \text{ plf unit shear wall shear at SW2}$$

The force applied to the shear wall at grid line A from the moment produced by the eccentricity is acting to the right. The resisting shears of the wall act to the left.

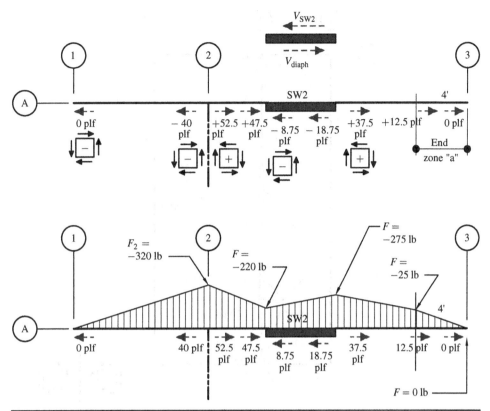

FIGURE 6.16 Transfer shears and strut force diagram at grid line A.

The net shear at the shear wall is equal to

$$v_{net} = +47.5 - 56.25 = -8.75 \text{ plf, net shear at left end of SW2, diaphragm}$$
$$\text{shear minus the shear wall shear}$$

$$v_{net} = +37.5 - 56.25 = -18.75 \text{ plf, net shear at right end of SW2}$$

Construction of the strut force diagram at grid line A (see Fig. 6.16): The direction of the diaphragm shears transferred into the strut is shown in the figure.

Starting at grid line 1:

$$F_2 = \frac{(-40 + 0)}{2}(16) = -320 \text{ lb tension force at grid line 2}$$

$$F = -320 + \frac{(52.5 + 47.5)}{2}(2) = -220 \text{ lb tension at the start of SW2}$$

$$F = -220 - \frac{(8.75 + 18.75)}{2}(4) = -275 \text{ lb tension at the end of SW2}$$

$$F = -275 + \frac{(37.5 + 12.5)}{2}(10) = -25 \text{ lb tension at start of end-zone pressure}$$

$$F_3 = -25 + \frac{(12.5 + 0)}{2}(4) = 0 \text{ lb at grid line 3}$$

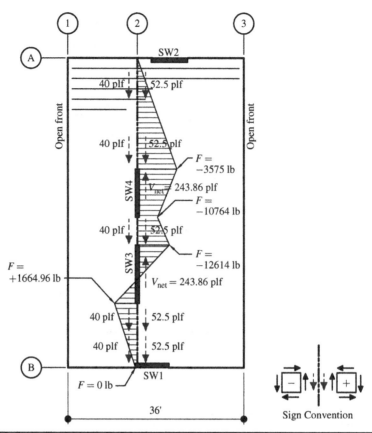

FIGURE 6.17 Collector force diagram at grid line 2.

Therefore, the diagram closes. The construction of the force diagram at grid line B is the same.

Construction of the collector force diagram at grid line 2 (see Fig. 6.17): The direction of the diaphragm shears transferred into the strut is shown in the figure.

Starting at grid line A:

$$v_{\text{SW3, SW4}} = \frac{7400}{(10 + 12)} = +336.4 \text{ plf unit shear in SW3 and SW4}$$

$$v_{\text{net}} = +336.4 - 52.5 - 40 = +243.86 \text{ plf net shear at shear walls}$$

$$F = -(52.5 + 40)(38) = -3575 \text{ lb compression at the start of SW4}$$

$$F = -3575 + 243.86(10) = -1076.4 \text{ lb compression at end of SW4}$$

$$F = -1076.4 - (52.5 + 40)(2) = -1261.4 \text{ lb compression at start of SW3}$$

$$F = -1261.4 + 243.86(12) = +1664.96 \text{ lb tension at end of SW3}$$

$$F_{\text{B}} = +1664.96 - (40 + 52.5)(18) = 0 \text{ lb at grid line B}$$

Therefore, the diagram closes. ▲

6.7 Cantilever Diaphragm with Back Span—Boundary Member Forces (Fig. 6.18)

The prescriptive requirements for cantilever diaphragms with back spans were covered in 2008 edition of SDPWS Section 4.2.5.2 and as referenced in Figure 4A(b) of that publication. The 2015 and 2021 editions of SDPWS no longer proportion the cantilever back span length for cantilever diaphragms as "the length of the cantilever back span shall be less than or equal to 25 ft or two thirds of the width, whichever is less." The current edition has eliminated the "cantilever" diaphragm section, combining the two terms into "open-front" diaphragms only, even though the term cantilever is still used in the standard. The maximum cantilever length shall not exceed 35 ft. For loads applied parallel to the numbered grid lines, the diaphragm back span boundaries occur between grid lines A and D from 1 to 2. The cantilever portion boundaries are between grid lines 2 to 3 from A to D. For loads applied parallel to the lettered grid lines, the diaphragm is a simple span diaphragm, and the boundaries occur between grid lines A and D from 1 to 3. In accordance with code requirements, each boundary line must include boundary members consisting of drag struts or collectors, or other lateral resisting elements such as a shear walls or frames.

Example 6.3: Cantilever Diaphragm with Back Span—Boundary Member Forces

The diaphragm shown in Fig. 6.18 has an overall length of 30 ft and a width of 60 ft. An 18 ft, a 12 ft, and a 4 ft shear walls are located along grid line 1. A 12-ft shear wall is located along grid line 2, and 5-ft shear walls are located at grid lines A and D as shown in the figure. The location of the interior wall line creates a 20-ft back span with a 10-ft cantilever. The diaphragm will be assumed to be rigid. The boundary members along grid lines A and B act as standard chord members. A full-length collector is installed on each side of the interior support wall at grid line 2. The wall at grid line 3 has no stiffness. A uniform load of 200 plf is applied to the diaphragm as shown in the figure.

RDA Results (reference Fig. 6.18):

$$F_{v1} = 4435.1 \text{ lb direct shear at grid line 1}$$

$$F_{T1} = 760.6 \text{ lb rotational shear at grid line 1}$$

$$F_{T2} = 940.8 \text{ lb rotational shear at grid line 2}$$

$$F_{TA} = F_{TD} = 755.8 \text{ lb rotational shear at grid lines A and D}$$

Construction of the collector force diagram at grid line 1 (see Fig. 6.19):

$$v_{SW} = \frac{R_1}{\Sigma L_{SW}} = \frac{1500}{34} = +44.11 \text{ plf}$$

Line 1, $v_{net} = 4435.1 - 760.6 = 3674.4$ lb reaction at grid line 1

Line 2, $v_{net} = 1565.3 + 940.8 = 2506$ lb reaction at grid line 2

$$v_D = \frac{3674.4}{60} = 61.23 \text{ plf unit diaphragm shear}$$

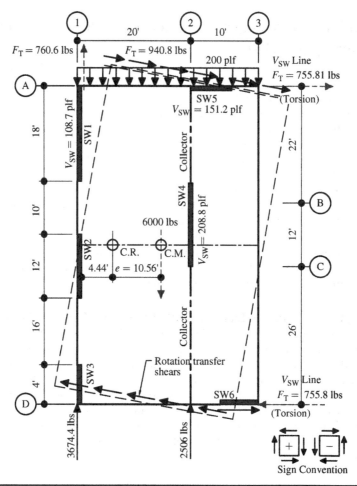

Figure 6.18 Cantilever diaphragm with back span.

$$v_{\text{sw}} = \frac{3674.4}{34} = 108.59 \text{ plf unit shear wall shear}$$

$$v_{\text{net}} = 108.59 - 61.23 = 47.36 \text{ plf net shear at shear wall}$$

Starting at grid line A:

$F = 47.36(18) = 852.5$ lb tension at the end of SW1

$F = 852.5 - 61.23(10) = +240.18$ lb tension at the start of SW2

$F = 240.18 + 47.36(12) = 808.5$ lb tension at the end of SW2

$F = 808.5 - 61.23(16) = -171.18$ lb compression at the start of SW3

$F = -171.18 + 47.36(4) = 0$ lb at grid line D

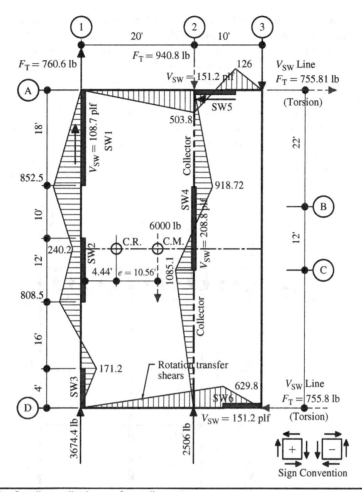

Figure 6.19   Cantilever diaphragm force diagrams.

Construction of the collector force diagram at grid line 2 (see Fig. 6.19):

Line 2, $v_{net} = 1565.3 + 940.8 = 2506$ lb reaction at grid line 2

$$v_D = \frac{2506}{60} = 41.76 \text{ plf unit diaphragm shear}$$

$$v_{SW} = \frac{2506}{12} = 208.83 \text{ plf unit shear wall shear}$$

$$v_{net} = 208.83 - 41.76 = 167.06 \text{ plf net shear at shear wall}$$

Starting at grid line A:

$F = -(41.76)(22) = -918.72$ lb compression at the start of SW4

$F = -918.72 + 167.07(12) = 1086.1$ lb tension at the end of SW4

$F = 1086.1 - 41.76(26) = 0$ lb at grid line D

Therefore, the diagram closes.

The resolution of the chord forces at grid lines A and D is done in a similar manner. Note that those force diagrams only reflect the rotational effects. Chord bending and the influence of the shear wall on the chord forces should be considered. ▲

As demonstrated in previous examples, it is not always practical or desirable to provide full length collectors at grid line 2 if the framing is oriented in a direction perpendicular to the shear wall. One possible solution is to install transfer diaphragms above and below shear wall SW4 as shown in Fig. 6.20. The transfer diaphragms act as propped cantilever beams, which are supported by the shear walls located at grid lines 1 and 2. Collectors are installed off each end of shear wall SW4 that embed the full depth of each transfer diaphragm. Transverse collectors are also installed at grid lines B, C, D, and E, which act as the transfer diaphragm chords. The diaphragm

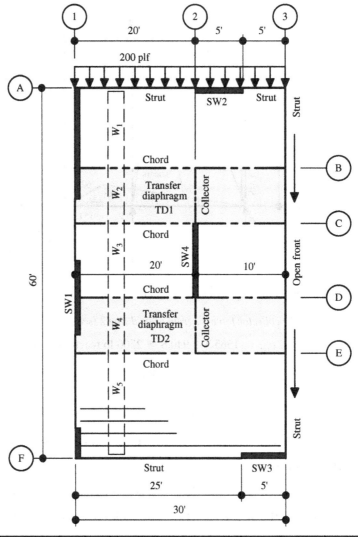

Figure 6.20 Cantilever diaphragm—partial length collectors at grid line 2.

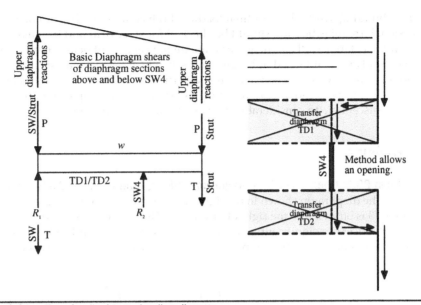

Figure 6.21   Transfer diaphragm loading diagram.

sections between grid lines A and B and between E and F can be assumed to act as individual simple span diaphragms that are supported at grid lines 1 and 3. The struts at grid line 3 transfer the reactions of the diaphragm sections as concentrated loads to the ends of the transfer diaphragm cantilevers. The uniform load applied to the diaphragm can be broken down into unit loads that are proportional to the diaphragm section widths as shown in the figure. The load diagram for each transfer diaphragm is shown in Fig. 6.21. The transfer diaphragm layout provides an opportunity to include an opening or recess in the diaphragm as shown at the right of the figure. For loading in the transverse direction, this would create a diaphragm with an intermediate offset, which was covered in Chap. 4, still providing a complete load path across the offset. An even simpler solution would be to locate a single transfer diaphragm at the shear wall between grid lines C and D as shown in Prob. P6.2.

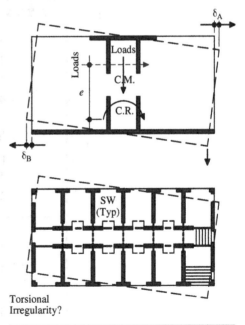

Figure 6.22   Plans with potential torsional irregularities.

The lower plan in Fig. 6.22 shows a floor plan layout typically used for multistory, multi-family structures. The upper floor plan would be representative of a typical office building. Under longitudinal loading, both could have Type 1a or 1b horizontal torsional irregularities and should be checked for story drift and irregularities.

If the exterior walls of the lower plan are not capable of being shear walls, then an open-front and cantilever diaphragm occurs, in which case an analysis similar to Sec. 6.3 should be done. Open-front and cantilever structures can be very complex and demand the utmost respect and attention from the designer. Even simple open-front structures can be susceptible to considerable damage if improperly designed.

6.8   Problems

Problem P6.1: Open-Front Diaphragm Both Sides, Interior Shear Walls Offset

Given: The diaphragm shown in the figure is the same as Example 6.2, except that shear wall SW3 is offset 2 ft to the right of grid line 2. This variation is presented to show that the wall offset can significantly affect the shear distribution and collector forces within the diaphragm. The diaphragm will be analyzed with the same wind loads as the

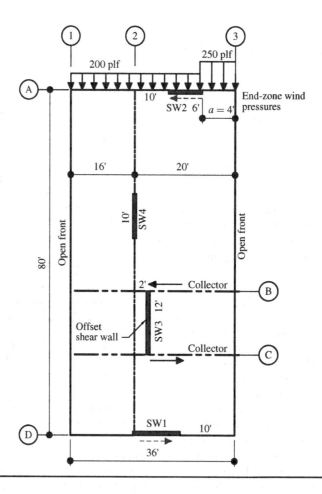

FIGURE P6.1

previous example. The shear walls along grid line 2 are assumed to act in the same line of resistance.

Find: The force diagrams for the collector at grid line 2 determine the strut forces at grid lines B and C.   ▲

Problem P6.2: Cantilever Diaphragm—No Collector at Interior Shear Wall

Given: The diaphragm shown has a more direct solution utilizing a single transfer diaphragm located at shear wall SW4. The diaphragm and shear wall configuration is the same as Fig. 6.20, with the exception of a single transfer diaphragm. A uniform load of 200 plf is applied to the diaphragm.

Find: All the collector, chord, and strut forces.   ▲

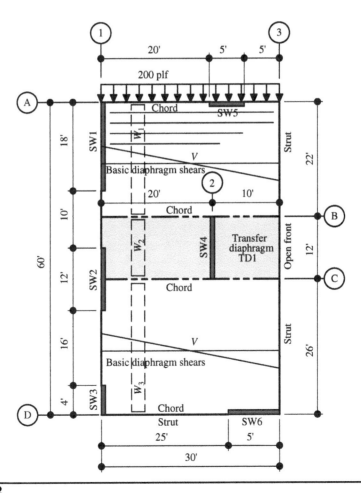

FIGURE **P6.2**

6.9   References

1. American Wood Council (AWC), *Special Design Provisions for Wind and Seismic with Commentary (SDPWS-21)*, AWC, Leesburg, VA, 2021.
2. American Society of Civil Engineers (ASCE), *ASCE/SEI 7-16 Minimum Design Loads for Buildings and Other Structures*, ASCE, New York, 2016.
3. International Code Council (ICC), *International Building Code, 2021 with Commentary*, ICC, Whittier, CA, 2021.
4. Malone, Breneman, *A Design Example of a Cantilever Wood Diaphragm*, Wood Products Council—WoodWorks, 2019.

CHAPTER 7

Diaphragms with Vertical Offsets

7.1 Introduction

Vertical offsets in the diaphragm often occur in floor and roof diaphragms. Whenever a full height interior shear wall or lateral resisting element cannot be installed at the location of the vertical offset, it will cause a discontinuity or vertical offset in the diaphragm web and chords, creating a Type 3 Horizontal irregularity, an abrupt discontinuity or change in stiffness in the diaphragm. This type of offset, in the authors' opinion, can create a questionable performing structure that requires considerable engineering judgment and 3D Finite Element Analysis (FEA) modeling. Very little has been written on this type of discontinuity. ATC-7 briefly addressed the effects of diaphragms with vertical offsets but did not provide any details on how to design for this type of irregularity.[1] A brief discussion and example for the analysis of vertical offset diaphragms can be found in a publication written by Edward F. Diekmann.[2] His example referenced cantilever diaphragm solutions supporting simple-span diaphragms with long cantilever diaphragm lengths, which was written before cantilever diaphragm length limitations were implemented. Earlier editions of SDPWS[3] allowed cantilever lengths more than 25 ft provided it could be shown that the drift at the ends of the diaphragms could meet allowable story drift and not pose instability problems. Since this would be difficult to achieve because special connections are required at the offset and the current edition of SDPWS does not allow cantilever lengths beyond 35 ft, a different solution is advised. These types of disruptions or discontinuities create very complicated load paths that require innovative solutions. Steps in roof and floor diaphragms are usually caused by the desire to create a clerestory wall with windows or other architectural features as shown in Figs. 7.1 and 7.2. The cross section in Fig. 7.2 shows the typical condition where an interior mezzanine might occur between grid lines 1 and 3. In this case, the roof shears at grid line 3 are not transferred down into the mezzanine below or down to the foundation. The figure shows that the magnitude of the uniform loads that are applied to the roof diaphragm sections will vary, partly because of the presence of the mezzanine or partly because of the difference in the wall heights. Assuming that the placement of a transverse interior shear wall is not possible at grid line 3, the challenges become:

- How to laterally support the edge of the upper and lower diaphragms along grid line 3.

231

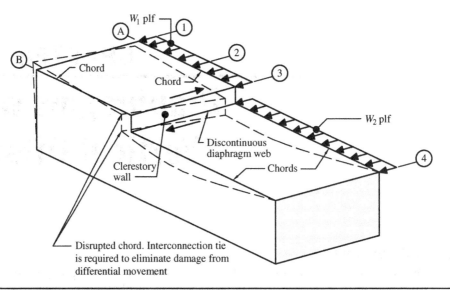

Figure 7.1   Isometric of vertically offset diaphragm.

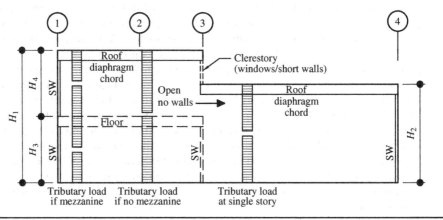

Figure 7.2   Section through vertically offset diaphragm.

- How to transfer the diaphragm shears across the vertical offset for transverse loads.
- How to effectively splice or transfer the disrupted diaphragm chords or struts across the vertical offset for transverse and longitudinal loading, respectively.
- How to limit the diaphragm deflection and drift to acceptable limits and prevent instability problems in the diaphragm supporting elements.
- How to calculate the deflection for this type of structure.
- How to avoid significant damage under higher seismic modes during a seismic event as shown in Fig. 7.3.

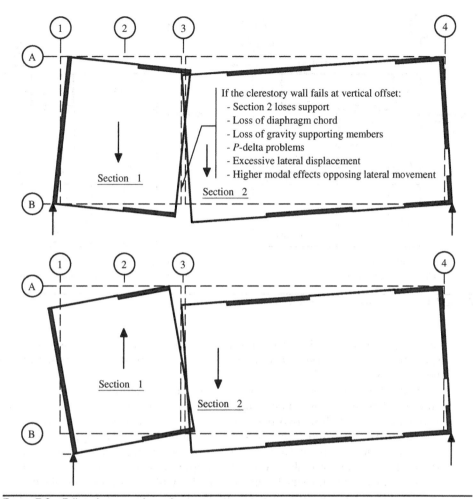

If the clerestory wall fails at vertical offset:
- Section 2 loses support
- Loss of diaphragm chord
- Loss of gravity supporting members
- P-delta problems
- Excessive lateral displacement
- Higher modal effects opposing lateral movement

FIGURE 7.3 Failure issues—dynamic responses.

If any of these design challenges cannot be resolved, the diaphragm will be subject to excessive damage or possible failure. Figure 7.3 shows a diaphragm with unequal sections on each side of the vertical offset. Although section 1 might comply with the requirements for an open front diaphragm, it would not be advisable to take that approach, because in the event of a failure at the vertical offset, section 2 of the diaphragm would lose support and the building would become unstable. Notice the warping at the edges located at grid lines 3A and 3B. Both diaphragm sections should act as a combined, single simple-span diaphragm spanning from grid lines 1 to 4, and continuity should be maintained across the discontinuous chords at the vertical offset.

7.2 Method of Analysis

The analysis of vertically offset diaphragms is greatly affected by the location of the vertical offset. As the vertical offset moves toward or away from the center of the diaphragm, the diaphragm will respond differently to the applied loads. The entire

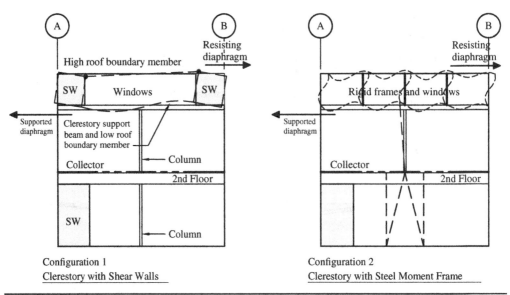

FIGURE 7.4 Shear transfer options at vertical offset.

diaphragm must be designed assuming that both sections act jointly as a single simple-span diaphragm. The possible dynamic forces and displacements applied to each section are shown in Fig. 7.3.

Shear transfer across the vertical offsets can be complicated. Architectural constraints often limit what can and cannot be done to the clerestory wall. Configurations 1 and 2 of Fig. 7.4 show two possible methods for transferring the shears when full height shear walls cannot be installed from the foundation to the upper roof diaphragm. In configuration 1, short shear wall sections are located at grid lines A and B in the clerestory. The discontinuous shear walls are supported by beams below. Interior columns can be installed to reduce the beam spans which will provide additional stiffness to the system, and reduce the drift caused by rotation of the shear panels. Since the shear panels are discontinuous at the supporting beams, the supporting beams and columns must be designed in accordance with ASCE 7-16[4] Section 12.3.3.3, which requires the application of an overstrength factor for any supporting element in SDCs B through F. The shear wall shown at the first-floor level provides lateral support for the mezzanine or second floor. In configuration 2, wall panels at the clerestory are not allowed due to architectural constraints. In that event, a steel rigid frame or braced frame could be used. The installation of interior columns and a lateral resisting element at the first floor accomplishes the same effects as that described for configuration 1. The deflection of the shear panels or frame must comply with allowable code drift limits to avoid damage to the framing and connections along the clerestory wall line. Regardless of what configuration is used, the clerestory framing should be significantly stiff enough to limit differential movement between the upper and lower roofs.

One of the reasons an open-front/simple-span diaphragm configuration should not be used is that section 2 would be supported at grid line 3 and assumed to act as a simple support. As such, theoretically, there would be no moment or chord force at grid line 3. The diaphragm chords of each section would then be assumed to act

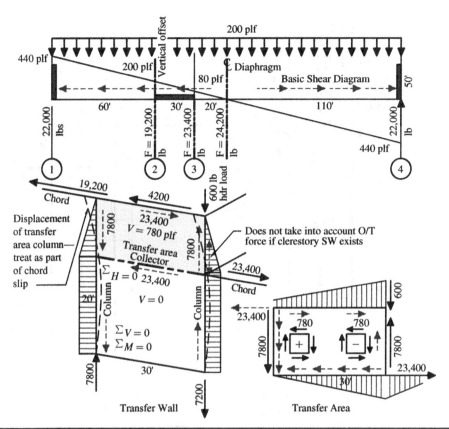

FIGURE 7.5   Transfer wall, member force diagrams, chord forces.

independently. As a minimum, ASCE 7-16 Section 12.1.3 requires that both sections of the diaphragm must be interconnected to prevent separation or tearing at the offset. Because chord continuity is required at the offset, there is no reason to use a cantilever diaphragm approach. Continuity in the diaphragm chords can be accomplished by installing full height vertical transfer walls between grid lines 2 and 3, at both grid lines A and B that extend from the foundation to the upper roof diaphragm with transfer areas located between the offset chords as shown in Figs. 7.5 and 7.6. The walls can be sheathed full height to act as shear walls for longitudinal loading or only in the transfer area to accomplish the chord splice.

The method of distributing the forces into the transfer wall and across the vertical offset shows that the wall area between the upper and lower roof acts more like a transfer area than a transfer diaphragm, where the eccentric chord forces causing rotation of that area are resisted by the columns placed at each end of the wall. Since the columns transfer their forces directly into the foundation, there is no shear in the wall section below the collector. The force at the top of that transfer wall is equal to the accumulated transfer shears into the diaphragm chord from grid lines 1 to 3. The direction of the diaphragm shears being transferred into the chord show that the member is in tension. The chord force applied at the top of the wall is equal to the chord force at the low roof at the vertical offset.

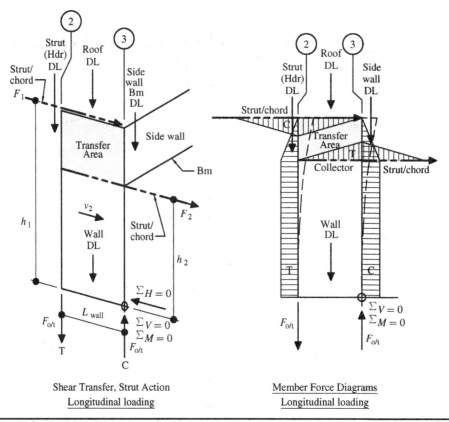

FIGURE 7.6 Transfer wall, member force diagrams, strut forces.

Figure 7.5 shows an elevation of the wall and transfer area under transverse loading (chord action).

$$v = \frac{23,400}{30} = 780 \text{ plf, unit wall shear between the upper and lower chords}$$

$$v_{\text{Rotation}} = \frac{23,400(10)}{30} = 7800 \text{ lb, rotational resisting force at transfer area}$$

$$F_{o/t} = \frac{23,400(30 - 20)}{30} = 7800 \text{ lb, compression force at grid line 2}$$

Reaction from clerestory support beam = 600 lb

$$T = 7800 - 600 = 7200 \text{ lb tension force at grid line 3}$$

This shows that the overturning force does not increase below the collector, but is transferred directly through the column down to the foundation. Also, note that the shear in the sheathing below the collector is zero; therefore, no additional force is applied to the low roof collector, showing that the transfer wall acts like a transfer area and not a transfer diaphragm.

All the applicable gravity loads should be included in the calculation of the overturning forces with the appropriate load factors and load combinations outlined in ASCE 7-16 Section 2.3 or 2.4. The chord and boundary member force diagrams are plotted using the method described in Chap. 3 and are shown in Figs. 7.5 and 7.6.

Every component and connection along the vertical offset undergo displacements in multiple directions. A finite element analysis was performed on a similar structure to verify the interaction of the two roof sections. The displaced shape of the diaphragm sections was identical to that of Fig. 7.7, and revealed the areas where special attention should be given. The computer output showed several interesting results. The first is the warping of the clerestory wall, which confirms that significant displacements are occurring along that line. All the members and connections along this line should be designed and detailed to account for this movement. Because of the critical nature of the discontinuity of the chords and that the wall is a transfer wall, the discontinuous chord force should be amplified by the overstrength factor. The transfer wall vertical

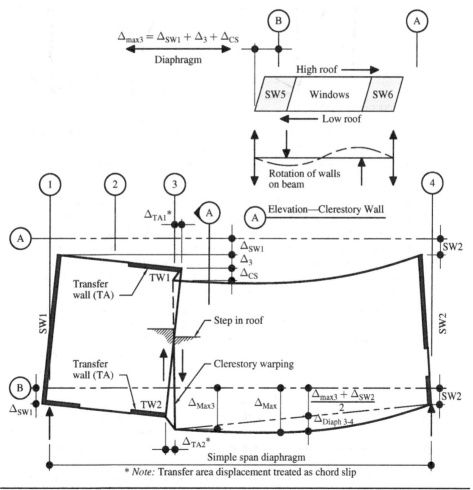

$$\Delta_{max3} = \Delta_{SW1} + \Delta_3 + \Delta_{CS}$$

Diaphragm

High roof →

SW5 — Windows — SW6

← Low roof

Rotation of walls on beam

Elevation—Clerestory Wall

$\Delta_{TA1}*$

Δ_{SW1}
Δ_3
Δ_{CS}

Transfer wall (TA)

TW1

Step in roof

Transfer wall (TA)

Clerestory warping

TW2

Δ_{Max3} — Δ_{Max}

$$\frac{\Delta_{max3} + \Delta_{SW2}}{2}$$

$\Delta_{Diaph\ 3-4}$

Δ_{SW1}

$\Delta_{TA2}*$

Simple span diaphragm

* *Note:* Transfer area displacement treated as chord slip

FIGURE 7.7 Deflected shape.

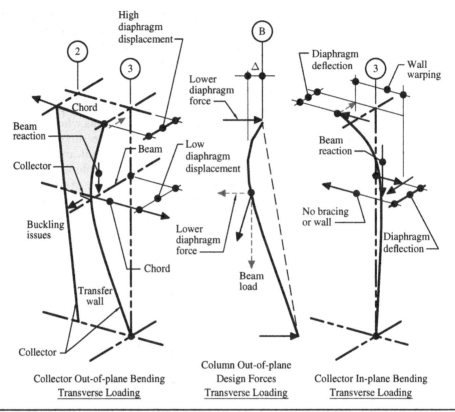

FIGURE 7.8 Transfer wall displacements.

boundary members (vertical columns/collectors) can undergo in-plane and out-of-plane bending due to the diaphragm displacements along grid line 3 as shown in Fig. 7.8. The out-of-plane displacement of the vertical column/collector at the low roof level can be significant. In lieu of a transfer wall, a single column could be used which could also include the clerestory rigid frame option. The full height column would resist bending in both directions and would have an additional deformation at the clerestory height. This would contribute to the bi-axial loading. The use of a steel column can be considered; however, doing so can be challenging and could create a high demand on a single element, reducing redundancy. Figures 7.6 to 7.8 show that solving the discontinuity issues across the vertical offset is not only challenging but can become very involved and time consuming. Deflection and story drift at the offset are likely to control most of the diaphragm design and components. The total deflection along grid line 3 shown in Fig. 7.7 is equal to the combined bending, shear deflection, chord slip, and nail slip contribution of both sections of the diaphragm plus the deflection of the clerestory shear walls caused by the opposing upper and lower roof forces and their rotation caused by the flexible supporting beam under bending. The standard diaphragm deflection equations cannot be used because of the discontinuities at the offset.

The deflection along grid line 3 must include the effects of chord slip along the upper and lower boundary members at the vertical offset and contribution of transfer area deflection/rotation (if used). The maximum diaphragm deflection may not occur at grid line 3. The contribution of the longer diaphragm mid-span deflection must also be considered. The total displacements at grid line 3 and at mid-span of diaphragm section 2 as shown in Fig. 7.7 are critical locations to check for compliance with allowable story drifts and the ability of the main lateral transfer and gravity load supporting column at grid line 3 to perform under the deformations and potential local P-delta effects.

$$\delta_x = \frac{C_d \delta_{xe}}{I_e} \leq \text{allowable story drift, ASCE 7-16 Table 12.12-1}$$

$$\theta = \frac{P_x \Delta I_e}{V_x h_{sx} C_d} \leq 0.10, \ P\text{-delta effects, ASCE 7-16 Section 12.8.7}$$

For loading in the longitudinal direction (strut action), the free body of the wall is shown in Fig. 7.7.

7.3   Single Vertical Offset in Diaphragm

The building shown in Fig. 7.9 is a classic example of a vertically offset diaphragm with sloped roof sections. From observations, the load path to transfer lateral loads from the low roof through the clerestory back into rest of the structure is not apparent. Perhaps rigid columns were used as transfer elements in the construction; however, the basic building configuration indicates that a complete load path may be lacking. The structure is located in SDC D. The front entry wall is solid windows and door and there are no interior shear walls below the clerestory wall. The clerestory wall is composed of all windows, no shear panels, or apparent rigid frames. The chance that this structure will survive a moderate seismic event without significant damage is questionable.

Example 7.1: Vertical Offset with Sloped Roof

The structure shown in Fig. 7.10 is 150 ft long by 50 ft wide with a double sloped roof. The high roof slope is a 3:12 roof pitch and the low roof is a 4:12 roof pitch. The roof is offset 6 ft vertically at grid line 3. The clerestory wall is supported by two 5-1/8″ by 24″

Figure 7.9   Photo vertically offset diaphragm with sloped roof.

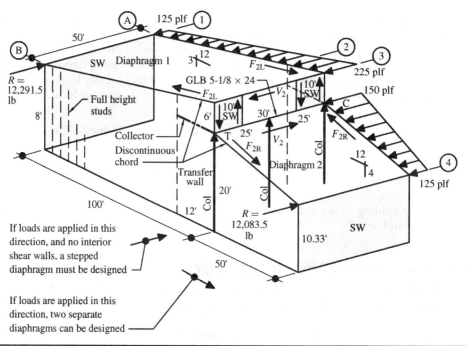

FIGURE 7.10   Isometric of vertically offset sloped roof.

by 25′-0″ glu-lam beams. The beams are supported at each exterior wall and at mid-width of the diaphragm by wood or steel columns. The diaphragm shears from the high roof are transferred across the vertical offset by two 10-ft-long shear walls (panels). Transfer walls are placed between grid lines 2 and 3 along lines A and B. The diaphragm chords of the lower roof are extended into these transfer walls with collectors. The exterior walls are constructed with full height studs, which apply a linear varying load at the sloped roof. Assume that calculated wind loads applied to the high roof vary linearly from 125 plf to 225 plf, and the low roof loads vary linearly from 125 plf to 150 plf. The diaphragm must be designed as a simple-span diaphragm. The shears and moments at the critical locations can be determined by a computer analysis or by hand as follows, referencing Fig. 7.11:

$$R_1 = V_1 = \left[0.625(33.33) + 5(83.33) + \frac{0.125(150)^2}{2} \right] \frac{1}{150} = 12.3 \text{ k left reaction}$$

$$v_1 = \frac{12{,}300}{50} = +246 \text{ plf diaphragm unit shear at grid line 1}$$

$$R_4 = V_4 = \left[5(66.67) + 0.625(116.67) + \frac{0.125(150)^2}{2} \right] \frac{1}{150} = 12.1 \text{ k right reaction}$$

Verify maximum moment location by free body 1:

$$V = 0 \text{ at, } 0 = 12.3 - 0.125x - \frac{0.001x^2}{2}, \; x = 76'$$

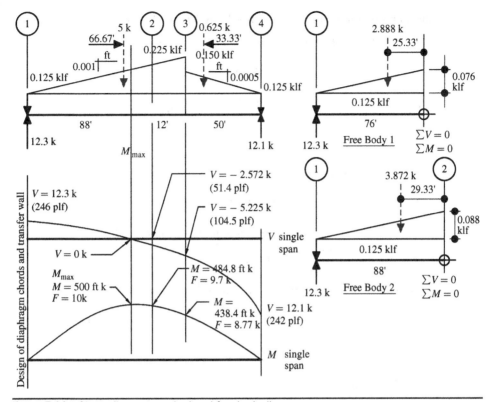

FIGURE 7.11 Shear diagrams and related free-body diagrams.

$$V_{76'} = 12.3 - 0.125(76) - 2.888 = 0 \text{ k, verified}$$

$$M_{max} = 12.3(76) - \frac{0.125(76)^2}{2} - 2.888(25.33) = 500 \text{ ft-k} \quad \therefore \text{ OK}$$

$$F_{max} = \frac{500}{50} = 10 \text{ k}$$

Shear and moment at grid line 2, free body 2 (max. V_2, M_2):

$$V_2 = 12.3 - 0.125(88) - 3.872 = -2.572 \text{ k}$$

$$v_2 = \frac{2572}{50} = -51.4 \text{ plf}$$

$$M_2 = 12.3(88) - \frac{0.125(88)^2}{2} - 3.872(29.33) = 484.8 \text{ ft-k}$$

$$F_2 = \frac{484.8}{50} = 9.696 \text{ k}$$

Shear and moment at grid line 3, free body 3 (max. V_3, M_3):

$$V_3 = 12.1 - 0.125(50) - 0.625 = -5.225 \text{ k}$$

$$v_3 = \frac{-5225}{50} = -104.5 \text{ plf}$$

$$M_3 = 12.1(50) - \frac{0.125(50)^2}{2} - 0.625(16.67) = 438.4 \text{ ft-k}$$

$$F_{chord} = \frac{438.4}{50} = 8.768 \text{ k}$$

Clerestory and diaphragm shears:

$$v_3 = \frac{5225}{50} = 104.5 \text{ plf, diaphragm shear at grid line 3}$$

$$V_1 = 12.3 \text{ k, controls diaphragm shear at grid line 1}$$

$$v_1 = \frac{12,300}{50} = 246 \text{ plf}$$

Shear distribution into clerestory shear walls at grid line 3 (see Fig. 7.12):

$$V_{SW\ CS} = \frac{5.225}{2} = 2.613 \text{ k each shear wall at clerestory}$$

$$v_{SW\ CS} = \frac{2613}{10} = 261.3 \text{ plf}$$

Right shear wall:

$$F_A = [-2.613(6) + 2.63(5)]\frac{1}{10} = 0.387 \text{ k tension}$$

$$F_{A-10'} = [2.613(6) + 2.63(5) + 3(10)]\frac{1}{10} = 5.748 \text{ k compression}$$

Left shear wall:

$$F_{B+10'} = [-2.613(6) + 2.63(5) + 3(10)]\frac{1}{10} = 2.612 \text{ k compression}$$

$$F_B = [2.613(6) + 2.63(5)]\frac{1}{10} = 2.748 \text{ k compression}$$

The overturning force of the clerestory walls at grid lines A and B is transferred directly into the transfer wall boundary member by nailing. A 200-plf dead load is applied at the upper and lower roof levels. The hold-down connections for these walls should be determined using the controlling load combination. The resulting dead loads of the upper roof plus wall and header load to the walls are shown in the figure. The shear and moment diagrams for the supporting beams are determined as follows:

Beam A.5 to B:

$$R_B = \left|2.612(15) + \frac{0.2(25)^2}{2}\right|\frac{1}{25} = 4.07 \text{ k left reaction}$$

$$R_{A.5} = \left|2.612(10) + \frac{0.2(25)^2}{2}\right|\frac{1}{25} = 3.54 \text{ k right reaction}$$

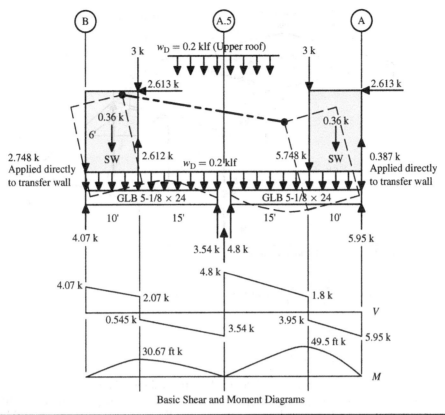

Basic Shear and Moment Diagrams

Figure 7.12 Design of clerestory walls and supporting beams.

Beam A to A.5:

$$R_{A.5} = \left[5.748(10) + \frac{0.2(25)^2}{2} \right] \frac{1}{25} = 4.8 \text{ k left reaction}$$

$$R_A = \left[5.748(15) + \frac{0.2(25)^2}{2} \right] \frac{1}{25} = 5.95 \text{ k right reaction}$$

The shear and moment diagrams are shown in the figure. The displacements of the diaphragm at grid line 3 are shown in Fig. 7.13, as previously discussed, and shown in Fig. 7.7. A complete example for the determination of the deflection produced by shear wall rotation on a supporting glue-lam beam can be referenced in SEAOC's *Seismic Design Manual, Vol. 2.*[5] Figure 7.14 summarizes the method outlined in that manual for a single wall located at the end of a supporting beam.

$$\Delta_v = \frac{Pa^2b^2}{3EIL}, \text{ where } P = F_{o/t}, \text{ deflection at any point } x \text{ due to concentrated load}$$

$$F_{o/t} = \frac{Vh}{b} \text{ overturning force, neglecting gravity loads}$$

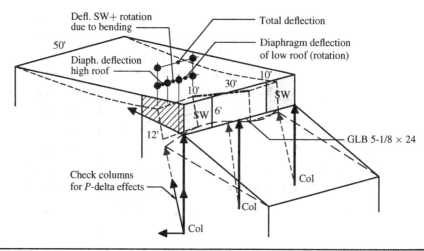

FIGURE 7.13   Displaced shape at offset.

$$\tan \theta = \frac{\Delta_v}{b} = \frac{\Delta_{\text{rot}}}{h} \text{ from similar triangles}$$

$$\Delta_{\text{rot}} = \frac{h\Delta_v}{b}$$

$$\Sigma\Delta_{\text{wall}} = \Delta_{\text{rot}} + \Delta_{\text{SW}}$$

Figure 7.14 also shows the displaced shape of the wall and its support beam. When multiple walls, multiple loading conditions, and/or wall placement complicate the analysis, a more direct method of determining the rotation deflection would be to perform a computer analysis.

It is necessary to resolve the forces applied to the transfer wall as shown in Fig. 7.15. The diaphragm chord forces (horizontal force component previously calculated) at grid lines 2 and 3 are broken down into vertical, horizontal, and resultant force components as shown in the figure. The unit diaphragm shears at the transfer wall are shown in the plan view in the figure. The vertical and horizontal components of the resultant shears applied along the wall from the roof diaphragm are also shown in the figure and are equal to

$$V_{\text{H}} = \frac{(51.4 + 104.5)(12)}{2} = 0.935 \text{ k, horizontal component}$$

$$V_{\text{V}} = \frac{(0.935)(3)}{12} = 0.234 \text{ k, vertical component}$$

The resultant component forces are placed at the center of the top plate. Summing horizontal forces along the upper diaphragm chord, the resulting force at grid line 3 is equal to

$$F_3 = 9.696 - \frac{(51.4 + 104.5)(12)}{2} = 8.761 \text{ k} \cong 8.768 \text{ k} \qquad \therefore \text{ OK}$$

$$v_{\text{wall}} = \frac{8.768}{12} = 0.7307 \text{ klf at area between the upper and lower chords}$$

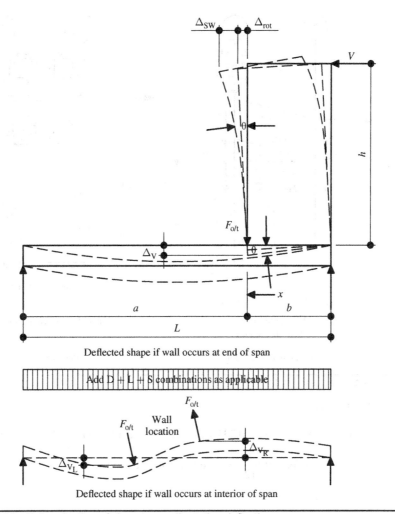

FIGURE 7.14 Determination of clerestory wall rotation deflection.

By adding the shears transferred into the upper chord from the diaphragm and shear wall, the upper chord force will close to zero at grid line 3.

$$\sum V_3 = 2.748 + 4.07 + 2.923 = 9.741 \text{ k, sum of applied loads at grid line 3}$$

$$\sum M_2 = 0$$

$$F_3 = [-9.696(23) - 0.234(6) + 0.18(8) + 2.4(6) + 9.741(12) + 0.935(24.5)$$

$$+ 8.768(20)]\frac{1}{12} = 8.881 \text{ k compression}$$

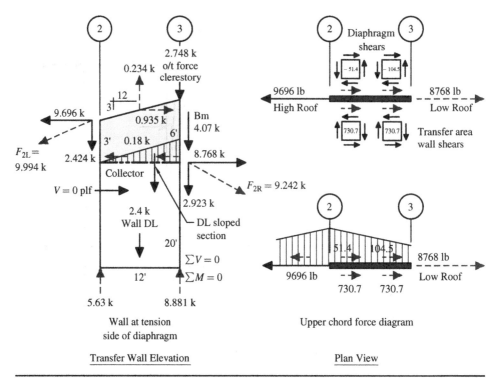

FIGURE 7.15 Transfer wall forces.

$$\Sigma M_3 = 0$$

$$F_2 = \left[-9.696(23) + 0.234(6) - 0.18(4) - 2.424(12) - 2.4(6) + 0.935(24.5)\right.$$

$$\left. + 8.768(20)\right]\frac{1}{12} = -5.629 \text{ k compression}$$

$$\Sigma V = 0$$

$$V = 5.629 + 8.881 + 0.234 - 2.424 - 0.18 - 2.4 - 2.748 - 4.07 - 2.923 = 0$$

Therefore, OK. Hold-down connections are not required; however, minimal strap connections should be installed.

The diaphragm nailing should be based on the maximum unit shear calculated and shown in Fig. 7.11. Consideration should be given to stiffening up the supporting beams and clerestory walls by incorporating a sill section between the walls, which will reduce the deflection taking place along the vertical offset. A possibility of modifying the clerestory wall is shown in Fig. 7.16. ▲

7.4 Multiple Vertical Offsets in Diaphragm

It is important to have a good understanding of how multi-leveled vertically offset diaphragm structures respond to seismic forces. Computer software programs currently available to the engineer provide the ability to easily perform advanced methods of analysis, which will show the displaced shape of the structure, providing a key to areas

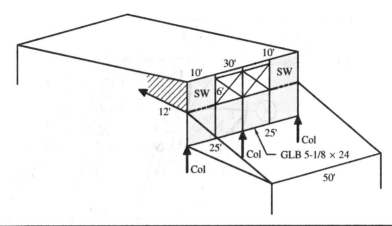

FIGURE 7.16 Alternate stiffened wall at offset.

that require special attention. Most complex structures are modeled using a 3D computer analysis program showing the complete response of the structure. Occasionally, portions of complex structures are analyzed by using a plane frame program along each line of resistance to avoid the time required to model a 3D structure. The problem of using a plane frame analysis for this type of structure is that the analysis does not capture the response of the structure as a whole during higher seismic modes.

Figure 7.17 is a photograph of an actual project that has low roof diaphragms on each side of a higher vertically offset section. The entire lateral resisting system consists of moment resisting frames. The results of a modal analysis for the structure shown in Fig. 7.18 demonstrate that higher modes can impose additional forces into the structural frames than what a plane frame analysis can capture.

FIGURE 7.17 Photo vertically offset diaphragms—both sides.

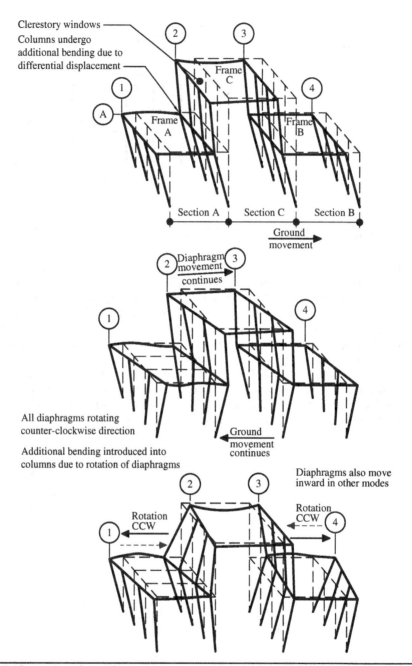

FIGURE 7.18 Mode 1, 2, and 3—mode shapes.

Example 7.2: Vertical Offset Diaphragm-Raised Center Section

The structure shown in Fig. 7.19 has a vertically offset section located at the interior of the diaphragm, horizontally offset to the left from the centerline of the structure as shown in Fig. 7.20. The configuration allows for a clerestory on each side of the pop-up section. The structure is 280 ft long by 80 ft wide. Transverse frames are not allowed at

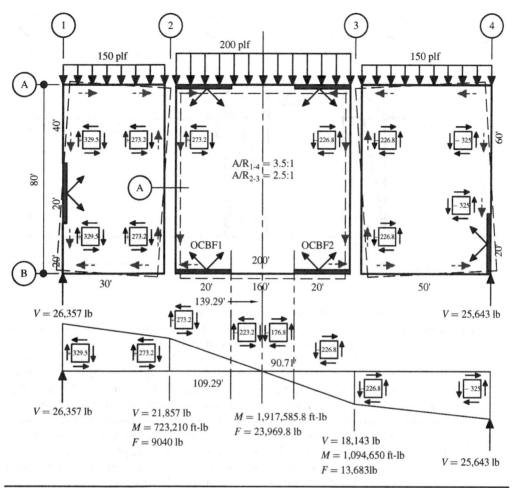

FIGURE 7.19 Vertically offset diaphragm—raised center section.

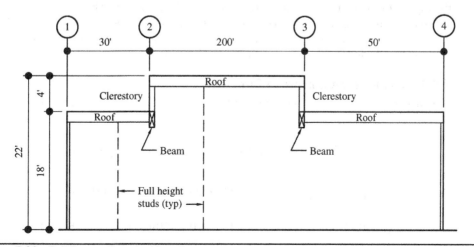

FIGURE 7.20 Building section at mid-section.

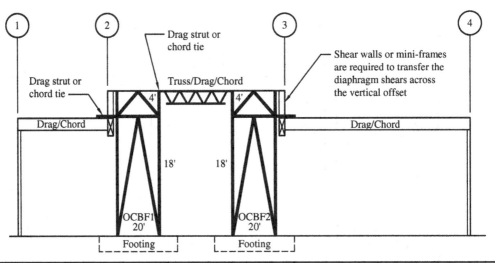

FIGURE 7.21 Building elevation at side walls.

grid lines 2 or 3 due to architectural constraints. The exterior walls are constructed with full height studs, which produce higher uniform loads to the roof diaphragm at the center section than at the sections with lower walls. Figure 7.21 shows ordinary concentric braced frames (OCBF) oriented in the longitudinal direction between grid lines 2 and 3, which also serve as transfer frames for the vertically offset diaphragm strut/chords. The uniform loads applied to the diaphragm are shown in Figs. 7.19 and 7.22. Chord and drag strut configurations and locations can also be seen in those figures. The low roof sections do not have longitudinal frames in either of the low roof sections and must rely on the braced frames in the center section for stability in the longitudinal direction. The diaphragm must be designed as a single-span diaphragm spanning from grid lines 1 to 4 in the transverse direction and from A to B in the longitudinal direction, with multiple discontinuities in the chords and struts. The low roof drags or chord elements can consist of wood or steel beams. The high roof drags or chord members are the top beams of the frames plus a preengineered steel truss, steel beam, or wood beam. The maximum drag or chord force if applied to a truss must be called out on the drawings for the truss manufacturer. Note that the following calculations do not reflect the overstrength factor for collectors, which must be added.

7.4.1 Loads in the Transverse Direction

Construction of the shear diagram can be seen in Fig. 7.22, which is determined as follows:

$$R_1 = [0.15(50)(25) + 0.2(200)(150) + 0.15(30)(265)]\frac{1}{280} = 26.357 \text{ k}$$

$$v_1 = \frac{R_1}{W} = \frac{26,357}{80} = +329.5 \text{ plf diaphragm shear at grid line 1}$$

$$R_4 = [0.15(30)(15) + 0.2(200)(130) + 0.15(50)(255)]\frac{1}{280} = 25.643 \text{ k}$$

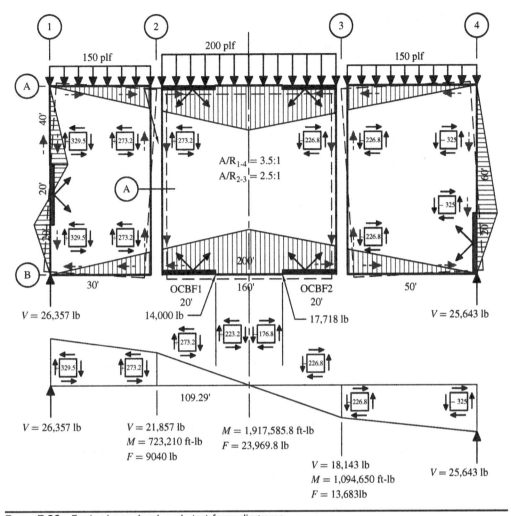

FIGURE 7.22 Basic shear, chord, and strut force diagrams.

$$v_4 = \frac{R_4}{W} = \frac{25,643}{80} = -325 \text{ plf diaphragm shear at grid line 4}$$

$$V_2 = R_1 - wx = 26.375 - 0.15(30) = 21.857 \text{ k shear at grid line 2}$$

$$v_2 = \frac{V_2}{W} = \frac{21,857}{80} = +273.2 \text{ plf diaphragm shear at grid line 2}$$

$$V = 0 \text{ at } 139.29 \text{ ft to the right of grid line 1}$$

$$V_3 = V_2 - wx = 21.857 - 0.2(200) = -18.143 \text{ k shear at grid line 3}$$

$$v_3 = \frac{V_3}{W} = \frac{-18,143}{80} = -226.8 \text{ plf diaphragm shear at grid line 3}$$

Chord forces (see Fig. 7.22):

$$M_2 = 26.357(30) - \frac{0.15(30)^2}{2} = 723.21 \text{ ft-k at grid line 2}$$

$$F_2 = \frac{M_2}{W} = \frac{723.21}{80} = 9.04 \text{ k chord force at grid line 3}$$

$$M_{max} = 26.357(139.29) - 0.15(30)(124.29) - \frac{0.2(109.29)^2}{2} = 1917.59 \text{ ft-k}$$

$$F_{max} = \frac{M_{max}}{W} = \frac{1917.59}{80} = 23.97 \text{ k maximum chord force}$$

$$M_3 = 25.643(50) - \frac{0.15(50)^2}{2} = 1094.65 \text{ ft-k at grid line 3}$$

$$F_3 = \frac{M_3}{W} = \frac{1094.65}{80} = 13.683 \text{ k chord force at grid line 3}$$

Transverse strut force diagrams at grid line 1 (see Fig. 7.22):

$$v_{frame} = \frac{26,357}{20} = 1318 \text{ plf unit shear along the frame}$$

$$v_{net} = v_{frame} - v_{diaph} = 1318 - 329.5 = 988.4 \text{ plf net shear along the frame}$$

Starting at grid line B:

$$F = -0.3295(20) = -6.59 \text{ k tension at the start of the frame}$$

$$F = -6.59 + 0.9884(20) = +13.18 \text{ k compression at the end of the frame}$$

$$F = 13.18 - 0.3295(40) = 0 \text{ k at grid line A}$$

Therefore, the diagram closes.

Transverse strut force diagrams at grid line 4:

$$v_{frame} = \frac{25,643}{20} = +1282 \text{ plf unit shear along the frame}$$

$$v_{net} = v_{frame} - v_{diaph} = 1282 - 325 = 957.2 \text{ plf net shear along the frame}$$

Starting at grid line B:

$$F = +0.9572(20) = +19.14 \text{ k compression at the end of the frame}$$

$$F = 19.14 - 0.325(60) = 0 \text{ k at grid line A}$$

Therefore, the diagram closes.

Chord forces diagrams at frames at grid line B (see Fig. 7.22):

$$v_{frame} = \frac{V_2 - wL_{frame}}{W} = \frac{21,857 - 200(20)}{80} = +223.2 \text{ plf unit shear at end of OCBF1}$$

$$v_{frame} = \frac{V_2 - wx}{W} = \frac{21,857 - 200(180)}{80} = -176 \text{ plf unit shear at start of OCBF2}$$

Starting at grid line 1, taking the area of shear diagram:

$$F_2 = \frac{(329.5 + 273.2)}{2}(30) = 9040 \text{ lb tension at grid line 2}$$

$$F = 9040 + \frac{(273.2 + 223.2)}{2}(20) = 14,000 \text{ lb tension at the end of the frame}$$

$$F_{max} = 14,000 + \frac{(223.2 + 0)}{2}(109.29 - 20) = 23,969 \text{ lb tension, maximum}$$

$$F = 23,969 - \frac{(176.8 + 0)}{2}(180 - 109.29) = 17,718 \text{ lb tension, at the start of OCBF2}$$

$$F_3 = \frac{(325 + 226.8)}{2}(50) = 13,683 \text{ lb tension at grid line 3}$$

The chord forces along grid line A are the same magnitude but act in compression.

The resultant hold-down forces of braced frames OCBF1 and OCBF2 are shown in Fig. 7.23, neglecting gravity loads.

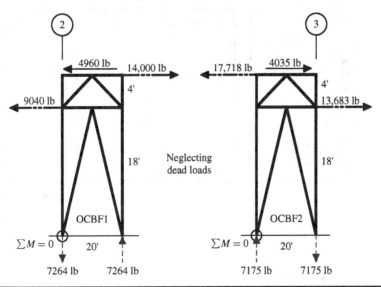

FIGURE 7.23 Transfer frame forces—chord forces.

7.4.2 Loading in the Longitudinal Direction (see Fig. 7.24)

Assume that the calculated seismic unit strip load is equal to 560 plf. The uniform load per section is distributed in accordance with the width of the sections.

$$w_{1-2} = \frac{wW_{1-2}}{W} = \frac{560(30)}{280} = 60 \text{ plf}$$

$$w_{2-3} = \frac{wW_{2-3}}{W} = \frac{560(200)}{280} = 400 \text{ plf}$$

$$w_{3-4} = \frac{wW_{3-4}}{W} = \frac{560(50)}{280} = 100 \text{ plf}$$

For wind loading in the longitudinal direction, the wind loading should be applied in accordance with Chapters 26 through 30 of ASCE 7-16. The magnitude of the unit load to the section between grid lines 1 and 2 will be much higher than the other sections because of the magnitude of the windward pressures. The combined windward and leeward wind load to the high roof will be small because of the small vertical offset, and the wind load to the section between grid lines 3 and 4 will be that equal to the leeward pressures. Each diaphragm section acts as an individual diaphragm in this direction due to the vertical offset.

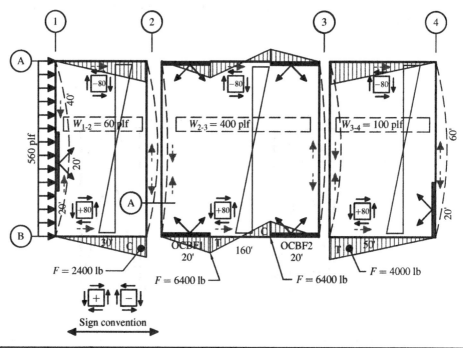

Figure 7.24 Longitudinal strut force diagrams.

Diaphragm section between grid lines 1 and 2 (see Fig. 7.24):

$$V_A = V_B = \frac{wL}{2} = \frac{60(80)}{2} = 2400 \text{ lb reactions at grid lines A and B}$$

$$V_{A-B} = \frac{V_A}{W_{1-2}} = \frac{2400}{30} = 80 \text{ plf diaphragm unit shear}$$

Since there are no shear walls or frames between grid lines 1 and 2, the full strut force is transferred into frame OCBF1.

Diaphragm section between grid lines 2 and 3:

$$V_A = V_B = \frac{wL}{2} = \frac{400(80)}{2} = 16,000 \text{ lb reactions at grid lines A and B}$$

$$V_{A-B} = \frac{V_A}{W_{2-3}} = \frac{16,000}{200} = 80 \text{ plf diaphragm unit shear}$$

$$V_{frames} = \frac{16,000}{2(20)} = 400 \text{ plf unit shear acting on frames}$$

$$V_{net} = v_{frame} - v_{diaph} = 400 - 80 = 320 \text{ plf net shear at frames}$$

$$F = 320(20) = 6400 \text{ lb tension at the end of OCBF1}$$

$$F = 6400 - 80(160) = -6400 \text{ lb compression at the start of OCBF2}$$

By inspection, the force at grid line 3 is zero.

Diaphragm section between grid lines 3 and 4:

$$V_{A-B} = \frac{wL}{2} = \frac{100(80)}{2} = 4000 \text{ lb reaction at grid lines A and B}$$

$$v_{A-B} = \frac{V_A}{W_{3-4}} = \frac{4000}{50} = 80 \text{ plf diaphragm unit shear}$$

$$F = 80(50) = 4000 \text{ lb at the end of OCBF2}$$

Since there are no shear walls or frames between grid lines 3 and 4, the full strut force is transferred into frame OCBF2. The resultant forces for braced frames OCBF1 and OCBF2 from chord are shown in Fig. 7.23, neglecting gravity loads. The resultant forces for braced frames OCBF1 and OCBF2 from strut forces are shown in Fig. 7.25, neglecting gravity loads. Comparing the reactions at the base of the frames with the reactions from the transverse loading, loading in the longitudinal direction controls the design of the frames and footings. ▲

7.5 Alternate Configurations

The longitudinal walls along grid lines A and B can be of any configuration as suggested in Figs. 7.26 and 7.27. The configurations shown, which represent structures commonly encountered, can have shear wall, braced frames, or cantilever steel columns

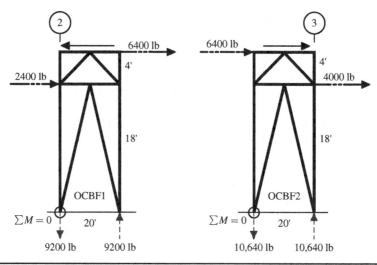

FIGURE 7.25 Transfer frame forces—strut forces.

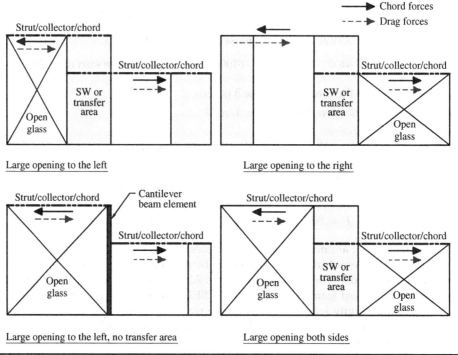

FIGURE 7.26 Alternate side wall configurations.

as the transfer elements. There is no redundancy in any of the configurations shown, which would suggest that a reasonable factor of safety should be used in the design. The tools necessary for analyzing these structures have already been presented. Special attention should be given to how the individual sections of these structures will respond to and affect each other, as a whole. The design objectives should include verification of

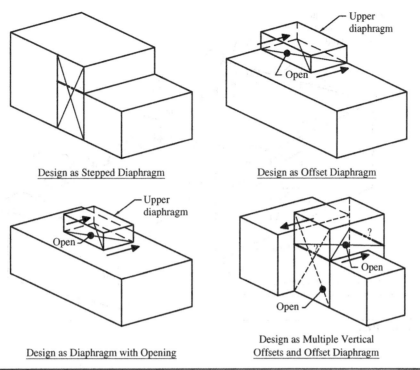

Design as Stepped Diaphragm   Design as Offset Diaphragm

Design as Diaphragm with Opening

Design as Multiple Vertical
Offsets and Offset Diaphragm

FIGURE 7.27   Alternate configurations.

complete load paths, a built-in redundancy, drift, possible irregularities, and an assurance that the sections will function as a unit.

7.6   Problem

Problem P7.1: Vertical Offset with Flat Roof

Given: The structure shown in Fig. P7.1 is 320 ft long by 80 ft wide with a flat roof and an aspect ratio of the diaphragm of 4:1. The roof is offset 4 ft vertically at grid line 3. The diaphragm is treated as a simple-span diaphragm from grid lines 1 to 4 and the shears from the low roof are transferred across the vertical offset to the high roof by a continuously sheathed clerestory wall. The clerestory wall at line 3 is supported by steel beams and columns. Transfer walls are located between grid lines 2 and 3 at lines A and B. The discontinuous diaphragm chords of the lower roof are extended into the transfer walls for continuity. The exterior walls are constructed with full height studs, which apply a larger uniform wind force to the high roof than to the low roof, assuming there is no interior mezzanine in the higher area. The calculated factored wind load applied to the high roof is 200 plf, and the uniform load to the low roof is 150 plf. Shear walls SW2 and SW3 are used as transfer walls (transfer diaphragms) to connect the diaphragm chords across the vertical offset.

Find: Design the transfer wall on the tension side of the building, including shears and collector/chord and overturning forces.   ▲

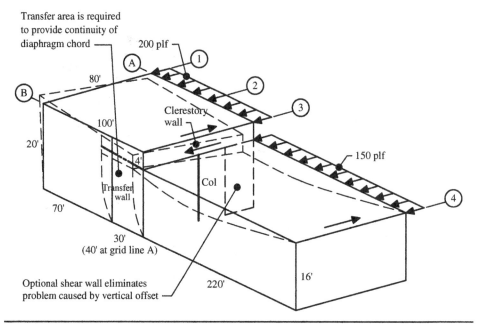

Transfer area is required
to provide continuity of
diaphragm chord

200 plf

A

B

80'

100'

20'

Clerestory
wall

150 plf

4'

Transfer
wall

Col

70'

30'
(40' at grid line A)

16'

Optional shear wall eliminates
problem caused by vertical offset

220'

FIGURE P7.1

7.7   References

1. Applied Technology Council (ATC), *Guidelines for Design of Horizontal Wood Diaphragms*, ATC-7, ATC, Redwood, CA, 1981.
2. Diekmann, E. F., "Design of Wood Diaphragms," *Journal of Materials Education*, 1982, Fourth Clark C. Heritage Memorial Workshop, Wood Engineering Design Concepts University of Wisconsin, WI.
3. American Wood Council (AWC), *Special Design Provisions for Wind and Seismic with Commentary (SDPWS-21)*, AWC, Leesburg, VA, 2021.
4. American Society of Civil Engineers (ASCE), *ASCE/SEI 7-16 Minimum Design Loads for Buildings and Other Structures*, ASCE, New York, 2016.
5. Structural Engineers Association of California (SEAOC), *IBC Structural/Seismic Design Manual*, Vol. 2. SEAOC, CA, 2018.

CHAPTER **8**

Complex Diaphragms with Combined Openings and Offsets

8.1 Introduction

During the schematic design phase of a project, there is usually some flexibility in the layout and placement of the shear walls or frames in the lateral-force-resisting system. Site constraints and architectural goals place limits on the choices. Maximum architectural flexibility is provided by using a minimum number of shear walls with the minimum lengths required. Unfortunately, such an approach also provides minimal redundancy and the resulting structure can have very complex or questionable load paths. An example of these marginal systems will be thoroughly examined in the diaphragm design in this chapter. In most cases, a reasonably redundant and good performing system can be developed without much effort. The objective of the engineer should be to break down the plan into simple rectangular sections, when possible, that are supported at all boundary edges, thereby removing any offsets problems or irregularities. The rectangular sections can then be individually analyzed as simple span diaphragms.

8.2 Loads and Load Paths

Figures 8.1 through 8.3 show the framing plans of a structure that is representative of many custom residential and commercial structures currently being designed and built. The structure is two stories and contains several vertical and horizontal offsets and other irregularities. The first-floor plan, Fig. 8.1, extends from grid line 1 to 5, while the second floor extends from grid lines 1.3 to 5. Problems associated with the layout of this structure are as follows:

- There are several second-floor shear walls at grid lines C.8, 1.3, and D that are discontinuous to the foundation. See Figs. 8.1 and 8.2.

- There are several second-floor shear walls at grid lines C, D, and 2 that do not line up and are of different lengths creating in-plane offset shear walls.

- The second-floor shear walls that occur at grid line 1.3 are horizontally offset from grid line 1 creating discontinuous shear walls and causing the diaphragm in that area to be a transfer diaphragm.

259

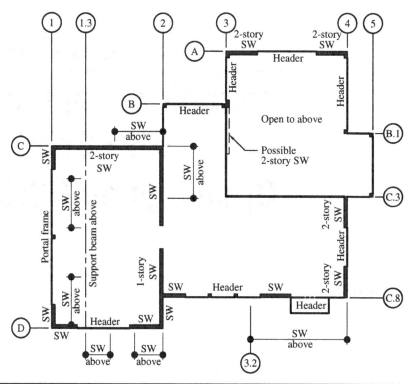

FIGURE 8.1 First-floor plan—complex structure.

- The minimal number of walls between grid lines 2 and 5 from A to D places a high demand on the lateral force resisting system. For the sake of the example, it is assumed that their location and placement are governed by architectural constraints.
- The second-floor diaphragm between grid lines 2 and 4 has a horizontal offset that is greater than 50 percent of the depth of the diaphragm, creating a notched diaphragm in both the transverse and longitudinal directions. See Fig. 8.2.
- A horizontal offset at grid line 4 from B.1 to C.3 creates a disruption in the strut at grid line 4, which complicates the transfer of wind loads acting on the wall along grid line A.

8.2.1 Lateral System in the Transverse Direction

Roof Diaphragms (See Fig. 8.3)

Starting at the roof and working downward, the roof is divided into two sections, diaphragm 1 and diaphragm 2. Diaphragm 1 is a simple rectangular diaphragm that is supported by the shear walls located at grid lines 1.3 and 2. The width of the diaphragm is from grid line C to grid line D. Diaphragm 2 is supported at grid lines 2 and 4 and can be designed as an offset diaphragm or as rectangular diaphragm bounded by grid lines 2 to 4 from B to C.8. The smaller section between grid lines 3 and 4 from A to B can be designed as a separate diaphragm that is supported at grid line 3 by the collector that is embedded into diaphragm 2, and by the strut at grid line 4. The roof diaphragm can be further simplified if an interior shear wall is added at grid line 3. This would create

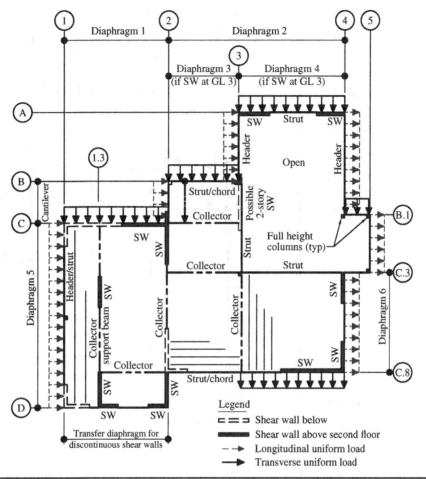

Figure 8.2 Second floor plan.

three simple rectangular diaphragms. The collectors that are required to provide complete load paths are shown in the figure.

Second-Floor Diaphragms (See Fig. 8.2)

Support walls for the second-floor diaphragms are located at grid lines 1, 2, and 4, which create two separate diaphragms, diaphragm 1 and diaphragm 2. The second-floor shear walls at grid line 1.3 support the roof diaphragm and are discontinuous to the foundation, creating a discontinuous shear wall. These walls are vertically supported by a beam at the second floor. The lateral forces from these walls are transferred through diaphragm 1 as a concentrated load and are distributed into the first-floor portal frame that is located along grid line 1 and the first-floor shear walls at grid line 2. Diaphragm 2 can be designed as an offset diaphragm, or as a simple rectangular diaphragm that occurs between grid lines 2 and 4 from C.3 to C.8. The simple rectangular diaphragm option is valid if the section between grid lines C.3 and C.8 meets the allowable aspect ratio. The right side of the small floor section between grid lines 2 and 3 from B to C.3 is supported by the strut at grid line 3 that is embedded into the

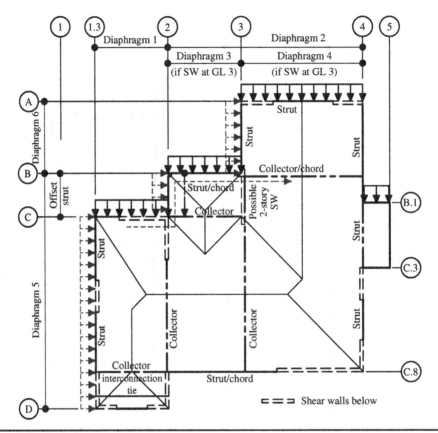

Figure 8.3   Roof plan.

rectangular diaphragm. A collector must be installed along grid line C.3 from 2 to 3 to act as part of the diaphragm chord for the rectangular diaphragm section, or to transfer the disrupted chord force into the transfer diaphragm if an offset diaphragm analysis is to be used. The second-floor diaphragm can be simplified in the same manner as the roof diaphragm by adding a shear wall at grid line 3, which would create diaphragms 3 and 4. All the sections would then become simple span rectangular diaphragms. The wall along grid line A from 3 to 4 must have full height studs that span to the roof. Those lateral forces applied to that wall are transferred back into the shear walls between grid lines C.3 and C.8 through the continuous roof strut.

8.2.2   Lateral System in the Longitudinal Direction

Roof Diaphragms (See Fig. 8.3)
In the longitudinal direction, the diaphragm is supported by walls at grid lines A, C, C.8, and D. Collectors should be installed at grid lines B, C, and C.8 as shown in the figure, thereby defining the boundary of diaphragm 5 between grid lines B and D, and diaphragm 6 between lines A to B. Diaphragm 6 is supported off the shear walls at grid line A and the collector at grid line B. Its reaction at grid line B must be transferred through the collector and drag element across the offset to the shear wall at grid line C. The shear walls at grid lines C.8 and D can be designed to act in the same line of resistance.

Second-Floor Diaphragms (See Fig. 8.2)

In the longitudinal direction, the second-floor diaphragm is broken down into two sections, diaphragm 5 and diaphragm 6. Diaphragm 5 is supported by shear walls at grid lines C and D from 1 to 3, with a cantilever that extends from grid line C to line B. The cantilever section is supported by the collector at grid line C that is tied to the shear wall at that location. The opposite end of the diaphragm is supported by the shear wall and drag strut at grid line C.8 and the shear wall at grid line D. A collector is installed at grid line C.8 from 1.3 to 2 to allow the shear walls to be designed to act in the same line of resistance. Diaphragm 6 is supported by the shear wall at grid line C.8 and by the drag strut at grid line C.3 that is embedded into diaphragm 5. Although the complexity of the lateral system has not necessarily been simplified, all boundaries of the diaphragms at each level are supported by shear walls or other boundary elements, and complete load paths have been provided.

Example 8.1: Complex Diaphragm Analysis

The complex diaphragm shown in Fig. 8.4 is presented to show how various parts of the diaphragm are broken down into segments to simplify the analysis. Transverse and longitudinal wind loads are applied to the various sections of the diaphragm as shown in Fig. 8.4. Seismic load distribution strips are shown in the longitudinal direction as an

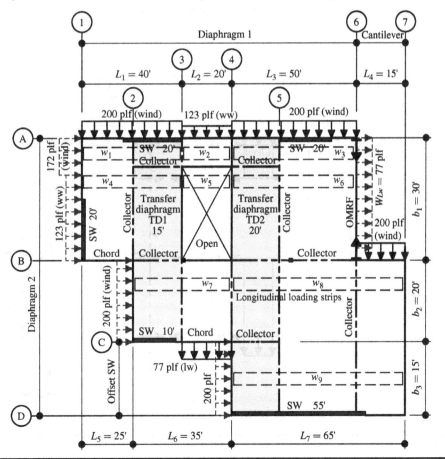

Figure 8.4 Load distribution in a complex diaphragm.

example only to demonstrate the distribution of forces based on width of the diaphragm sections. Looking at the diaphragm plan in the transverse direction, the diaphragm is supported by a shear wall at grid line 1 and by an ordinary moment resisting frame (OMRF) at grid line 6. The diaphragm has a horizontal offset between grid lines 1 and 2 and between 2 and 4, and a large opening located between lines 3 and 4. The section between grid lines 6 and 7 cantilevers from the main diaphragm. In the longitudinal direction, the diaphragm is supported at grid line A by in-line shear walls. The opposite side of the diaphragm is supported by offset shear walls at grid lines C and D, which are assumed to act in the same line of lateral force resistance.

Figure 8.5 shows the diaphragm breakdown for transverse loading. The sections between grid lines 1 and 6 act like a double offset diaphragm, where the disrupted chord force at grid line 2B is transferred into transfer diaphragm TD1, and the disrupted chord force at grid line 4C is transferred into TD2. The section between grid lines 3 and 4 is designed like the method described in Chap. 5 for a diaphragm with an opening. Transfer diaphragm TD1 receives three discontinuous forces, one from the bottom of the panel above the opening, one from the top of the panel below the opening, and one from the disrupted chord at grid line 2C. Transfer diaphragm TD2 receives four forces, one from the bottom of the panel above the opening, one from the top of the panel below the opening, one from the disrupted chord at grid line 4C and one from the top chord of the cantilever section that is transferred back to TD2.

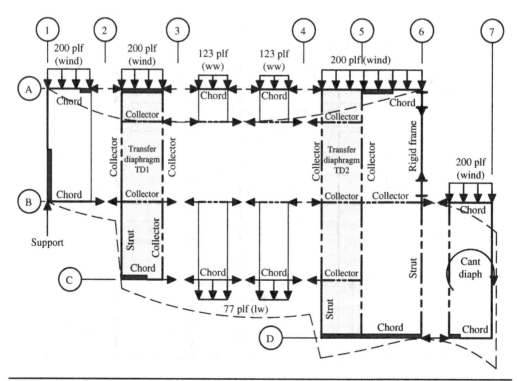

Figure 8.5 Segmentation of the diaphragm for transverse loading.

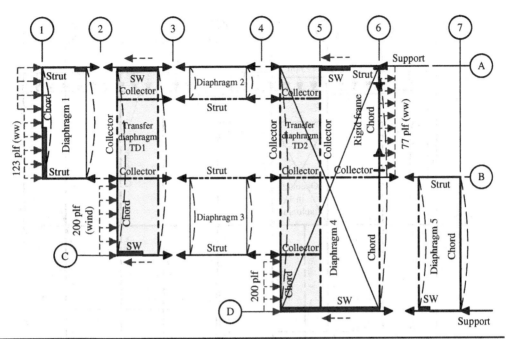

FIGURE 8.6 Segmentation of the diaphragm for longitudinal loading.

Figure 8.6 shows the diaphragm breakdown for longitudinal loading. The diaphragm can be broken down into several simple rectangular sections as shown in the figure. Diaphragm section 1 is supported by shear walls at grid line A and by the strut at grid line B. The strut at grid line B is embedded into transfer diaphragm TD1, which then transfers that force into the shear walls at grid lines A and C. Diaphragm sections 2 and 3 are supported by struts at the edges of the opening in the diaphragm, and by the shear walls located at grid lines A and C. The strut forces are transferred into transfer diaphragms TD1 and TD2. Diaphragm section 4 is supported by shear walls located at grid lines A and D. The chords for diaphragm 4 are located along grid lines 4 and 6. Diaphragm 5 is supported by shear walls at grid line D and by the strut at grid line B. The strut is connected to a collector that is embedded into the transfer diaphragm 2. The shear walls at grid lines C and D are offset and assumed to act in the same line of resistance. ▲

8.3 Complex Diaphragms with Interior Shear Walls—Analysis in the Transverse Direction

The diaphragm shown in Fig. 8.7 is 190 ft long by 90 ft wide. The shear wall at grid line 1, SW1, is 20 ft long and contains an opening. Shear walls SW2 and SW3, located at grid line 6, are 25 ft and 10 ft long, respectively. The lateral-force-resisting element at grid line 10 is a 40-ft-long moment resisting steel frame. If wind controls the design, a uniform load of 200 plf, which is the combined windward and leeward wind loads, can be applied to one side of the diaphragm sections that do not contain an opening. The wind load of 200 plf is broken down into windward and leeward wind loads of 123 plf

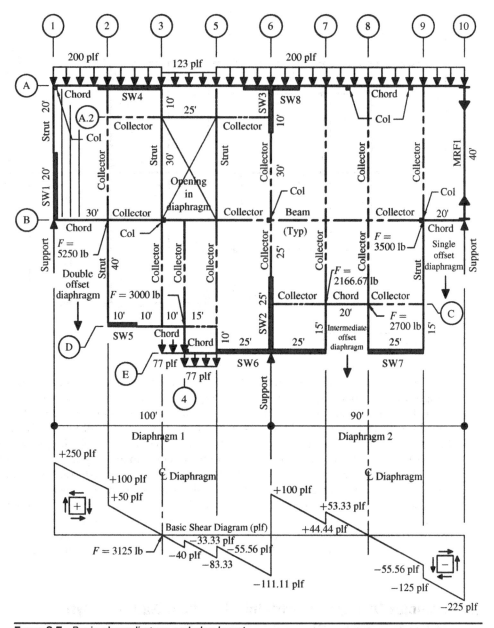

Figure 8.7 Basic shear diagram and plan layout.

and 77 plf, respectively, and are distributed to the appropriate sides of the diaphragm at the opening. It is important during the analysis to pay attention to the magnitude of the chord and collector forces and where they occur. A better sense of the critical nature of these elements can be obtained by this exercise.

Construction of the Basic Shear Diagram (See Fig. 8.7)   The diaphragm unit shears must be determined at each strut, collector, and change in diaphragm depth locations.

Diaphragm 1:

$$R_1 = R_6 = \frac{wL}{2} = \frac{200(100)}{2} = 10{,}000 \text{ lb}$$

$$v_1 = \frac{R_1}{W} = \frac{10{,}000}{40} = +250 \text{ plf, diaphragm unit shear at grid line 1}$$

$$V_2 = R_1 - wx = 10{,}000 - 200(30) = 4000 \text{ lb, shear at grid line 2}$$

$$v_{2L} = \frac{V_2}{W} = \frac{4000}{40} = +100 \text{ plf, diaphragm unit shear, left side of grid line 2}$$

$$v_{2R} = \frac{V_2}{W} = \frac{4000}{80} = +50 \text{ plf, diaphragm unit shear, right side of grid line 2}$$

$$V_3 = V_2 - wx = 4000 - 200(20) = 0 \text{ lb, shear at grid line 3, } v_3 = 0 \text{ plf}$$

$$V_4 = V_3 - wx = 0 - 200(10) = -2000 \text{ lb, shear at grid line 4}$$

$$v_{4L} = \frac{V_4}{W} = \frac{-2000}{50} = -40 \text{ plf, diaphragm unit shear, left side of grid line 4}$$

$$v_{4R} = \frac{V_4}{W} = \frac{-2000}{60} = -33.33 \text{ plf, diaphragm unit shear, right side of grid line 4}$$

$$V_5 = V_4 - wx = -2000 - 200(15) = -5000 \text{ lb, shear at grid line 5}$$

$$v_{5L} = \frac{V_5}{W} = \frac{-5000}{60} = -83.33 \text{ plf, diaphragm unit shear, left side of grid line 5}$$

$$v_{5R} = \frac{V_5}{W} = \frac{-5000}{90} = -55.56 \text{ plf, diaphragm unit shear, right side of grid line 5}$$

$$V_6 = -10{,}000 \text{ lb, shear at grid line 6}$$

$$v_6 = \frac{V_6}{W} = \frac{-10{,}000}{90} = -111.11 \text{ plf, diaphragm unit shear at grid line 6}$$

Diaphragm 2:

$$R_6 = R_{10} = V_6 = \frac{wL}{2} = \frac{200(90)}{2} = 9000 \text{ lb}$$

$$v_6 = \frac{V_6}{W} = \frac{9000}{90} = +100 \text{ plf, diaphragm unit shear at grid line 6}$$

$$V_7 = V_6 - wx = 9000 - 200(25) = 4000 \text{ lb, shear at grid line 7}$$

$$v_{7L} = \frac{V_7}{W} = \frac{4000}{90} = 44.44 \text{ plf, diaphragm unit shear, left side of grid line 7}$$

$$v_{7R} = \frac{V_7}{W} = \frac{4000}{75} = +53.33 \text{ plf, diaphragm unit shear, right side of grid line 7}$$

$$V_8 = V_7 - wx = 4000 - 200(20) = 0 \text{ lb, } v_8 = 0 \text{ plf}$$

$$V_9 = V_8 - wx = 0 - 200(25) = -5000 \text{ lb, shear at grid line 9}$$

$$v_{9L} = \frac{V_9}{W} = \frac{-5000}{90} = -55.56 \text{ plf, diaphragm unit shear, left side of grid line 9}$$

$$v_{9R} = \frac{V_9}{W} = \frac{-5000}{40} = -125 \text{ plf, diaphragm unit shear, right side of grid line 9}$$

$$v_{10} = \frac{R_{10}}{W} = \frac{-9000}{40} = -225 \text{ plf, diaphragm unit shear at grid line 10}$$

Chord Forces (See Fig. 8.7)

$$M_2 = R_1 - \frac{wx^2}{2} = 10{,}000(30) - \frac{200(30)^2}{2} = 210{,}000 \text{ ft-lb}$$

$$F_2 = \frac{M_2}{W} = \frac{210{,}000}{40} = 5250 \text{ lb}$$

$$M_4 = R_6 - \frac{wx^2}{2} = 10{,}000(40) - \frac{200(40)^2}{2} = 240{,}000 \text{ ft-lb}$$

$$F_4 = \frac{M_4}{W} = \frac{240{,}000}{80} = 3000 \text{ lb}$$

$$M_7 = R_6 - \frac{wx^2}{2} = 9000(25) - \frac{200(25)^2}{2} = 162{,}500 \text{ ft-lb}$$

$$F_7 = \frac{M_7}{W} = \frac{162{,}500}{75} = 2166.67 \text{ lb}$$

$$M_8 = R_6 - \frac{wx^2}{2} = 9000(45) - \frac{200(45)^2}{2} = 202{,}500 \text{ ft-lb}$$

$$F_8 = \frac{M_8}{W} = \frac{202{,}500}{75} = 2700 \text{ lb}$$

$$M_9 = R_{10} - \frac{wx^2}{2} = 9000(20) - \frac{200(20)^2}{2} = 140{,}000 \text{ ft-lb}$$

$$F_9 = \frac{M_9}{W} = \frac{140{,}000}{40} = 3500 \text{ lb}$$

Transfer Diaphragms (See Fig. 8.8)   Five transfer diaphragms, TD1 through TD5, have been created to handle the disrupted chord forces. The directions of the forces acting at the areas of discontinuity are shown in the figure and will be confirmed later in the

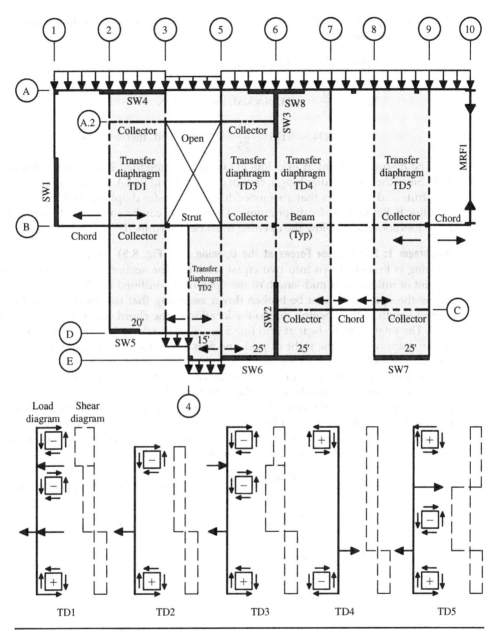

Figure 8.8 Transfer diaphragm layouts.

analysis. TD2 is a little unconventional because the upper end does not tie into a shear wall. However, tying into the continuous collector at grid line B is acceptable if the forces resolve and the force diagram closes to zero. It will simply change the collector forces at that location. The anticipated transfer diaphragm loading, and shear diagrams are shown at the bottom of the figure, along with the direction of the shears acting on the sheathing elements. Before continuing the analysis, the transfer diaphragm aspect ratios must be checked.

$$TD1 = \frac{80}{20} = 4, \text{ blocked, therefore OK}$$

$$TD2 = \frac{50}{15} = 3.33, \text{ blocked, therefore OK}$$

$$TD3 = TD4 = TD5 = \frac{90}{25} = 3.6, \text{ blocked, therefore OK}$$

The disrupted chord at grid line 4D must be embedded into transfer diaphragm TD2. Since the transfer diaphragm cannot extend to grid line A, it must be supported by the struts and collectors that are embedded into transfer diaphragms TD1 and TD3 at grid line B. Before transfer diaphragms TD1 and TD3 can be analyzed, the forces in the collectors above and below the opening must be determined.

Diaphragm 1: Panel Shear Forces at the Opening (See Fig. 8.9)   The section above the opening is broken down into two equal sections. The sections are assumed to have a point of inflection at mid-length of the opening as outlined in Chap. 5. The sections below the opening cannot be broken down assuming that its inflection point occurs at mid-length of the opening due to the location of the chord offset and occurrence of TD2. The total vertical shear at grid line 5 is −5000 lb acting downward on the section of the diaphragm to the right of grid line 5 as determined by placing the sheathing element symbols adjacent to the transfer diaphragm as shown in the free-body diagram at the upper right of the figure, in accordance with the sign convention legend. The total vertical shear is distributed to the upper and lower section at the opening is in proportion to their width as in Chap. 5. The direction of the shear acting on the right edge of the sections above and below the opening is acting in the upward direction, also shown in the figure. Applying the sheathing element symbols at all edges of the sections determine if the unit shears are positive or negative.

$$V_{upper} = \frac{V_5 W_{A-A.2}}{\Sigma W} = \frac{-5000(10)}{60} = -833.33 \text{ lb, shear applied to upper section}$$

$$v_{upper} = \frac{V_{upper}}{W_{A-A.2}} = \frac{-833.33}{10} = -83.33 \text{ plf, unit shear of upper section}$$

$$V_{lower} = \frac{V_5 W_{B-E}}{\Sigma W} = \frac{-5000(50)}{60} = -4166.67 \text{ lb, shear applied to lower section}$$

$$v_{lower} = \frac{V_{lower}}{W_{B-E}} = \frac{-4166.67}{50} = -83.33 \text{ plf, unit shear of lower section}$$

Starting at grid line 5 at the upper panel, summing to the left (see free-body Fig. 8.9):

$$V_5 = 833.33 \text{ lb, acting upward}$$

$$V_{left} = V_4 = V_5 - wx = 833.33 - 123(12.5)$$
$$= -704.2 \text{ lb, acting upward on the left edge}$$

$$v_4 = \frac{V_{left}}{W} = \frac{704.2}{10} = +70.42 \text{ plf, unit shear at the left edge of the section}$$

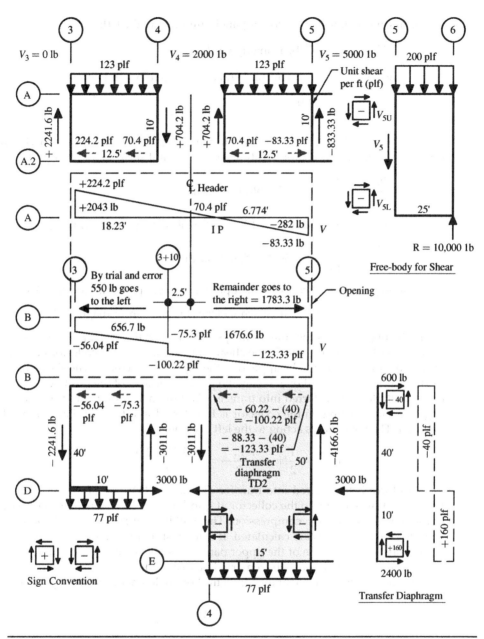

Figure 8.9 Shear panel forces at opening—diaphragm 1.

The shear on the right side of the left panel (near grid line 4) is equal to 704.2 lb and is acting in the downward direction. The shear at grid line 3 is equal to

$$V_3 = V_4 - wx = -704.2 - 123(12.5) = -2241.6 \text{ lb acting upward}$$

$$v_3 = \frac{V_3}{W} = \frac{2241.6}{10} = +224.2 \text{ plf, unit shear.}$$

Shears at grid line 5 at the lower panel, summing to the left:

$$V_5 = -4166.67 \text{ lb, acting upward}$$

$$V_{\text{left}} = V_5 - wx = 4166.67 - 77(15) = +3011 \text{ lb, acting downward}$$

$$v_{\text{left}} = \frac{V_{\text{left}}}{W} = \frac{3011}{50} = -60.22 \text{ plf, unit shear}$$

The shear on the right side of the left panel at grid line 4 is equal to 3011 lb and is acting in the upward direction creating a negative unit shear in the sheathing.

$$v_{\text{right}} = \frac{V_{\text{right}}}{W} = \frac{3011}{40} = -75.3 \text{ plf, unit shear}$$

$$V_3 = V_{\text{right}} - wx = 3011 - 77(10) = +2241 \text{ lb, acting downward at grid line 3}$$

$$v_3 = \frac{V_3}{W} = \frac{2241.6}{40} = -56.04 \text{ plf, unit shear}$$

Summing the total shear from the upper and lower panels at grid line 3 as a check:

$$\Sigma V = +2241.6 - 2241.6 = 0, \text{ which matches the basic shear diagram. Therefore, OK.}$$

In the upper panels the unit shear goes from a positive 224.2 plf at grid line 3 to a negative 83.33 plf at grid line 5. The shears transferred into the collector at the bottom of the upper panels should be applied to the figure for reference. The shear diagram of the transferred shears is shown below the panels. The area of the positive shears acting on the collector is transferred into transfer diaphragm TD1 on the left. The area of the negative shears acting on the collector is transferred into transfer diaphragm TD3 on the right. The force that is acting to the left becomes

$$F_{3,A.2} = \frac{1}{2}bh = \frac{v_3 L_{3-4}}{2} = \frac{224.2(18.23)}{2} = 2043 \text{ lb}$$

This force is transferred into transfer diaphragm TD1. Observing the direction of the shears transferred into the collector at grid line 5, A.2, the force applied to transfer diaphragm TD3 is also in compression. This will be confirmed when the longitudinal collector force diagrams are calculated. It is interesting to note that the force, $F = 0$, in the member at the bottom of the upper panel does not occur at the assumed inflection point location (i.e., halfway between grid lines 3 and 5).

The force in the collector at the top of the bottom left panel between grid lines 3 and $3 + 10'$ is

$$F = \frac{(56.04 + 75.3)}{2}(10) = 656.7 \text{ lb}$$

This force is transferred into transfer diaphragm TD1.

Before the force between grid lines 4 and 5 can be determined, transfer diaphragm TD2 shears must be calculated because the transfer diaphragm receives the disrupted chord force at grid line 4D, and transfers part of that force into its support at grid line B. Part of the combined basic diaphragm shear and transfer diaphragm shear forces that are applied into the collector at grid line B are shared by transfer diaphragms TD1 and TD3.

$$F_{\text{chord}} = 3000 \text{ lb}$$

Transfer Diaphragm Shears (Fig. 8.9, Lower Right)

$$V_B = \frac{3000(10)}{50} = 600 \text{ lb}, \quad v_B = \frac{V_B}{W_{TD}} = \frac{600}{15} = -40 \text{ plf negative shear at grid line B}$$

$$V_E = \frac{3000(40)}{50} = 2400 \text{ lb}, \quad v_E = \frac{V_E}{W_{TD}} = \frac{2400}{15} = +160 \text{ plf positive shear at grid line B}$$

Add the basic diaphragm shears with the transfer diaphragm shears to determine the net shears within TD2.

Grid line 4, from B to D $v = -60.22 - (40) = -100.22 \text{ plf}$
Grid line 5, from B to D $v = -83.33 - (40) = -123.33 \text{ plf}$
Grid line 4, from D to E $v = -60.22 + (160) = +99.78 \text{ plf}$
Grid line 5, from D to E $v = -83.33 + (160) = +76.67 \text{ plf}$

The force in the collector between grid lines $3 + 10'$ and 5 is

$$F = \frac{(100.22 + 123.33)}{2}(15) = 1676.6 \text{ lb tension}$$

Applying these forces to the transfer diaphragms and completing the longitudinal chord/collector force diagrams will show that the diagrams at grid lines B, D, and E do not quite close to zero. This is partly due to the configuration of the section below the opening which differs from a constant width diaphragm with an opening. By trial and error, 550 lb goes to TD1 and 1783.3 lb goes to TD3. The force applied to TD1 is in compression and the force to TD3 is in tension.

Transfer Diaphragm TD1 (See Fig. 8.10)   The disrupted chord force and collector forces at the edge of the opening are applied to transfer diaphragm TD1 as shown in the figure.

$$\Sigma M_D = 0$$

$$V_A = [(5250 + 550)(40) + 2043(70)]\frac{1}{80} = 4687.63 \text{ lb}$$

$$v_A = \frac{V_A}{W_{TD}} = \frac{4687.63}{20} = -234.4 \text{ plf negative unit shear}$$

$$\Sigma M_A = 0$$

$$V_D = [(5250 + 550)(40) + 2043(10)]\frac{1}{80} = 3155.4 \text{ lb}$$

$$v_D = \frac{V_D}{W_{TD}} = \frac{3155.4}{20} = +157.8 \text{ plf positive unit shear}$$

Summing at grid line B:

$$V_B = -3155.4 + (5250 + 550) = 2644.6 \text{ lb}$$

$$v_B = \frac{V_B}{W_{TD}} = \frac{2644.6}{20} = -132.2 \text{ plf negative unit shear from grid line A.2 to B}$$

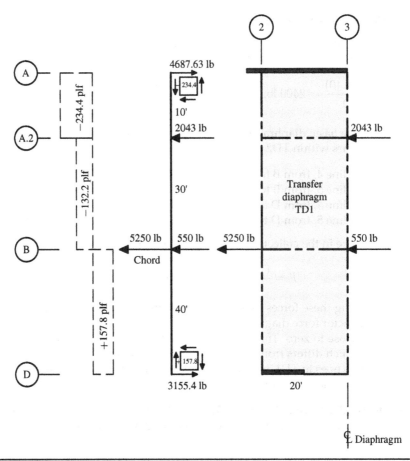

Figure 8.10 Transfer diaphragm forces at left side.

Transfer Diaphragm TD3 (See Fig. 8.11) The disrupted chord force and collector forces at the edge of the opening are applied to transfer diaphragm TD3 as shown in the figure.

$$\sum M_D = 0$$

$$V_A = \left[1783.3(50) - 282(80)\right]\frac{1}{90} = 740.1 \text{ lb}$$

$$v_A = \frac{V_A}{W_{TD}} = \frac{740.1}{25} = -29.6 \text{ plf negative unit shear}$$

$$\sum M_A = 0$$

$$V_E = \left[1783.3(40) - 282(10)\right]\frac{1}{90} = 761.24 \text{ lb}$$

$$v_E = \frac{V_E}{W_{TD}} = \frac{761.24}{25} = +30.45 \text{ plf positive unit shear}$$

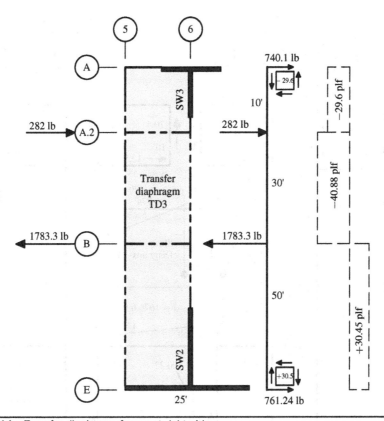

Figure 8.11 Transfer diaphragm forces at right side.

Summing at grid line B:

$$V_B = -761.24 + 1783.3 = 1022.1 \text{ lb}$$

$$v_B = \frac{V_B}{W_{TD}} = \frac{1022.1}{25} = -40.88 \text{ plf negative unit shear from grid line B to A.2}$$

Determination of Diaphragm Net Shears (See Fig. 8.12) The transfer diaphragm shears are placed on the figure for easy reference. The basic diaphragm unit shears are placed at the top of the diaphragm near the grid line callouts for reference.

Location (Grid Lines)	Net Shears
1 from A to B	$= +250$ plf
2 (left) from A to B	$= +100$ plf
2 (right) from A to A.2	$v = +50 - (234.4) = -184.4$ plf
2 (right) from A.2 to B	$v = +50 - (132.2) = -82.2$ plf
2 (right) from B to D	$v = +50 + (157.8) = +207.8$ plf
3 (left) from A to A.2	$v = +0 - (234.4) = -234.4$ plf
3 (left) from A.2 to B	$v = +0 - (132.2) = -132.2$ plf
3 (left) from B to D	$v = +0 + (157.8) = +157.8$ plf

(Continued on next page)

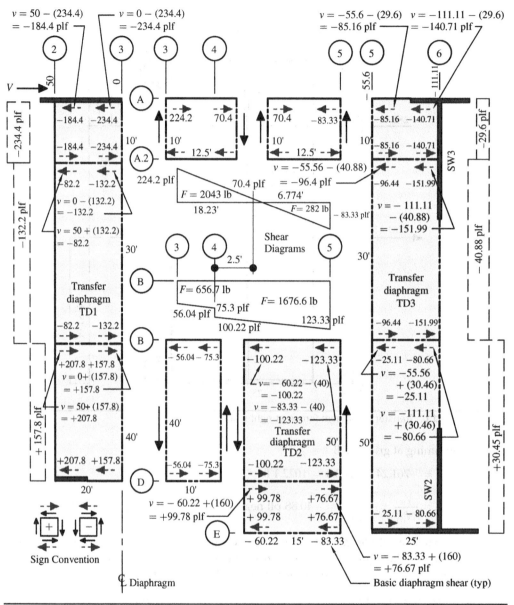

FIGURE 8.12 Net shears.

Location (Grid Lines)	Net Shears
3 (right) from A to A.2	$= +224.2$ plf
3 (right) from B to D	$= -56.04$ plf
4 (left) from B to D	$= -75.3$ plf
4 (right) from B to D	$v = -60.22 - (40) = -100.22$ plf
4 (right) from D to E	$v = -60.22 + (160) = +99.78$ plf

(Continued on next page)

5 (left) from A to A.2	$= -83.33$ plf
5 (left) from B to D	$v = -83.33 - (40) = -123.33$ plf
5 (left) from D to E	$v = -83.33 + (160) = +76.67$ plf
5 (right) from A to A.2	$v = -55.56 - (29.6) = -85.16$ plf
5 (right) from A.2 to B	$v = -55.56 - (40.88) = -96.44$ plf
5 (right) from B to E	$v = -55.56 + (30.46) = -25.11$ plf
6 (left) from A to A.2	$v = -111.11 - (29.6) = -140.71$ plf
6 (left) from A.2 to B	$v = -111.11 - (40.88) = -151.99$ plf
6 (left) from B to E	$v = -111.11 + (30.46) = -80.66$ plf

Determination of Longitudinal Collector/Chord Forces (See Fig. 8.13)   The net shears and the direction of their transfer into the boundary members are placed on the diaphragm sections for reference.

Force diagrams at grid line A.2:

$$F_3 = \frac{[(184.4 - 82.2) + (234.4 - 132.2)]}{2}(20) = 2043 \text{ lb}$$

Force at grid line 3 plus 12.5':

$$F_{3+12.5'} = 2043 - \frac{(224.2 + 70.4)}{2}(12.5) = 201.75 \text{ lb}$$

The shear in the right section changes from a $+70.4$ plf to a -83.33 plf. The forces of the positive and negative shear areas are shown above the section.

$$F_5 = +201.75 - 201.5 + 282 = 282.5 \text{ lb, equals previous calculation}$$

$$F_6 = 282 - \frac{[(96.44 - 85.16) + (151.99 - 140.71)]}{2}(25) = 0 \text{ lb}$$

Therefore, the diagram closes.

Force diagrams at grid line B:

$$F_2 = -5250 \text{ lb chord tension force}$$

$$F_3 = -5250 + \frac{[(82.2 + 207.8) + (132.2 + 157.8)]}{2}(20) = +550 \text{ lb}$$

$$F_4 = +550 - \frac{(56.04 + 75.3)}{2}(10) = +106.7 \text{ lb}$$

$$F_5 = +106.7 - \frac{(100.22 + 123.33)}{2}(15) = -1783.3 \text{ lb}$$

$$F_6 = -1783.3 + \frac{[(96.44 - 25.11) + (151.99 - 80.66)]}{2}(25) = 0 \text{ lb}$$

Therefore, the diagram closes.

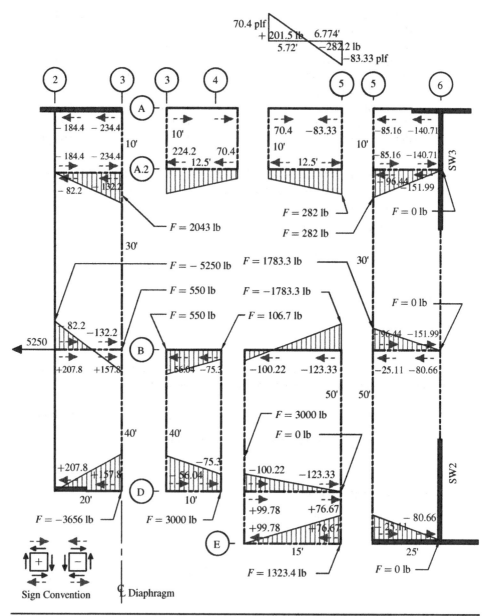

Figure 8.13 Longitudinal collector force diagrams—diaphragm 1.

Force diagrams at grid line D:

$$F_3 = -\frac{(207.8 + 157.8)}{2}(20) = -3656 \text{ lb}$$

$$F_4 = -3656 + \frac{(56.04 + 75.3)}{2}(10) = -3000 \text{ lb}$$

$$F_5 = -3000 + \frac{[(100.22 + 99.78) + (123.33 + 76.67)]}{2}(15) = 0 \text{ lb}$$

Therefore, the diagram closes.

Force diagrams at grid line E:

$$F_5 = -\frac{(99.78 + 76.67)}{2}(15) = -1323.4 \text{ lb}$$

$$F_6 = -1323.4 + \frac{(25.11 + 80.66)}{2}(25) = 0 \text{ lb}$$

Therefore, the diagram closes.

Determination of Transverse Collector/Chord Forces (See Fig. 8.14)

Force diagrams at grid line 2, starting at grid line D:

$$F_B = -207.8(40) = -8312 \text{ lb}$$

$$F_{A.2} = -8312 + (100 + 82.2)(30) = -2846 \text{ lb}$$

$$F_A = -2846 + (184.4 + 100)(10) = 0 \text{ lb} \pm$$

Therefore, the diagram closes.

Force diagrams at grid line 3, starting at grid line D:

$$F_B = (56.04 + 157.8)(40) = 8553.6 \text{ lb}$$

$$F_{A.2} = 8553.6 - 132.2(30) = 4587.6 \text{ lb}$$

$$F_A = 4587.6 - (234.4 + 224.2)(10) = 0 \text{ lb} \pm$$

Therefore, the diagram closes.

Force diagrams at grid line 4, starting at grid line E:

$$F_D = -99.78(10) = -997.8 \text{ lb}$$

$$F_B = -997.8 + (100.22 - 75.3)(40) = 0 \text{ lb}$$

Therefore, the diagram closes.

Force diagrams at grid line 5, starting at grid line E:

$$F_D = (25.11 + 76.67)(10) = 1017.8 \text{ lb}$$

$$F_B = 1017.8 - (123.33 - 25.11)(40) = -2911 \text{ lb}$$

$$F_{A.2} = -2911 + 96.44(30) = -17.8 \text{ lb}$$

$$F_A = -17.8 + (85.16 - 83.33)(10) = 0 \text{ lb} \pm$$

Therefore, the diagram closes.

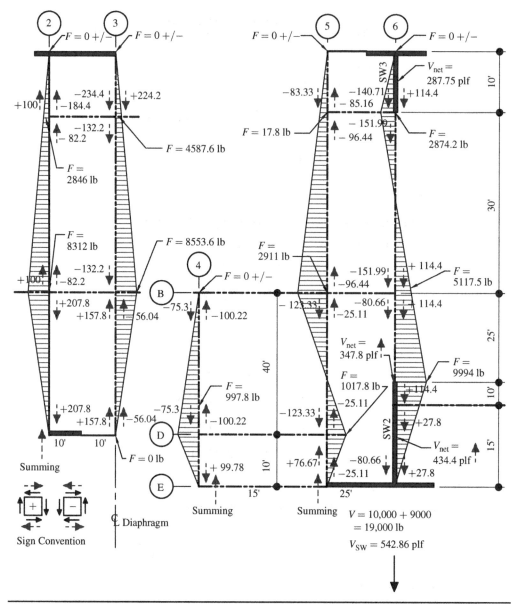

Figure 8.14 Transverse collector force diagrams—diaphragm 1.

Force diagrams at grid line 6, starting at grid line E:

$$V_6 = 10,000 + 9000 = 19,000 \text{ lb reaction from left and right diaphragm}$$

$$v_{SW} = \frac{V_6}{\sum L_{SW2,SW3}} = \frac{19,000}{10 + 25} = +542.86 \text{ plf}$$

$$v_{net} = v_{SW} - v_{diaph}$$

$$v_{net} = +542.86 - (140.71 + 114.4) = +287.75 \text{ plf net shear wall SW3}$$

Note that the diaphragm shear values on the right side of SW2 change in the lower 15 ft of the wall; therefore, the net shears at the wall will also change.

$$v_{net} = +542.86 - (80.66 + 27.8) = +434.4 \text{ plf net shear at } 15' \text{ SW2}$$

$$v_{net} = +542.86 - (80.66 + 114.4) = +347.8 \text{ plf net shear at } 10' \text{ SW2}$$

Starting at grid line E:

$$F = 434.4(15) + 347.8(10) = +9994 \text{ lb, end of SW2}$$

$$F_B = 9994 - (80.66 + 114.4)(25) = 5117.5 \text{ lb}$$

$$F_{A.2} = 5117.5 - (151.99 + 114.4)(30) = 2874.2 \text{ lb}$$

$$F_A = -2874.2 + 287.75(10) = 0 \text{ lb} \pm$$

Therefore, the diagram closes.

Diaphragm 2 (See Fig. 8.15)   This diaphragm has an intermediate offset.

Determine chord forces at areas of discontinuity:

$$R_6 = R_{10} = \frac{wL}{2} = \frac{200(90)}{2} = 9000 \text{ lb}$$

$$M_9 = R_{10}x - \frac{wx^2}{2} = 9000(20) - \frac{200(90)^2}{2} = 140,000 \text{ ft-lb}$$

$$F_{9B} = \frac{M_9}{W} = \frac{14,000}{40} = 3500 \text{ lb, chord force at grid line 9B}$$

$$M_8 = \frac{wL^2}{8} = \frac{200(90)^2}{8} = 202,500 \text{ ft-lb (occurs at centerline of diaphragm)}$$

$$F_{8C} = \frac{M_8}{W} = \frac{202,500}{75} = 2700 \text{ lb, chord force at grid line 8C}$$

$$M_7 = R_6x - \frac{wx^2}{2} = 9000(25) - \frac{200(25)^2}{2} = 162,500 \text{ ft-lb}$$

$$F_{7C} = \frac{M_7}{W} = \frac{162,500}{75} = 2166.67 \text{ lb, chord force at grid line 7C}$$

Determine Transfer Diaphragm TD4 Shears (See Fig. 8.15)

$$\sum M_E = 0$$

$$V_A = \frac{2166.67(15)}{90} = 361.1 \text{ lb}$$

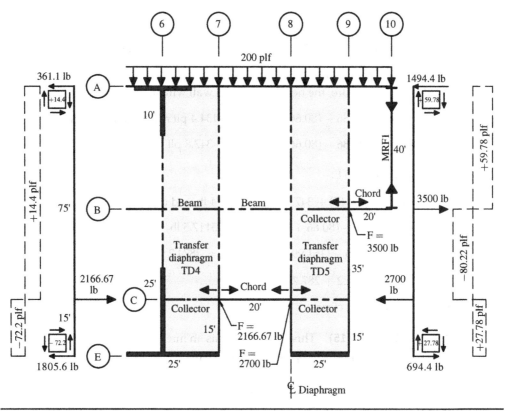

FIGURE 8.15   Transfer diaphragm shear forces—diaphragm 2.

$$v_A = \frac{V_A}{W_{TD}} = \frac{361.1}{25} = +14.4 \text{ plf positive shear}$$

$$\sum M_A = 0$$

$$V_E = \frac{2166.67\,(75)}{90} = 1805.6 \text{ lb}$$

$$v_E = \frac{V_E}{W_{TD}} = \frac{1805.6}{25} = -72.2 \text{ plf negative shear}$$

Determine Transfer Diaphragm TD5 Shears

$$\sum M_E = 0$$

$$V_A = [-2700(15) + 3500(50)]\frac{1}{90} = 1494.4 \text{ lb}$$

$$v_A = \frac{1494.4}{25} = +59.78 \text{ plf positive shear}$$

$$\sum M_A = 0$$

$$V_E = [-3500(40) + 2700(75)]\frac{1}{90} = 694.4 \text{ lb}$$

$$v_E = \frac{694.4}{25} = +27.78 \text{ plf positive shear}$$

Sum shears at grid line B:

$$V_C = -694.4 + 2700 = 2005.6 \text{ lb}$$

$$v_C = \frac{2005.6}{25} = -80.22 \text{ plf negative shear}$$

Determination of Diaphragm Net Shears (See Fig. 8.16)   The basic diaphragm shears, transfer diaphragm shears, and resulting net shears are placed on the illustration for easy reference. The direction of the transfer shears should be placed on the plan as they are determined.

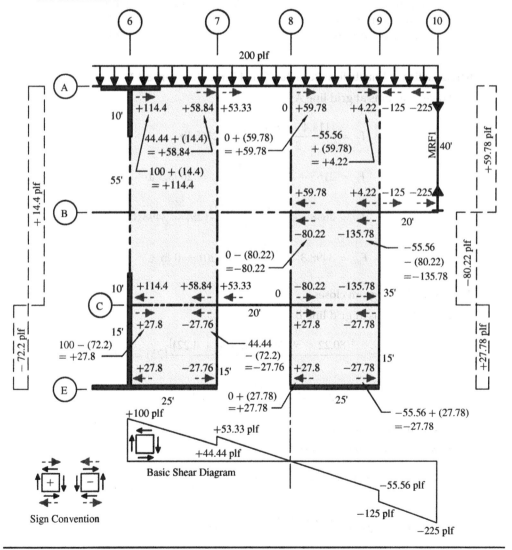

FIGURE 8.16   Net shears—diaphragm 2.

Location (Grid Lines)	Net Shears
6 from A to C	$v = 100 + (14.4) = +114.4$ plf
6 from C to E	$v = 100 - (72.2) = +27.8$ plf
7 (left) from A to C	$v = +44.44 + (14.4) = +58.84$ plf
7 (left) from C to E	$v = +44.44 - (72.2) = -27.76$ plf
7 (right) from A to C	$= +53.33$ plf
8 (left) from A to C	$= 0$ plf
8 (right) from A to B	$v = 0 + (59.78) = +59.78$ plf
8 (right) from B to C	$v = 0 - (80.22) = -80.22$ plf
8 (right) from C to E	$v = 0 + (27.78) = +27.78$ plf
9 (left) from A to B	$v = -55.56 + (59.78) = +4.22$ plf
9 (left) from B to C	$v = -55.56 - (80.22) = -135.78$ plf
9 (left) from C to E	$v = -55.56 + (27.78) = -27.78$ plf
9 (right) from A to B	$= -125$ plf
10 (left) from A to B	$= -225$ plf

Determination of Longitudinal Collector/Chord Forces (See Fig. 8.17)

Force diagrams at grid line A:

$$F_7 = \frac{(114.4 + 58.8)}{2}(25) = 2165 \text{ lb}$$

$$F_8 = 2165 + \frac{(53.33 + 0)}{2}(20) = 2698.3 \text{ lb}$$

$$F_9 = 2698.3 + \frac{(59.78 + 4.22)}{2}(25) = 3498.3 \text{ lb}$$

$$F_{10} = 3498.3 - \frac{(125 + 225)}{2}(20) = 0 \text{ lb} \pm$$

Therefore, the diagram closes.

Force diagrams at grid line B:

$$F_9 = -\frac{[(80.22 + 59.78) + (135.78 + 4.22)]}{2}(25) = -1750 \text{ lb}$$

$$F_{10} = -1750 + \frac{(125 + 225)}{2}(20) = 0 \text{ lb}$$

Therefore, the diagram closes.

Force diagrams at grid line C:

$$F_7 = -\frac{[(114.4 - 27.8) + (58.84 + 27.76)]}{2}(25) = -2165 \text{ lb}$$

$$F_8 = -2165 - \frac{(53.33 + 0)}{2}(20) = -2698.3 \text{ lb}$$

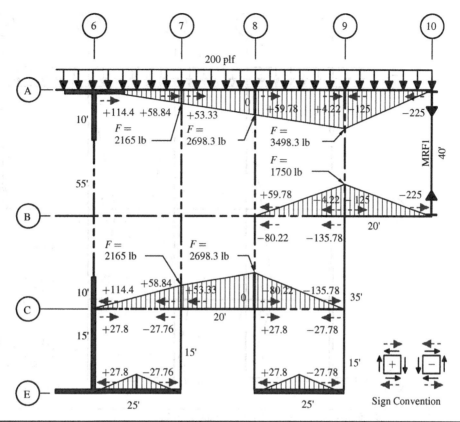

Figure 8.17   Longitudinal collector forces—diaphragm 2.

$$F_9 = -2698.3 + \frac{[(80.22 + 27.8) + (135.78 - 27.76)]}{2}(25) = 0 \text{ lb } \pm$$

Therefore, the diagram closes.

Determination of Transverse Collector/Chord Forces (See Fig. 8.18)

Force diagrams at grid line 7, starting at grid line E:

$$F_C = -27.76(15) = -416.4 \text{ lb}$$

$$F_A = -416.4 + (58.84 - 53.33)(75) = 0 \text{ lb } \pm$$

Therefore, the diagram closes.

Force diagrams at grid line 8, starting at grid line E:

$$F_C = -27.76(15) = -416.4 \text{ lb}$$

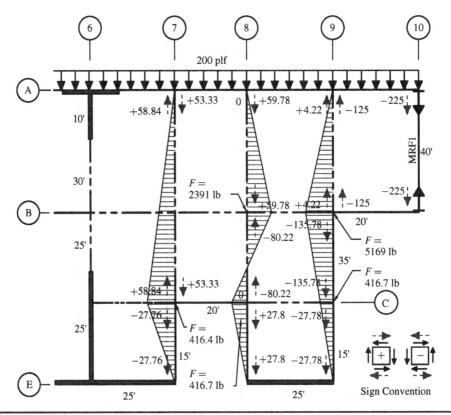

Figure 8.18 Transverse collector forces—diaphragm 2.

$$F_B = -416.4 + 80.22(35) = +2391 \text{ lb}$$

$$F_A = +2391 - 59.78(40) = 0 \text{ lb} \pm$$

Therefore, the diagram closes.

Force diagrams at grid line 9, starting at grid line E:

$$F_C = -27.78(15) = -416.7 \text{ lb}$$

$$F_B = -416.7 - 135.78(35) = -5169 \text{ lb}$$

$$F_A = -5169 + (4.22 + 125)(40) = 0 \text{ lb}$$

Therefore, the diagram closes.

There are several ways to break the diaphragms down into sections to simplify the analysis. No single approach is more correct than another. Diaphragm 2 can be somewhat simplified as shown in Fig. 8.19, by making the section between grid lines 6 and 10 from A to B as the main diaphragm section and the remainder supported off that section at grid line 9.

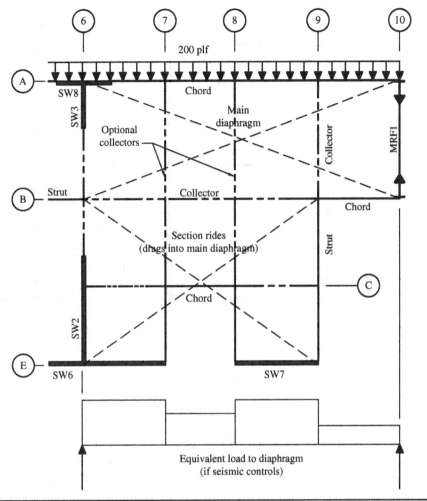

Figure 8.19   Optional layout for diaphragm 2.

8.4   Complex Diaphragms with Interior Shear Walls—Analysis in the Longitudinal Direction

For loading in the longitudinal direction, the diaphragm is supported by shear walls at grid lines A, D, and E as shown in Fig. 8.20. Shear walls 5 and 6 at grid lines D and E are offset but considered to act in the same line of lateral force resistance. Shear wall 7 is not connected to shear wall 6 by a strut and therefore acts independently. There are several ways to break the diaphragm down into smaller sections to simplify the analysis, one of which is shown in Fig. 8.21. The easiest method is to break the diaphragm around the opening into smaller sections, which will be supported off the beams (collectors) located along grid line B and off the shear walls at the exterior wall lines. The beams at grid line B collect the reactions from the individual sections and transfer the accumulated forces as a concentrated force into the main diaphragm located between

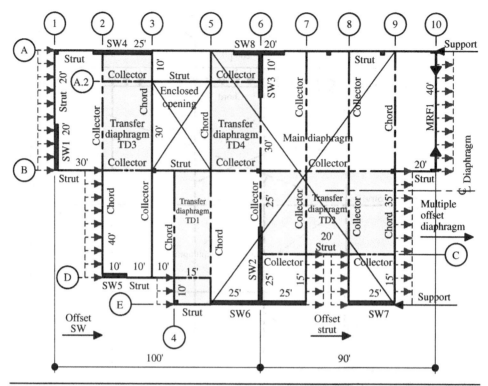

Figure 8.20   Analysis in longitudinal direction diaphragm loading.

grid lines 6 and 9 from A to E and also supports the left side of diaphragm 6. The main diaphragm has an intermediate horizontal offset at its left support, grid line E. The diaphragm section between grid lines 1 and 6 is broken down into seven separate segments: diaphragm 1, diaphragm 2, diaphragm 3, and diaphragm 4, and transfer diaphragms TD1, TD3, and TD4.

The strut coming off SW5 is embedded into transfer diaphragm TD1 where the discontinuous force is then transferred into the shear walls along grid line E and off the struts and collectors along grid line B. Load sharing at this area is complicated. All of the diaphragm sections and transfer diaphragm reactions that are distributed into grid line B between grid lines 1 to 6 can easily be calculated with the exception of transfer diaphragm 1. The reaction of transfer diaphragm 1 at grid line B depends on the effects of the discontinuous strut force of SW 5, which in turn determines the whole force that is applied as a concentrated force into the main diaphragm.

The main diaphragm, diaphragm 5, is supported by 50-ft SW6 and 25-ft SW7 connected by the reduced width of the sections between C and E designated as sections A and B, respectively. This section of the diaphragm can be analyzed like the intermediate end offset in Example 4.3 without a strut and Prob. 4.3. Since section A is connected to SW6, its reaction effects the resulting shears acting along grid lines D and E from 2 to 7, which also depends on the reaction of transfer diaphragm TD1 at grid line E. Complicated to say the least, which requires guessing what the final reactions are for TD1 then taking an iterative approach until closure of all force diagrams. Transfer diaphragm

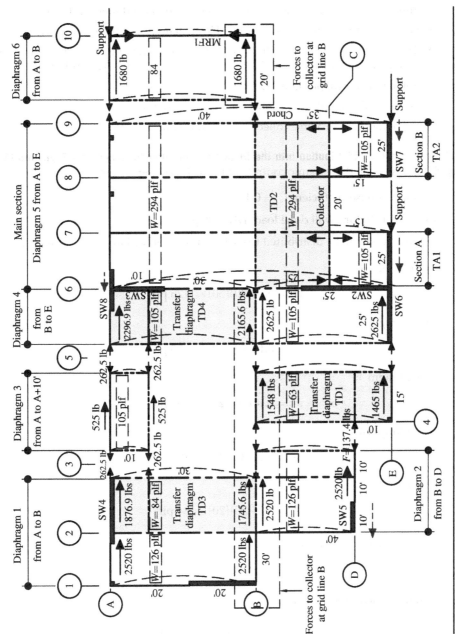

Figure 8.21 Longitudinal load distribution and segmentation.

TD2 is used to resist and transfer the tension and compression chord forces from sections A and B at the corners of the offset. Approach taken

- Solve simple diaphragms 1, 2, 3, 4, and 6 reactions and their unit shears.
- Solve transfer diaphragms TD3 and TD4 reactions and their unit shears.
- The resolution of TD1 and diaphragm 5 requires an iterative approach because of load sharing between SW5, TD1, and the collector force at grid line B.
- Solve transfer diaphragm TD2.
- Solve for the longitudinal collector and strut forces.
- Solve for the transverse collector force diagrams.

Determine the Load Distribution into the Diaphragm and Transfer Diaphragm Segments (See Fig. 8.21)   Assume that seismic controls with the following design criteria:

Seismic response coefficient $C_s = 0.14$ (ASD)

Roof plus tributary wall dead load, DL = 30 psf

The strip load will be distributed to each section in accordance with its width.

Diaphragm 1:

$$w = 0.14(30)(30) = 126 \text{ plf}$$

Diaphragm 2:

$$w = 0.14(30)(30) = 126 \text{ plf}$$

Diaphragm 3:

$$w = 0.14(30)(25) = 105 \text{ plf}$$

Diaphragm 4:

$$w = 0.14(30)(25) = 105 \text{ plf}$$

Diaphragm 6:

$$w = 0.14(30)(20) = 84 \text{ plf}$$

Transfer Diaphragm TD1:

$$w = 0.14(30)(15) = 63 \text{ plf}$$

Transfer Diaphragm TD3:

$$w = 0.14(30)(20) = 84 \text{ plf}$$

Transfer Diaphragm TD4:

$$w = 0.14(30)(25) = 105 \text{ plf}$$

Main Diaphragm 5

Grid line A to C:

$$w = 0.14(30)(70) = 294 \text{ plf from grid line 6 to 9}$$

Section A:

$$w = 0.14(30)(25) = 105 \text{ plf}$$

Section B:

$$w = 0.14(30)(25) = 105 \text{ plf}$$

Construction of the Basic Shear Diagrams (See Figs. 8.22 and 8.23)

Diaphragm 1:

$$R_A = -R_B = \frac{wL}{2} = \frac{126(40)}{2} = 2520 \text{ lb}$$

$$v_A = -v_B = \frac{R_A}{W} = \frac{2520}{30} = 84 \text{ plf}$$

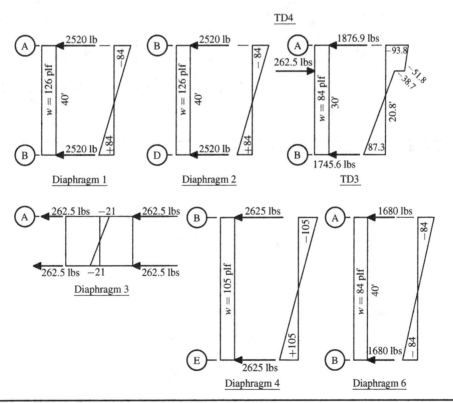

FIGURE 8.22 Basic shear diagrams.

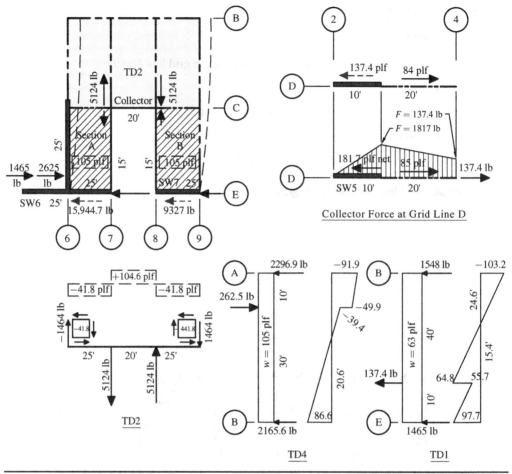

FIGURE 8.23 Main diaphragm loading and shear diagrams.

Diaphragm 2:

$$R_B = -R_D = \frac{126(40)}{2} = 2520 \text{ lb}$$

$$v_B = -v_D = \frac{2520}{30} = 84 \text{ plf}$$

Diaphragm 3:

$$w = C_s(\text{DL})W = 0.14(30)(25) = 105 \text{ plf}$$

$$V_{\text{support}} = \frac{wL}{2} = \frac{105(10)}{2} = 525 \text{ lb}$$

$$v = \frac{525}{25} = 21 \text{ plf}$$

It can reasonably be assumed that this shear is shared equally between TD1 and TD4.

$$F = \frac{525}{2} = 262.5 \text{ lb}$$

Diaphragm 4:

$$R_B = -R_E = \frac{105(50)}{2} = 2625 \text{ lb}$$

$$v_B = -v_E = \frac{2625}{25} = 105 \text{ plf}$$

Diaphragm 6:

$$R_A = -R_B = \frac{84(40)}{2} = 1680 \text{ lb}$$

$$v_A = -v_B = \frac{1680}{20} = 84 \text{ plf}$$

Transfer diaphragm TD1(Fig. 8.23): Because of the iterative process described above an initial guess for the TD reactions at grid lines B and E must be made. Assume the reactions are based on the uniform loads only, ignoring the discontinuous strut force at grid line D4.

$$R_B = R_E = \frac{63(50)}{2} = 1575 \text{ lb}$$

$$v_B = v_E = \frac{1575}{15} = 105 \text{ plf}$$

Transfer diaphragm TD3 (Fig. 8.22):

$$R_A = \left[\frac{84(40)^2}{2} + 262.2(30) \right] \frac{1}{40} = 1876.9 \text{ lb}$$

$$v_A = \frac{1876.9}{20} = 93.8 \text{ plf}$$

$$R_B = \left[\frac{84(40)^2}{2} + 262.2(10) \right] \frac{1}{40} = 1745.6 \text{ lb}$$

$$v_B = \frac{1745.6}{25} = 87.3 \text{ plf}$$

Transfer diaphragm TD4 (Fig. 8.23):

$$R_A = \left[\frac{105(40)^2}{2} + 262.2(30) \right] \frac{1}{40} = 2296.9 \text{ lb}$$

$$v_A = \frac{2296.9}{25} = 91.9 \text{ plf}$$

$$R_B = \left[\frac{105(40)^2}{2} + 262.2(10) \right] \frac{1}{40} = 2165.6 \text{ lb}$$

$$v_B = \frac{2165.6}{25} = 86.6 \text{ plf}$$

Diaphragm 5 (see Fig. 8.24)

Construction of the Basic Diaphragm Shears (See Fig. 8.24)

From initial guess of TD1, sum of forces to grid line B:

$$\Sigma V_B = 2520 + 1745.6 + 1575 + 2520 + 2625 + 2165.6 + 1680 = 14{,}831.2 \text{ lb}$$

$$R_E = \left[\frac{294(75)^2}{2} + 1481.2(40) + 210(15)(82.5) \right] \frac{1}{90} = 18{,}666.6 \text{ lb}$$

$$v_E = \frac{1866.6}{2(25)} = 373.3 \text{ plf}$$

$$R_A = R_B = 9333.3 \text{ lb}$$

$$V_C = V_E - wx = 9333.3 - 105(15) = 7758.3 \text{ lb}$$

$$v_C = \frac{V_C}{W} = \frac{7758.3}{25} = +310.3 \text{ plf}$$

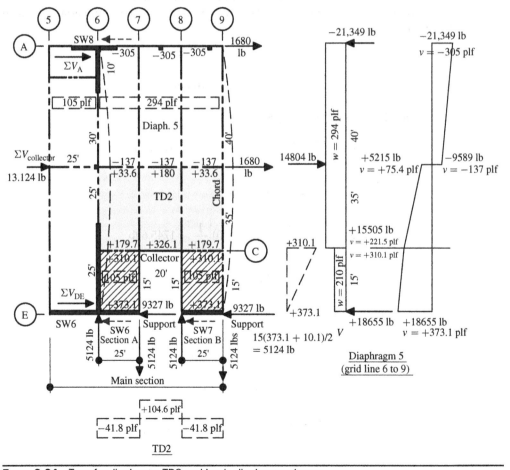

FIGURE 8.24   Transfer diaphragm TD2 and basic diaphragm shears.

Total assumed shear to SW5 and $6 = 2520 + 1575 + 2625 + 9333.3 = 16,053.3$ lb

At this point, the iteration process starts:

- Determine unit shears to SW5 and 6.
- Determine net shear at SW5.
- Determine discontinuous force transferred to TD1 at grid line D4 (see Fig. 8.3).
- Calculate the new reactions of TD1 to grid lines B and E.
- Sum forces to collector at grid line B.
- Start over until closure.

Final forces:

- Sum of forces to collector at grid line B (see Figs. 8.21 and 8.24).
- Main diaphragm 5: $R_A = 21,349$ lb, $R_E = 18,655$ lb (9327 lb sections A and B), $v_A = -305$ plf, $v_E = 373.1$, $v_C = 310.1$ plf, $v_B = -137$ plf
- TD1 (see Fig. 8.23): $R_B = 1548$ lb, $R_E = 1465$ lb, $v_B = -103.2$ plf, $v_E = +97.7$ plf
- Sum of shear to SW5 and $6 = 15,944.7$ lb, $v_{SW} = 265.7$ plf
- Discontinuous force at D4, V_{net} SW5 $= 265.7 - 84 = 181.7$ plf, $F_{D4} = 137.4$ lb (see Fig. 8.23)

Chord forces of sections A and B at grid lines 7 and 8 = sum of shears from E to C (reference Fig. 8.23).

$$F_{T/C} = \frac{(373.1 + 310.1)}{2}(15) = 5124 \text{ lb}$$

Transfer Diaphragm TD2 (See Figs. 8.23 and 8.24)   The chord forces from sections A and B are applied to the transfer diaphragm as shown.

$$V_9 = \frac{5124(20)}{70} = 1464 \text{ lb}$$

$$v_{6,9} = \frac{V_9}{W_{TD}} = \frac{1464}{35} = -40.8 \text{ plf negative unit shear}$$

Summing shears at grid line 7:

$$V_7 = -1464 + 5124 = +3660 \text{ lb}$$

$$v_7 = \frac{V_7}{W_{TD}} = \frac{3660}{35} = +104.6 \text{ plf positive unit shear}$$

Determination of Net Shears in TD2 (See Fig. 8.24)

Location (Grid Lines)	Net Shears
B from 6 to 7 and 8 to 9	$v = -41.8 + 75.4 = +33.6$ plf
B from 7 to 8	$v = +104.6 + 75.4 = +180$ plf
C from 6 to 7 and 8 to 9	$v = -41.3 + 221.5 = +179.7$ plf
C from 7 to 8	$v = +104.6 + 221.5 = +326.1$ plf

Determination of Longitudinal Collector/Chord Forces (See Figs. 8.25 to 8.27)

Force diagrams at grid line A (see Fig. 8.25):

Total shear to wall line A:

$$\sum V_A = 2520 + 1876.9 + 525 + 2296.9 + 21,349 + 1680 = 30,247.8 \text{ lb}$$

Force at 2 minus $5' = 84(25) = 2100 \text{ lb}$

$$v_{SW} = \frac{\sum V_A}{\sum L_{SW}} = \frac{30,247.8}{45} = 672.2 \text{ plf}$$

Net shears at SW4 from grid line 2 minus $5'$ to grid line 2:

$$v_{SW4} = 672.2 \text{ plf}$$
$$v_{diaph} = 84 \text{ plf}$$

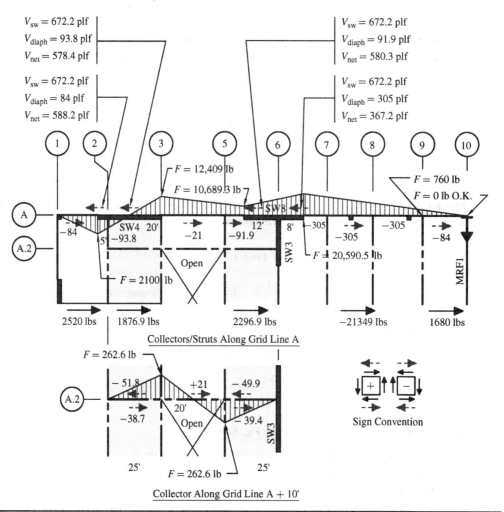

FIGURE 8.25 Diaphragm net shears and collector forces at grid line A.

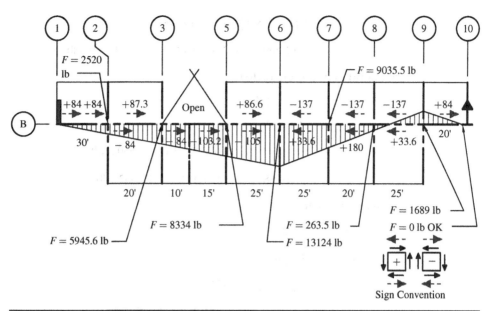

Figure 8.26 Collector forces at grid line B.

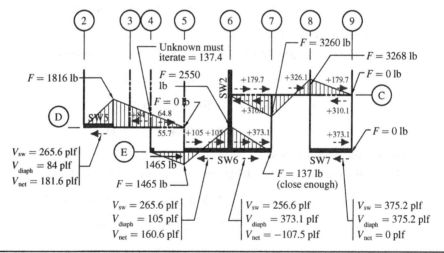

Figure 8.27 Collector forces at grid lines C, D, and E.

$$v_{net} = v_{SW} - v_{diaph} = 672.2 - 84 = 588.2 \text{ plf}$$

$$F_2 = 2100 - 588.2(5) = 841 \text{ lb}$$

SW4 grid line 2 to 3:

$$v_{SW4} = 672.2 \text{ plf}$$

$$v_{diaph} = 93.8 \text{ plf}$$

$$v_{net} = v_{SW} - v_{diaph} = 672.2 - 93.8 = 578.4 \text{ plf}$$

$$F_3 = 841 - 578.4(20) = -12{,}409 \text{ lb}$$

$$F_{6 \text{ minus } 12'} = -12{,}409 + 21(25) + 91.9(13) = -10{,}689.3 \text{ lb}$$

SW8 grid line 6 minus 12' to 6:

$$v_{SW4} = 672.2 \text{ plf}$$

$$v_{\text{diaph}} = 91.9 \text{ plf}$$

$$v_{\text{net}} = v_{SW} - v_{\text{diaph}} = 672.2 - 91.9 = 580.3 \text{ plf}$$

$$F_6 = -10{,}689.3 - 580.3(12) = -17{,}652.9 \text{ lb}$$

SW8 grid line 6 to 6 + 8':

$$v_{SW4} = 672.2 \text{ plf}$$

$$v_{\text{diaph}} = 305 \text{ plf}$$

$$v_{\text{net}} = v_{SW} - v_{\text{diaph}} = 672.2 - 305 = 367.2 \text{ plf}$$

$$F_{6 \text{ plus } 8'} = -17{,}652.9 - 367.2(8) = -20{,}590.5 \text{ lb tension}$$

$$F_{10} = -20{,}590.5 + 305(17 + 20 + 25) + 84(20) = 0$$

Therefore, the diagram closes.

Force diagrams at grid line A + 10' (see Fig. 8.25 for net shears):

$$F_3 = -(51.8 - 38.7)(20) = -262.4 \text{ lb}$$

$$F_5 = -262.2 + 21(25) = +262.6 \text{ lb}$$

$$F_6 = +262.6 - (49.9 - 39.4)(25) = 0 \text{ lb}$$

Therefore, the diagram closes.

Force diagrams at grid line B Starting at grid line 1 (see Fig. 8.26):

$$F_2 = 84(30) = +2520 \text{ lb compression at grid line 2}$$

$$F_3 = +2520 + (87.28 + 84)(20) = +5945.6 \text{ lb}$$

$$F_5 = +5945.6 + 84(10) + 103.3(15) = +8334 \text{ lb}$$

$$F_6 = +8334 + (86.6 + 105)(25) = +13{,}124 \text{ lb}$$

$$F_7 = +13{,}124 - (137 + 33.6)(25) = +9035.5 \text{ lb}$$

$$F_8 = +9035.5 - (137 + 180)(20) = +263.5 \text{ lb}$$

$$F_9 = +263.5 - (137 + 33.6)(25) = -1680 \text{ lb}$$

$$F_{10} = -1680 + 84(20) = 0 \text{ lb}$$

Therefore, the diagram closes.

Note that the collector forces from grid line 1 to 6 are in compression while being transferred into the main diaphragm as a concentrated force. The collector between grid lines 6 and 9 resists that force and is also in compression. The strut between grid lines 9 and 10 hangs off of the main diaphragm and is in tension.

Force diagrams at grid line C (see Fig. 8.27):

$$F_7 = -(310.1 - 179.7)(25) = -3260 \text{ lb}$$

$$F_8 = -3260 + 326.4(20) = +3268 \text{ lb}$$

$$F_9 = +3268 - (310.1 - 179.7)(25) = 8 \text{ lb}$$

Therefore, the diagram closes.

Force diagrams at grid line D, see Fig. 8.27, previously calculated by iteration:

$$F_{2+10'} = 1816 \text{ lb, previously determined}$$

$$F_{4'} = -137.4 \text{ lb, previously determined}$$

$$F_5 = -137.4 + (64.8 - 55.7)(15) = 0 \text{ lb}$$

Therefore, the diagram closes.

Force diagrams at grid line E (see Fig. 8.27):

$$F_5 = +1465 \text{ lb, previously calculated}$$

$$v_{SW6} = 265.6 \text{ plf}$$

$$v_{diaph} = 105 \text{ plf}$$

$$v_{net} = v_{SW} - v_{diaph} = 265.6 - 105 = 160.6 \text{ plf}$$

$$F_6 = +1465 - 160.6(25) = -2550 \text{ lb}$$

SW6 grid line 6 to 7:

$$v_{SW6} = 265.6 \text{ plf}$$

$$v_{diaph} = 373.1 \text{ plf}$$

$$v_{net} = v_{SW} - v_{diaph} = 265.6 - 310.1 = -107.5 \text{ plf}$$

$$F_7 = -2550 + 107.5(25) = +137.5 \text{ lb due to iteration}$$

Close enough because additional iteration required.

Note: SW7 shears equal diaphragm shears, therefore are no net shear to create a force.

Determination of Transverse Collector/Chord Forces (See Fig. 8.28)

Force diagrams at grid line 2, starting at grid line A:

$$F_{A+10'} = -\frac{[(93.8 - 84) + (51.8 - 42)]}{2}(10) = -98 \text{ lb}$$

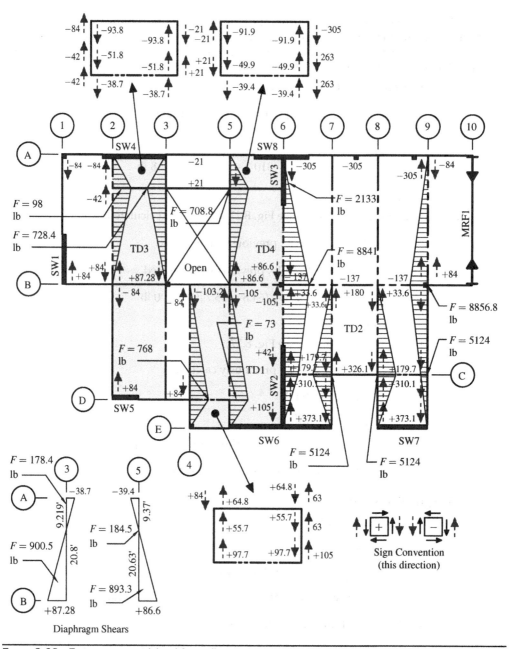

Figure 8.28   Transverse strut/chord force diagrams.

$$F_B = -98 + \frac{[(42 - 38.7) + (87.3 - 84)]}{2}(30) = 0 \text{ lb}$$

Therefore, the diagram closes.

Force diagrams at grid line 3, starting at grid line A:

The shear on the sheathing elements from grid line A+10′ to B, at 3 and 5, change from negative to positive values. The forces along grid lines 3 and 5 due to this change are calculated by similar triangles and are shown at the bottom left of the figure.

$$F_{A.2} = -\frac{[(93.8 - 21) + (51.8 + 21)]}{2}(10) = -728 \text{ lb}$$

Summing shears from similar triangles:

$$F_B = -728 - 178.4 - 900.5 = 5.9 \text{ lb close enough}$$

Therefore, the diagram closes.

Force diagrams at grid line 5, starting at grid line A:

$$F_{A.2} = \frac{[(91.9 - 21) + (49.9 + 21)]}{2}(10) = 709 \text{ lb}$$

Adding similar triangle forces:

$$F_B = 709 + 184.5 - 893.3 = 0 \text{ lb}$$

Therefore, the diagram closes.

Force diagrams at grid line 4, starting at grid line B:

$$F_D = -\frac{[(103.2 - 84) + (84 - 64.8)]}{2}(40) = -768 \text{ lb compression}$$

$$F_E = -768 + \frac{(55.7 + 97.7)}{2}(10) = 0 \text{ lb}$$

Therefore, the diagram closes.

Force diagrams at grid line 5 from B to E, starting at grid line B:

$$F_D = \frac{-[(105 - 103.2) + (64.8 - 63)]}{2}(40) = -72 \text{ lb}$$

$$F_E = -72 + \frac{[(63 - 55.7) + (105 - 97.7)]}{2}(10) = -1 \text{ lb}$$

Therefore, the diagram closes.

Force diagrams at grid line 6, starting at grid line A:

$$F_{A.2} = -\frac{[(305 - 91.9) + (263 - 49.9)]}{2}(10) = -2133 \text{ lb}$$

$$F_B = -2133 - \frac{[(263 - 39.4) + (137 + 86.6)]}{2}(30) = -8841 \text{ lb}$$

$$F_C = -8841 + \frac{[(105 + 33.6) + (179.7 - 42)]}{2}(35) = -4005.8 \text{ lb}$$

$$F_E = -4005.8 + \frac{[(310.1 - 42) + (373.1 - 105)]}{2}(15) = 15.8 \text{ lb}$$

Therefore, the diagram closes.

Force diagrams at grid line 7, starting at grid line E:

$$F_C = \frac{(584.92 + 521.92)}{2}(15) = 8301.3 \text{ lb tension}$$

$$F_B = 8301.3 - \frac{[(324.48 - 87.3) + (177.48 + 59.7)]}{2}(35) = 0 \text{ lb}$$

Therefore, the diagram closes.

Force diagrams at grid lines 7 and 8, starting at grid line E:

$$F_E = \frac{(373.1 + 310.1)}{2}(15) = -5124 \text{ lb}$$

$$F_B = -5124 + \frac{[(326.1 - 179.7) + (180 - 33.6)]}{2}(35) = 0 \text{ lb}$$

Therefore, the diagram closes. Collector forces are same at grid line 7.

Force diagrams at grid line 9, starting at grid line E:

$$F_C = -5124 \text{ lb}$$

$$F_B = -5124 - \frac{(179.7 + 33)}{2}(35) = -8856.8 \text{ lb}$$

$$F_A = -8856.8 + \frac{[(137 + 84) + (305 - 84)]}{2}(40) = 16.7 \text{ lb}$$

Therefore, the diagram closes.

All the force diagrams in the longitudinal and transverse direction close to zero; therefore, the analysis has been verified.

This example demonstrates that designing minimal lateral-force-resisting systems in the structure requires a significant increase in design time and detailing. Establishing complete load paths in this diaphragm can be difficult, especially when alternative solutions are required once the consequences of the design are known. It also demonstrates that boundary members and collectors can develop large forces and should always be checked. Although by reducing the number of shear walls in the structure, it might seem that the client could save money; in actuality, the costs are simply shifted from the deleted walls into other parts of the diaphragm. A large part of those costs is from the additional collectors and connections required to complete the complicated load paths. The costs of such complicated load path can be greater than installing more walls to eliminate or lessen the impact of the irregularities which has the benefit of increasing redundancy.

Figure 8.29 shows the same diaphragm with wind loads applied. The reader is encouraged to solve this problem so that a comparison can be made with seismic loading.

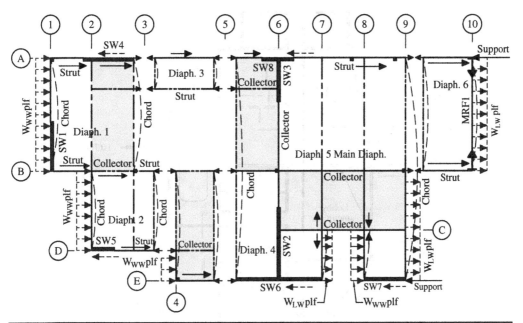

FIGURE 8.29 Wind load distribution into the main diaphragm.

8.5 Alternate layouts

The configuration of the shear walls, opening, and offsets in the diaphragm can make it difficult to eliminate all irregularities with additional walls. However, there are several possible wall arrangements that can simplify the lateral system. These options should be explored in the schematic phase of the design and reviewed with the client. An excellent set of exercises would be to analyze the diaphragm shown in Fig. 8.29 for each of the following modifications:

1. Add a 12-ft shear wall at grid lines 2 and 8—transverse loading.
2. Add an 8-ft vertical offset at grid line 5—transverse loading.
3. Add a shear wall along grid line B from 6 to 8.
4. Remove shear walls SW6 and SW7 and add a shear wall at grid line C from 7 to 8.
5. Remove shear walls SW4 and SW8 and add a shear wall at grid line A.2 from 3 to 5.
6. Make your own configurations.

Each of these exercises will provide its own special lesson regarding the development of complete load paths, and the transfer of forces across discontinuities.

8.6 Multi-Family Projects

Multi-family, multistory structures are increasingly popular and commonly have multiple offsets and other irregularities. The example plan shown in Fig. 8.30 with multiple offsets at the exterior walls is not that unusual and shows that it can be difficult to

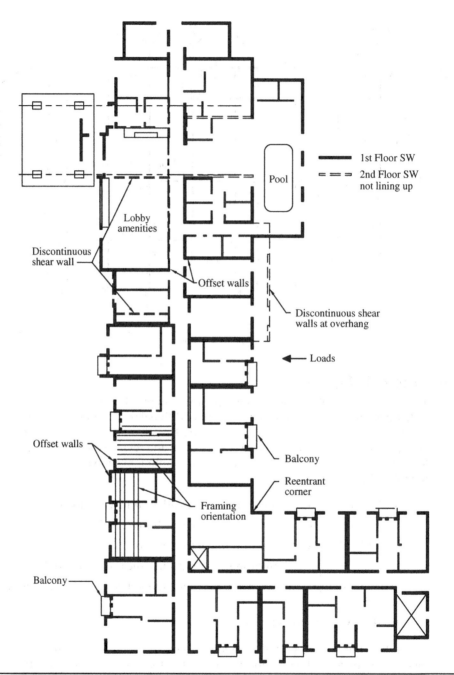

FIGURE 8.30 Discontinuities in multi-family plans.

establish clean lines of the diaphragm chords, struts, and collectors. Transfer of forces across these discontinuities can be complicated and requires special analytical tools to analyze. Another frequent challenge is when there are few opportunities for shear walls at the exterior wall lines, which can lead to an open-front diaphragm condition, as described in Chap. 6. Even when no adequate shear walls are located at an exterior wall line, structural continuity at the boundary lines is still required.

The orientation of the floor or roof framing determines if framing can be used as boundary or collector elements, or if transfer diaphragms or transfer areas must be used. If the floor or roof framing runs from demising wall to demising wall, there are several opportunities to use framing members that can act as interior chords or collectors, simplifying the diaphragm segments into simple rectangular sections. If the framing runs from the exterior wall to the corridor wall, it becomes much more difficult because there is no opportunity to use a continuous framing member oriented parallel with the exterior wall to act as a chord or strut. Continuous steel straps and flat blocking could be used which might require long strap lengths and the possible use of transfer diaphragms. Potential methods of analyzing offsets at the exterior walls will be examined in greater detail in Chap. 12.

CHAPTER 9

Standard Shear Walls

9.1 Introduction

The analysis and design of standard shear walls without openings can be found in numerous publications.[1-4] Most of the coverage is fairly comprehensive and has been common knowledge to practicing professional for decades. Prior versions of the IBC, and legacy codes prior to the IBC, governed the design of diaphragms and shear walls and provided all the necessary tables and information for their design. However, recent editions of the IBC[5] have removed most of the design criteria and tables for the design of diaphragms and shear walls and now refer to the SDPWS[1] for the design and construction of wood diaphragms and shear walls in either strength design (LRFD) or in allowable stress design (ASD). Detailed coverage of the requirements and adjustment factors for strength design and allowable stress design can be found in the current edition of SDPWS. Computer software programs and spreadsheets have become readily available on the market or have been created in-house. A basic review of the methods of analyzing simple shear walls without openings is presented here for reference and to help provide a complete coverage on the topic of shear walls.

Throughout the examples in this book, dead loads are applied to shear walls to demonstrate their potential effects on overturning forces, wall deflections, and stiffness, and to offer discussions on the pros and cons for their inclusion in the design of shear walls. Few shear walls have been tested with dead loads applied, and many textbook examples do not cover how to include them. Consequently, many believe their effects are not important to be included in the design and their omission is often considered appropriate in standard practice. The application of dead loads in the analysis can be a matter of preference or engineering judgment. However, it is the authors' opinion that there are appropriate and inappropriate limits to ignoring or including vertical loads when designing wall shears.

For medium and high aspect ratio shear walls, the assumption that walls act rigidly in-plane and can redistribute gravity loads to reduce the uplift forces is reasonable and widely performed as in the examples of this book. Not using the gravity loads to reduce uplift forces is a valid conservative choice. Not increasing the compression forces due to the wall rotation is unconservative. In high aspect ratio shear walls, particularly when stacked in multi-story structures, ignoring the potential increase in compression forces is not recommended. The outstanding questions are how much dead load can be applied to resist overturning uplift force and how much must be added to the compression force? Common sense should suggest that long walls will not act as a rigid body element over their entire length when resisting overturning forces, while short walls can. Four methods of handling gravity loading on shear walls are used in practice at some level and are presented for consideration:

1. Neglect the addition of dead loads.
   - Pros: Simplifies calculations and is considered as acceptable in standard practice. Matches conditions of most tested shear walls.
   - Cons: Can lead to larger hold-downs than required, can reduce wall stiffness, can result in unconservative compression forces in high aspect ratio and multi-story walls.

2. Analyze as a rigid body with dead load included.
   - Pros: Can reduce the size of hold-downs, can increase wall stiffness, help reduce wall rotation, can more accurately determine the increased shears acting in the header sections of FTAO shear walls, can account for the distribution of dead loads by headers and continuous rim joists across a segmented or an FTAO shear wall.
   - Cons: will increase bearing stresses in the wall bottom plate on the compression side of the wall, might require consideration of actual wall stiffness to transfer dead loads to boundary elements, require more calculations.

3. Analyze as a beam on an elastic foundation.
   - Pros: Could be more accurate.
   - Cons: Requires a level of calculations beyond standard practice.

4. Use judgment-based rules to distribute part of the dead load to the boundary elements. Example rules used by others on the length of the wall which distributes dead load to the ends:
   - One stud bay, or one-half of a stud bay.
   - A length based on an allowable aspect ratio for a shear wall. For example, 2-ft of an 8-ft tall wall if using a 4:1 rule, 4 ft of an 8-ft tall wall if using a 2:1 rule.
   - An arbitrary distance based on engineering judgment. For example, using last 10 ft of a 50-ft-long perforated shear wall instead of the whole length.

Because there are currently no standards or guidance in this regard, engineering judgment must be used.

9.2   Shear Wall Basics—Standard Walls

Shear walls are vertical lateral-force-resisting elements that act as supports for roof and floor diaphragms, cantilevered from and transferring their forces down into foundations or other support below. Shear walls can be full length walls or short wall sections connected by a strut/collector that can occur at any location along a line of lateral-force-resistance as shown in Fig. 9.1. The wall sheathing can consist of any of the materials listed below the figure, but most commonly consist of wood structural panels (WSP), such as plywood or oriented strand board (OSB), or gypsum wallboard. When a diaphragm is loaded by wind or seismic forces, the loads are transferred from the

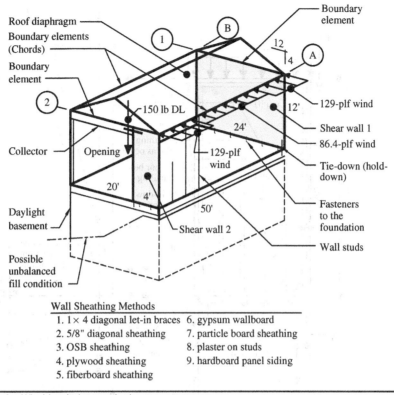

Roof diaphragm
Boundary elements (Chords)
Boundary element
Collector
Opening
Daylight basement
Possible unbalanced fill condition
150 lb DL
20'
4'
129-plf wind
24'
50'
12'
B
1
2
A
12
4
Boundary element
129-plf wind
Shear wall 1
86.4-plf wind
Tie-down (hold-down)
Fasteners to the foundation
Wall studs
Shear wall 2

Wall Sheathing Methods

1. 1 × 4 diagonal let-in braces   6. gypsum wallboard
2. 5/8" diagonal sheathing   7. particle board sheathing
3. OSB sheathing   8. plaster on studs
4. plywood sheathing   9. hardboard panel siding
5. fiberboard sheathing

Figure 9.1 Wind loads into a single-story structure.

diaphragm into the top of the supporting shear walls or other boundary elements at each end of the diaphragm, as discussed in Chap. 2.

A typical shear wall elevation with lateral forces and gravity loads applied is shown in Fig. 9.2. The dead loads applied to the wall consist of roof or floor dead loads, the dead load of the wall, and dead loads applied from headers that frame into the ends of the wall. It is the authors' opinion that dead loads should be applied resulting in more efficient hold-downs, and more accurately determine shear wall stiffness and wall rotation. Horizontal forces applied at the top of the wall causes the wall to rotate and overturn. The applied dead loads counteract and resist the overturning moment. Whenever the overturning moment exceeds the dead load resisting moment, pre-manufactured hold-down anchors are required to be installed on the boundary elements to resist the tension force caused by the net overturning moment.

Multiple studs or posts are installed at each end of the wall that acts as the shear wall boundary elements or chords. The number of multiple studs or solid post sizes is determined by design or as specified as a minimum by the hold-down manufacturer, whichever is greater. Anchor bolts and/or nailing in the bottom plate of the wall resist the wall shears and keep the wall from sliding off the framing or foundation below.

There are several methods commonly used to establish the length of the overturning moment resisting arm used for calculating the hold-down force, depending upon which textbook is used as a standard. Some use the full length of the wall. Some use the

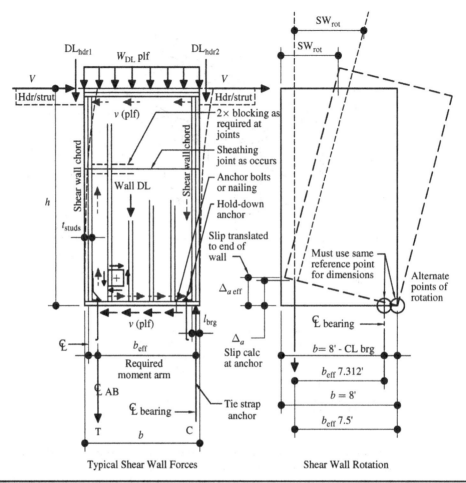

Figure 9.2   Shear wall elevation and rotation.

end of the wall to centerline of the hold-down anchor rod, and others use the center of bearing to the center of the hold-down anchor rod. For low aspect ratio (wide) walls, using the full width, b, to calculate the rotational contribution to shear wall deflection is a reasonable engineering assumption. However, for high aspect ratio (narrow) walls, the difference between the location of the edge of wall and the anchorage location can be a significant portion of the wall width. A more precise method to account for this is not described in older editions of SDPWS. The latest edition of SDPWS now refines how the deflection from anchorage slip and wall rotation is calculated. For high aspect ratio walls, the rotational term $\dfrac{h\Delta_a}{b}$, must be modified considering the actual location of the hold-down anchor bolt, where Δ_a is taken. If discrete hold-downs or continuous tie rods are used, the hold-down location is inboard from the end of the wall at the centerline of the anchor rod where Δ_a occurs, reducing the moment resisting arm from "b" to an effective width, b_{eff}. Figure 9.2 shows what has been included in the 2021 edition of SDPWS. Either translate the elongation of the anchorage, $\Delta_{a'}$ out to the end of the wall and use b minus the centerline of bearing for the moment resisting arm length, or use the full elongation, Δ_a, and the distance from the centerline of the anchor to the centerline of bearing as the moment resisting arm length, which is now noted as b_{eff}. Either approach will produce the same wall rotation. As an option, the code will allow the overturning point of rotation to be taken from the end of the wall in lieu of center of bearing based on engineering judgment. Wall rotation due to anchor slip can be calculated as follows:

$$SW_{rot} = \frac{h(\Delta_a)}{b_{eff}} \ or \ \frac{h(\Delta_{aeff})}{b}$$

To get the total vertical elongation, $\Delta_{a'}$ and wall rotation, the effects of shrinkage and bearing perpendicular to grain crushing must be added.

9.2.1 Shrinkage Effects

Wood has a large amount of naturally occurring moisture. As the lumber dries to the ambient climate environment, known as its equilibrium moisture content (EMC), shrinkage also occurs. There are several ways to calculate the amount of shrinkage in wood members. One simple calculation is to assume a dimensional change of 0.0025 in per inch of cross-sectional dimension for every 1 percent change in MC. This loss of moisture results in volumetric changes to the lumber, which is often referred to as shrinkage.

$$\text{Shrinkage} = 0.0025(D)(\text{Starting MC} - \text{End MC})$$

where D is the dimension of the member in the direction under consideration (in), in this case the thickness of a wall plate.

Estimating a 15 percent initial MC at the time of construction and 10 percent equilibrium MC results in the following shrinkage for a $2\times$ bottom plate:

$$\Delta_{shrinkage} = 0.0025(1.5)(15 - 10) = 0.019 \text{ in}$$

The effect of shrinkage is commonly considered when designing continuous rod tie-down systems. IBC 2304.3.3 also requires consideration of the effects of shrinkage of wood framing on MEP systems when wood framing supports more than two stories and a roof. This is a good threshold beyond which consideration of the effects of shrinkage on shear wall deformations is advised.

9.2.2 Bearing Perpendicular-to-Grain Crushing

Crushing contributes to wall rotation which is caused by perpendicular-to-grain bearing stresses between the wall compression chord members and bottom plate. These stresses must be adjusted in accordance with NDS Section 4.2.6.[6] The allowable bearing stresses noted therein are based on steel plate to wood plate contact. For wood-to-wood contact, an adjustment factor $= 1.75$ for parallel to perpendicular grain wood contact must be used. Reference NDS C4.2.6.

Boundary values for bearing perpendicular-to-grain stresses and crushing using DFL:

$$F_{c\perp 0.02} = 0.73 F'_{c\perp} = 0.73(625) = 456.3 \text{ psi lower boundary}$$

$$F_{c\perp 0.04} = F'_{c\perp} = 625 \text{ psi upper boundary}$$

When $f_{c\perp} \leq F_{c\perp 0.02}$ in

$$\Delta_{\text{crush}} = 0.02\left(\frac{f_{c\perp}}{F_{c\perp 0.02}}\right)$$

When $F_{c\perp 0.02} \leq f_{c\perp} \leq F_{c\perp 0.04}$

$$\Delta_{\text{crush}} = 0.04 - 0.02\left(1 - \frac{f_{c\perp}}{F'_{c\perp}}\right)/0.27$$

When $f_{c\perp} > F_{c\perp 0.04}$

$$\Delta_{\text{crush}} = 0.04\left(\frac{f_{c\perp}}{F_{c\perp 0.04}}\right)^3$$

The analysis procedure for the wall overturning force is as follows:

$v_{\text{SW}} = \dfrac{V}{b}$, unit wall shear (plf), gravity loads are typically not included when

calculating the unit shear in the wall

$b_{\text{eff}} = b - \left[t_{\text{studs}} + \text{CL} + \dfrac{l_{\text{brg}}}{2}\right]\dfrac{1}{12}$, (ft), moment resisting arm where

b = length of wall (ft)

CL = distance from the face of the hold-down to the centerline of the anchor bolt (in), provided by the hold-down manufacturer

l_{brg} = length of bearing area (in)

t_{studs} = thickness of end studs (in)

$F_{\text{o/t}} = \dfrac{Vh \pm \sum M_{\text{DL}}}{b_{\text{eff}}}$, overturning force, using the appropriate load case factors

V = applied lateral force (lb or kips)

h = height of wall (ft)

$\sum M_{\text{DL}}$ = sum of resisting dead load moments

Shear walls are required to be designed for the strength load combinations of ASCE7-16 Section 2.3.1 or Allowable Stress Design, Section 2.4.1 to determine the maximum tension and compression forces in the shear wall boundary elements, understanding that they do not occur simultaneously or in the same load combination.[7]

The calculation of shear wall deflections is important for the determination of story drifts, the design of open-front diaphragms, determining wall rigidities and in irregular shaped diaphragm design. The shear wall deflection equations have been well documented in many publications. The four-term equation below was developed from static load tests on shear wall assemblies with aspect ratios of 2:1 or less. The APA[2] noted that the equation was not as accurate for walls with aspect ratios greater than 2:1 and is a nonlinear solution.

$$\Delta_{\text{SW}} = \frac{8vh^3}{EAb} + \frac{vh}{Gt} + 0.75he_n + \frac{\Delta_a h}{b_{\text{eff}}} \qquad \text{SDPWS-21 Eq.C4.2.2-1}$$

where $\dfrac{8vh^3}{EAb}$ = bending deflection

$$\frac{vh}{Gt} = \text{shear deflection}$$

$$0.75he_n = \text{nail slip contribution}$$

$$\frac{\Delta_a h}{b_{eff}} = \text{wall rotation}$$

$v =$ the unit wall shear (plf)

$h =$ the wall height (ft)

$b =$ the wall width (ft)

$b_{eff} =$ Effective wall length resisting wall rotation

$E =$ modulus of elasticity of the studs/chords (psi)

$\Delta_a =$ hold-down slip and elongation, shrinkage and crushing

$e_n =$ nail slip per SDPWS Table C4.2.2D) (in)

$G_t =$ panel rigidity through the thickness (lb/in) per SDPWS Table C4.2.2A

$A =$ area of the boundary elements (sq. in)

As an alternate, the deflection can be calculated using the three-term equation provided in Section 4.3.4 in SDPWS-21. The three-term equation is a linear simplification of the four-term equation.

$$\delta_{SW} = \frac{8vh^3}{EAb} + \frac{vh}{1000G_a} + \frac{h\Delta_a}{b} \qquad \text{SDPWS Eq.4.3-1}$$

$\Delta_a =$ total vertical displacement of wall anchorage system (including fastener slip, device elongation, rod elongation, etc.)

A graphical comparison between the two equations can be referenced in SDPWS commentary Figure C4.3.4B and in Fig. 9.11.

Example 9.1: Single-Story Shear Wall Overturning Analysis

The one-story structure shown in Fig. 9.1 is 50-ft long by 24-ft wide with 12-ft high walls. The roof consists of 4:12 double pitched trusses spaced at 24″ o.c. A 4-ft shear wall with a drag strut occurs at grid line 2 and a full-length shear wall occurs at grid line 1. Wind loads control the design (see Figs. 9.1 and 9.2).

Given:

$h = 12$ ft

Diaphragm end-zone pressure $0.6W = 129$ plf

Diaphragm interior zone pressure $0.6W = 86.4$ plf

$V = 2385$ lb, force at top of the wall

$W_{DL} = 135$ plf, includes wall DL

$DL_{hdr} = 150$ lb

4-ft shear wall (see Fig. 9.2):

$$\text{A.R.} = \frac{12}{4} = 3 < 3.5, \quad \text{OK}$$

$V = 2385$ lb diaphragm reaction to SW's including the effects of end-zone wind pressure.

$$v_{SW} = \frac{2385}{4} = 596.3 \text{ plf}$$

Note, the selected nominal shear capacity must be reduced in accordance with SDPWS Section 4.3.3 because the aspect ratio exceeds 2:1. For WSP shear walls, the reduction factor $= 1.25 - 0.125(h/b)$ multiplied by the nominal shear capacity, divided by 2 for ASD wind $> v_{SW}$.

Calculate the effective moment resisting arm:

CL $= 1.5$ in, distance from face of hold-down to centerline of anchor bolt

$$l_{brg} = 3 \text{ in}$$

$$b_{eff} = b - \left[t_{studs} + CL + \frac{l_{brg}}{2}\right]\frac{1}{12} = 4 - \frac{\left(3 + 1.5 + \frac{3}{2}\right)}{12} = 3.5 \text{ ft}$$

Load case $0.6D + 0.6W$:

$$T = \left[2385(12) - 0.6(4)(135)(1.875) - 0.6(150)(3.75)\right]\frac{1}{3.5} = 7907.2 \text{ lb}$$

Load case $D + 0.6W$:

$$C = \left[2385(12) - 150(0.25) + 135(4)(1.625)\right]\frac{1}{3.5} = 8417.2 \text{ lb}$$

24-ft shear wall:

$$\text{A.R.} = \frac{12}{24} = 0.5 < 3.5, \text{ therefore OK, no reduction required}$$

$$V = 2385 \text{ lb}$$

The shear applied to grid line 1 is the same as the load applied to grid line 2 because the wind end-zone pressure is required to be applied to each corner separately. The end-zone wind pressure is applied to account for accidental torsion.

$$v_{SW} = \frac{V}{b} = \frac{2385}{24} = 99.4 \text{ plf}$$

$$b_{eff} = b - \left[t_{studs} + CL + \frac{l_{brg}}{2}\right]\frac{1}{12} = 24 - \frac{\left(3 + 1.5 + \frac{3}{2}\right)}{12} = 23.5 \text{ ft}$$

Load case $0.6D + 0.6W$ maximum tension force:

$$T = \left[2385(12) - 0.6(24)(135)(11.875)\right]\frac{1}{23.5} = 235.5 \text{ lb, uplift}$$

Note, $T = 221$ lb if calculated using the full width of the shear wall instead of considering actual location of T/C locations.

Load case $D + 0.6W$ maximum compression force:

$$C = \left[2385(12) + 135(24)(11.625)\right]\frac{1}{23.5} = 2820.6 \text{ lb} \quad \blacktriangle$$

9.3 Hold-Down Anchors and Boundary Members

Solid posts or multiple studs are typically used for shear wall boundary members (chords). A variety of hold-down installation options have been used when attaching discrete hold-down anchors to the boundary members, as shown in Fig. 9.3. Questions arise about how the boundary edge nailing should be applied when multiple stud boundary members are used. Option A of the figure shows where edge nailing is only installed at the outer stud location. Under this condition, stitch nailing should be installed connecting each of the 2× studs together capable of transferring the full vertical shear at that location. If this option is used, footnote (10) of the 2021 SDPWS Table 4.3A requires an 8 percent

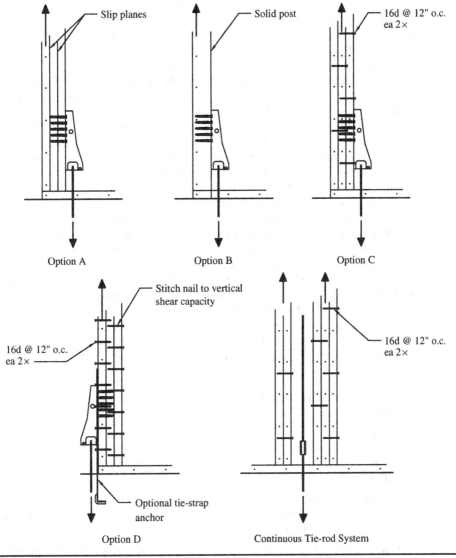

Figure 9.3 Typical hold-down installations.

reduction in shear capacity if 10d common nails are used in WSP sheathing, regardless of the nail spacing used. This is due to testing on the effects on eccentric hold-down connections. Option B shows the anchor installation on a solid 4× or 6× post, with a single or multiple rows of nailing (as required). Option C shows the condition where edge nailing is installed in each of the 2× stud members. For this condition, member to member fastening can be at a larger spacing, preferably a maximum spacing of 12″ o.c. Option D shows the anchor installed at the outside face of the wall. This creates virtually little or no eccentricity. Footnote (10) only refers to the use of 10d common nails in shear walls with WSP sheathing and does not specifically address the type of boundary element or its nailing. However, the reduction factor was based on testing of option A. For continuous tie-rod systems, all 2× stud members of the stud pack should be edge nailed.

The APA—The Engineered Wood Association conducted tests[8] comparing the performance of 3× framing members versus stitch-nailed double 2× members at adjoining panel edges and end members. The results from cyclic shear wall testing showed that the shear walls with double 2×'s stitch-nailed together performed about the same as those with a single 3× by all measures, except the shear walls with double 2× framing had increased displacement capacity and ductility, which is a desirable characteristic for seismic performance. Photographs from the tests showing the end post anchorage failures provided interesting observations. Several of the failures suggested that critical grain-slope and knot placement near the hold-downs contributed to the failure at the anchor locations. This type of failure is less likely to occur if multiple 2× members are used for the end posts. The result is similar to glu-lam beam member fabrication and design, where the same imperfections occurring in each piece of lumber at the same connection is unlikely.

Hold-down anchor options have increased greatly in recent years. The development of new hold-down anchors and improvements made to existing ones has been the result of testing and experience gained from field applications. Figures 9.4 and 9.5 provide details showing the installation of discrete single and double hold-down anchors. Typically, these types of anchors are fastened to solid posts or multiple stitch-nailed studs with bolts or more ductile screw attachments. The hold-down anchors are connected to the foundation with an embedded anchor bolt, or the embedment of the steel strap hold-down itself. Upper floor shear walls are typically connected to walls or beams below with standard hold-downs or with steel straps, as shown in Fig. 9.5. All single and double anchors must be installed in strict accordance with the manufacturers' requirements. When two anchors are installed on the same members, the tabulated loads for the hold-downs may be combined provided that

- The allowable loads for a single hold-down is based on a minimum wood member thickness. When two hold-downs are used, the number of members required to resist the tension or compression forces must be checked.

- When hold-downs are installed with screws on the opposite sides of a member, the post must be large enough to prevent opposing hold-down screw interference, which could potentially cause splitting. As an alternate, the hold-downs can be vertically offset in accordance with the manufacturer's recommendations. See Fig. 9.3.

- When hold-downs are installed with bolts on the opposite sides of a member, the number of members required to resist the tension or compression forces and the net section for tension must be checked.

- Multiple studs used as boundary members or at interior joints must be stitch-nailed or screwed together to act as a single member.

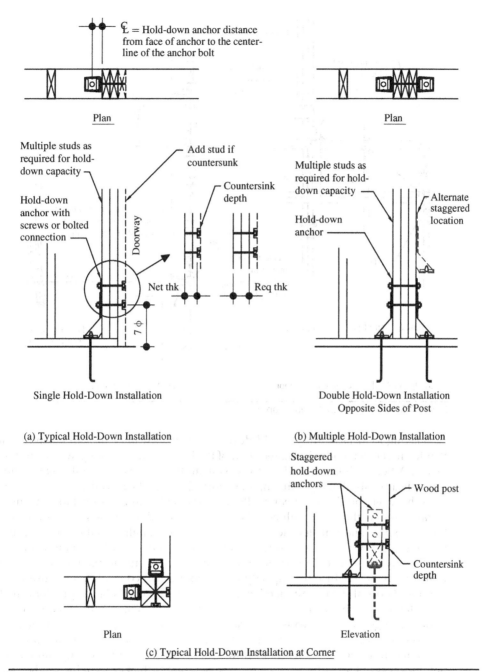

FIGURE 9.4   Typical hold-down installations.

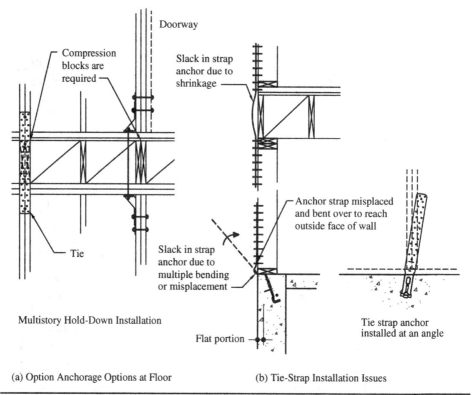

Doorway

Compression blocks are required

Slack in strap anchor due to shrinkage

Tie

Slack in strap anchor due to multiple bending or misplacement

Anchor strap misplaced and bent over to reach outside face of wall

Flat portion

Multistory Hold-Down Installation

Tie strap anchor installed at an angle

(a) Option Anchorage Options at Floor

(b) Tie-Strap Installation Issues

FIGURE 9.5 Improperly installed strap anchors.

Having to install multiple hold-downs on a shear wall chord can be an indication that too much demand is being required of the lateral-force-resisting system and its elements. A better choice would be to reevaluate the proposed lateral system and add some shear walls, thereby reducing the demand on the shear walls.

Whenever multistory shear walls are designed, compression blocking must be installed below the shear wall chords to prevent buckling of the framing members in the floor space and transfer the compression forces brought about by shear wall overturning. Figure 9.5 shows the condition at a story-to-story transfer of hold-down forces.

Failure to install hold-down anchors in accordance with the manufacturer's instructions can greatly reduce the capacity of the hold-down connection or cause localized failures of critical shear wall members. Several common installation problems that can cause failures can be seen in Figs. 9.5(b) and 9.6. A common installation error occurs at door openings when the heads of the hold-down bolts are concealed by countersinking the bolts into the stud as shown in Fig. 9.4(a). The anchor manufacturer requires a minimum thickness for a specified hold-down capacity. The countersinking can reduce the net thickness of the multiple studs to an unacceptable thickness, reducing the capacity of the anchor. An additional stud can be installed to accommodate for the countersinking. Another common installation error occurs when the hold-down anchor bolt is misplaced in the foundation. Figure 9.6 shows two conditions where the anchor bolts have been misplaced. The detail on the left side of the figure shows the condition where the hold-down anchor is installed with an in-plane offset. The common response is to bend

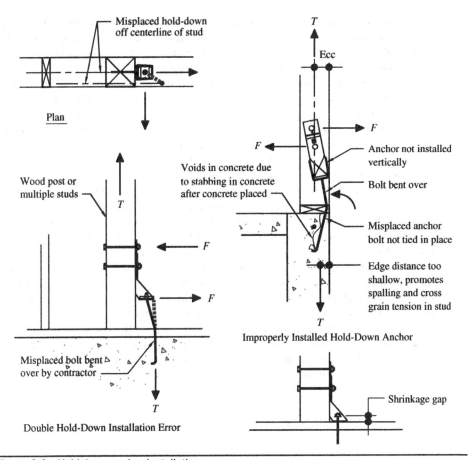

Misplaced hold-down
off centerline of stud

Plan

Wood post or
multiple studs

Voids in concrete due
to stabbing in concrete
after concrete placed

Anchor not installed
vertically

Bolt bent over

Misplaced anchor
bolt not tied in place

Edge distance too
shallow, promotes
spalling and cross
grain tension in stud

Improperly Installed Hold-Down Anchor

Misplaced bolt bent
over by contractor

Double Hold-Down Installation Error

Shrinkage gap

Figure 9.6 Hold-down anchor installation errors.

the anchor bolt extension to enable connection to the hold-down. The detail on the right side of the figure shows the condition where the anchor bolt is installed horizontally offset (out-of-plane) from the centerline of the stud. Whenever a tension force is applied to either of these connections, the hold-down tries to straighten out causing cross-grain tension to occur at the bolted connection or localized bending in the boundary member. Failure to tie the anchor bolt in place properly or stabbing it into the wet concrete can also cause voids in the concrete around the bolt severely reducing the capacity or rendering it ineffective. If the anchor bolt is placed too close to the edge or end of the foundation wall, an inadequate edge/end distance can cause spalling of the concrete or cause a reduced capacity of the anchor. Figure 9.5(b) shows a strap anchor installation at the foundation wall. Common installation problems occur when the anchor is misplaced or intentionally bent over to accommodate the installation of the wall framing.

Manufacturers may allow one bend cycle for certain products to aid in installation, but this is not advisable and should be verified with the manufacturer's installation instructions. Even when allowed by the manufacturer, when a strap is bent back into position for nailing to the wall, a set kink in the strap can occur which prevents it from functioning until the slack is removed. Whenever discrete hold-downs are used, shrinkage can occur in the framing, which will leave a gap between the bottom plate of the anchor and the nut as shown in Fig. 9.6; or, whenever a tie strap is installed from floor to floor, shrinkage can cause the strap to buckle as shown in Fig. 9.5(b). In both conditions, the slack must be taken out before the finish wall covering is installed, which can be very difficult to do due to time constraints. Occasionally, flat blocking is placed behind the sheathing where the strap does not lap onto the studs to capture the unfilled nail holes. Installations that weaken or compromise the hold-down or other member in the lateral-force-resisting system should be corrected before proceeding further. An engineered solution is typically required to fix these types of problems. The anchor manufacturers have limited options for correcting these installation errors but typically do have set procedures to correct most of the installation errors noted above. The manufacturer should be contacted to see if the installation can be salvaged.

Automatic tensioning anchors as shown in Figs. 9.7 and 9.8 compensate for shrinking of the framing and settlement within the structure and are gaining in popularity for use in mid-rise structures. A continuous rod tie-down system utilizes a combination of threaded rods with bearing plates and take-up devices at each level to transfer the forces to the foundation. In addition to providing a complete system for overturning, they can also help reduce the drift at the top of the wall due to the restraint from overturning at each floor. Design catalogs and on-line software are available from the manufacturers that will help customize a tie-down system and make it more efficient. Specifying the tie-down requirements for each wall on the construction drawings and conducting structural observations on all lateral-force-resisting systems, as a standard of practice, can reduce the number of field installation errors and ensure that the lateral system designed is installed to perform under the loads for which it was designed.

Figure 9.7 Automatic tensioning hold-down testing.

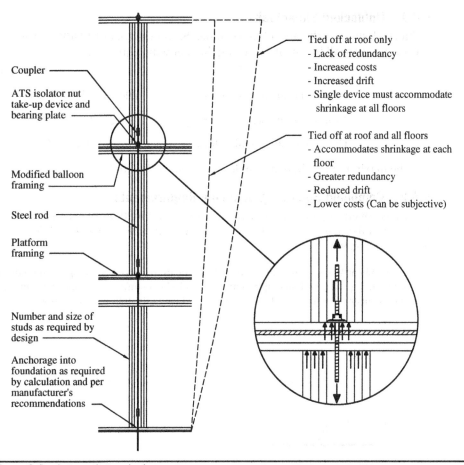

Coupler

ATS isolator nut
take-up device and
bearing plate

Modified balloon
framing

Steel rod

Platform
framing

Number and size of
studs as required by
design

Anchorage into
foundation as required
by calculation and per
manufacturer's
recommendations

Tied off at roof only
- Lack of redundancy
- Increased costs
- Increased drift
- Single device must accommodate
  shrinkage at all floors

Tied off at roof and all floors
- Accommodates shrinkage at each
  floor
- Greater redundancy
- Reduced drift
- Lower costs (Can be subjective)

FIGURE 9.8 Automatic tensioning system.

9.4   Shear Wall Sheathing Combinations

The current version of SDPWS has significantly modified the nominal unit shear capacities for wood-framed diaphragms and shear wall in accordance with Section 4.1.4,
respectively. Previous versions of Tables 4.2A-4.2D for diaphragms and 4.3A-4.3D for
shear walls separated the nominal unit seismic shear capacities from wind capacities
into different table values, with the wind capacities increased by a factor of 1.4. The current edition has only one set of nominal shear values per each table. Allowable unit
shear capacities are now determined by modifying the nominal shear value as follows:

- Seismic: $\text{ASD} = \dfrac{v_{\text{nom}}}{2.8}$, $\text{LRFD} = 0.5 v_{\text{nom}}$

- Wind: $\text{ASD} = \dfrac{v_{\text{nom}}}{2.0}$, $\text{LRFD} = 0.8 v_{\text{nom}}$

No further increases are allowed.

9.4.1 Unblocked Shear Walls

Unblocked shear walls are allowed under Section 4.3.5.3 of SDPWS. The allowable shear capacity shall be calculated using the following equation.

$$V_{n(ub)} = v_{n(b)}C_{(ub)}$$ SDPWS Eq.4.3-2

where $v_{n(ub)}$ = Nominal unit shear capacity (plf) for unblocked shear walls

$v_{n(b)}$ = Nominal unit shear capacity (plf) from Table 4.3A

$C_{(ub)}$ = Unblocked shear wall adjustment factor from Table 4.3.5.3

The maximum height of an unblocked shear wall is 16 ft.

9.4.2 Combined Sheathing Summing Requirements

The rules governing the summing of shears for combinations of sheathing materials on the same wall or within a line of walls are covered in SDPWS Section 4.3.5.4 and as summarized in Fig. 9.9.

4.3.5.4.1 Shear Walls with Similar Shear Wall Sheathing Systems on Opposing Sides of Common Framing: The combined apparent shear wall shear stiffness, $G_{a(c)}$ and the combined nominal unit shear capacity, $v_{n(c)}$, shall be determined using the following equations:

$$G_{a(c)} = G_{a(1)} + G_{a(2)}$$

$$v_{n(c)} = K_{min}G_{a(c)}$$

$$where \ K_{min} = \frac{v_1}{G_{a(1)}} \ or \ \frac{v_2}{G_{a(2)}}, \text{minimum ratio}$$

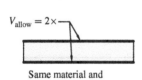

$V_{allow} = 2\times$

Same material and construction

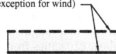

$V_{allow} = 2\times$ the smaller nominal unit shear capacity or the larger unit shear capacity, whichever is greater (see exception for wind)

Dissimilar materials

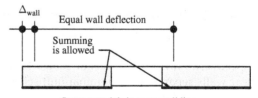

Δ_{wall} Equal wall deflection

Summing is allowed

Same materials in same wall line

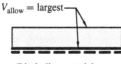

$V_{allow} = $ largest

Dissimilar materials

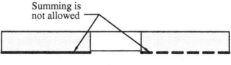

Summing is not allowed

Dissimilar materials in same wall line

FIGURE 9.9 Summing shear capacities.

4.3.5.4.2 Shear Walls with Dissimilar Shear Wall Sheathing Systems on Opposing Sides of Common Framing: The combined nominal unit shear capacity, $V_{n(c)'}$ shall be taken as either two times the smaller nominal unit shear capacity or as the larger nominal unit shear capacity, whichever is greater. The combined apparent shear wall shear stiffness, $G_{a(c)'}$ shall be associated with the nominal unit shear capacity used in design and shall be either two times the apparent shear stiffness of the shear wall sheathing system with the smaller nominal unit shear capacity or the apparent shear stiffness of the shear wall sheathing system with the larger nominal unit shear capacity.

Exception: For wind design, the combined nominal unit shear capacity, $V_{n(c)'}$ of shear walls sheathed with a combination of wood structural panels, hardboard panel siding, or structural fiberboard on one side and gypsum wallboard on the opposite side shall equal the sum of the sheathing capacities of each side.

4.3.5.4 Where the shear wall sheathing system on one side of the shear wall is a blocked shear wall and the shear wall sheathing system on the other side of the shear wall is an unblocked shear wall, provisions of 4.3.5.5.2 apply.

The SDPWS does not address two layers of WSP sheathing on one side of a shear wall. APA Report TT-115 has demonstrated that two layers of WSP sheathing applied to one side of the wall develops a capacity to be twice that of a single layer applied to one side of the wall, but has a nail spacing limitation of not less than 4″ o.c. This application is best suited for a retrofit project where only one side is exposed to access.[9]

9.5   Shear Wall Aspect Ratios and Reduction Factors

9.5.1   Required Aspect Ratios

Wind and seismic loads and seismic design categories vary between geographical areas. However, the need to provide a complete load path does not. Additionally, applicable seismic detailing requirements apply even when wind controls the design forces as specifically stated in IBC Section 1604.10. Regardless, failing to follow through with a complete review of the controlling forces in shear walls and collectors for all load cases can lead to an overstress in one of the lateral-force-resisting elements. Earlier editions of SDPWS had different aspect ratio requirements for wind and seismic that were embedded in the footnotes of the aspect ratio table. The current edition requires the aspect ratio from the table to be applied to all shear walls regardless of whether wind or seismic controls the design. The maximum allowable shear wall ratios can be referenced in SDPWS Table 4.3.3, shown in Fig. 9.10.

9.5.2   Required Capacity Reduction Factors

Wood Structural Panel, WSP, Segmented Shear Walls

SDPWS Section 4.3.3.2 notes that walls with aspect ratios (h/b) greater than 2:1, the nominal shear capacity shall be multiplied by the Aspect Ratio Factor (WSP) $= 1.25 - 0.125\,h/b$. For structural fiberboard shear walls with aspect ratios (h/b) greater than 1:1, the nominal shear capacity shall be multiplied by the Aspect Ratio Factor (fiberboard) $= 1.09 - 0.09\,h/b$. Reference SDPWS Figure 4C.

Perforated Shear Walls

Section 4.3.3.4, Aspect Ratio of Segments: The aspect ratio limitations of Table 4.3.3 shall apply to perforated shear wall full height segments within a perforated shear wall as

AF & PA SDPWS Table 4.3.3 Maximum Shear Wall Dimension Ratios

Diaphragm Sheathing Type	Maximum h/b_s ratios
Wood structural panel, unblocked	2:1
Wood structural panel, blocked	3.5:1
Particleboard, blocked	2:1
Diagonal sheathing, conventional	2:1
Gypsum board	2:1 (1)(2)
Portland cement plaster	2:1 (1)
Structural fiberboard	3.5:1

Footnotes
1. Walls having aspect ratios exceeding 1.5:1 shall be blocked shear walls.
2. For wind design of wood structural panel shear walls or structural fiberboard shear walls, with gypsum wallboard on the opposite side, maximum h/b for gypsum wallboard shall be permitted to be 3.5:1.

FIGURE 9.10 Maximum shear wall aspect ratios. (*Courtesy, American Wood council, Leesburg, VA.*)

illustrated in Figure 4E. Portions of walls with aspect ratios exceeding 3.5:1 shall not be considered in the sum of shear wall segments, Σb_i. In the design of perforated shear walls, the length of each perforated shear wall segment with an aspect ratio greater than 2:1 shall be multiplied by $2b_i/h$ for the purposes of determining b_i and Σb_i. The provisions of Section 4.3.3.2 and the exceptions to Section 4.3.5.5.1 shall not apply to perforated shear wall segments.

Force-Transfer Shear Walls

Section 4.3.3.3 notes that the aspect ratio limitations of Table 4.3.3 shall apply to the overall shear wall including openings and to each wall pier at the sides of openings as illustrated in Figure 4D. The height of a wall pier with an opening on one side shall be defined as the clear height of the pier at the side of the opening. The height of a wall pier with an opening on each side shall be defined as the larger of the clear heights of the pier at the sides of the openings. The length of a wall pier shall be defined as the sheathed length of the pier. Wall piers with aspect ratios exceeding 3.5:1 shall not be considered as portions of force-transfer shear walls.

9.6   Shear Wall Rigidity and Nominal Shear Wall Stiffness

Current architectural designs can become very complex and certain wood framed structures have configurations which may behave more like rigid diaphragms than flexible diaphragms. Because of this, there have been many discussions on when a rigid diaphragm analysis for wood roofs is more appropriate than the classic flexible diaphragm approach. NEHRP Seismic Design Brief 10[10] and ASCE 7-16 commentary both state that "The diaphragms in most buildings braced by wood light-frame shear walls are semi-rigid." Section 1604.4 of the IBC requires that the lateral forces to be distributed to the various vertical elements of the lateral force resisting system in proportion to their rigidities, considering the rigidity of the horizontal bracing system or diaphragm. ASCE 7 Section 12.3.1 notes that the structural analysis shall consider the relative stiffnesses of diaphragms and the vertical elements of the seismic force-resisting system. To simplify the implementation of a structural analysis, ASCE 7 and the IBC give conditions on when a diaphragm can be idealized as flexible and when a diaphragm can be idealized as rigid.

The rigidity, k, of a shear wall is based on the deformation of the wall at a given load. The process of determining the wall rigidity can be an iterative process because of potential nonlinearities in the hold-down slip and elongation, crushing and shrinkage of the wall bottom plate, and nail slip. For an applied lateral force, F, and the corresponding calculated horizontal deflection, δ, a linear stiffness, K, is calculated by

$$K = F/\delta_{sw}$$

Wood structural panel shear wall deflections and stiffness can be determined using either the SDPWS three-term deflection Eq. 4.3-1 or four-term deflection Eq. C4.3.2-1. The differences between the effects of the two equations are shown in Fig. 9.11. The three-term equation is a linear simplification of the four-term equation through combining the panel shear and nail slip terms into one term. The tabulated terms, G_a, are based on the nail slip term calculated at the ASD wind capacity of the shear wall nailing. As demonstrated in Fig. 9.11 for a specific wall, at shear forces less than the ASD wind capacity, the three-term equation will produce a larger deflection than the four-term equation, which would result in less stiffness.

$$\delta_{SW} = \frac{8vh^3}{EAb} + \frac{vh}{1000G_a} + \frac{h\Delta_a}{b_{eff}} \qquad \text{SDPWS Eq. 4.3-1}$$

When vertical dead loads are used to resist overturning, commonly accepted deflection calculation methods have some intrinsic complexities. At low levels of horizontal loading, the vertical gravity loads alone can be enough to prevent net uplift from occurring at the boundary elements at the "tension" side of the shear wall. However, once the horizontal forces overcome the resisting moment due to gravity forces, net uplift occurs and any slip and elongation of the hold-down adds to wall rotation of the shear wall, increasing the deflection and decreasing stiffness. This basic model to calculate shear wall deflections creates a nonlinear relationship between lateral load and horizontal deflection where the calculated stiffness can vary with both the vertical and lateral loading on the wall. An engineer could attempt to calculate a deflection and wall stiffness consistent with each independent load combination and direction of loading applicable to the structural design; however, coupled with rigid diaphragm analysis where the lateral loads on a wall depend on the wall stiffness, such a process would be a herculean effort, likely accomplished successfully only in a nonlinear structural analysis program, with little demonstrated structural benefit in normal equivalent lateral force seismic design.

Combining rigid diaphragm analysis and shear wall deflection calculations is problematic due to nonlinearities, which can affect the distribution of loads to the shear walls and will affect the shear wall deflections. This can lead to a different set of stiffness values that may not be consistent, whenever changing:

- Load combinations
- Vertical or lateral loads

- Direction of loading
- Redundancy or
- Accidental torsion

Taking this approach requires an iterative search for the point of convergence, which is not practical for multistory structures.

Sources of nonlinearities:

- Hold-down slip at uplift (e.g., shrinkage gap)
- Hold-down system tension and elongation
- Compression crushing. Nonlinear in NDS
- Shrinkage
- Four-term deflection equation

Therefore, the following suggestions are provided in the WoodWorks publication *A Seismic Design Example of a Cantilever Wood Diaphragm*[11] as one possible rational approach to shear wall stiffness calculations when performing a rigid diaphragm analysis. The approach, based on stiffness at the maximum lateral capacity, is to

- Use a single vertical and lateral load combination, $1.0D + Q_e$ with $\rho = 1.0$ and $A_x = 1.0$, while ignoring vertical seismic load effects, to calculate deflections and nominal shear wall stiffnesses.
- Use the deflection calculated at the 1.4 times the maximum ASD lateral capacity of the wall using the three-term equation to define a nominal stiffness of the wall using:

$$K = \frac{F_{max}}{\delta_{SW\,max}}$$

- Use the nominal stiffness for all load combinations and drift checks, where needed, including deflection checks

$$\delta_{SW} = \frac{F}{K}$$

The maximum wall capacity is equal to the maximum allowable shear or hold-down capacity, whichever is less.

This method allows having only one set of nominal stiffness values for one set of construction details for a shear wall. If the nailing or hold-down hardware changes, the maximum capacity will change, and the nominal stiffness will need to be recalculated.

9.6.1 Vertical Loading for Wall Stiffness Calculations

An expected gravity loading of $1.0D$ is used for shear wall deflections and stiffness calculations. This represents a single "nominal" gravity loading on which to base the deflection calculations to perform an RDA. This approach can be generalized to multistory buildings where it is common to ignore the presence of any live loading to resist overturning.

Special situations such as high snow or storage loads may prompt a designer to consider a high and low gravity loading as separate conditions for which different shear wall stiffness values are calculated. This is similar to the gravity loads used for nonlinear procedures in ASCE 41-13 Eq. 7-3 and the simplification of gravity loading in ASCE 7-16 Section 16.3.2 for nonlinear response history analysis procedures. Similarly, the vertical seismic effects ($E_v = +/- 0.2\ S_{DS}D$) are not considered for wall stiffness calculations. This approach may not be valid for structures with significant vertical discontinuities in the seismic or gravity load-resisting systems. Under the proposed ASCE 7-22 provisions balloting, public comment version:

12.8.6.1 Minimum Base Shear and Load Combination for Computing Displacement and Drift: The elastic analysis of the seismic force-resisting system for computing displacement and drift shall be made using the prescribed seismic design forces of Section 12.8 using a load factor of 1.0 on E_h in the presence of expected gravity loads. Expected gravity loads shall be taken as no less than $1.0D + 0.5L$. L shall be taken as $0.8L_o$ for live loads that exceed 100 lb/ft² (4.79 kN/m²) and $0.4L_o$ for all other live loads, and L_o is the unreduced design live load (see Table 4-1).

9.6.2 Lateral Loading for Wall Stiffness Calculations

A similar approach is used in consideration of the magnitude of the lateral loading when calculating nominal wall stiffnesses. SDPWS Section C4.3.2 (Fig. 9.11) shows how SDPWS Eq. C4.3.2-2 is a simplification of the four-term shear wall deflection Eq. C4.3.2-1 by calibrating the three-term equation to match the nonlinear four-term equation at the applied lateral load of 1.4 times the ASD shear wall capacity. This simplification removes the nonlinear behavior of the nail slip term, e_n. A similar approach is used for the wall at the wall capacity. Given a wall design that has the required strength, the stiffness of the wall used in this example is calculated based on a strength level wall capacity limit, not based on the actual applied loads.

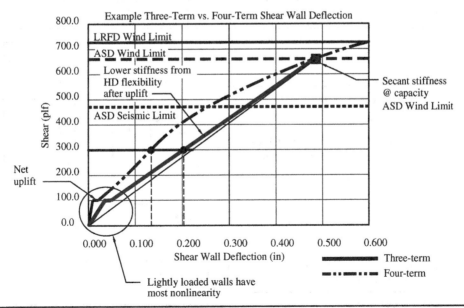

Figure 9.11 Three-term versus four-term equation comparison.

One could follow the approach of only considering the shear capacity of the wall sheathing. However, once a hold-down has been selected it could be prudent to also check the shear capacity of the wall based on the hold-down capacity. Although not mandatory, both capacities will be checked for this example and the lowest force at the top of the wall will be used to calculate the wall stiffness. For shear walls not requiring hold-downs, the lateral wall force used to calculate stiffness can be determined at the point where uplift occurs at the tension side of the wall.

9.6.3   Detailed Shear Wall Deflections and Nominal Stiffness Calculations (See Fig. 9.12)

Below is an example of detailed deflection and stiffness calculations using previously selected shear wall details with a vertical applied loading of 1.0D and lateral loading of the selected lateral capacity of the wall. Because wall capacities are used to calculate

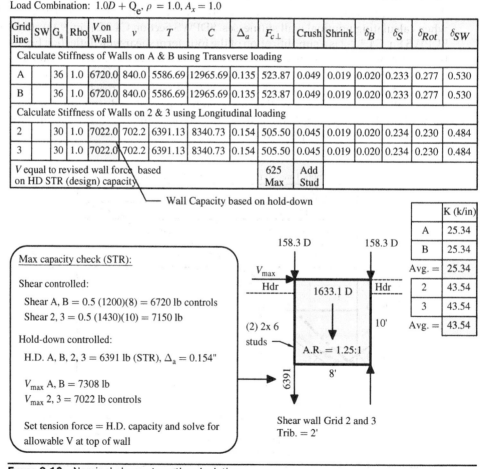

Load Combination: $1.0D + Q_e$, $\rho = 1.0$, $A_x = 1.0$

Grid line	SW	G_a	Rho	V on Wall	v	T	C	Δ_a	$F_{c\perp}$	Crush	Shrink	δ_B	δ_S	δ_{Rot}	δ_{SW}
Calculate Stiffness of Walls on A & B using Transverse loading															
A		36	1.0	6720.0	840.0	5586.69	12965.69	0.135	523.87	0.049	0.019	0.020	0.233	0.277	0.530
B		36	1.0	6720.0	840.0	5586.69	12965.69	0.135	523.87	0.049	0.019	0.020	0.233	0.277	0.530
Calculate Stiffness of Walls on 2 & 3 using Longitudinal loading															
2		30	1.0	7022.0	702.2	6391.13	8340.73	0.154	505.50	0.045	0.019	0.020	0.234	0.230	0.484
3		30	1.0	7022.0	702.2	6391.13	8340.73	0.154	505.50	0.045	0.019	0.020	0.234	0.230	0.484
V equal to revised wall force based on HD STR (design) capacity						625 Max	Add Stud								

Wall Capacity based on hold-down

Max capacity check (STR):

Shear controlled:

Shear A, B = 0.5 (1200)(8) = 6720 lb controls
Shear 2, 3 = 0.5 (1430)(10) = 7150 lb

Hold-down controlled:

H.D. A, B, 2, 3 = 6391 lb (STR), $\Delta_a = 0.154"$

V_{max} A, B = 7308 lb
V_{max} 2, 3 = 7022 lb controls

Set tension force = H.D. capacity and solve for allowable V at top of wall

158.3 D   158.3 D

V_{max}

Hdr   1633.1 D   Hdr

(2) 2x 6 studs

A.R. = 1.25:1

6391

8'

10'

Shear wall Grid 2 and 3
Trib. = 2'

	K (k/in)
A	25.34
B	25.34
Avg. =	25.34
2	43.54
3	43.54
Avg. =	43.54

Figure 9.12   Nominal shear strength calculation.

corresponding deflections, design load amplification factors do not apply ($\rho = 1.0$ and $A_x = 1.0$).

Shear Wall Maximum Capacity: See Fig. 9.12 for applied dead loads and calculated wall geometry. These values and spreadsheet are taken from the cantilever diaphragm example paper for reference. Determine the maximum force (LRFD) that can be applied at the top of the wall. Seismic forces govern:

$$L_{sw} = 8 \text{ ft}, h_{sw} = 10 \text{ ft}$$

$$\text{A.R.} = 1.25{:}1 < 3.5{:}1$$

Since the aspect ratio does not exceed 2:1, no reduction is required per SDPWS Section 4.3.4.

$SW_{A,B}$—Maximum nailing/shear capacity:

$$15/32'' \text{ structural 1 WSP w/10d @3'' o.c., } v_{nom} = 1860 \text{ plf, } G_a = 37$$

$$V_{A,B} = 0.5(1860 \text{ plf}) (8 \text{ ft}) = 6720 \text{ lb}$$

$SW_{A,B}$—Use the hold-down capacity to determine corresponding lateral force capacity at top of wall. Set the overturning force equal to the maximum hold-down capacity at strength level, then solve for the maximum force at the top of the wall:

$$T = 6391 = [V(10 \text{ ft}) - 1836(7.687 \text{ ft}) - 3248(3.812 \text{ ft})$$
$$+ 2295(0.063 \text{ ft})]/7.312, V_{allow} = 7308 \text{ lb}$$

$V = 6720$ lb controls lateral capacity. Use for deflection calculations.

A software spreadsheet was developed to aid in the calculation of the nominal stiffness. The maximum wall capacity was inserted directly into the spreadsheet under the "V on wall" tab, then calculated for the resulting nominal wall stiffness. Once the nominal shear stiffness, k, is determined, any force that is applied to the top of the wall from a rigid diaphragm analysis, the wall deflection can be determined by dividing that force by the nominal wall stiffness instead of resorting back to the deflection equation.

$$\delta_{sw} = \frac{F}{K}$$

9.7 Sloped Shear Walls

Shear walls with sloped top plates are a common occurrence. These walls require an additional consideration in the application of the loads to the wall. The biggest effect of the slope is that the vertical and horizontal components of the strut forces and diaphragm shears applied along the top plate of the shear wall increase the overturning force. The strut on the right side of the wall shown in Fig. 9.13 also increases the overturning moment because of the increased height where the force is applied. The classic approach of designing sloped diaphragms is to analyze it as though it is a flat horizontal beam, regardless of whether it is sloped or not. The diaphragm shears and strut or collector forces calculated are assumed to be applied horizontally to the shear wall. The shear wall shown in the figure has struts installed on each end of the wall. The strut forces are broken down into vertical and horizontal components. The sum of the unit diaphragm shears applied along the top plate of the wall are also broken down into vertical and horizontal components and applied at the centerline of the wall top plate as shown in the figure.

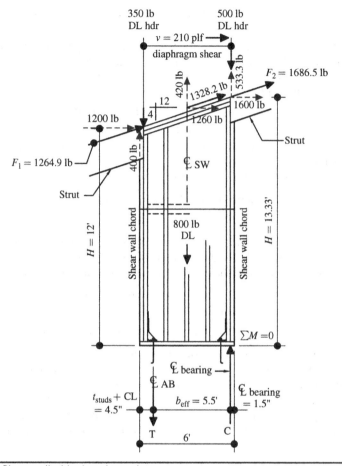

FIGURE 9.13   Shear wall with sloped top plate.

Example 9.2: Sloped Shear Wall Analysis (See Fig. 9.13)

This example assumes that strut forces applied to the wall governs over diaphragm chord forces. The slope of the wall top plate is a 4:12 roof pitch. The horizontal component of the calculated strut force on the left side of the wall is 1200 lb and the horizontal force component on the right side is 1600 lb. A uniform diaphragm unit shear of 210 plf is applied along the length of the wall, which is the horizontal component of the unit shear applied to the wall plate. All lateral forces are wind forces and have already been factored by 0.6 for ASD design.

$$b = 6 \text{ ft } 0 \text{ in, length of shear wall}$$

$$H_1 = 12 \text{ ft, height on the left end of the shear wall}$$

$$H_2 = 13.33 \text{ ft, height on the right end of the shear wall}$$

$$DL = 800 \text{ lb, dead load of wall and roof above}$$

$$P_{DL1} = 350 \text{ lb, dead load of header on the left side of the wall}$$

$$P_{DL2} = 500 \text{ lb, dead load of header on the right side of the wall}$$

Aspect ratio taken at average wall height $= \dfrac{(12 + 13.33)}{2(6)} = 2.11{:}1$

$$b_{\text{eff}} = b - \left[t_{\text{studs}} + \text{CL} + \frac{l_{\text{brg}}}{2} \right] \frac{1}{12} = 6 - \frac{(4.5 + 1.5)}{12} = 5.5 \text{ ft}$$

Sum of horizontal components of diaphragm unit shears:

$$V_{\text{H}} = v_{\text{diaph}} b = 210(6) = 1260 \text{ lb}$$

Sum of vertical components of diaphragm unit shears:

$$V_{\text{V}} = \frac{1260(4)}{12} = 420 \text{ lb}$$

$F_{\text{H1}} = 1200$ lb, horizontal force component of the left strut

$$F_{\text{V1}} = \frac{1200(4)}{12} = 400 \text{ lb, vertical force component of left strut}$$

$$F_1 = \sqrt{F_{\text{H1}}^2 + F_{\text{V1}}^2} = \sqrt{(400)^2 + (1200)^2} = 1264.9 \text{ lb, resultant axial force, left}$$

$F_{\text{H2}} = 1600$ lb, horizontal force component of the right strut

$$F_{\text{V2}} = \frac{1600(4)}{12} = 533.3 \text{ lb, vertical force component of right strut}$$

$$F_2 = \sqrt{F_{\text{H2}}^2 + F_{\text{V2}}^2} = \sqrt{(533.3)^2 + (1600)^2} = 1686.5 \text{ lb, resultant axial force, right}$$

$\Sigma M = 0$ right

Load case 0.6D + 0.6W:

$$T = \left[1200(12) + \frac{1260(12 + 13.33)}{2} + 1600(13.33) + 400(5.875) - 0.6(350)(5.75) \right.$$
$$\left. + 420(2.875) - 0.6(800)(2.875) - 0.6(533.3)(0.125) \right] \frac{1}{5.5}$$
$$= 9566.5 \text{ lb}$$

$\Sigma M = 0$ left

Load case D + 0.6W:

$$C = \left[1200(12) + \frac{1260(12 + 13.33)}{2} + 1600(13.33) + 500(5.5) - 350(0.25) \right.$$
$$\left. - 420(2.625) + 400(0.375) - 533.3(5.625) + 800(2.625) \right] \frac{1}{5.5}$$
$$= 9544.7 \text{ lb}$$

$\Sigma H = 0$

$\Sigma V_{\text{H}} = 1200 + 1260 + 1600 = 4060 \text{ lb}$

$$v_{\text{SW}} = \frac{\Sigma V_{\text{H}}}{b} = \frac{4060}{6} = 676.7 \text{ plf}$$

Since the aspect ratio is greater than 2:1, a shear capacity reduction is required. ▲

9.8   Multistory Shear Wall Analysis

There are a wide variety of commercially available software programs and in-house spreadsheets that have been created for multistory shear wall analysis. Some of these programs examine a single story at a time. Figure 9.14 provides an example of how to analyze the wall shears and overturning forces for a three-story shear wall with a flat roof. The unit shear in the wall is equal to the roof shear, V_r, divided by the width

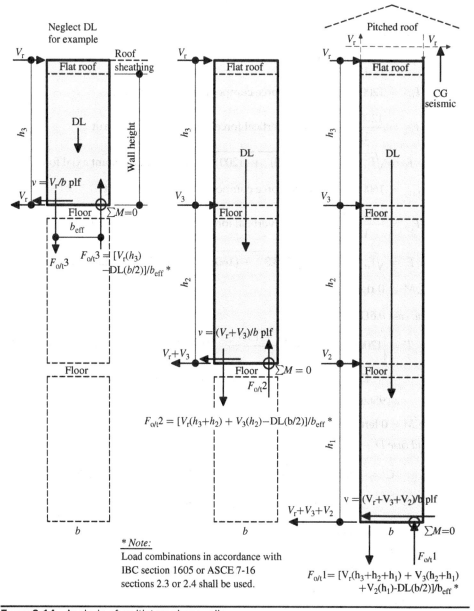

Note:
Load combinations in accordance with IBC section 1605 or ASCE 7-16 sections 2.3 or 2.4 shall be used.

FIGURE 9.14 Analysis of multistory shear walls.

of the wall. The overturning force is equal to the roof shear times the distance from the base of the wall to the lateral force which is equal to the wall height plus the depth of the flat roof, h_3. Applying the shear force at the top of the wall and ignoring the depth of the roof or floor framing is a common mistake, as will be demonstrated. The resisting moment is equal to the sum of the dead load times their respective distances from the point about which the moments are being summed. All loads and forces must be factored in accordance with the appropriate load combination for wind or seismic.

Top Story:

$h_3 = h_{wall} + h_{roof}$, overturning moment arm

V_r = The factored roof shear force that is applied to the wall segment

D = The factored dead load that is applied to the wall segment

b = Width of wall segment

$v = \dfrac{V_r}{b}$, unit shear in the wall

b_{eff} = Distance between the center of bearing to the center of the hold-down anchor

$F_{o/t} = \dfrac{[V_r h_3 \pm \sum M_D]}{b_{eff}}$

Second Floor:

V_r = Factored shear force at the roof level

V_3 = Factored shear force at the third-floor level

D = The factored dead load that is applied to the wall segment

$h_3 = h_{wall} + h_{roof}$

$h_2 = h_{wall} + h_{floor}$

b = Width of wall segment

$v = \dfrac{V_r + V_3}{b}$, unit shear in the wall at the second floor

$F_{o/t} = \dfrac{\left[V_r\left(h_3 + h_2\right) + V_3 h_2 \pm \sum M_D\right]}{b_{eff}}$

First Floor:

V_r = Factored shear force at the roof level

V_3 = Factored shear force at the third-floor level

V_2 = Factored shear force at the second-floor level

D = The factored dead load that is applied to the wall segment

$h_3 = h_{wall} + h_{roof}$

$h_2 = h_{wall} + h_{floor}$

$h_1 = h_{wall} + h_{floor}$

b = Width of wall segment

$$v = \frac{V_r + V_3 + V_2}{b}, \text{ unit shear in the wall at the first floor}$$

$$F_{o/t} = \frac{\left[V_r (h_3 + h_2 + h_1) + V_3 (h_2 + h_1) + V_2 h_1 \pm \Sigma M_D \right]}{b_{eff}}$$

For sloped roofs, the distance to the center of gravity (C.G.) of the roof is added for seismic loading.

As an option, the diaphragm shears that occur at the roof or floor sheathing can be transferred down through the roof or floor framing into the top of the wall as discussed in detail in Sec. 9.10. Assume that the tension overturning force from the wall above is equal to 2400 lb and the horizontal shear force applied to the wall is equal to 4000 lb as shown in Fig. 9.15. The design is controlled by wind. Determine the tension forces only.

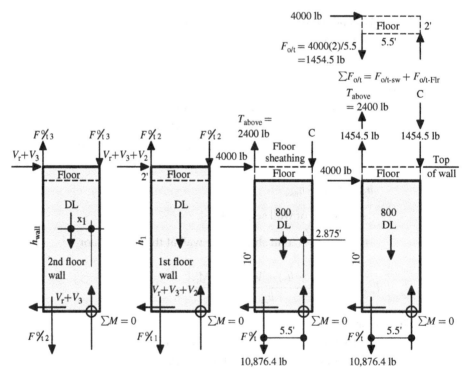

Forces Applied at the Floor Sheathing Level   Forces Transposed to the Top of the Wall

Note:
Load combinations in accordance with
IBC section 1605 or ASCE 7-16
sections 2.3 or 2.4 shall be used.

FIGURE 9.15 Alternate method—transposed forces.

Case 1: Shear applied at the floor sheathing level:

$$h_{wall} = 10 \text{ ft}$$

$$h_{floor} = 2 \text{ ft}$$

$$b = 6 \text{ ft}$$

$$b_{eff} = b - \left[t_{studs} + \text{CL} + \frac{l_{brg}}{2} \right]\frac{1}{12} = 6 - \frac{3" + 1.5" + \frac{3"}{2}}{12} = 5.5 \text{ ft,}$$

where $(2)2\times$ studs are used at each end of the wall and the hold-down $\text{CL} = 1\text{-}\frac{1}{2}$ in

$$V = 4000 \text{ lb}$$

$$D = 800 \text{ lb, total factored dead load both walls}$$

$$T = 2400 \text{ lb, overturning forces from the wall above}$$

$$F_{o/t} = \frac{[4000(10+2) + 2400(5.5) - (800)(2.875)]}{5.5} = 10{,}709.1 \text{ lb}$$

Case 2: Shear applied at the top of the wall:

The process of transferring the horizontal shear across the depth of the floor framing effectively increases the overturning forces by

$$F = \frac{Vh_{floor}}{b} = \frac{4000(2)}{5.5} = 1454.5 \text{ lb}$$

Transposing these forces to the top of the wall produces an overturning force equal to

$$F_{o/t} = \frac{[4000(10) + 2400(5.5) + 1454.5(5.5) - (800)(2.875)]}{5.5} = 10{,}709.1 \text{ lb}$$

Therefore, the resulting forces are the same.

If the transfer across the depth of the floor framing had been ignored and the lateral force was applied at the top of the wall, the hold-down capacity would have been under-designed by 1454 lb.

Example 9.3: Multistory Shear Wall Analysis

The three-story shear wall examined in this example has two 8-ft wide shear wall sections which shares the total shear force to the wall line. The wind load distribution to the end wall is shown at the right of Fig. 9.16. Dead loads of the wall panels and header reactions to the wall panels are also shown in the figure. The analysis is based on allowable stress design. Wind loads have been factored for the load combination governing the design.

Controlling Load Combinations ASD:

$$D + 0.6W$$

$$0.6D + 0.6W$$

$$h_3 = 10 \text{ ft height at third floor which equals the wall plus roof height}$$

$$h_2 = 10 \text{ ft height at second floor which equals the wall plus floor height}$$

$$h_1 = 12 \text{ ft height at first floor which equals the wall plus floor height}$$

$$h_{floor} = 1 \text{ ft all floors and roof}$$

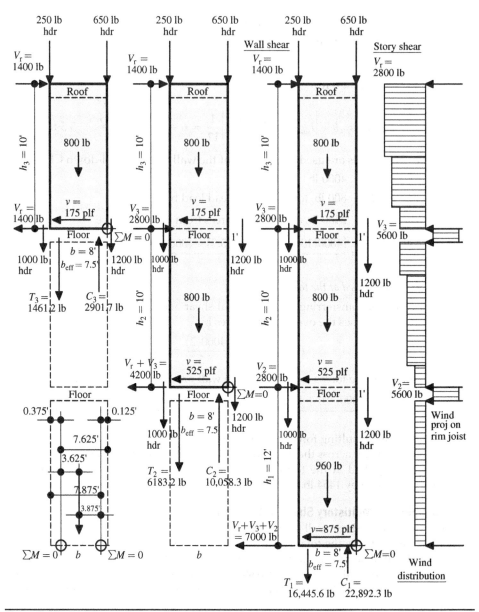

FIGURE 9.16 Multistory shear walls—wind controls.

$b = 8$ ft all floors

$$b_{\text{eff}} = b - \left[t_{\text{studs}} + \text{CL} + \frac{l_{\text{brg}}}{2} \right] \frac{1}{12} = 8 - \frac{3''+1.5'' + \dfrac{3''}{2}}{12} = 7.5 \text{ ft}$$

where (2)2× studs are used at each end of the wall and the hold-down

CL = 1-½ in, distance from the face of the hold-down to the centerline of the bolt

$$V_r = \frac{2800}{2} = 1400 \text{ lb, load per wall panel at the roof}$$

$$V_3 = \frac{5600}{2} = 2800 \text{ lb, load per wall panel at the third floor}$$

$$V_2 = \frac{5600}{2} = 2800 \text{ lb, load per wall panel at the second floor}$$

The typical distances to the forces from the point of summing moments are shown at the lower left of the figure.

Third floor:

$$v_{wall} = \frac{V_r}{b} = \frac{1400}{8} = 175 \text{ plf, unit wall shear at third floor}$$

$$T_3 = [1400(10) - 0.6(800)(3.875) - 0.6(250)(7.875)]\frac{1}{7.5} = 1461.2 \text{ lb tension}$$

$$C_3 = [1400(10) + (800)(3.625) + (650)(7.625) - 250(0.375)]\frac{1}{7.5}$$
$$= 2901.7 \text{ lb compression}$$

Second floor:

$$v_{wall} = \frac{V_r + V_3}{b} = \frac{1400 + 2800}{8} = 525 \text{ plf, unit wall shear at the second floor}$$

$$T_2 = [1400(20) + 2800(10) - 0.6(800)(2)(3.875) - 0.6(250 + 1000)(7.875)]\frac{1}{7.5}$$
$$= 6183.2 \text{ lb tension}$$

$$C_2 = [1400(20) + 2800(10) + (800)(2)(3.625) + (650 + 1200)(7.625)$$
$$- (250 + 1000)(0.375)]\frac{1}{7.5} = 10{,}058.3 \text{ lb compression}$$

Compression perpendicular to the grain should be checked for two studs bearing on the wall bottom plate.

First floor:

$$v_{wall} = \frac{V_r + V_3 + V_2}{b} = \frac{1400 + 2800 + 2800}{8} = 875 \text{ plf, unit shear at the first floor}$$

$$T_1 = [1400(32) + 2800(22 + 12) - 0.6(1600 + 960)(3.875)$$
$$- 0.6(250 + 2000)(7.875)]\frac{1}{7.5} = 16{,}465.6 \text{ lb tension}$$

$$C_1 = [1400(32) + 2800(22 + 12) + (1600 + 960)(3.625) + (650 + 2400)(7.625)$$
$$- (250 + 2000)(0.375)]\frac{1}{7.5} = 22{,}892.3 \text{ lb compression}$$

$$f_{c\perp} = \frac{22{,}892.3}{16.5(1.13)} = 1278 \text{ psi}, \text{ where } C_b = 1.13 \text{ bearing area factor for 3 in width.}$$

Compression perpendicular to the grain has been exceeded for two studs bearing on the first-floor wall bottom plate. The allowable tension stress in the boundary studs should also be checked at the net section. The addition of studs will reduce the resisting moment arm length, b_{eff}, which will increase the overturning force. The forces on the wall are approaching the high end of the demand that should be required of the wall. The wall width should be increased if the forces increase. ▲

9.9 Multistory Shear Wall Effects

In recent years, engineers have been questioning the traditional practice of analyzing shear walls from floor to floor as separate wall elements when calculating deflections, thinking that there could be more interaction between the shear walls from floor to floor than previously thought. When looking at deflections, the accumulated moments are traditionally ignored vis-à-vis the first term of the SDPWS deflection equation which only has the applied shear and not an applied bending moment at the top applied by the walls above. Alternatively, the wall from floor to roof may act as a single tall shear wall where accumulated moments going down the wall increase the flexural deformations at lower levels, and accumulated rotations and horizontal displacements going up the wall increase the total displacements at upper levels. Upper walls can be considered to rotate and translate horizontally in direct response to rotation of the top of the supporting wall below. This effect would cause an added lateral displacement at each floor above. This process is repeated for each wall above. Consideration of multistory shear wall effects results in larger calculated displacements, and lower wall stiffnesses, than traditional methods.

Current research suggests that the traditional method of shear wall analysis might be more appropriate for low-rise structures. Multistory shear walls greater than three stories should consider flexure and wall rotation from walls below and the wall rotation and moment from the walls above, as shown on the right side of Fig. 9.17.

The connectivity of the walls is caused by the out-of-plane stiffness of the diaphragm. The wall elevation on the left side of the figure shows two framing conditions at the floor framing, one is platform framed and the other is semi-balloon framed. Both are considered flexible for out-of-plane loading on the diaphragm. A good analogy is to visualize a sheet of paper acting as a diaphragm out-of-plane, it is very thin and flexible. Any load out-of-plane would cause it to bend. If turned 90 degrees and loaded in-plane, it is much stiffer because of the increased depth. If the floor diaphragm is flexible, it can easily transfer the moment and rotation from the wall above to the wall below, causing connectivity between the walls. A prescriptive guidance is that if the out-of-plane stiffness of the diaphragm is flexible, analyze the entire wall as a tall wall. If the out-of-plane stiffness (concrete or steel beam of significant stiffness) is stiff, analyze the entire wall as traditional floor to floor.

At the present time, some research has been done by the Canadian Wood Council and FPInnovations, producing reports on a mechanics-based approach for stacked multistory shear walls,[12] but no full-scale testing has been done. At the time of this printing, a task group in the United States has been set up to review this concept, but to date, no conclusions have been drawn and the concept is currently not in the standards used in the United States. It is presented here for the readers' consideration only.

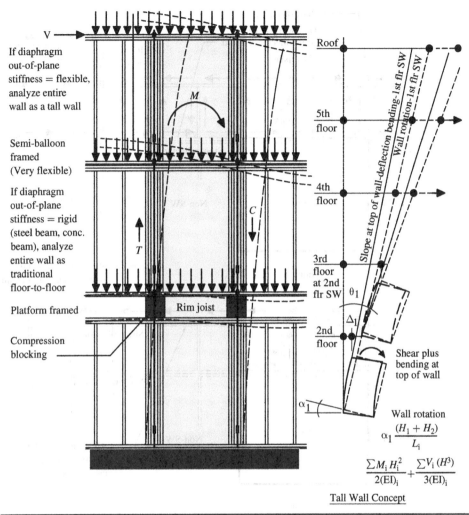

Semi-balloon framed (Very flexible)

If diaphragm out-of-plane stiffness = rigid (steel beam, conc. beam), analyze entire wall as traditional floor-to-floor

Platform framed

Compression blocking

Rim joist

If diaphragm out-of-plane stiffness = flexible, analyze entire wall as a tall wall

V

M

T

C

Roof

5th floor

4th floor

3rd floor at 2nd flr SW

2nd floor

$\theta 1$

Δ_1

α_1

Slope at top of wall-deflection

Wall bending-1st flr SW

Wall rotation-1st flr SW

Shear plus bending at top of wall

Wall rotation

$$\alpha_1 \frac{(H_1 + H_2)}{L_i}$$

$$\frac{\sum M_i H_i^2}{2(EI)_i} + \frac{\sum V_i (H^3)}{3(EI)_i}$$

Tall Wall Concept

Figure 9.17 Multistory shear wall effects.

9.10 Interior Shear Walls

A basic introduction to interior shear walls was presented in Chap. 2. The distribution of lateral forces into interior shear walls that are oriented parallel to the trusses is relatively easy because the trusses can be used as the collector elements. However, walls that are installed perpendicular to the trusses create framing conditions that make the installation of a continuous collector very difficult. Interior shear walls are usually required when the allowable diaphragm aspect ratio has been exceeded, or it is desirable to reduce the load to other shear walls. A typical wall elevation is shown in Fig. 9.18. The installation of the interior shear wall, in effect, creates two separate diaphragms. The interior wall acts as a support for both diaphragms and therefore defines a diaphragm boundary. A full-length collector must be installed to support the remaining width of the diaphragms on both sides of the wall and transfer the

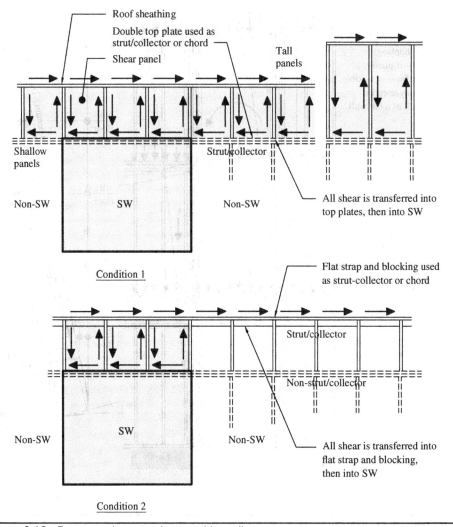

FIGURE 9.18   Forces on shear panels at corridor walls.

diaphragm forces to the interior shear wall as shown in Fig. 9.21, in accordance with the code requirements.

When trusses or rafters are oriented perpendicular to exterior shear walls, the transfer of the diaphragm forces to the shear walls is typically accomplished with full height solid blocking between the trusses. However, at interior shear walls or stub-heel trusses at an exterior wall, the distance from the roof sheathing to the top plate of the shear wall to the roof sheathing prevents the use of solid blocking. Whenever this condition occurs, other methods must be used to complete the load path.

The transfer of diaphragm forces at the floor or roof sheathing level down into the shear wall top plate is a topic that has not been covered in much detail in the past. In some cases, the transfer of forces is automatically assumed to occur. For some wall configurations that can be a reasonable assumption, but in other cases, it might require a more thorough evaluation as to what can be done to provide a complete load path down into the shear wall. The requirement for a detailed review depends on a factor of things:

- What is being used as a collector: Steel strap and flat blocking at the roof level, a beam or built-up light-framed collector at the roof level, or are the shear forces transferred down into collectors located at the wall top plate level?
- What is the height of the diaphragm framing that is being transverse?
- What is the stiffness of the diaphragm framing and its ability to transfer forces as a rigid body down to the shear wall?
- Will a continuous rim joist, shear panels, or blocking have to be used?

Figure 9.18 shows two possible conditions for collector locations and framing for transferring shears from the diaphragm to the top of the shear wall.

Condition 1 uses the wall double top plate as the collector with shear panels installed between the trusses. Shear panels act as mini-shear walls where shear forces are applied at the top of the panel and are resisted at the bottom at the top of the wall. The lateral forces produce overturning forces at each end of the panels. These rotational forces are equal in magnitude and act in opposite directions across the truss. As long as the rotational forces can effectively be transferred across the trusses, the forces cancel out to zero and hold-downs are not required at the ends of the panels. This method allows the diaphragm shears to be transferred directly to the double top plate which acts as the collector.

In condition 2, the collector is located at the roof elevation and can be a steel strap and flat blocking, a built-up light-framed collector, or a solid beam. Transfer of shear forces are the same as condition 1. If shear panels are only installed over the wall, a tie strap might be required at each end of the wall if the dead load of a truss is not large enough to cancel out the panel overturning force. In this case, the shear panels act like a second story of a two-story stacked shear wall. One optional method is to install pre-engineered truss blocking panels between each truss designed to resist shear and overturning.

The designer of the lateral-force-resisting system is responsible to provide the shear forces to the truss manufacturer and typically places the loads applied to the panels on the construction drawings. The shear panels shown in Figs. 9.18 to 9.20 are typically site-built that consist of plywood or OSB sheathing nailed to lumber blocking between the trusses. Figures 9.24 through 9.26 show typical framing options for these panels that will be discussed and designed in Example 9.4. The shear panels act as mini-shear walls that resist shear and overturning. In extreme cases, the height of the panels can exceed 12 ft in height if the trusses are sloped and the shear wall is located near the ridge line. The analysis of the shear wall and panels can follow the procedure that was covered in Sec. 9.9 by applying the lateral shear force at the roof sheathing level and analyzing it as a two-story shear wall, or by transposing the forces down to the top of the wall. The shear panels making up the upper wall section must be connected to the lower shear wall just like a second-floor shear wall, including overturning hold-down anchors or straps as required. The shear panels shown in the photograph in Fig. 9.19 is 2 ft wide (22.5-in-wide net) by 8 ft high. If a uniform shear of 200 plf is applied at the top of the panels, the counteracting shear force acting vertically on each end of each panel, neglecting dead load, is equal to 1600 lb. This overturning force must be resisted by an equal

FIGURE 9.19 Photo of typical shear panels.

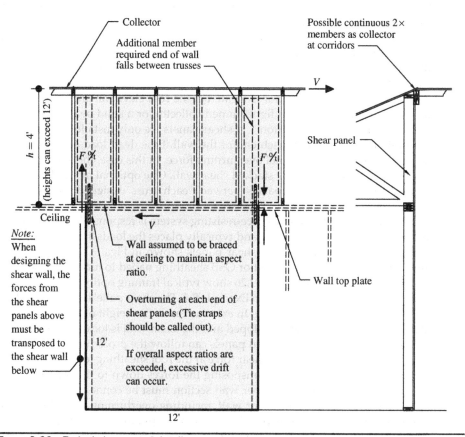

FIGURE 9.20 Typical shear panel detail.

and opposite shear force across each truss, or a hold-down connected to the wall below is required if a net overturning force exists. The wall below the shear panels in the photograph is a bearing wall, but not a shear wall. The shear wall is located to the right of the picture. In this case, the shear panels transfer their forces into the wall double top plate which acts as the collector transferring the force to the shear wall. Where there is no wall located below the trusses, the collector would have to be installed at the underside of the roof sheathing.

Daylight can be seen between the shear panels at the truss interface, indicating that vertical truss blocking members have not been installed. This creates a problem that will be solved in Example 9.4. The end shear panels at the ends of the shear wall have no opposing force and therefore require tie straps to resist the overturning forces of the panels unless the dead load of the end truss is enough to counteract the overturning force.

Example 9.4: Analysis of Interior Shear Walls, Trusses Perpendicular to Wall

The diaphragm shown in Fig. 9.21 has a length of 60 ft and a width of 24 ft. A 12-ft interior shear wall is installed at grid line 2. The roof trusses are oriented perpendicular to

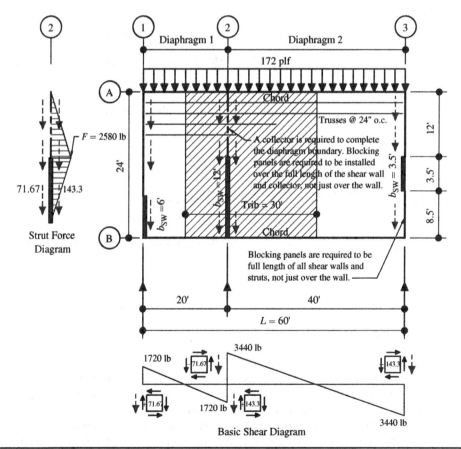

FIGURE 9.21 Interior shear wall, trusses perpendicular to wall.

the interior shear wall. A uniform load of 172 plf is applied at the roof diaphragm. Assume the wind load has been factored to 0.6W and controls, ASD.

Construction of the basic shear diagram (see Fig. 9.21):

$$R_1 = \frac{wL_1}{2} = \frac{172(20)}{2} = 1720 \text{ lb}, \quad v_1 = v_{2L} = \frac{R_1}{W} = \frac{1720}{24}$$

$$= 71.67 \text{ plf diaphragm unit shear}$$

$$R_2 = \frac{w(L_1 + L_2)}{2} = \frac{172(20 + 40)}{2} = 5160 \text{ lb}$$

$$R_3 = \frac{wL_2}{2} = \frac{172(40)}{2} = 3440 \text{ lb}, \quad v_3 = v_{2R} = \frac{R_3}{W} = \frac{3440}{24}$$

$$= 143.3 \text{ plf diaphragm unit shear}$$

$$v_{SW2} = \frac{R_2}{2} = \frac{5160}{12} = 430 \text{ plf shear wall unit shear}$$

$$v_{net} = v_{SW} - v_{diaph} = 430 - 71.67 - 143.3 = +215 \text{ plf net unit shear at shear wall}$$

Strut force diagram:

$$F_{strut} = L_{strut} \sum v_{diaph} = 12(-71.67 - 143.3) = -2580 \text{ lb compression}$$

$$F_B = F_{strut} - v_{net}b_{SW} = -2580 + 215(12) = 0 \text{ lb, therefore the diagram closes}$$

Shear panel forces (reference Figs. 9.22 and 9.23):

$$v_{panel} = 430 \text{ plf, same as the shear wall}$$

$$V_{panel} = 430(2) = 860 \text{ lb}$$

Assume the dead load of the roof is equal to 15 psf.

$$DL_{truss} = \frac{L(DL)L_{panel}}{2} = \frac{60(15)(2)}{2} = 900 \text{ lb per truss}$$

$$V_{side} = \frac{V_{panel}h_{panel}}{L_{panel}} = \frac{860(4)}{2} = 1720 \text{ lb overturning force at each end of the panel}$$

Load combination 0.6D+0.6W:

$$T = [5160(4) - 0.6(2 + 4 + 6 + 8 + 10 + 12)(900) - 0.6(480)(6)]\frac{1}{12} = -314 \text{ lb}$$

Therefore, there is no overturning and tie straps are not required at the ends of the shear panels. This would not be the case if the panel height is greater than 4 ft.   ▲

Three types of shear panels are commonly used and included in most typical detail books. Each of these details will be checked for capacity versus demand for the previous panel forces calculated. Figure 9.24 shows the condition where 2× flat blocking is installed at the top and bottom of the panel with the roof sheathing nailed to the blocking and the bottom blocking nailed into the wall below. Vertical side members are not installed, which leaves the edges of the sheathing unsupported and prone to buckling.

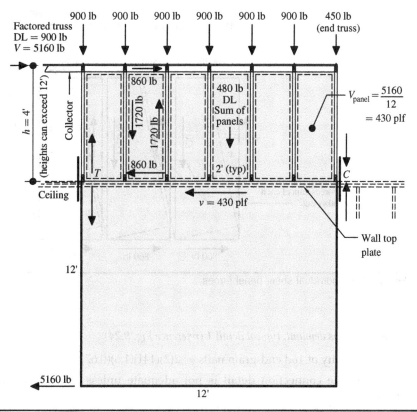

FIGURE 9.22 Forces on shear panels.

As previously mentioned, the panels act as mini-shear walls and therefore must have boundary members installed at all edges per code. Sheathing alone cannot act as boundary members. The truss chords are nailed into the flat blocking with (2) 16d end-grain nails. The only connections resisting the overturning force (F_2) are the end-grain nails. Assuming that sheathing buckling is not an issue:

$$F_1 = \text{shear force applied at the top of the panel}$$

$$h_{\text{panel}} = \text{height of the shear panel}$$

$$L_{\text{panel}} = \text{width of the shear panel}$$

$$F_2 = \frac{F_1 h_{\text{panel}}}{L_{\text{panel}}} = \text{overturning force at the side of the panel}$$

Framing is Douglas Fir-Larch.

$$C_D = 1.6 \text{ load duration factor per NDS 2.3.2}$$

$$C_{\text{eg}} = 0.67 \text{ end-grain factor per NDS 11.5.2}$$

16d common nails are used.

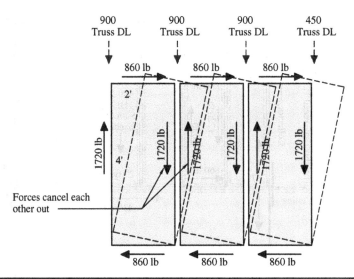

FIGURE 9.23 Individual shear panel forces.

Capacity versus demand, typical detail 1 (reference Fig. 9.24):

Capacity of 16d end-grain nails = 2(2)(141)(1.6)(0.67) = 604 lb < 1720 lb

Therefore, the connection detail is not adequate unless smaller lateral forces are applied.

Figure 9.25 shows a panel that is framed the same as the previous panel except that full height vertical edge members are installed on each end of the panels to brace the

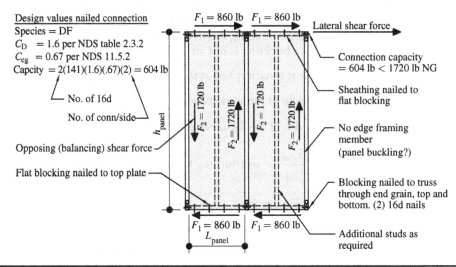

FIGURE 9.24 Typical detail 1—shear panel capacity versus demand.

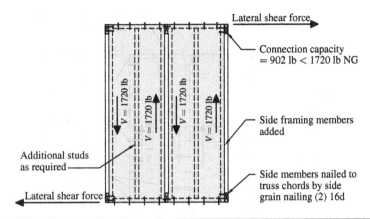

FIGURE 9.25 Typical detail 2—shear panel capacity versus demand.

edges of the sheathing. Nailing is installed through the top and bottom truss chords into the vertical members with (2) 16d side-grain nails. The vertical shear forces (F_2) at the interior truss are resisted by these nails. At the end panels, where there is no adjoining panel, there is no opposing force. A tie strap anchor must be installed to resist the overturning force if required by calculation.

Capacity versus demand, typical detail 2 (reference Fig. 9.25):

$$\text{Capacity of 16d side-grain nails} = 2(2)(141)(1.6) = 902 \text{ lb} < 1720 \text{ lb}$$

Therefore, the detail connection is no good.

Figure 9.26 shows a method for the construction of shear panels that will resist a much larger force. Full height vertical edge members are installed on each end of the panels. Vertical fillers or blocking members are also installed between the webs in the plane of the trusses at the panel locations. Nailing is installed into the blocking full height of the vertical members with 16d side-grain nails. The vertical shear forces (F_2) at the interior trusses are resisted by these nails. The transfer of the vertical forces into the trusses will allow using the dead load of the trusses to resist overturning of the shear panels. At the end panels, where there is no adjoining panel, there is no opposing force, and a tie strap anchor may be required.

Capacity versus demand, typical detail 3 (reference Fig. 9.26):

$$\text{Capacity (8) 16d side-grain nails at 6'' o.c.} = (8)(141)(1.6) = 1804.8 \text{ lb} > 1720 \text{ lb}$$

Therefore, the detail connection is OK.

$$\text{Nail spacing to shear wall} = \frac{141(1.6)(12)}{430} = 6.3'' \text{ o.c., use 16d @ 6'' o.c.}$$

$$\text{Nail roof sheathing to shear panels or blocking} = \frac{74(1.6)(12)}{430}$$
$$= 3.3'' \text{ o.c., use 8d @ 3'' o.c.}$$

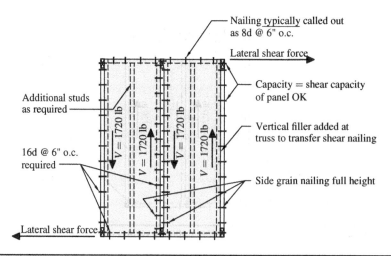

Nailing typically called out
as 8d @ 6" o.c.

Lateral shear force

Capacity = shear capacity
of panel OK

Additional studs
as required

Vertical filler added at
truss to transfer shear nailing

16d @ 6" o.c.
required

Side grain nailing full height

$V = 1720$ lb

Lateral shear force

Figure 9.26   Typical detail 3—shear panel capacity versus demand.

where

$Z = 74$ lb, NDS Table 12Q with $G = 0.5$

15/32" plywood sheathing

$C_\text{D} = 1.6$

In addition to the panels, a collector that provides the remaining part of the diaphragm boundary element from the end of the shear wall across to grid line A is required to transfer the diaphragm forces into the shear wall. Under some conditions, a continuous wall is present the full width of the diaphragm along grid line 2, but only part of the wall is used as a shear wall. In this case, diaphragm shears can simply be transferred through shear panels into the wall double top plate which can be used as the collector transferring the forces back to the shear wall. Whenever the trusses are continuous over the wall but in absence of a continuous wall below on either side of the shear wall, an alternate approach might be required to create the collector. Figures 9.27 through 9.29 show the possible construction of a continuous collector projecting through open web trusses. The collector can consist of multiple continuous 2× members placed below the truss top chord. The diaphragm shear is transferred into these members by solid blocking placed between the truss top chords as shown in section A. The multiple 2× member joints are spliced by a steel strap splice connection. The multiple members are nailed together to act as a single composite member to form the collector. Far too often, collectors and drag struts are designed for tension forces only. The collector must be designed to resist both tension and compression forces. The top member of the multiple collector elements should be nailed to the truss top chords at each truss crossing and into the blocking to help transfer the shear. This allows a side grain nailing condition instead of toe nailing. These connections allow the truss chords to brace the collector element in both directions, which eliminates buckling issues. Additional shear clips can be added to the blocking as required. Special nailing of the sheathing along the strut or collector line should be called out on the drawings. The forces applied to the blocking

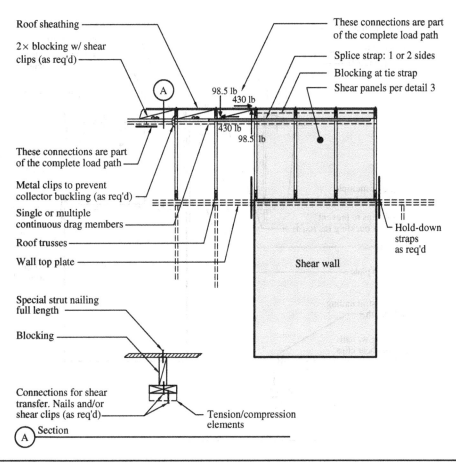

Roof sheathing

2× blocking w/ shear clips (as req'd)

These connections are part of the complete load path

Splice strap: 1 or 2 sides

Blocking at tie strap

Shear panels per detail 3

A

98.5 lb
430 lb

430 lb
98.5 lb

These connections are part of the complete load path

Metal clips to prevent collector buckling (as req'd)

Single or multiple continuous drag members

Roof trusses

Wall top plate

Hold-down straps as req'd

Shear wall

Special strut nailing full length

Blocking

Connections for shear transfer. Nails and/or shear clips (as req'd)

Tension/compression elements

A Section

FIGURE 9.27 Drag strut framing—option 1.

between the truss top chords acts in the same manner as the shear panels (i.e., the eccentric lateral forces produce opposing vertical forces to maintain static equilibrium). Connection of the blocking to the truss top chords requires fasteners to resist these vertical forces.

In the event that a continuous wall top plate is not available on either side of the shear wall to act as a collector and the collector is formed at the top chord of the trusses, the connection of the strut or collector to the shear wall can be accomplished as shown in Figs. 9.27 through 9.29. In Fig. 9.27, the shear panels are framed up to the underside of the roof sheathing. The connection of the collector to the wall is accomplished by installing 2× blocking in the shear blocking panels then overlapping the panels and collector with a steel strap to make the splice connection. The shear panel sheathing must be nailed to the blocking with (2) rows of 8d @ 6" o.c. to meet the 430 plf shear demand. In Fig. 9.28, the collector is continued through the shear wall over the shear blocking panels. The shear from the diaphragm sheathing to the shear panels is the same method as shown in section A. The diagonal web members of the truss

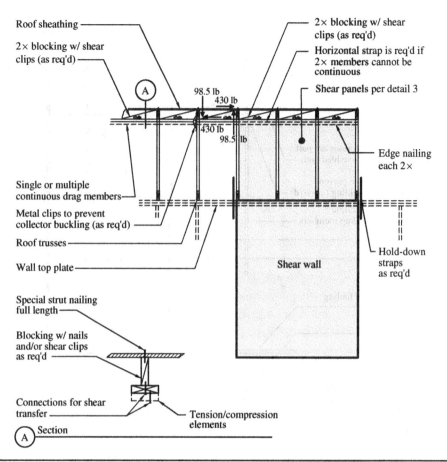

Roof sheathing

2× blocking w/ shear clips (as req'd)

2× blocking w/ shear clips (as req'd)

Horizontal strap is req'd if 2× members cannot be continuous

Shear panels per detail 3

A

98.5 lb
430 lb

430 lb
98.5 lb

Edge nailing each 2×

Single or multiple continuous drag members

Metal clips to prevent collector buckling (as req'd)

Roof trusses

Wall top plate

Shear wall

Hold-down straps as req'd

Special strut nailing full length

Blocking w/ nails and/or shear clips as req'd

Connections for shear transfer

Tension/compression elements

A Section

FIGURE 9.28 Drag strut framing—option 2.

frequently prevent alignment of the strut with the wall. In that event, the collector can be offset as shown in Fig. 9.29. The offset produces a couple that must be resisted by opposing forces developed in the trusses at each end of the wall. The sheathing in the offset area must be designed for the shear from the collector plus the basic diaphragm shear.

9.11 Shear Wall Foundation Issues

When checking for overturning stability, foundation reinforcement, and soil bearing pressure at the bottom of the foundation, make sure the lateral force is applied at the diaphragm sheathing level, not the top of the shear wall unless all forces are transposed from the top of the shear panels, at the diaphragm level, to the top of the shear wall below. Lateral and gravity load paths are not complete until all forces have been safely transferred into the soil.

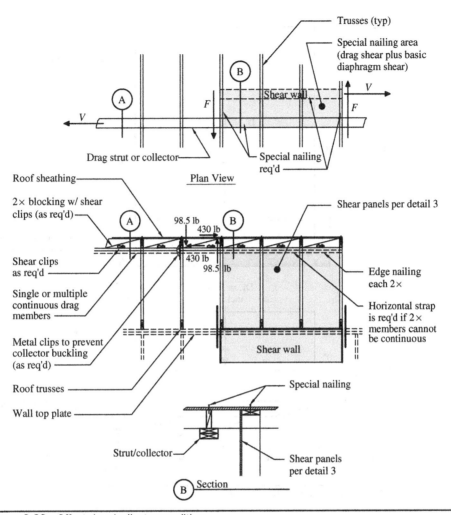

FIGURE 9.29 Offset drag/collector condition.

Example 9.5: Shear Wall Foundation Design

A foundation will be designed for the shear wall in the previous example. Assume that a 2-ft 0-in-wide footing, extending 1 ft on each end of the shear wall, is desired as shown in Fig. 9.30. The total dead load of the roof, wall, and footing has been calculated to be 11,820 lb centered on the wall and footing.

Allowable soil bearing pressure = 2000 psf,   1.33 increase allowed for wind or seismic per the soils report.

Check for load combination 0.6D+0.6W, IBC Eq. 16-2 or ASCE 7-16 Eq. 7 for this example. All load combinations must be checked for governing load case.

$$F_{wind} = 0.6(8600) = 5160 \text{ lb}$$

$$M_o = Vh_{wall+fdn+ftg} = 5160(19) = 98,040 \text{ ft-lb, overturning moment}$$

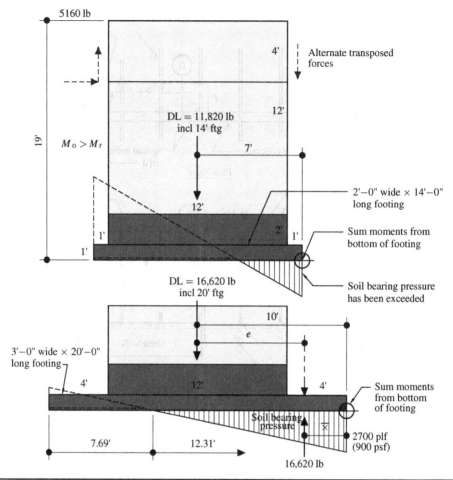

Figure 9.30 Shear wall foundation design.

$$M_R = DL\frac{L_{ftg}}{2} = 0.6(11,820)(7) = 49,644 \text{ ft-lb, resisting moment}$$

$$M_o > 0.6M_R$$

Therefore, the wall/foundation is unstable.

Increase the footing size to increase the dead load and the dead load moment resisting arm. The adjusted footing size shown at the bottom of the figure is increased to 3 ft 0 in wide by 1 ft thick by 20 ft 0 in long. The footing length is increased to a 4-ft extension each side of the shear wall. The revised calculated dead load is 16,620 lb.

$$M_o = 5160(19) = 98,040 \text{ ft-lb, overturning moment}$$

$$M_R = 16,620(10) = 166,200 \text{ ft-lb, resisting moment}$$

$$M_R > M_o$$

Therefore, the wall/foundation is stable.

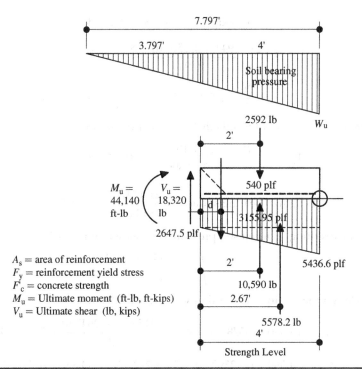

FIGURE 9.31 Strength level soil pressure for footing design.

Check the soil bearing pressure using LC 7, $1.0D + 0.6W$:

$$\bar{x} = \frac{M_R - M_o}{\sum DL} = \frac{166,200 - 98,040}{16,620} = 4.1 \text{ ft}$$

$$e = \frac{L_{ftg}}{2} - \bar{x} = \frac{20}{2} - 4.1 = 5.9 \text{ ft} > \frac{L}{6} = \frac{20}{6} = 3.33 \text{ ft}$$

Therefore, the resultant is outside the kern of the footing.

$$\text{SBP}_{max} = \frac{2P}{3bx} = \frac{2(16,620)}{3(3)(4.1)} = 900 \text{ psf} < 2000 \text{ psf}, \quad \therefore \text{ OK}$$

$$\text{Length of bearing surface} = \frac{16,620(2)}{900(3)} = 12.31 \text{ ft}$$

The footing is to be designed for the ultimate forces shown in Fig. 9.31.

Check the latest edition of ACI 318 (strength) to design the footing, for this example. All load combinations must be checked to determine the governing load case.

$$L_r = 2.5 \text{ kips}, \qquad 0.5L_r = 1.25 \text{ k}$$

$$D = 16.62 \text{ kips}, \qquad 1.2D = 19.944 \text{ k}$$

$$W = 5.16 \text{ kips}, \qquad 1.6W = 8.256 \text{ k}$$

$$\bar{x} = \frac{(19.944 + 1.25)(10) - 8.256(19)}{19.944 + 1.25} = 2.599 \text{ ft}$$

$$e = \frac{20}{2} - 2.599 = 7.401 \text{ ft} > \frac{L}{6} = \frac{20}{6} = 3.33 \text{ ft}$$

Therefore, the resultant is outside the kern of the footing.

$$\text{SBP}_{\text{max}} = \frac{2P}{3bx} = \frac{2(21,194)}{3(3)(2.599)} = 1812.2 \text{ psf} < 2000 \text{ psf}, \quad \therefore \text{ OK}$$

$$\text{Length of bearing surface} = \frac{21,194(2)}{1812.2(3)} = 7.797 \text{ ft}$$

The resulting factored soil pressures are shown in Fig. 9.31.

$$d = 8.75 \text{ in} = 0.729 \text{ ft}$$

$$b = 36 \text{ in}$$

Design the footing in accordance with ACI 318 latest edition.

$$V_{\text{U}} = 10,590 + 5578.2 - 2592 - \frac{(3155.95 + 2647.5)}{2}(0.729) = 11,460 \text{ lb}$$

$$M_{\text{U}} = 10,590(2) - 2592(2) + 5578.2(2.67) = 30,889.8 \text{ ft-lb} \quad \blacktriangle$$

Example 9.6: Grade Beam Design

Occasionally, the overturning moment is large enough that a standard short footing does not have sufficient dead load to maintain stability of the shear wall, or the soil bearing pressures exceed the allowable values. A solution to this problem is to tie the shear wall to a grade beam thereby increasing the moment resisting arm lengths. The shear wall and grade beam configuration shown in Fig. 9.32 will be examined for loads applied in both directions.

Loads acting to the right (Fig. 9.32): A two-story 8-ft wide shear wall, with 8-ft story heights is located at the right end of a concrete grade beam. The foundation footing is 2 ft 0 in wide by 1 ft thick by 26 ft 0 in long, with a 1-ft extension each side of the foundation wall. The foundation wall is 2 ft 0 in high with a wall thickness of 8 in. The calculated factored lateral forces and dead loads are shown in the figure.

$$M_{\text{o}} = 4600(20) + 6000(12) = 164,000 \text{ ft-lb, overturning moment}$$

$$M_{\text{R}} = 8400(5) + (2400 + 3600)(8) + 16,200(13) = 300,600 \text{ ft-lb, resisting moment}$$

$$0.6M_{\text{R}} = 0.6(16,620) = 99,720 \text{ ft-lb} > M_{\text{o}}$$

Therefore, the wall/foundation is stable.

Check the soil bearing pressure:

$$\bar{x} = \frac{300,600 - 164,000}{30,600} = 4.464 \text{ ft}$$

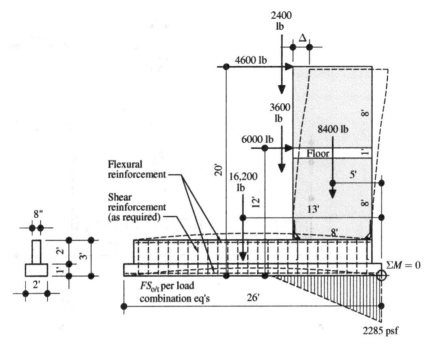

Use of the foundation as a grade beam to increase resistance to overturning

FIGURE 9.32   Grade beam design—loads right.

$$e = \frac{26}{2} - 4.464 = 8.536 \text{ ft} > \frac{L}{6} = \frac{26}{6} = 4.33 \text{ ft}$$

Therefore, the resultant is outside the kern of the footing and part of the footing is in tension.

$$\text{SBP}_{\text{max}} = \frac{2P}{3bx} = \frac{2(30,600)}{3(2)(4.464)} = 2285 \text{ psf} < 2000 \times 1.33 = 2667 \text{ psf}, \ \therefore \text{ OK}$$

Loads acting to the left (Fig. 9.33):

$$M_o = 4600(20) + 6000(12) = 164,000 \text{ ft-lb, overturning moment}$$

$$M_R = 8400(21) + (2400 + 3600)(17) + 16,200(13)$$
$$= 489,000 \text{ ft-lb, resisting moment}$$

$$0.6M_R = 0.6(489,000) = 293,400 \text{ ft-lb} > M_o$$

Therefore, the wall/foundation is stable.

Check the soil bearing pressure:

$$\bar{x} = \frac{489,000 - 164,000}{30,600} = 10.62 \text{ ft}$$

$$e = \frac{26}{2} - 10.62 = 2.379 \text{ ft} > \frac{L}{6} = \frac{26}{6} = 4.33 \text{ ft}$$

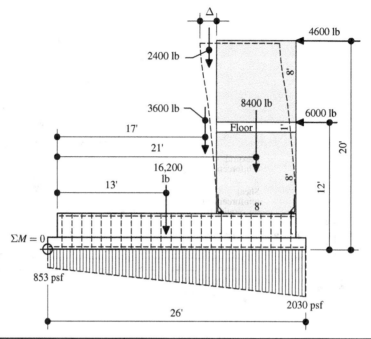

Figure 9.33 Grade beam design—loads left.

Therefore, the resultant is inside the kern of the footing.

$$A_{ftg} = 2(26) = 52 \text{ sq. ft}$$

$$S_{ftg} = \frac{2(26)^2}{6} = 225.33 \text{ ft}^3$$

$$\frac{P}{A} = \frac{30,600}{52} = 588.5 \text{ psf}$$

$$\frac{Pe}{S} = \frac{30,600(10.62)}{225.33} = 1442 \text{ psf}$$

$$SBP_{max} = 2030 \text{ psf} < 2667 \text{ psf, therefore, OK}$$

$$SBP_{min} = 853 \text{ psf} < 2667 \text{ psf, therefore, OK}$$

Therefore, the proposed grade beam provides stability and keeps the soil bearing pressures within allowable limits. The foundation reinforcement should be designed for flexural and shear reinforcement.   ▲

9.12   References

1. American Wood Council (AWC), *Special Design Provisions for Wind and Seismic with Commentary*, AWC, Leesburg, VA, 2021.
2. APA—The Engineered Wood Association, *Design/Construction Guide-Diaphragms and Shear Walls*, APA Form L350, APA—The Engineered Wood Association, Engineering Wood Systems, Tacoma, WA, 2007.
3. Structural Engineers Association of California (SEAOC), *IBC Structural/Seismic Design Manual, Vol. 2.*, SEAOC, CA, 2018.
4. Breyer, D. E., Martin, Z., and Cobeen, K. E., *Design of Wood Structures ASD*, 8th ed., McGraw-Hill, New York, 2020.
5. International Code Council (ICC), *International Building Code, 2021 with Commentary*, ICC, Whittier, CA, 2021.
6. American Wood Council (AWC), *National Design Specification for Wood Construction and Supplement*. Leesburg, VA, 2018.
7. American Society of Civil Engineers (ASCE), *ASCE/SEI 7-16 Minimum Design Loads for Buildings and Other Structures*, ASCE, New York, 2016.
8. APA—The Engineered Wood Association, *APA Report T2003-22, Shear Wall Lumber Framing: double 2xs vs. Single 3x's at Adjoining Panel Edges*, APA Report No. T2003-22, APA—The Engineered Wood Association, Engineering Wood Systems, Tacoma, WA, 2003.
9. *Single Sided Double Sheathed Shear Wall with Wood Structural Panel Sheathing*. APA—The Engineered Wood Association, Engineering Wood Systems. Tacoma, WA.
10. Cobeen, K. E., Dolan, J. D., Thompson, D., van de Lindt, J. W., *Seismic Design of Wood Light-Frame Structural Diaphragm Systems—A Guide for Practicing Engineers*, National Institute of Standards and Technology (NIST), NEHRP Seismic Design Technical Brief No. 10, U.S. Department of Commerce, 2014
11. Malone, R. T., and Breneman, S., *A Seismic Design Example of a Cantilever Wood Diaphragm*, Wood Products Council, Woodworks, 2019.
12. Newfield, G., Ni, C., Wang, J., *A Mechanics-Based Approach for Determining Deflections of Stacked Multi-Storey Wood-Based Shear Walls*, FPInnovations, Vancouver, BC, and Canadian Wood Council, Ottawa, ON, 2013.

CHAPTER 10

Shear Walls with Openings

10.1 Introduction

It is not uncommon to utilize wall sections that contain openings as shear walls. There are four basic methods of analyzing standard shear walls with openings. The first method utilizes a segmented shear wall approach ignoring the portions of wall with the opening and include only the full height wall segments that are sheathed like a standard shear wall, as shown in Fig. 10.1. The header and sill sections above and below the opening are not considered as contributing to resisting lateral forces. Each full height segment must meet the aspect ratio requirements of SDPWS[1] Table 4.3.3 and Figure 4D. Since each segment is considered as an independent shear wall, hold-downs would be required at each end of each full height segment if required by the design. The wall analysis is the same as that described for the solid shear walls presented in Chap. 9. Of the four methods described herein, the advantage of this method is that fewer calculations are required. The obvious disadvantage of this approach is that a larger number of hold-downs and increased nailing may be required because it is a more conservative design. The other methods of analysis are as follows:

- *Perforated shear walls (PSW) (Fig. 10.2):* Perforated shear walls can be analyzed using an empirical analysis based on tests. This method uses the entire length of wall as the shear wall. A reduction in the allowable strength is taken as a ratio of the total area of the wall divided by the total area of the full height sheathed sections regardless of whether they comply with the required aspect ratio and the total area of the openings. The advantage to this method is that, if the design complies with SDPWS requirements, hold-downs are only required at each end of the wall since the entire portion of wall is considered the shear wall. The disadvantages are that the wall has a reduced capacity to resist shear forces, and intermediate tie-down anchorage is required.

- *Force transfer around an opening (FTAO) (Fig. 10.5):* The FTAO method is an engineered solution based on a rational analysis in accordance with well-established principles of mechanics. The advantages to this method are that, like the perforated shear wall, hold-downs are only required at each end of the wall and intermediate tie-downs are not required. In this method, the sections above and below the opening are designed to transfer forces across an opening. The

359

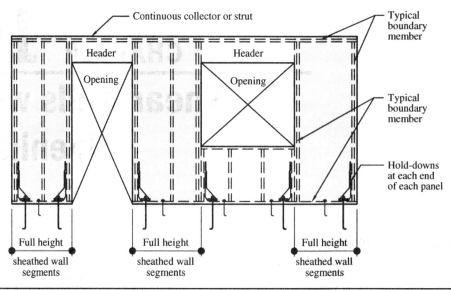

Figure 10.1 Segmented shear wall elevation.

disadvantages are that tie straps and blocking must be installed at each corner of the opening, special detailing is required, and the number of calculations is increased.

- *Cantilevered wall segments (Fig. 10.26):* The cantilevered method is an engineered solution developed by Edward Diekmann[7] based on a rational analysis in accordance with well-established principles of mechanics. This method can be used where the depth of the header section above the window opening is too shallow to meet reasonable allowable aspect ratios for those individual sections of the wall. This shallow condition prevents the header section from effectively distributing forces across the opening to the adjacent full height wall sections. The advantages of this method are that hold-downs are only required at each end of the wall, intermediate tie-downs are not required, the wall can resist higher shear forces than the perforated shear walls, and it requires fewer calculations than the FTAO method. The disadvantages are that tie straps and blocking must be installed at the lower corner of each opening and special detailing is required. To date, no testing has been done on this type of wall.

10.2   Perforated Shear Walls

The perforated shear wall method of analysis has changed slightly in the 2021 edition of SDPWS from that of previous versions. This method is based on tests and uses empirically determined shear resistance reduction factors. In contrast to the segmented shear wall approach, the entire length of the wall is used to resist shear. The wall can be sheathed on one or both sides with wood structural panels only; except in the case where wind-controls design, wood sheathing on the exterior side and GWB on the interior side is allowed. The sheathed sections above and below the opening provide some local restraint at each end of the section and add some strength and stiffness to the wall.

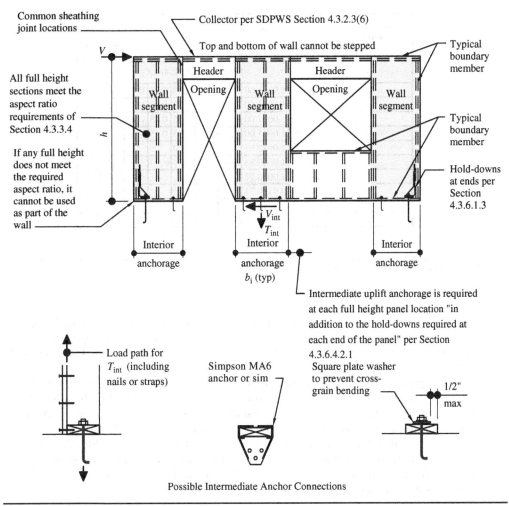

Common sheathing joint locations

Collector per SDPWS Section 4.3.2.3(6)

Top and bottom of wall cannot be stepped

V

Header

Opening

Wall segment

Wall segment

Header

Opening

Wall segment

Typical boundary member

Typical boundary member

Hold-downs at ends per Section 4.3.6.1.3

All full height sections meet the aspect ratio requirements of Section 4.3.3.4

h

If any full height does not meet the required aspect ratio, it cannot be used as part of the wall

V_{int}

T_{int}

Interior anchorage

Interior anchorage

Interior anchorage

b_i (typ)

Intermediate uplift anchorage is required at each full height panel location "in addition to the hold-downs required at each end of the panel" per Section 4.3.6.4.2.1

Load path for T_{int} (including nails or straps)

Simpson MA6 anchor or sim

Square plate washer to prevent cross-grain bending

1/2" max

Possible Intermediate Anchor Connections

FIGURE 10.2 Perforated shear wall elevation.

However, since these sections are not tied to the full height sections on either side of the opening, they cannot fully develop forces across the opening. Consequently, the shear capacity of the wall as a whole is reduced by a shear capacity reduction factor as noted in SDPWS Table 4.3.5.6.

The American Plywood Association—The Engineered Wood Association (APA) conducted limited testing on this type of shear wall with un-reinforced openings, the results of which can be referenced in APA Research Report 157.[2] The APA tests were based on locating hold-downs only at the ends of the wall to resist overturning and uplift forces. Intermediate anchorage connections were not provided at the full height sheathed sections in the test samples. The results of the tests were consistent with the research that was done by Dr. Hideo Sugiyama in Japan. Sugiyama's research resulted in the development of a simplified empirical method of calculating an effective shear stiffness/strength ratio used to modify the shear stiffness and strength of a fully

Percentage of full height sheathing $\dfrac{A_{fhs}}{A_{wall}}$	Percentage of Wall Area Openings $\dfrac{A_o}{A_{wall}}$									
	0%	10%	20%	30%	40%	50%	60%	70%	80%	90%
	Shear Capacity Ratio, C_o									
10%	1.00	1.00	1.00	1.00	0.77	0.63	0.53	0.45	0.40	0.36
20%	1.00	1.00	1.00	0.91	0.71	0.59	0.50	0.43	0.38	-
30%	1.00	1.00	1.00	0.83	0.67	0.56	0.48	0.42	-	-
40%	1.00	1.00	1.00	0.77	0.63	0.53	0.45	-	-	-
50%	1.00	1.00	0.91	0.71	0.59	0.50	-	-	-	-
60%	1.00	1.00	0.83	0.67	0.56	-	-	-	-	-
70%	1.00	1.00	0.77	0.63	-	-	-	-	-	-
80%	1.00	0.91	0.71	-	-	-	-	-	-	-
90%	1.00	0.83	-	-	-	-	-	-	-	-
100%	1.00	-	-	-	-	-	-	-	-	-

1. Definitions of A_o, A_{fhs}, and A_{wall} are provided in Equation 4.3-6.

FIGURE 10.3 Table 4.3.5.6—shear capacity adjustment factor, C_o.

sheathed wall to account for the openings. In the Sugiyama method of analysis, the shear ratio was based on the ratio of the total area of window and door openings to the total area of the wall, and the ratio of the full height wall segments to the total length of the wall.

Subsequent to the testing conducted in APA Research Report 157, modifications to the IBC were made to provide some conservatism to the method, in which the shear capacity of a perforated shear wall is to be taken as the tabulated nominal unit shear capacity multiplied by the sum of the shear wall segment lengths, b_i, and the appropriate shear capacity adjustment factor, C_o. In the current 2021 edition of SDPWS, Section 4.3.5.6 and Table 4.3.5.6, the shear capacity adjustment factor has been significantly revised. The C_o factor can be taken from either Table 4.3.5.6, Fig. 10.3, or calculated using SDPWS equation 4.3-6:

$$V_n = v_n C_o \Sigma b_i \text{ nominal shear capacity} \qquad \text{SDPWS Eq. 4.3-5}$$

where

v_n = Nominal unit shear capacity, lb/ft
C_o = Shear capacity adjustment factor from Table 4.3.5.6 or calculated using the following equation

$$C_o = \frac{A_{wall}}{(3A_o + A_{fhs})} \leq 1.0 \qquad \text{SDPWS Eq. 4.3-6}$$

where

A_{fhs} = Total area sheathed with full height sheathing, ft², regardless of whether individual wall segments meet the aspect ratio limits in 4.3.3.4
A_{wall} = Total area of a perforated shear wall equal to the length of the perforated shear wall times its height
A_o = Total area of openings in the perforated shear wall where individual opening areas are calculated as the opening width times the clear opening height, ft²

$\Sigma b_i =$ Sum of perforated shear wall segment lengths b_i, ft. Lengths of perforated shear wall segments with aspect ratios greater than 2:1 shall be adjusted in accordance with SDPWS Section 4.3.3.4

$h =$ Height of the perforated shear wall

Method of analysis:

1. Determine the % of full height sheathing $= \dfrac{A_{fhs}}{A_{wall}}$.

2. Determine the % of wall openings $= \dfrac{A_o}{A_{wall}}$.

3. The shear resistance adjustment factor, C_o, is determined from SDPWS Table 4.3.5.6 or calculated in accordance with Eq. 4.3-6. Interpolation is allowed.

4. The allowable nominal shear capacity v_n is taken from SDPWS Table 4.3A and is to be adjusted for aspect ratio as required. The maximum aspect ratio, (h/b_i), of individual wall piers shall not exceed 3.5:1, Table 4.3.3, and Figure 4E. Where any segment exceeds 2:1, the unadjusted shear resistance shall be multiplied by $\dfrac{2b_i}{h}$.

5. Adjusted shear resistance $= C_o v_n$.

6. $V_n = C_o v_n \Sigma b_i$, nominal shear capacity of the wall.

7. Solve for hold-down forces at ends of wall, T, C.

8. Solve for intermediate uplift anchorage t full height sheathing in accordance with Section 4.3.6.4.2.1 or by rational analysis.

Limitations—SDPWS Section 4.3.2.3: Framing members, blocking, and connections around openings are not designed.

1. Full height segments must be located at each end of the perforated shear wall.

2. The aspect ratio limitations of Section 4.3.3.3 and Fig. 10.4a shall apply.

3. The nominal unit shear capacity shall not exceed 2435 plf as given in Table 4.3A.

4. Where out-of-plane offsets occur, portions of the wall on each side of the offset shall be considered as separate perforated shear walls.

5. Collectors for shear transfer of forces between the diaphragm and the shear wall shall be the full length of the perforated shear wall. This can be the wall top plate(s) or continuous rim joist acting as the collector.

6. A perforated shear wall shall have a uniform top of wall and bottom of wall elevation, no vertical offsets or slopes. Perforated shear walls not having uniform plate elevations shall be designed by other methods.

7. The maximum wall height shall not exceed 20 ft.

8. All sheathed areas of the perforated shear wall shall be constructed with sheathing and sheathing attachment associated with the tabulated nominal unit shear capacity in Section 4.3.5.6 for the selected shear wall configuration.

Exception: Sheathed areas constructed with sheathing and sheathing attachment associated with a tabulated nominal unit shear capacity less than required in Section 4.3.2.3(9) are permitted, provided they are included in the total area of openings, A_o.

Hold-downs designed for a tension force, T, and a compression force, C, are required to be installed at each end of the wall. Each end of each perforated shear wall segment shall be designed for a compression force, C, in each segment in accordance with SDPWS Section 4.3.6.1.3, where

$$T = C = \frac{Vh}{C_o \Sigma b_i}$$

SDPWS Eq. 4.3-8

where

T = Tension chord uplift force, lb
C = Compression chord force, lb
V = Total shear force in perforated shear wall, lb
h = Wall height, ft
Σb_i = Sum of the lengths of the full height segments, ft
C_o = Shear capacity adjustment factor from SDPWS Table 4.3.5.6, as shown in Fig. 10.4

If multiple studs are used for the chord members of the wall, the studs must be fastened together to act as a single member to transfer the shear forces that the shear wall is being designed for.

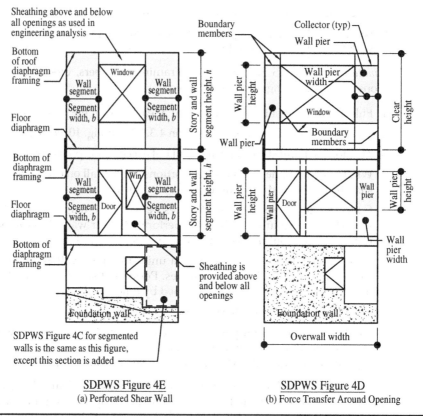

SDPWS Figure 4E
(a) Perforated Shear Wall

SDPWS Figure 4D
(b) Force Transfer Around Opening

FIGURE 10.4 Allowable shear wall aspect ratios—SDPWS Figures 4D and 4E.

A commonly overlooked component of a perforated shear wall is the requirement for uplift anchorage of the bottom plate at the full height sheathing sections. Section 4.3.6.4.2.1 requires these anchors in addition to the hold-downs at the ends of the wall. The uplift force for the additional anchors, t, is equal to the unit shear, v_{max}, times the spacing of the anchors. Examples of uplift anchors can be seen in Fig. 10.2. The objective of installing these anchors is to prevent cross-grain bending and splitting in the bottom plate of the full height wall segment when that section of the wall tries to rotate about its base. The figure shows that the uplift force, t, is transferred through the sheathing into the bottom plate and then into the anchor bolt or nailing to the framing below.

$$v_{max} = \frac{V}{C_o \Sigma b_i} \qquad \text{SDPWS Eq. 4.3-9}$$

where

v_{max} = Unit shear force, in plf
V = Total shear force in perforated shear wall, lb
Σb_i = Sum of the lengths of the full height segments, ft
C_o = Shear capacity adjustment factor

Perforated shear wall deflection can be calculated using the three-term deflection Eq. 4.3-1, where v is equal to v_{max} and b is taken as Σb_i.

Most examples in textbooks and technical publications only address single-story conditions. For multistory perforated shear walls, adequate transfer of end wall hold-downs and intermediate uplift anchorage must be made to the wall below. The forces from the intermediate anchorage of the wall above can be transferred down to the wall below by two different methods. If a solid continuous rim joist exists, the rim joist can be sized to resist the intermediate uplift anchorage forces. If blocking exists over the lower wall, tie straps will be required to transfer the intermediate uplift anchorage forces across the floor framing to the wall below.

10.3 Shear Walls with Force Transfer Around an Opening

10.3.1 History and Modifications

The force transfer around an opening (FTAO) method of analysis was developed to provide a solution for large openings in diaphragms and later applied to exterior walls to reduce the number of hold-downs within a structure. The original method was developed by Edward Diekmann in the early 1980s, which was based on a rational analysis like a design method described in ATC-7, *Guidelines for the Design of Wood Sheathed Diaphragms*.[3] The method is based on the concept of rigid body behavior assuming that the wall, as a whole, will act like a Vierendeel truss or frame. Diekmann's original method assumed that the sections above, below and on each side of the opening should comply with the allowable aspect ratio as shown in Fig. 10.5 and are stiff enough to contribute as members of the Vierendeel truss. The points of inflections were originally assumed to be located at mid-length of the header and sill sections, and at mid-height of the piers. Previous design examples included only one opening in the wall.

Recent modifications to the method of analysis have simplified the calculation process and have been verified by full scale testing by the APA. Among these revised

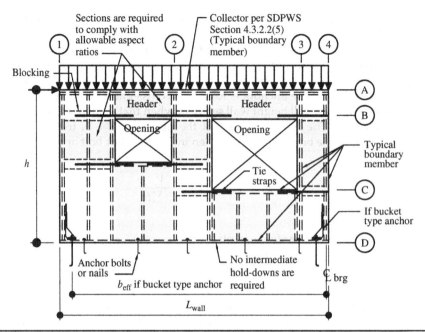

Figure 10.5 Shear wall elevation with force transfer around opening (FTAO).

versions are the commonly used Thompson method and the APA/Diekmann simplified method as published in Technical Note T-555B.[4]

Commonality:

- Both are variations of the original Diekmann method.

- Both determine the location of the inflection point or tributary width to a pier section at an opening is by assuming the relative stiffness between the header and adjacent pier sections, based on their respective depths and widths. As an example, the location of the inflection point, near the upper left corner of an opening would be determined by multiplying the left pier width by the opening length divided by the sum of the left and right pier widths. This would be like distributing forces to a shear wall based on their length (relative stiffness).

- Both assume the vertical shears acting along the wall length are uniform; and that, the distribution of those shears at an opening is in accordance with the relative stiffness of the header and sill sections, determined by their depths.

- Neither of the methods, their design examples or tested wall assemblies included the application of gravity loads, which is deemed appropriate from the standpoint of determining wall shears, but not acceptable for determining the wall stiffness or hold-down capacity, as noted in Sec. 9.6 and Chap. 6. The Thompson method did note that some engineers argue that gravity loads need to be applied for accuracy.

- Both methods assume that whenever different height openings occur, the deepest opening is assumed for all openings, uniform depth.

- Both allow multiple openings.
- None of the methods, including SDPWS Figure 4D, require the header or sill sections as having to comply with any aspect ratio.

Prior to the development of the APA simplified method, extensive testing by the APA, Report M410,[5] compared the test results with design examples using several different methods of FTAO analysis and found that the Diekmann method more closely matched the test results than the other methods. The testing was able to determine the tie strap forces at the corners of the opening and wall deflection but was not able to determine the location of the inflection point or localized shear forces due to the limitations of the test equipment. The original Diekmann method broke (segmented) the pier, header, and sill down into rectangular sections at the location of their points of inflection, then using free body diagrams, solved for the unknown forces by summing forces about each corner of the segments. The new simplified APA method without gravity loads assumes the wall shears are uniformly distributed across the wall; and that, the vertical shears above and below the opening are distributed to each section in proportion to their depth. The corner forces are determined by the collected shears based on their tributary width of the header and sill sections.

10.3.2   Code Requirements

The design requirements for FTAO shear walls are covered in SDPWS Section 4.3.2.2 and SDPWS Figure 4D, as shown in Fig. 10.4b. SDPWS Section 4.3.2.2 sets limitations for force transfer around an opening (FTAO) shear wall, which are as follows:

1. Framing members, blocking, and connections around the openings are designed.
2. The length of each wall pier shall not be less than 2 ft.
3. A full height wall segment shall be located at each end of the force transfer wall.
4. Where out-of-plane offsets occur, portions of the wall on each side of the offset shall be considered as separate FTAO shear walls.
5. Collectors for the transfer of shear forces between the diaphragm and the shear wall shall not be less than the full length of the force-transfer wall.

Some confusion can occur between SDPWS Figure 4D and current design examples. Many current design examples only consider a wall with a full height section on each side of the window opening and ignore the door pier. SDPWS Figure 4D does not show a full height section on the left side of the window opening as having to comply with the required aspect ratio, which can cause some confusion as to what constitutes the true opening, and what represents the left member of the Vierendeel truss. The full height section to the left of the door opening is shaded suggesting it can be part of the FTAO wall which would not replicate a Vierendeel truss. SEAOC's *IBC Structural/Seismic Design Manual, Vol. 2*[6] notes, "Building structures that have sliding glass doors to balconies or air-conditioner wall-mounted units below the windows, such as those in hotels, usually cannot utilize force transfer around openings because there is not enough wall width (depth) above and below the openings. In these cases, a segmented shear wall with continuous tie-downs may be necessary." Section 4.3.2.2 notes that the design shall be based on the results of a rational analysis, so its inclusion combined with good engineering judgment could be reasonable.

The location of hold-downs also requires some consideration. Current examples show hold-down anchorage installed at each end of the wall section containing the window opening. If the door section is included in the design, engineering judgment would be required for the placement of the hold-downs in locations that would provide the best performance of the wall. Hold-down placement can significantly affect how the wall will transfer forces within itself.

Limitation 5 requires a collector to be installed the full length of the wall. This has long been interpreted as requiring a full-length strap and blocking across the full length of the wall above and below the openings. This common misconception has been reinforced by design examples, photographs, and many presentations suggesting it is required. Conversations with AWC and the APA clarified that their intent was that that requirement only refers to the double top plate of the wall and not across the top and bottom of the openings. The tie straps are only required to distribute corner forces into the adjacent full height sections a sufficient distance to effectively create rigid member end connections and providing a continuous load path across the wall. The previous assumption, as misinterpreted, would require very long straps on large multi-family type structures. An alternate method will be presented in Examples 10.1 to 10.3 to limit the length of these straps and still provide complete load paths and distribution of forces into the wall. Example 10.3 was presented as Example 10.2 in the previous edition of this book showing that this is not a new concept. The most recent version of the APA example paper, testing, and the 2021 SDPWS Commentary Figure C4.3.2.2 have clarified that shorter strap lengths are allowed.

10.3.3 Sill and Header Sections Aspect Ratios

In accordance with SDPWS Section 4.3.3.3 and Figure 4D, only the shaded pier sections are required to comply with the allowable aspect ratios in Table 4.3.3, showing that the header and sill sections do not need to comply. However, the absence of further specific guidance on this issue can result in not meeting the underlying stiffness assumptions of the FTAO method. SDPWS Section 4.3.5.2 requires that FTAO designs are to be based on a rational analysis, leaving the decision of header and sill aspect ratios up to engineering judgment. This lack of specific direction has resulted in designs with window openings up to 20 ft in length, which is neither a reasonable nor a rational design. To comply with rigid body behavior, the wall elements above and below the opening should meet some reasonable aspect ratio. If not, these members could become too flexible to effectively transfer forces across the opening and different results can occur as shown in Fig. 10.7. One concern regarding setting an aspect ratio for the header and sill section has been setting an arbitrary limit on what is stiff enough to comply with the basic assumptions of rigid body behavior and the basic assumptions of the method of analysis, yet accommodating larger standard commercial window sizes. However, the APA has empirical evidence as published in report M410[5] which is based on a number of tests, has noted that an A.R. = 6.5:1 (7.5 ft/1.17 ft) for a header or sill section is stiff enough to perform adequately. That publication noted that header sections less than 12-in in depth are not applicable to FTAO walls. Exceeding this aspect ratio would require engineering judgment and possibly the approval or rejection by the authority having jurisdiction (AHJ). A basic summary of the maximum opening sizes and test samples that would meet this A.R. requirement is shown in Fig. 10.6 based on an allowable aspect ratio of 6.5:1 or less.

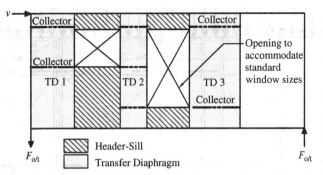

APA Test Walls, Form M410

Wall height = 8'
Maximum opening width = 7.5'
Minimum pier width = 2.25, A.R. = 3.56:1
Typical header depth = 1.167', A.R. = 6.43:1
Minimum sill depth = 1.83, A.R. = 4.1:1

APA notes that a header or sill section with an A.R. = 6.5:1 is stiff enough to work in an FTAO shear wall.

Maximum Header/Sill Length

A.R.	3.5:1	4:1	4.5:1	5:1	5.5:1	6:1
d_{header}						
1'	3.5'	4'	4.5'	5'	5.5'	6'
1.33'	4.67'	5.33'	6'	6.67'	7.33'	8'
2'	7'	8'	9'	10'	11'	12'

Maximum Transfer Diaphragm Widths

A.R.	3.5:1	4:1	4.5:1	5:1	5.5:1	6:1
$h = 12'$, b min:	3.5'	3'	2.67'	2.33'	2.25'	2'
$h = 10'$, b min:	2.83'	2.5'	2.25'	2'	-	-
$h = 9'$, b min:	2.5'	2.25'	2'	-	-	-
$h = 8'$, b min:	2.25'	2'	-	-	-	-

1. h = Wall Height
2. Minimum Width 2'−0" per SDPWS Section 4.3.2.2 (2)

FIGURE 10.6 Header and sill aspect ratio limitations.

10.3.4 Transfer Diaphragm Aspect Ratios

Diekmann noted when assigning aspect ratios to transfer diaphragms in a horizontal diaphragm that the transfer diaphragm should be similar to the aspect ratios of the main diaphragm so that they would have a similar stiffness. One could reason that the flexibility in assigning aspect ratios to the headers and sill sections in shear walls could also be applied to transfer diaphragms in FTAO shear walls, meaning that the aspect ratio for transfer diaphragms could be up to a maximum of 6.5:1, similar to the conclusions of the APA. The minimum width of a pier section complying with SDPWS limitation Section 4.3.2.2(2) is 2 ft 0 in, which would automatically set the minimum transfer diaphragm width. Figure 10.6 shows several tables that show allowable transfer diaphragm widths versus wall heights.

Any wall pier, header, sill, or transfer diaphragm section that exceeds an aspect ratio greater than 2:1 must have the nominal shear capacity adjusted by an aspect ratio factor in accordance with SDPWS Section 4.3.3.

10.3.5 Method of Analysis—Transfer Diaphragm Method

Most current design examples only show openings at the same height. If different opening heights occur, current methods of analysis have noted that the deepest opening is assumed to be used at each opening for simplicity of the calculations. This could affect the actual stiffness of the wall. The other issue as previously noted is the omission of dead loads, which can change the wall stiffness and hold-down anchorage.

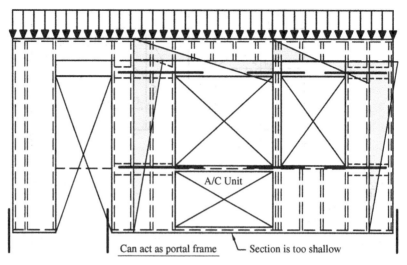

(a) Force Transfer If Shallow Sill Sections

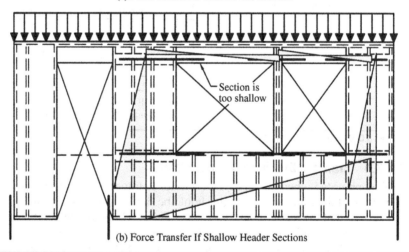

(b) Force Transfer If Shallow Header Sections

Figure 10.7 Force transfer if shallow sill or header section.

Example 10.3, which was also included in the first edition of this book, is an example where transfer diaphragms can be used to allow partial length collectors and tie straps. Applying that methodology to the whole wall can provide many benefits. These benefits not only allow any number and combination of different opening heights and asymmetrical piers, but also allow the application of gravity loads and provide more flexibility in complex design wall layouts. The method will hereafter be referred to as the transfer diaphragm method.

Concept

The full height transfer diaphragms placed on each side of each opening are assumed to act similar to standard segmented shear walls. The rotation of these sections pushes against and pulls away from the header and sill sections as shown in Fig. 10.8. The

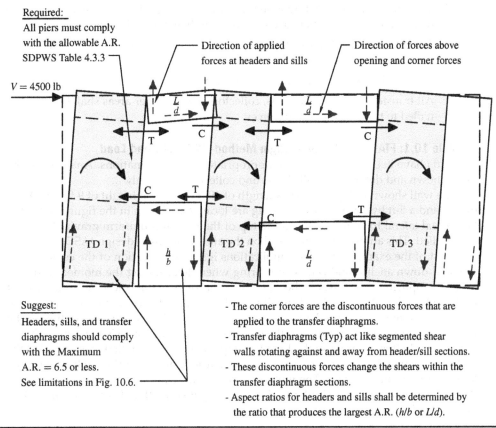

Required:
All piers must comply
with the allowable A.R.
SDPWS Table 4.3.3

Direction of applied
forces at headers and sills

Direction of forces above
opening and corner forces

$V = 4500$ lb

Suggest:
Headers, sills, and transfer
diaphragms should comply
with the Maximum
A.R. = 6.5 or less.
See limitations in Fig. 10.6.

- The corner forces are the discontinuous forces that are
  applied to the transfer diaphragms.
- Transfer diaphragms (Typ) act like segmented shear
  walls rotating against and away from header/sill sections.
- These discontinuous forces change the shears within the
  transfer diaphragm sections.
- Aspect ratios for headers and sills shall be determined by
  the ratio that produces the largest A.R. (h/b or L/d).

Figure 10.8 FTAO transfer diaphragm method concept.

connectivity to the header and sill sections will cause concentrated collector forces to be applied into the transfer diaphragm once the transfer diaphragms rotate. The pier portion of the transfer diaphragms between the header and sill sections must comply with the minimum width and allowable aspect ratio in SDPWS Table 4.3.3, but the main section of the transfer diaphragm could reasonably follow the test result's extended aspect ratios limitations for the headers previously discussed, as shown in Fig. 10.6, and as prescribed in other applications in this book.

Limitations and Assumptions

All FTAO shear walls are required to comply with SDPWS Section 4.3.2.2, Table 4.3.3, and Figure 4D. Other limitations and requirements for the transfer diaphragm method are as follows:

1. The wall design shall be based on a rational analysis.

2. All wall layouts shall only include members that are stiff enough to exhibit rigid body behavior and shall comply with acceptable aspect ratios previously discussed. Opinion: Although transfer diaphragms could have an aspect ratio up to 6.5:1, better performance would result if the ratio were held to A.R. = 4.5:1 or less. However, this is left up to engineering judgment.

3. Vertical shear forces acting along the wall openings shall be distributed to the header and sill sections in accordance with their respective depths.

4. The tributary width across the opening is based on the relative stiffness between the header/sill sections and the adjacent pier sections.

5. Only WSPs shall be used for the wall sheathing.

6. All transfer diaphragm chords, collectors, and transfer areas shall be properly nailed to transfer the design forces.

Example 10.1: FTAO Transfer Diaphragm Method—Without Dead Load

All loads have been factored for the appropriate load combinations. Analyze for the loads shown and determine wall shears and collector forces only.

The wall shown in Fig. 10.9 has a length of 19 ft 6 in and a height of 9 ft 0 in. A 3-ft by 3-ft and a 5-ft by 5-ft window opening are located as shown in the figure. A 4500-lb horizontal lateral force is applied at the top of the wall and uniform gravity loads are not applied. This analysis will be based on the allowable stress design (ASD).

Most of the examples in other publications ignore the location of the centerline of the hold-down anchor and center of bearing when determining the moment resisting arm for simplicity. This is usually because the eccentricity of the hold-down with respect to the outer edge of the wall is difficult to address when determining shear. However, when calculating the required anchor capacity, if discrete bucket type or continuous tie rod anchors are used, code requires the use of b_{eff} as the moment resisting arm as shown in Fig. 10.5. In this example, embedded tie strap anchors are installed at each end of the wall as shown in the figure to eliminate the eccentricity and centerline of bearing problem to simplify the analysis. The method of distributing and determining forces up to the transfer diaphragms is similar to the APA simplified method because of the lack of gravity loads.

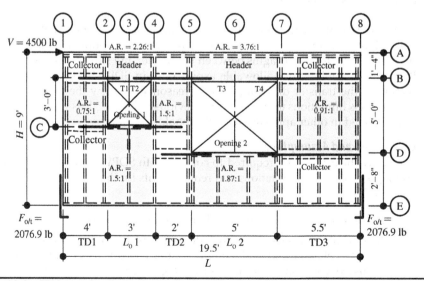

FIGURE 10.9 FTAO transfer diaphragm method—without dead loads.

Check aspect ratios (see Fig. 10.9): For shallow headers or sills, it is suggested to use the aspect ratio for that section that produces the higher value, L/d or h/d:

Allowable A.R.:

- Piers per SDPWS Section and Table 4.3.3, A.R. $= 3.5{:}1$ max
- Headers and sills 6.5:1 max
- Transfer diaphragms 6.5:1 max., preferable 4.5:1

$$\text{Overall wall} = \frac{9}{19.5} = 0.46{:}1 \text{ OK}$$

$$\text{Header}_{Lo1} = \frac{3}{1.33} = 2.26{:}1 \text{ OK}$$

$$\text{Sill}_{Lo1} = \frac{4.67}{3} = 1.56{:}1 \text{ OK}$$

$$\text{Pier}_{TD1} = \frac{3}{4} = 0.75{:}1 \text{ OK}$$

$$\text{Pier}_{TD2} = \frac{3}{2} = 1.5{:}1 \text{ OK}$$

$$\text{Header}_{Lo2} = \frac{5}{1.33} = 3.76{:}1 \text{ OK}$$

$$\text{Sill}_{Lo2} = \frac{5}{2.67} = 1.87{:}1 \text{ OK}$$

$$\text{Pier}_{TD3} = \frac{5}{5.5} = 0.91{:}1 \text{ OK}$$

$$\text{TD1} = \frac{9}{4} = 2.25{:}1 \text{ OK}$$

$$\text{TD2} = \frac{9}{2} = 4.5{:}1 \text{ OK}$$

$$\text{TD3} = \frac{9}{5.5} = 1.64{:}1 \text{ OK}$$

Determine the overturning force (see Fig. 10.10):

$$T = C = \frac{4500(9)}{19.5} = 2076.9 \text{ lb}$$

Determine the distribution of shear above and below the opening (see Fig. 10.10): It is assumed that the shears applied to the header and sill section are distributed in proportion to their depths, and because no dead loads are applied, that distribution of the vertical shear is the same across the length of the wall openings. The unit shears in the full height sections are equal to the overturning force divided by the height of the wall.

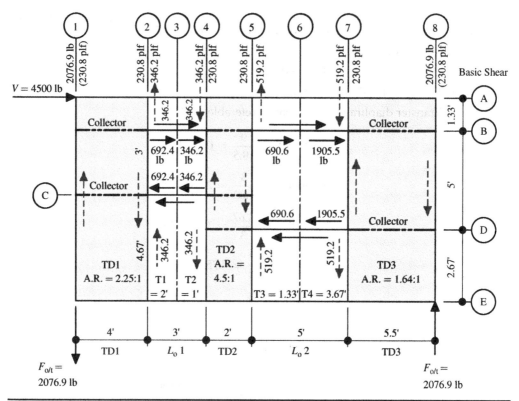

FIGURE 10.10   Shear above and below opening, tributary widths, and boundary forces at window openings.

Distribution of shear at the 3-in opening:

$$V_{AB} = \frac{2076.9(1.33)}{(4.67 + 1.33)} = 460.4 \text{ lb}, \; v = \frac{460.4}{1.33} = 346.2 \text{ plf}$$

$$V_{DE} = \frac{2076.9(4.67)}{(4.67 + 1.33)} = 1616.5 \text{ lb}, \; v = \frac{1616.5}{4.67} = 346.2 \text{ plf}$$

Distribution of vertical shear forces at the 5-in opening:

$$V_{AB} = \frac{2076.9(1.33)}{(2.67 + 1.33)} = 690.6 \text{ lb}, \; v = \frac{690.6}{1.33} = 519.2 \text{ plf}$$

$$V_{DE} = \frac{2076.9(2.67)}{(2.67 + 1.33)} = 1386.3 \text{ lb}, \; v = \frac{1386.3}{2.67} = 519.2 \text{ plf}$$

The unit shear at the edges of the full height sections:

$$v = \frac{2076.9}{9} = 230.8 \text{ plf}$$

Determine the tributary width of the openings (see Fig. 10.10): Tributary widths are determined in proportion to the pier widths on each side of the opening and the opening length.

$$T1 = \frac{TD1(L_{o1})}{(TD1 + TD2)} = \frac{4(3)}{(4+2)} = 2 \text{ ft}$$

$$T2 = \frac{TD2(L_{o1})}{(TD1 + TD2)} = \frac{2(3)}{(4+2)} = 1 \text{ ft}$$

$$T3 = \frac{TD2(L_{o2})}{(TD2 + TD3)} = \frac{2(5)}{(2+5.5)} = 1.33 \text{ ft}$$

$$T4 = \frac{TD1(L_{o1})}{(TD1 + TD2)} = \frac{5.5(5)}{(2+5.5)} = 3.67 \text{ ft}$$

Determine the corner forces at the openings (see Fig. 10.10): The corner forces above the opening act to the right. The force below the openings is the same magnitude as the forces above the openings but acts in the opposite direction. The corner forces are equal to the unit shear in the header multiplied by its tributary width.

$$F_{T1} = 346.2(2) = 692.4 \text{ lb}$$

$$F_{T2} = 346.2(1) = 346.2 \text{ lb}$$

$$F_{T3} = 519.2(1.33) = 690.6 \text{ lb}$$

$$F_{T4} = 519.2(3.67) = 1905.5 \text{ lb}$$

The direction of the transfer shear forces applied to the transfer diaphragm boundary members are shown as dashed lined arrows in Fig. 10.10.

Determine the transfer diaphragm unit shears and net shears (see Fig. 10.11): The forces acting on the transfer diaphragm are shown in the figure. The transfer diaphragm analysis is the same as described in Chap. 3.

Transfer diaphragm shears TD1:

$$V_{AB} = \frac{692.4(7.67) - 692.4(4.67)}{9} = 230.8 \text{ lb}, \quad v = \frac{230.8}{4} = 57.7 \text{ plf}$$

Placing the sheathing element symbols next to the reaction force per standard procedure shows that the shears from A to B and from C to E are negative in value and are equal to:

$$V_{BC} = -230.8 + 692.4 = 461.6 \text{ lb}, \quad v = \frac{461.6}{4} = +115.4 \text{ plf}$$

$$V_{CD} = 461.6 - 692.4 = -230.8 \text{ lb}, \quad v = \frac{-230.8}{4} = -57.7 \text{ plf}$$

Transfer diaphragm Net shears TD1:
Basic shear $= +230.8$ plf as previously calculated. The net shears are equal to the basic shear $+/-$ the transfer diaphragm shear. Once the net shear has been determined, the direction of the resulting net transfer shear is placed on the transfer diaphragm in accordance with the sheathing element symbol sign convention as shown in the figure.

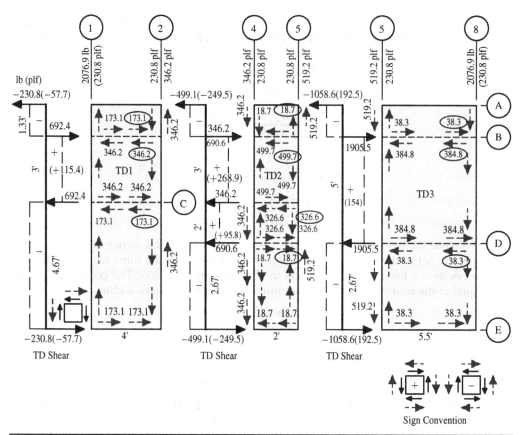

Figure 10.11   Transfer diaphragm shears and net shears.

Grid line	Net shear
1AB	$v_{net} = +230.8 - 57.7 = +173.1$ plf
1BC	$v_{net} = +230.8 + 115.4 = +346.2$ plf
1CD	$v_{net} = +230.8 - 57.7 = +173.1$ plf

Transfer diaphragm shear TD2:

$$V_{AB} = \frac{(346.2 + 690.6)(7.67) - 346.2(4.67) - 690.6(2.67)}{9} = -499.1 \text{ lb,}$$

$$v = \frac{-499.1}{2} = -249.5 \text{ plf}$$

$$V_{BC} = -499.1 + 690.6 + 346.2 = 537.7 \text{ lb,} \qquad v = \frac{537.7}{2} = +268.9 \text{ plf}$$

$$V_{CD} = 537.7 - 346.2 = 191.5 \text{ lb,} \qquad v = \frac{191.5}{2} = 95.8 \text{ plf}$$

$$V_{DE} = 191.5 - 690.6 = -499.1 \text{ lb,} \qquad v = \frac{-499.1}{2} = -249.5 \text{ plf}$$

Transfer diaphragm net shear TD2:

Grid line	Net shear
4AB	$v_{net} = +230.8 - 249.5 = -18.7$ plf
4BC	$v_{net} = +230.8 + 268.9 = +499.7$ plf
4CD	$v_{net} = +230.8 + 95.8 = +326.6$ plf
4DE	$v_{net} = +230.8 - 249.5 = -18.7$ plf

The net shears at grid line 5 is the same value.

Transfer diaphragm shear TD3:

$$V_{AB} = \frac{(1905.5)(7.67) - 1905.5(2.67)}{9} = -1058.6 \text{ lb}, \quad v = \frac{-1058.6}{5.5} = -192.5 \text{ plf}$$

$$V_{BD} = -1058.6 + 1905.5 = 846.9 \text{ lb}, \qquad\qquad v = \frac{846.9}{5.5} = +154 \text{ plf}$$

$$V_{DE} = 846.9 - 1905.5 = -1058.6 \text{ lb}, \qquad\qquad v = \frac{-1058.6}{5.5} = -192.5 \text{ plf}$$

Transfer diaphragm net shear TD3:

Grid line	Net shear
7AB	$v_{net} = +230.8 - 192.5 = +38.3$ plf
7BD	$v_{net} = +230.8 + 154 = 384.8$ plf
7DE	$v_{net} = +230.8 - 192.5 = +38.3$ plf

The net shears at grid line 8 are the same value. Note that the maximum unit shear forces in each area of the wall are circled in Fig. 10.11.

Determine the vertical force diagrams (see Fig.10.12): The development of the force diagrams isn't necessary, but useful in verifying that the analysis is correct, and that $\Sigma V = 0$ and $\Sigma H = 0$.

At grid line 1:

$$F_B = 173.1(1.33) = 230.2 \text{ lb}$$

$$F_C = 230.2 + 346.2(3) = 1268.6 \text{ lb}$$

$$F_E = 1268.6 + 173.1(4.67) = 2076.9 \text{ lb} = \text{tension force}$$

At grid line 2:

$$F_B = (346.2 - 173.1)(1.33) = +230.2 \text{ lb}$$

$$F_C = 230.2 - 346.2(3) = -808.4 \text{ lb}$$

$$F_E = -808.4 + (-173.1 + 346.2)(4.67) = 0 \text{ lb. closes}$$

At grid line 4:

$$F_B = (-346.2 - 18.7)(1.33) = -485.3 \text{ lb}$$

$$F_C = -485.3 + 499.7(3) = +1013.8 \text{ lb}$$

$$F_D = 1013.8 - (346.2 + 326.6)(2) = 974.5 \text{ lb}$$

$$F_E = 974.5 - (346.2 + 18.7)(2.67) = 0 \text{ lb. closes}$$

At grid line 5:

$$F_B = (519.2 + 18.7)(1.33) = +715.4 \text{ lb}$$

$$F_C = 715.4 - 499.7(3) = -783.6 \text{ lb}$$

$$F_D = -783.6 - 326.6(2) = -1436.9 \text{ lb}$$

$$F_E = -1436.9 + (18.7 + 519.2)(2.67) = 0 \text{ lb, closes}$$

At grid line 7:

$$F_B = (-519.2 + 38.3)(1.33) = -639.6 \text{ lb}$$

$$F_D = -639.6 + 384.8(5) = +1284.4 \text{ lb}$$

$$F_E = 1284.4 - (519.2 + 38.3)(2.67) = 0 \text{ lb, closes}$$

At grid line 8:

$$F_B = -38.3(1.33) = -50.9 \text{ lb}$$

$$F_D = -50.9 - 384.8(5) = -1974.9 \text{ lb}$$

$$F_E = -1974.9 - 38.3(2.67) = 2076.9 \text{ lb} = \text{compression force}$$

Determine the horizontal force diagrams (see Fig. 10.12):

Collector at grid line B, starting at grid line 1. Average shears at each end of collector multiplied by length of collector.

$$F_2 = \frac{(-346.2 + 173.1) + (-346.2 + 173.1)}{2}(4') = -692.4 \text{ lb} = \text{corner force}$$

$$F_5 = +346.2 \text{ corner force at line } 4 - \frac{(-499.7 - 18.7) + (-499.7 - 18.7)}{2}(2')$$

$$= -690.6 \text{ lb} = \text{corner force at line 5}$$

$$F_8 = +1905.5 \text{ corner force at line } 7 - \frac{(-384.8 + 38.3) + (-384.8 + 38.3)}{2}(5.5') = 0 \text{ lb}$$

Collector at grid line C, starting at grid line 1:

$$F_2 = \frac{(346.2 - 173.1) + (346.2 - 173.1)}{2}(4') = +692.4 \text{ lb} = \text{corner force}$$

$$F_5 = +346.6 \text{ corner force at line } 4 - \frac{(499.7 - 326.6) + (499.7 - 326.6)}{2}(2') = 0 \text{ lb}$$

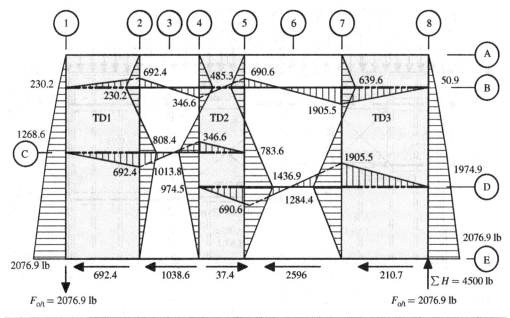

FIGURE 10.12 Collector force diagrams.

Collector at grid line D, starting at grid line 4:

$$F_5 = \frac{(326.6 + 18.7) + (326.6 + 18.7)}{2}(2') = 690.6 \text{ lb} = \text{corner force}$$

$$F_8 = 1905.5 \text{ corner force at line } 7 - \frac{(384.8 - 38.3) + (384.8 - 38.3)}{2}(5.5') = 0 \text{ lb}$$

Therefore, all force diagrams close to zero which verifies the analysis.

Check analysis, sum horizontal forces at bottom of wall (see Fig. 10.12):

$$\Sigma H = 173.1(4) + 346.2(3) - 18.7(2) + 519.2(5) + 38.3(5.5) = 4500 \text{ lb} = \text{lateral}$$
applied force, therefore, OK.

Notice the direction of the force at the bottom of TD2 caused by the rotation of the transfer diaphragm about the sill section. ▲

Example 10.2: FTAO Transfer Diaphragm Method with Dead Load

All loads have been factored for the appropriate load combinations. Analyze for the loads shown and determine wall shears and collector forces only.

The wall shown in Fig. 10.13 is identical to Example 10.1 except for the added uniform dead load. It is recommended to compare the results of this example with Example 10.1. The main difference with this example is that the vertical shears will vary linearly across the length of the wall due to the applied uniform gravity load.

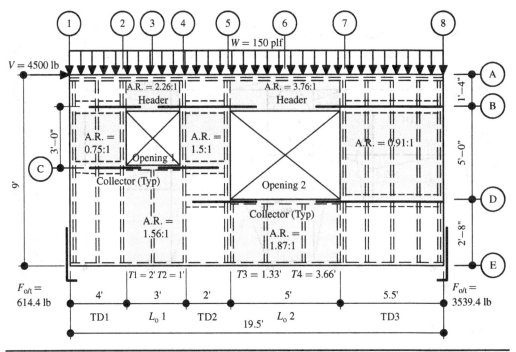

FIGURE 10.13 FTAO wall with DL using the transfer diaphragm method.

Given:

    $V = 4500$-lb lateral force
    $w = 150$-plf uniform DL
    All loads have been factored for the appropriate load combination.

All the tributary widths over the openings are determined in the same manner as Example 10.1, and are as follows:

    $T1 = 2'$, $T2 = 1'$ at opening 1
    $T3 = 1.33'$, $T4 = 3.66'$ at opening 2

Determine the aspect ratios of the wall and wall segments (see Fig. 10.13):

The aspect ratios for the overall wall, piers, headers, sills, and transfer diaphragms are the same as Example 10.1.

Determine the overturning force (see Fig. 10.13):

$$T = \left[4500(9) - \frac{150(19.5)^2}{2}\right]\frac{1}{19.5} = 614.4 \text{ lb}$$

$$C = \left[4500(9) + \frac{150(19.5)^2}{2}\right]\frac{1}{19.5} = 3539.4 \text{ lb}$$

Determine the vertical shear acting across the wall (see Fig. 10.14): The vertical shears above the wall will vary linearly due to the applied uniform dead load. Sum vertical forces starting at grid line 8 and proceeding to the left. For the case where dead loads are

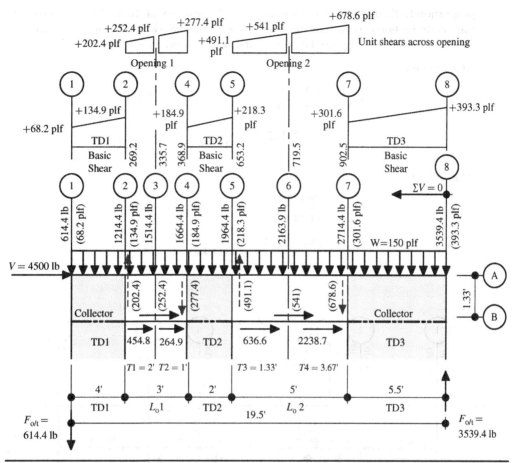

FIGURE 10.14 Distribution of dead loads to header.

applied, it is important to determine the vertical shears at the points of inflection to determine the corner forces. The vertical and unit shears at each end of each transfer diaphragms are equal to the vertical shear at that location divided by the wall height. The unit shears in the headers and sills and at points of inflection are equal to the vertical shear divided by the sum of the header and sill depths.

$$V_8 = 3539.4 \text{ lb, divided by wall height } v = \frac{3539.4}{9} = 393.3 \text{ plf}$$
$$V_7 = 3539.4 - 150(5.5) = 2714.4 \text{ lb}, v = 301.6 \text{ plf}$$
$$V_6 = 2714.4 - 150(3.67) = 2163.9 \text{ lb}$$
$$V_5 = 2163.9 - 150(1.33) = 1964.4 \text{ lb}, v = 218.3 \text{ plf}$$
$$V_4 = 1964.4 - 150(2) = 1664.4 \text{ lb}, v = 184.9 \text{ plf}$$
$$V_3 = 1664.4 - 150(1) = 1514.4 \text{ lb}$$
$$V_2 = 1514.4 - 150(2) = 1214.4 \text{ lb}, v = 134.9 \text{ plf}$$
$$V_1 = 1214.4 - 150(4) = 614.4 \text{ lb}, v = 68.2 \text{ plf}$$

Determine the vertical shear forces above and below the openings (see Figs. 10.14 and 10.15):
The distribution to the header and sill sections at all openings is assumed to be in

proportion to their depth as in Example 10.1. A summary of the unit shears across the full height sections and across the window openings are shown in Fig. 10.14. The unit shears in the sill sections are the same as the headers and are shown in Fig. 10.15.

At grid line 7:

$$V_{hdr} = \frac{2714.4(1.33)}{(1.33 + 2.67)} = 902.5 \text{ lb}, \ v = 678.6 \text{ plf}$$

$$V_{sill} = \frac{2714.4(2.67)}{4} = 1811.8 \text{ lb}, \ v = 678.6 \text{ plf}$$

At grid line 6:

$$V_{hdr} = \frac{2163.9(1.33)}{(1.33 + 2.67)} = 719.5 \text{ lb}, \ v = 541 \text{ plf}$$

$$V_{sill} = \frac{2163.9(2.67)}{4} = 1444.4 \text{ lb}, \ v = 541 \text{ plf}$$

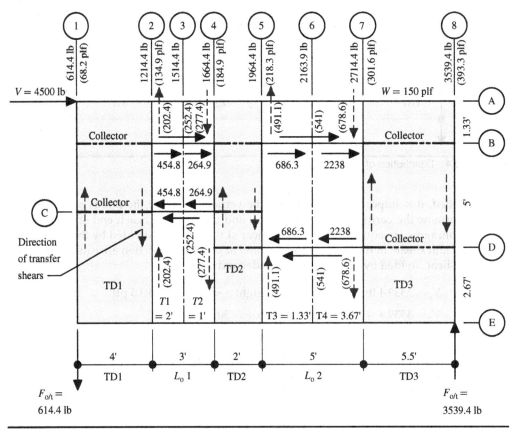

FIGURE 10.15 Tributary widths, corner forces, and shear forces.

At grid line 5:

$$V_{hdr} = \frac{1964.4(1.33)}{(1.33 + 2.67)} = 653.2 \text{ lb, } v = 491.1 \text{ plf}$$

$$V_{sill} = \frac{1964.4(2.67)}{4} = 1311.2 \text{ lb, } v = 491.1 \text{ plf}$$

At grid line 4:

$$V_{hdr} = \frac{1664.4(1.33)}{(1.33 + 4.67)} = 368.9 \text{ lb, } v = 277.4 \text{ plf}$$

$$V_{sill} = \frac{1664.4(4.67)}{6} = 1295.5 \text{ lb, } v = 277.4 \text{ plf}$$

At grid line 3:

$$V_{hdr} = \frac{1514.4(1.33)}{(1.33 + 4.67)} = 335.7 \text{ lb, } v = 252.4 \text{ plf}$$

$$V_{sill} = \frac{1514.4(4.67)}{6} = 1178.7 \text{ lb, } v = 252.4 \text{ plf}$$

At grid line 2:

$$V_{hdr} = \frac{1214.4(1.33)}{(1.33 + 4.67)} = 269.2 \text{ lb, } v = 202.4 \text{ plf}$$

$$V_{sill} = \frac{1214.4(4.67)}{6} = 945.2 \text{ lb, } v = 202.4 \text{ plf}$$

Determine the corner forces at the openings (see Figs. 10.14 and 10.15): Unlike the walls without gravity loads applied, the shears across the top of the wall are not uniform but vary linearly. Therefore, the varying shears acting along each tributary width of the header and sill section must be averaged and multiplied by their respective tributary lengths. The direction of the corner forces at grid line 2 at the bottom of the opening should be acting to the left because the rotation of the transfer diaphragm is pushing on the sill section which would cause compression. The direction of the corner forces at the top of the opening at that grid line should be acting in the opposite direction creating tension as shown in Fig. 10.15.

At header opening 1:

$$F_{T1} = \frac{(202.4 + 252.4)}{2}(2') = 454.8 \text{ lb}$$

$$F_{T2} = \frac{(252.4 + 277.4)}{2}(1') = 264.9 \text{ lb}$$

At header opening 2:

$$F_{T3} = \frac{(491.1 + 541)}{2}(1.33') = 686.3 \text{ lb}$$

$$F_{T4} = \frac{(541 + 678.6)}{2}(3.66') = 2238 \text{ lb}$$

Determine the transfer diaphragm shears and net shears (see Fig. 10.16): The shears within the transfer diaphragms will also vary due to the uniform DL that is applied. Using the transfer diaphragm method will determine the varying shear values. To the left of each transfer diaphragm is the calculated TD shears, caused by the opening corner forces. Above each transfer diaphragm are the individual basic shear diagrams with shear values. The analytical processes of calculating the transfer diaphragm shears, net shears, and the direction of applied transfer shears have been presented in the previous chapters and are shown in the figure. Additionally, the maximum shear values for nailing are circled in the figure.

Transfer diaphragm shears TD1:

$$V_{AB} = \frac{454.8(7.67) - 454.8(4.67)}{9} = -151.6 \text{ lb}, \quad v = \frac{-151.6}{4} = -37.9 \text{ plf}$$

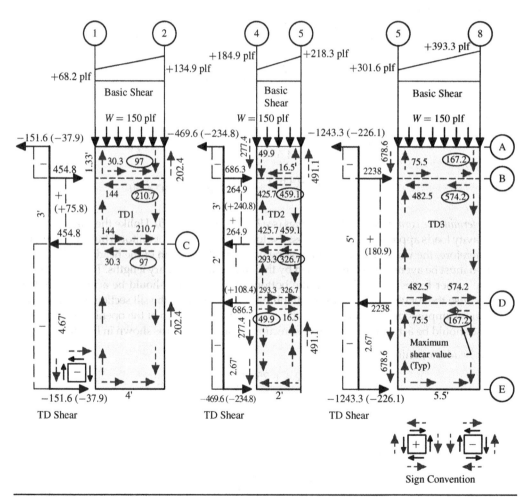

FIGURE 10.16 Transfer diaphragms.

Placing the sheathing element symbols next to the reaction force per standard procedure shows that the shears from A to B and from C to E are negative in value and are equal to

$$V_{BC} = -151.6 + 454.8 = 303.2 \text{ lb}, \quad v = \frac{303.2}{4} = +75.8 \text{ plf}$$

$$V_{CE} = 303.2 - 454.8 = -151.6 \text{ lb}, \quad v = \frac{-151.6}{4} = -37.9 \text{ plf}$$

Transfer diaphragm net shears TD1:

Grid line	Net shear
1AB	$v_{net} = +68.2 - 37.9 = +30.3$ plf
1BC	$v_{net} = +68.2 + 75.8 = +144$ plf
1CE	$v_{net} = +68.2 - 37.9 = +30.3$ plf
2AB	$v_{net} = +134.9 - 37.9 = +97$ plf
2BC	$v_{net} = +134.9 + 75.8 = +210.7$ plf
2CE	$v_{net} = +134.9 - 37.9 = +97$ plf

Transfer diaphragm shears TD2:

$$V_{AB} = \frac{(264.9 + 686.3)(7.67) - 264.9(4.67) - 686.3(2.67)}{9} = -469.6 \text{ lb},$$

$$v = \frac{-469.6}{2} = -234.8 \text{ plf}$$

$$V_{BC} = -469.6 + 686.3 + 264.9 = 481.6 \text{ lb}, \quad v = \frac{481.6}{2} = +240.8 \text{ plf}$$

$$V_{CD} = 481.6 - 264.9 = 216.7 \text{ lb}, \quad v = \frac{216.7}{2} = 108.4 \text{ plf}$$

$$V_{DE} = 216.7 - 686.3 = -469.6 \text{ lb}, \quad v = \frac{-469.6}{2} = -234.8 \text{ plf}$$

Transfer diaphragm net shears TD2:

Grid line	Net shear
4AB	$v_{net} = +184.9 - 234.8 = -49.9$ plf
4BC	$v_{net} = +184.9 + 240.8 = +425.7$ plf
4CD	$v_{net} = +184.9 + 108.4 = +293.3$ plf
4DE	$v_{net} = +184.9 - 234.8 = -49.9$ plf
5AB	$v_{net} = +218.3 - 234.8 = -16.5$ plf
5BC	$v_{net} = +218.3 + 240.8 = +459.1$ plf
5CD	$v_{net} = +218.3 + 108.4 = +326.7$ plf
5DE	$v_{net} = +218.3 - 234.8 = -16.5$ plf

Transfer diaphragm shears TD3:

$$V_{AB} = \frac{(2238)(7.67) - 2238(2.67)}{9} = -1243.3 \text{ lb}, \ v = \frac{-1243.3}{5.5} = -226.1 \text{ plf}$$

$$V_{BD} = -1243.7 + 2238 = 994.7 \text{ lb}, \ v = \frac{994.7}{5.5} = +180.9 \text{ plf}$$

$$V_{DE} = 994.7 - 2238 = -1243.3 \text{ lb}, \ v = \frac{-1058.6}{5.5} = -226.1 \text{ plf}$$

Transfer diaphragm net shears TD3:

Grid line	Net shear
7AB	$v_{net} = +301.6 - 226.1 = +75.5 \text{ plf}$
7BD	$v_{net} = +301.6 + 180.9 = 482.5 \text{ plf}$
7DE	$v_{net} = +301.6 - 226.1 = +75.5 \text{ plf}$
8AB	$v_{net} = +393.3 - 226.1 = +167.2 \text{ plf}$
8BD	$v_{net} = +393.3 + 180.9 = +574.2 \text{ plf}$
8DE	$v_{net} = +393.3 - 226.1 = +75.5 \text{ plf}$

Determine the vertical force diagrams (see Fig. 10.17): The development of the force diagrams isn't necessary, but useful in verifying that the analysis is correct, and that $\Sigma V = 0$. At grid line 1:

$$F_B = +30.3(1.33) = +40.3 \text{ lb}$$

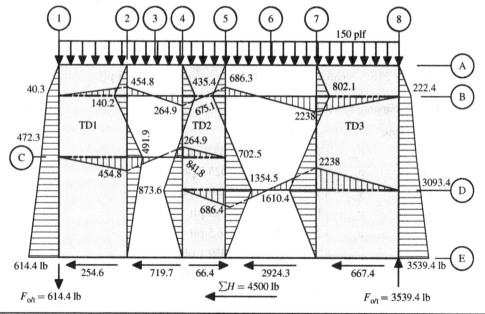

Figure 10.17 Force diagrams.

$$F_C = 40.3 + 144(3) = +472.3 \text{ lb}$$

$$F_E = 472.3 + 30.3(4.67) = 614 \text{ lb}$$

At grid line 2:

$$F_B = (202.4 - 97)(1.33) = +140.2 \text{ lb}$$

$$F_C = 140.2 - 210.7(3) = -491.9 \text{ lb}$$

$$F_E = -491.1 + (202.4 - 97)(4.67) = 0 \text{ lb}$$

At grid line 4:

$$F_B = (-277.4 - 49.9)(1.33) = -435.3 \text{ lb}$$

$$F_C = -435.3 + 425.7(3) = +841.8 \text{ lb}$$

$$F_D = +841.8 + (293.3 - 277.4)(2) = 873.6 \text{ lb}$$

$$F_E = 873.6 - (277.4 + 49.9)(2.67) = 0 \text{ lb}$$

At grid line 5:

$$F_B = (491.1 + 16.5)(1.33) = +675.1 \text{ lb}$$

$$F_C = 675.1 - 459.1(3) = -702.2 \text{ lb}$$

$$F_D = -702.2 - 326.7(2) = -1355.6 \text{ lb}$$

$$F_E = -1355.6 + (165 + 491.1)(2.67) = 0 \text{ lb}$$

At grid line 7:

$$F_B = (-678.6 - 75.5)(1.33) = -802.1 \text{ lb}$$

$$F_D = -802.1 + 482.5(5) = +1610.4 \text{ lb}$$

$$F_E = 1610.4 - (678.6 - 75.5)(2.67) = 0 \text{ lb}$$

At grid line 8:

$$F_B = -167.2(1.33) = -222.4 \text{ lb}$$

$$F_D = -222.4 - 574.2(5) = -3093.4 \text{ lb}$$

$$F_E = -3093.4 - 167.2(2.67) = 3539.4 \text{ lb}$$

Determine the horizontal force diagrams (see Fig. 10.17):

Collector at grid line B, starting at grid line 1:

$$F_2 = \frac{(-144 + 30.3) + (-210.7 + 97)}{2}(4') = 454.8 \text{ lb} = \text{corner tension force.}$$

$$F_5 = 264.9 \text{ corner force} - \frac{(-425.7 - 49.9) + (-459.5 - 16.5)}{2}(2') = 686.3 \text{ lb}$$

$$F_8 = 2238 \text{ corner force} - \frac{(-482.5 + 75.5) + (-574.2 + 167.2)}{2}(5.5') = 0 \text{ lb}$$

Collector at grid line C, starting at grid line 1:

$$F_2 = \frac{(144 - 30.3) + (210.7 - 97)}{2}(4') = 454.8 \text{ lb}$$

$$F_5 = 264.9 \text{ corner force} - \frac{(425.7 - 293.3) + (459.1 - 326.7)}{2}(2') = 0 \text{ lb}$$

Collector at grid line D, starting at grid line 4:

$$F_5 = \frac{(293.3 + 49.9) + (326.7 + 16.5)}{2}(2') = 686.4 \text{ lb}$$

Collector at grid line D, starting at grid line 7:

$$F_8 = 2238 - \frac{(482.5 - 75.5) + (574.2 - 167.2)}{2}(5.5') = 0 \text{ lb}$$

Summing the shears at the bottom of the wall must equal the applied lateral force. The shears at each segment are equal to the average of the unit shear in each segment (see Fig. 10.17).

$$\text{Example TD1} = \frac{(30.3 + 97)}{2}(4) = 254.6 \text{ lb}$$

Total resisting force = $254.6 + 719.7 - 66.4 = 2924.3 + 667.4 = 4500 \text{ lb}$

Therefore, all force diagrams close to zero which verifies the analysis. ▲

Shear Wall Deflection

Shear wall deflection for an FTAO wall is typically done by one of two methods. Most common is the APA/Diekmann approach which uses averaging of the deflection of the pier section from lateral loads being applied from both directions as shown in Fig. 10.18. The APA method closely matches the FTAO tests and was also verified by a computer analysis for accuracy.

Single Opening:

$$\Delta_{\text{aver.}} = \frac{\left(\Delta_{\text{pier 1}} + \Delta_{\text{pier 2}}\right)\text{left} + \left(\Delta_{\text{pier 1}} + \Delta_{\text{pier 2}}\right)\text{right}}{4}$$

Multiple Openings:

$$\Delta_{\text{aver.}} = \frac{\left(\Delta_{\text{pier 1}} + \Delta_{\text{pier 2}} + \Delta_{\text{pier 3}}\right)\text{left} + \left(\Delta_{\text{pier 1}} + \Delta_{\text{pier 2}} + \Delta_{\text{pier 3}}\right)\text{right}}{6}$$

The other method is the center strip method that is similar to a masonry wall. The difference with this method over that of a masonry wall is that the pier section deflection equations are not fixed at both ends. The APA method is preferred.

The transfer diaphragm method examples presented show that significant flexibility in wall layouts and loading can be achieved with this method. This is especially important if the loads and forces from multistory shear walls are considered, or the wall has a complex layout as shown in Fig. 10.19.

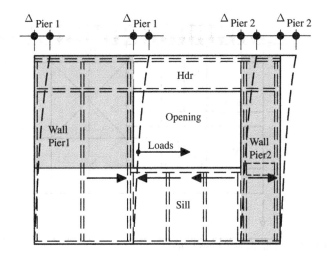

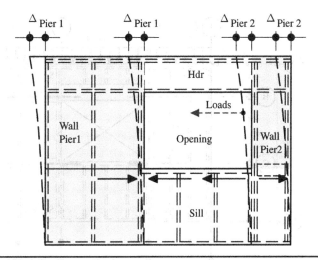

Figure 10.18 APA deflection method.

Example 10.3: FTAO Transfer Diaphragm Method with Dead Load—Partial Length Straps

All loads have been factored for the appropriate load combinations. Analyze for the loads shown and determine wall shears and collector forces only.

SDPWS Section 4.3.2.2(1) notes "Framing members, blocking, and connections around openings are designed" and nothing more. SDPWS Commentary Figure C4.3.2.2 shows partial length straps and refers back to that section. Therefore, it is left up to the engineer to properly design and determine the strap length by rational methods. Transfer diaphragms allow using shorter strap lengths while still providing complete alternate load paths as required by the code, provided the design is based on a rational analysis.

A sample wall will be examined with partial length straps, as shown in Fig. 10.20. A 2 ft 8 in transfer diaphragm has been installed on each side of the opening, designated

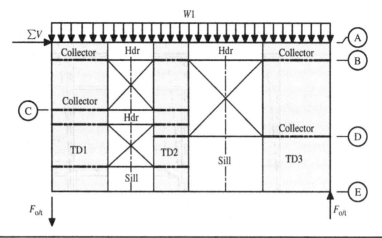

Figure 10.19   Complex wall possibilities.

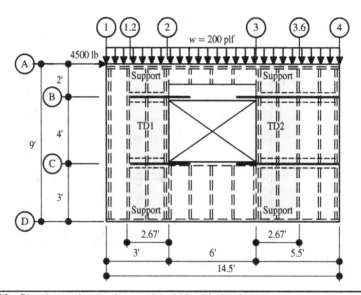

Figure 10.20   Blocking and strapping partial width with dead load.

as TD1 and TD2. These transfer diaphragms act as vertical beams that receive the disrupted forces on each end of the header and sill sections as shown in the figure and transfer them into the top and bottom plates of the wall. All the relevant shears and forces applied to the right wall pier are as shown in Fig. 10.21. They consist of the uniformly applied gravity loads, the forces that are applied to the transfer diaphragms and the vertical shear forces applied from the header section and sill section.

Check transfer diaphragm aspect ratios:

$$TD1 = TD2 = \frac{9}{2.67} = 3.37 < 6.5 \text{ max}$$

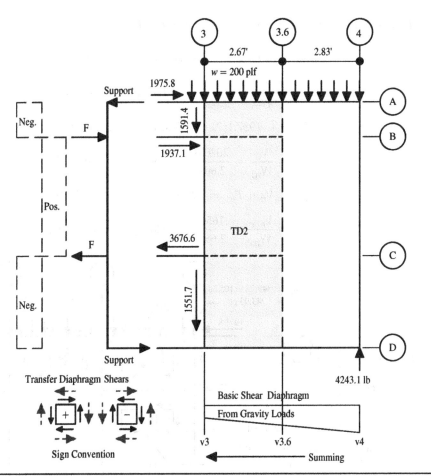

FIGURE 10.21 Pier section forces and loads.

The transfer diaphragm must be blocked. Adjust the allowable shears for nailing capacity since the A.R. is greater than 2:1.

Determine the basic shear diagram starting at the right wall reaction (see Fig. 10.21):

$R_4 = 4243.1$ lb

$v_4 = \dfrac{R_4}{h_{wall}} = \dfrac{4243.1}{9} = 471.46$ plf

$V_{3.6} = R_4 - wx_{4-3.6} = 4243.1 - 200(2.83) = 3677.1$ lb, summing to the left

$v_{3.6} = \dfrac{V_{3.6}}{h_{wall}} = \dfrac{3677.1}{9} = 408.57$ plf

$V_3 = V_{3.6} - wx_{3.6-3} = 3677.1 - 200(2.67) = 3143.1$ lb

Sum vertical shears at grid line 3:

$$\Sigma V = -1591.4 - 1551.7 + 3143.1 = 0 \quad \therefore \text{ OK}$$

Determine the transfer diaphragm net shears (see Fig. 10.22):

$$V_A = \left[1937.1(7) - 3676.6(3)\right]\frac{1}{9} = 281.1 \text{ lb}$$

$$v_A = \frac{V_A}{W_{TD}} = \frac{281.1}{2.667} = -105.3 \text{ plf, negative shear}$$

$$V_D = \left[-1937.1(2) + 3676.6(6)\right]\frac{1}{9} = 2020.6 \text{ lb}$$

$$v_D = \frac{V_D}{W_{TD}} = \frac{2020.6}{2.667} = -756.8 \text{ plf, negative shear}$$

$$V_{B-C} = V_D + F_{3C} = -2020.6 + 3676.6 = 1656 \text{ lb}$$

$$v_{B-C} = \frac{V_{B-C}}{W_{TD}} = \frac{1656}{2.667} = +620.2 \text{ plf, positive shear}$$

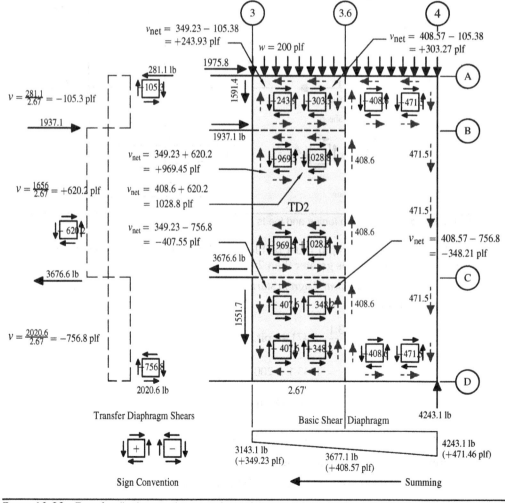

FIGURE 10.22 Transfer diaphragm shears and net shears.

Determine the net shears in the transfer diaphragm:

$$v_{net} = +349.23 - (105.3) = +243.93 \text{ plf, grid line 3 from A to B}$$

$$v_{net} = +349.23 + (620.2) = +969.45 \text{ plf, grid line 3 from B to C}$$

$$v_{net} = +349.23 - (756.8) = -407.55 \text{ plf, grid line 3 from C to D}$$

$$v_{net} = +408.57 - (105.3) = +303.27 \text{ plf, grid line 3.6 from A to B}$$

$$v_{net} = +408.57 + (620.2) = +1028.8 \text{ plf, grid line 3.6 from B to C}$$

$$v_{net} = +408.57 - (756.8) = -348.21 \text{ plf, grid line 3.6 from C to D}$$

$$v_{net} = +408.57 \text{ plf grid line 3.6 from A to D, right side}$$

$$v_{net} = +471.46 \text{ plf grid line 4 from A to D}$$

Determine the horizontal collector forces (see Fig. 10.23):

$$F_A = \frac{(408.6 + 471.5)}{2}(2.83) + \frac{(243.9 + 303.3)}{2}(2.67) = 1975.8 \text{ lb, compression}$$

$$F_B = \frac{[(969.5 - 243.9) + (1028.8 - 303.3)]}{2}(2.67) = 1937.1 \text{ lb, compression}$$

$$F_C = \frac{[(969.45 + 407.55) + (1028.8 + 348.2)]}{2}(2.67) = 3676.6 \text{ lb, tension}$$

All the calculated forces match the applied forces, $\therefore$ OK

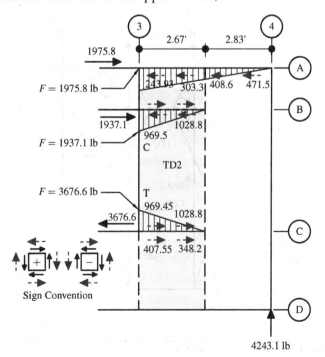

FIGURE 10.23 Horizontal collector forces.

Determine the vertical collector forces (see Fig. 10.24):

Starting at grid line 4A:

$$F_{4D} = 471.5(9) = 4243.1 \text{ lb compression}$$

Starting at grid line 3.6D:

$$F_{3.6C} = (348.2 + 480.6)(3) = 2270.3 \text{ lb compression}$$

$$F_{3.6B} = 2270.3 - (1028.8 - 480.6)(4) = -210.54 \text{ lb tension}$$

$$F_{3.6A} = -210.54 + (480.6 - 303.3)(2) = 0 \text{ lb, the diaphragm closes}$$

Starting at grid line 3D:

$$F_{3C} = -407.6(3) - 1551.7 = -2774.4 \text{ lb tension}$$

$$F_{3B} = -2774.4 + (969.5)(4) = +1103.45 \text{ lb compression}$$

$$F_{3A} = +1103.45 + (243.93)(2) - 1591.4 = 0 \text{ lb, the diaphragm closes}$$

The resulting wall hold-down forces and collector forces of Examples 10.1, 10.2, and this example should be compared to understand the significance of including gravity loads. It would also be of value for the reader to analyze the wall with the lateral force in the opposite direction plus the addition of gravity loads.

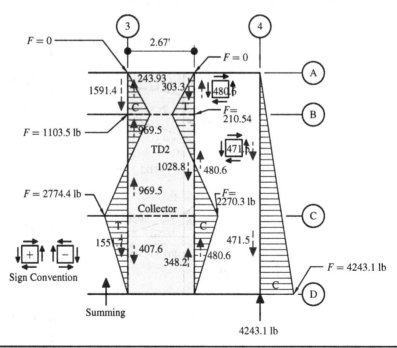

Figure 10.24   Vertical collector forces.

For a complete design of these walls, there are members, connections, and stresses that need to be checked in addition to the nailing requirements of the sheathing, the sizing of the tie straps, and the hold-downs at the ends of the wall. Testing has shown that failure typically occurred at the corner joints of the opening due to buckling, the transfer of shears and tension forces across the sheathing joints and bearing perpendicular to the grain issues. ▲

Transfer of the Horizontal Corner Forces

Figure 10.25 shows the forces acting at a typical corner joint and two possible framing conditions. The joint in the sheathing, in this case, is located at the inside face of the jamb stud, which is a common occurrence. The vertical shear from the header panel can be transferred across the joint through the light gauge hanger or shear clip connection as shown in framing configuration 1, or by a trimmer stud as shown in framing configuration 2.

Bearing Perpendicular to the Grain (See Fig. 10.25)

Bearing stresses perpendicular to the grain are produced by the header and the sill plates bearing on the vertical studs when the joint is in compression. The compression force is resisted by the horizontal blocking on the opposite side of the vertical studs, which also create bearing stresses. The number of $2\times$ blocking members must accommodate the width of the tie strap and reduce the bearing stresses to within allowable limits. The blocking members should be full depth of the wall studs to prevent local

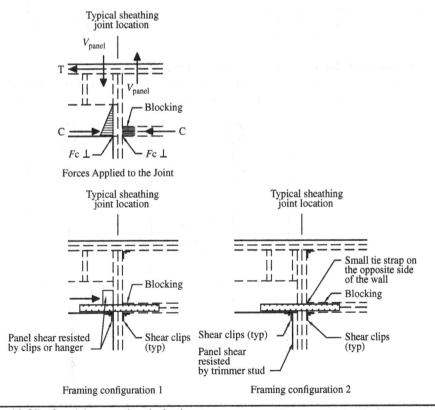

Figure 10.25 Special connection design issues.

buckling of the studs, not installed flat. Bearing stresses perpendicular to the grain should also be checked at the boundary element to bottom plate.

10.4 Shear Walls with Openings—The Cantilever Method

In the force transfer method, the sections above and below the opening are developed to their full capacity, because they are intended to be interconnected rigid elements that meet the allowable aspect ratio and can transfer the forces across the opening. Example 10.1 had a wall height of 9 ft, allowing a deeper header section, which may not always be the case. Whenever the wall is only 8 ft high or a window opening is higher, which is common for some residential and light commercial buildings, the depth of the header section above the opening might not meet the required aspect ratio for wider openings and should be ignored in the analysis because they are too flexible to effectively transfer forces across the opening. The reduced force distribution of this condition can be seen in Fig. 10.7. For this case, an alternate force transfer method of analysis exists. The method is based on simple statics, but only the full height sections and the sill sections are used as shown in Fig. 10.26, and the transfer of forces are required at the bottom of the opening only. The full height sections are treated as cantilevers which project above the sill section. The analysis of the wall is an engineered design where the sill section is used to resist the cantilever forces and make the wall act as a whole. The method was presented in a module of a workshop for the University of Wisconsin, by Edward F. Diekmann[7] published in 1982. The method requires hold-downs only at the ends of the wall as shown in the figure. The rules for blocking and strapping are the same as presented in the FTAO method.

Example 10.4: Cantilever Method—Single Opening

The wall shown in Fig. 10.26 has a height of 8 ft and a length of 16 ft. The sill section is 3 ft high. A 4-ft-high by 6-ft-wide opening is offset from the center of the wall as shown

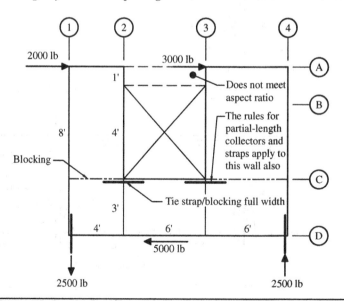

Figure 10.26   Cantilever method—single opening.

in the figure. The section above the opening does not comply with the allowable aspect ratios and therefore will not be used. Gravity loads will be ignored for this example, but when applied, can have the same effect as demonstrated in Example 10.2. A 5000-lb lateral force is applied at the top of the wall and is distributed to each pier in accordance with their respective widths.

Overturning force (see Fig. 10.26):

$$T = C = \frac{Vh}{b} = \frac{5000(8)}{16} = 2500 \text{ lb}$$

Shear force at top of the piers:

$$V_{1-2} = \frac{Vb_{1-2}}{b_{1-2} + b_{3-4}} = \frac{5000(4)}{10} = 2000 \text{ lb}$$

$$V_{3-4} = \frac{Vb_{3-4}}{b_{1-2} + b_{3-4}} = \frac{5000(6)}{10} = 3000 \text{ lb}$$

Left pier (see Fig. 10.27):

Summing moments about grid line 2D:

$$F_{2C} = [2000(8) - 2500(4)]\frac{1}{3} = 2000 \text{ lb}$$

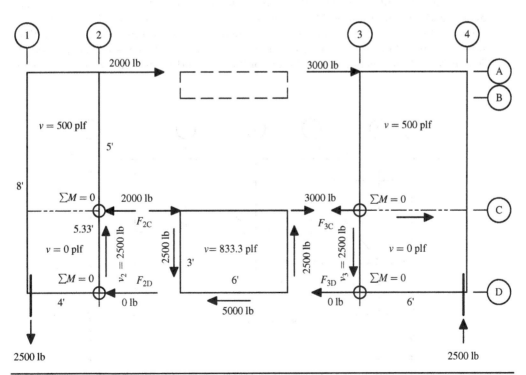

FIGURE 10.27 Wall section forces.

Summing moments about grid line 2C:

$$F_{2D} = \left[2000(5) - 2500(4)\right]\frac{1}{3} = 0 \text{ lb}$$

Summing vertical forces:

$$V_2 = 2500 \text{ lb, by inspection}$$

Right pier:

Summing moments about grid line 3D:

$$F_{3C} = \left[3000(8) - 2500(6)\right]\frac{1}{3} = 3000 \text{ lb}$$

Summing moments about grid line 3C:

$$F_{3D} = \left[3000(5) - 2500(6)\right]\frac{1}{3} = 0 \text{ lb}$$

Summing vertical forces:

$$V_3 = 2500 \text{ lb, by inspection} \quad \blacktriangle$$

Example 10.5: Cantilever Method—Double Opening

The wall shown in Fig. 10.28 has a height of 8 ft and a length of 23 ft. The sill section on the left has a height of 3 ft. The sill section on the right has a height of 5 ft. A 5-ft-high by 4-ft-wide opening and a 3-ft-high by 6-ft-wide opening are placed in the wall as shown in the figure. The sections above the opening do not comply with the allowable shear wall aspect ratios and therefore will not be used. Gravity loads will be ignored. A 5000-lb lateral force is applied at the top of the wall and is distributed to each pier in

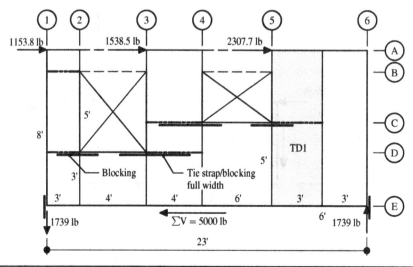

Figure 10.28   Cantilever method—double opening.

accordance with their respective lengths. A partial length collector and transfer diaphragm will be used at the right pier.

Overturning force (reference Fig. 10.28):

$$F = T = C = \frac{Vh}{b} = \frac{5000(8)}{23} = 1739 \text{ lb}$$

Shear force at top of piers:

$$V_{1-2} = \frac{Vb_{1-2}}{\Sigma b} = \frac{5000(3)}{13} = 1153.8 \text{ lb}$$

$$V_{3-4} = \frac{Vb_{3-4}}{\Sigma b} = \frac{5000(4)}{13} = 1538.5 \text{ lb}$$

$$V_{5-6} = \frac{Vb_{5-6}}{\Sigma b} = \frac{5000(6)}{13} = 2307.7 \text{ lb}$$

Left pier (see Fig. 10.29):

Sum moments about grid line 2D:

$$F_{2C} = [1153.8(8) - 1739(3)]\frac{1}{3} = 1337.9 \text{ lb}$$

Summing moments about grid line 2C:

$$F_{2D} = [1153.8(5) - 1739(3)]\frac{1}{3} = 184 \text{ lb}$$

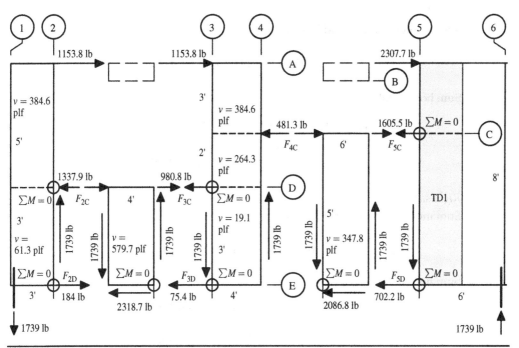

FIGURE 10.29 Wall section forces.

Sum vertical forces

$$V_2 = 1739 \text{ lb, by inspection}$$

Section shears:

$$v_{upper} = \frac{1153.8}{3} = 384.6 \text{ plf unit shear}$$

$$v_{lower} = \frac{1337.9 - 1153.8}{3} = 61.4 \text{ plf unit shear}$$

Right pier:

Sum moments about grid line 5D:

$$F_{5C} = \left[2307.7(8) - 1739(6)\right]\frac{1}{5} = 1605.5 \text{ lb}$$

Summing moments about grid line 5C:

$$F_{4D} = \left[2307.7(3) - 1739(6)\right]\frac{1}{3} = 702.2 \text{ lb}$$

Sum vertical forces:

$$V_4 = 1739 \text{ lb, by inspection}$$

Transfer the sill vertical shears to each section.

Left sill section:

Sum moments about grid line 3D:

$$F_{3C} = \left[1337.9(3) - 1739(4)\right]\frac{1}{3} = 980.8 \text{ lb}$$

Sum horizontal forces:

$$V_{bot} = 1337.9 + 980.8 = 2318.7 \text{ lb}$$

$$v_{sill} = \frac{V_{bot}}{b_{2-3}} = \frac{2318.7}{4} = 579.7 \text{ plf unit shear}$$

Right sill section:

Sum moments about grid line 4D:

$$F_{4C} = \left[1605.5(5) - 1739(6)\right]\frac{1}{5} = 481.3 \text{ lb}$$

Sum horizontal forces:

$$V_{bot} = 1605.5 + 481.3 = 2086.8 \text{ lb}$$

$$v_{sill} = \frac{V_{bot}}{b_{4-5}} = \frac{2086.8}{6} = 347.8 \text{ plf}$$

Center pier:

Sum moments about grid line 3D:

$$\Sigma M = 1538.5(8) - 481.3(5) - 980.8(3) - 1739(4) = 0, \quad \therefore \text{ OK}$$

Sum moments about grid line 3C:

$$F_{3D} = [1538.5(5) - 481.3(2) - 1739(4)]\frac{1}{3} = 75.4 \text{ lb}$$

Sum horizontal forces:

$$\Sigma H = 1538.5 - 75.4 - 980.8 - 481.3 = 0 \text{ lb}, \quad \therefore \text{ OK}$$

$$v_{\text{upper}} = \frac{V_{\text{upper}}}{b_{3-4}} = \frac{1538.5}{4} = 384.6 \text{ plf}$$

$$v_{\text{middle}} = \frac{V_{\text{upper}} - F_{4C}}{b_{3-4}} = \frac{1538.5 - 481.3}{4} = 264.3 \text{ plf}$$

$$v_{\text{lower}} = \frac{V_{\text{upper}} - F_{4C} - F_{3C}}{b_{3-4}} = \frac{1538.5 - 481.3 - 980.8}{4} = 19.1 \text{ plf}$$

Check transfer diaphragm (see Fig. 10.30): Note that in lieu of installing a strap full width of the right pier, a partial length strap and transfer diaphragm will be used.

$$\text{Right pier unit shear} = \frac{V_6}{h_{\text{wall}}} = \frac{1739}{8} = +217.38 \text{ plf}$$

Transfer diaphragm shears:

$$V_A = \frac{1605.5(5)}{8} = 1003.4 \text{ lb}$$

$$v_A = \frac{V_A}{W_{TD}} = \frac{1003.4}{3} = +334.5 \text{ plf, positive shear}$$

$$V_D = \frac{1605.5(3)}{8} = 602.1 \text{ lb}$$

$$v_D = \frac{V_D}{W_{TD}} = \frac{602.1}{3} = -200.7 \text{ plf, negative shear}$$

Net shears:

$$v_{\text{net}} = 217.38 + (334.5) = +551.9 \text{ plf from A to C}$$

$$v_{\text{net}} = 217.38 - (200.7) = +16.68 \text{ plf from C to D}$$

Horizontal collector at C:

$$F = (551.9 - 16.68)(3) = 1605.5 \text{ lb}, \quad \therefore \text{ OK}$$

Vertical collector at grid line 5, starting at grid line D:

$$\text{Unit shear from sill} = \frac{1739}{5} = +347.8 \text{ plf, left side of grid line 5}$$

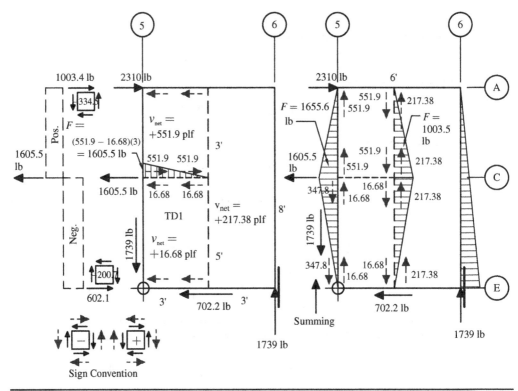

FIGURE 10.30 Transfer diaphragm design and collector forces.

$$F_C = -(347.8 - 16.68)(5) = -1655.6 \text{ lb tension}$$

$$F_A = -1655.6 + 551.9(3) = 0 \text{ lb, therefore the diagram closes}$$

Vertical collector at 5+3′, starting at grid line D:

$$F_C = (-16.68 + 217.38)(5) = 1003.5 \text{ lb compression}$$

$$F_A = 1003.5 - (551.9 - 217.38)(3) = 0 \text{ lb, therefore the diagram closes}$$

The different methods presented provide the tools to analyze most wall configurations. Each method has its unique advantages and disadvantages. Different factors such as the cost of construction, engineering fees required to provide the necessary design and detailing, familiarity, and preconceived opinions on the practicality of the various methods determine whether one method is used over another. ▲

10.5 Shear Walls with Small Openings

There have been an increased number of questions regarding the opening size within a shear wall, as to whether it can be neglected or must be engineered. Most of the inquiries have been in regard to plumbing or mechanical systems penetrating the shear walls. IBC Section 2305.1.1 requires: Openings in shear panels (diaphragms and shear walls) that materially affect their strength shall be fully detailed on the plans and shall have their edges adequately reinforced to transfer all shear stresses. Although that section does address what must be done, it does not provide guidance on how to determine if the opening size would trigger an engineered design.

Recently, FPInnovations[8] did extensive testing on diaphragms with large openings, as noted in Chap. 5. As a result, recommendations were made on how to determine if opening sizes can be ignored. Since diaphragms and shear walls can both be classified as shear panels, it seems appropriate to use the same rules to apply to shear walls. Repeated here for quick reference:

It is strongly recommended that analysis for a shear wall with an opening should be carried out except where all five of the following items are satisfied:

a. Depth no greater than 15 percent of shear wall height.

b. Length no greater than 15 percent of shear wall length.

c. Distance from shear wall edge to the nearest opening edge is a minimum of three times the opening dimension in the given direction.

d. The shear wall portion between opening and diaphragm edge satisfies the maximum aspect ratio requirement (all sides of the opening).

e. The shear does not exceed the nailing capacity of the wall without an opening.

Figure 10.31 demonstrates how to determine the effects of an opening size. A 4500-lb wind force is applied to the top of the wall. The wall is sheathed with 15/32" sheathing with 8d @ 6" o.c. edges with an ASD capacity of 365 plf. The unit shear forces can be derived by an in-house spreadsheet, APA FTAO calculator, or hardware manufacturer's software. The results show, for this example, that

Opening:

- $0.15H = 0.15(9) = 1.35' < 2'$

- $0.15L = 0.15(14.5) = 2.17' > 2'$ Violates opening height

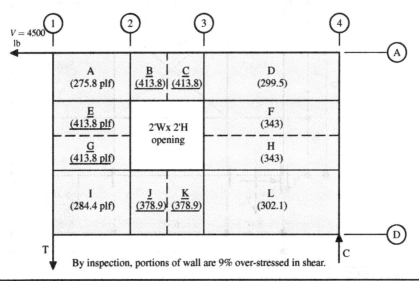

FIGURE 10.31   Walls with small openings.

Edges:

- Vertical $= 3(2) = 6' > 3'$ Violates edge distance
- Horizontal $= 3(2) = 6' > 3'$ Violates edge distance

10.6 Problems

Problem 10.1: Perforated Shear Wall with Double Door Opening

The perforated shear wall shown has a wall height of 8 ft 0 in and a length of 14 ft 6 in. A 4-ft-high by 6-ft-wide opening is placed in the wall as shown in the Fig. P10.1. A 4500-lb lateral wind force is applied at the top of the wall.

Given: Framing is D.F.

15/32" CDX sheathing with 10d nails

Find: Provide a complete design for the wall.   ▲

Problem 10.2: FTAO Shear Wall with Multiple Openings of Different Heights—Transfer Diaphragm Method with Dead Load

The wall has a height of 9 ft 0 in and a length of 24 ft 0 in. Three openings of different heights and widths are as shown in the Fig. P10.2. The pier widths are asymmetrical as shown. Straps and blocking are installed at the top and bottom of each opening as shown in the figure.

Given: A horizontal force of 6200 lb is applied to the top of the wall. A gravity load of 200 plf is applied to the top of the wall.

Find: Provide a complete force transfer around the opening analysis. Compare the results with Examples 10.1 and 10.2.   ▲

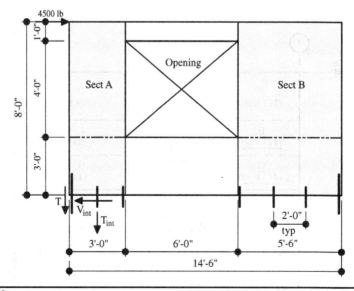

FIGURE P10.1

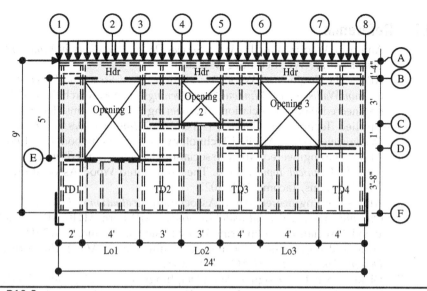

Problem 10.3: Cantilever Shear Wall Method with Single Opening

The wall Fig. P10.3 has a height of 8 ft and a length of 10 ft 6 in. The sill section has a height of 5.33 ft. A 1 ft 8 in high by 5 ft 0 in wide opening is offset from the center of the wall as shown. The section above the opening does not comply with the allowable aspect ratios and therefore will not be used. Gravity loads will be ignored.

Given: A 1000 lb lateral force is applied at the top of the wall and is distributed to each pier in accordance with their respective lengths.

Find: Analyze the wall using the cantilever method. ▲

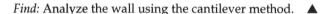

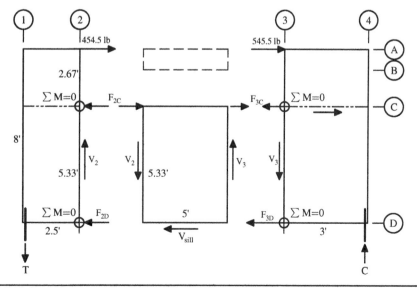

10.7 References

1. American Wood Council (AWC), *Special Design Provisions for Wind and Seismic with Commentary*, AWC, Leesburg, VA, 2021.
2. APA—The Engineered Wood Association, *Wood Structural Panel Shear Walls with Gypsum Wallboard and Window/Door Openings*, APA Research Report 157, APA—The Engineered Wood Association, Engineering Wood Systems, Tacoma, WA, 1996.
3. Applied Technology Council (ATC), *Guidelines for Design of Horizontal Wood Diaphragms, ATC-7*, Applied Technology Council, Redwood, CA, 1981.
4. APA—The engineered Wood Association, *Design for Force Transfer Around Openings (FTAO), Technical Topic T-555B*, The Engineered Wood Association, Engineering Wood Systems, Tacoma, WA, 2022.
5. APA—The Engineered Wood Association, *Evaluation of Force Transfer Around an Opening-Experimental and analytical Studies*, APA Form M410, APA—The Engineered Wood Association, Engineering Wood Systems, Tacoma, WA, 2011.
6. Structural Engineers Association of California (SEAOC), *IBC Structural/Seismic Design Manual, Vol. 2*, SEAOC, CA, 2018.
7. Diekmann, E. F., "Design of Wood Diaphragms," *Journal of Materials Education*, 1982, Fourth Clark C. Heritage Memorial Workshop, Wood Engineering Design Concepts University of Wisconsin, WI.
8. Neylon, B., Wang, J., Ni, C., *Design Example: Designing for Openings in Wood Diaphragm*, FPInnovations, Canadian Wood Council, CA, 2013.

CHAPTER 11

Discontinuous Shear Walls

11.1 Introduction

Discontinuous shear walls are a structural feature where careful consideration of the complete load path is needed; otherwise, serious load path deficiencies may occur. In-plane offsets or vertically discontinuous shear walls occur when shear walls in a multi-story structure do not line up vertically, or when shear walls are not continuous to the walls or foundation below, as shown in Fig. 11.1. The photo is of a mid-rise structure with multistories of wood framing over a concrete podium. The wall on the right side of the building has stacked shear walls that create easy, straight-lined load paths, but are discontinuous at the podium slab. The walls on the left side of the building are horizontally offset in-plane for architectural accents and are discontinuous at each floor level and at the podium. Developing complete load paths along this wall line can become very complicated. In ASCE 7 Chapter 12,[1] these discontinuities are classified as type 4 vertical irregularities. The discontinuous wall shears and overturning forces must be transferred through the floor framing into beams, columns, collectors, or walls below, and eventually find a load path down to the foundation. Failures can occur when lateral-force-resisting elements and their connections are not provided or properly designed to form a complete load path. Many floor plans do not provide ample opportunities for the placement of shear walls, and in some cases the installation of needed collectors. Minimal lateral-force-resisting systems often lead to more complex lateral load paths, which can increase the structural design time and construction cost due to the highly loaded lateral force-resisting elements and connections.

11.2 In-Plane Offset Shear Walls

The two-story shear wall configuration shown in Fig. 11.2 is an example of in-plane offset using segmented shear walls. Wall section A on the second floor is narrower in length than wall section C below. Wall section B is located over a portion of the first-floor wall that is not being used to resist lateral forces and is not directly continuous to the foundation. This configuration requires indirect load paths to be established, which can increase the difficulty in transferring the overturning forces and lateral shear forces of the upper wall sections to the foundation. Due to the discontinuity of wall section B, all supporting members (e.g., header, jamb studs, connections of the supporting

407

Figure 11.1   Photo of discontinuous load paths.

columns to the beam and hold-down anchors must be increased by the overstrength factor in accordance with ASCE 7 Section 12.3.3.3 for seismic loading if located in SDCs B through F.

Three options are available for the transfer of the overturning forces of wall section A down to the foundation. The first option is to install multiple studs in lower wall section C that line up with the interior end studs of upper wall section A, connect them together with hold-downs across the floor space, and then transfer the force directly into the foundation with a hold-down at location A. The disadvantage of this approach is that three hold-downs are required in wall section C. An advantage of this approach is a direct and continuous load path from the end of the upper wall to the foundation which is simple to design and eliminates a type 4 vertical irregularity.

The second option is not installing a hold-down at location A and treats lower wall section C as a transfer wall, similar to a transfer diaphragm, distributing the uplift from above through the lower wall sheathing to the hold-downs at each end of the lower wall. The basic wall shears of section C must be combined with the transfer diaphragm shears caused by the discontinuous force. The disadvantages with this option are that special detailing must be made to assure that the uplift force can be effectively transferred into the wall sheathing by nailing, the increase in shear and anchorage forces is addressed, and that the walls are properly constructed in accordance with the intended design. The special nailing can easily be overlooked if the nailing and detailing is not clearly defined in the drawings. It is also recommended that structural observations be required. The photo shown in Fig. 11.3 shows this exact condition. The double stud collector at the first floor and the tie strap at the second floor were properly installed. However, the nailing of the first-floor wall sheathing to the multiple studs was accomplished with only one row of nails at 12″ o.c. (field nailing), which in most cases will not

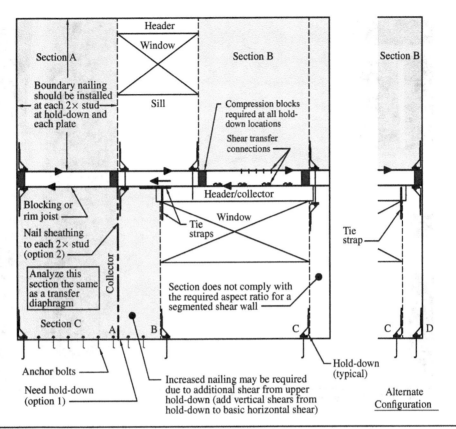

FIGURE 11.2 In-plane offset segmented shear walls.

provide enough capacity to resist the overturning force transferred into the wall. Each of the double 2× stud should have had special nailing as required by the overturning force. There was no hold-down installed at the bottom of the multiple stud location for a tie to the foundation. In such conditions, the additional calculation, detailing, construction, and inspection time to assure the construction matches the intended design are likely more costly than the cost of an additional hold-down.

For seismic loads, the second option can also be considered to create a type 4 vertical irregularity in the shear wall because the overturning forces are taken to the foundation through a discontinuous load path. Provided studs at location A can transfer the overturning compression loads directly into the foundation, the discontinuous load path is on the tensile overturning forces from above. This can be considered to trigger the requirements for discontinuous systems of ASCE 7 Section 12.3.3.3 where wall elements supporting the discontinuous load path are to be designed with seismic forces including overstrength factors. Although ASCE 7 commentary C12.3.3.3 notes that the intent of this section is to protect the gravity load carrying system (e.g., beams, columns, trusses, slabs, and walls) from overloads requiring it to be designed to resist the seismic load effects, including the overstrength factor. It further clarifies that walls that support isolated points loads from frame columns or discontinuous walls,

Figure 11.3   Photograph of in-plane offset shear walls.

perpendicular walls, or walls with significant vertical offsets as shown in Figures C12.4-3 and C12.4-4 can be subjected to the same type of failures caused by overload. The discontinuous force can also be classified as a transfer force similar to that discussed in Sections 12.10.1.1 and 12.10.3.3, which refers to diaphragms, because both systems must follow a similar transfer of discontinuous forces to provide a complete lateral load path. While the interpretation that ASCE 7 12.3.3.3 applies to transfer of discontinuous uplift forces through shear walls below acting as vertical transfer diaphragms may not be universal, it is a sound engineering practice recommended by the authors. If 12.3.3.3 is triggered, the seismic transfer force applied to the lower section is increased by the overstrength factor of Section 12.4.3 prior to adding it to the seismic load effects applied to the wall. If option 2 is selected, the multiple studs that are located below the end studs of the upper wall section act as a collector and transfer the upper wall overturning force into the sheathing of wall section C. The connections for shear and the overturning force of wall section A to lower wall C require only using the standard equations of section 2.3 or 2.4. The first-floor wall sheathing is required to be nailed into each member of the collector with special nailing equivalent to the overturning force amplified by the overstrength factor. The hold-downs at the end of the lower wall are sized to the loads including the amplified transfer loads.

The third option would be to design wall sections A and B as perforated or FTAO shear walls, bypassing the need to install a hold-down or transfer wall similar to that shown in Figs. 11.4 through 11.6.

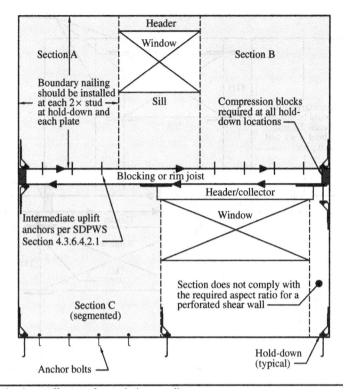

Section A

Header

Window

Section B

Boundary nailing
should be installed
at each 2× stud
at hold-down and
each plate

Sill

Compression blocks
required at all hold-
down locations

Blocking or rim joist

Header/collector

Window

Intermediate uplift
anchors per SDPWS
Section 4.3.6.4.2.1

Section does not comply with
the required aspect ratio for a
perforated shear wall

Section C
(segmented)

Anchor bolts

Hold-down
(typical)

Figure 11.4 In-plane offset perforated shear walls.

Upper wall section B, in Fig. 11.2, is vertically discontinuous to the foundation. Its tension and compression overturning forces on the left side of the wall must be transferred through the supporting header, into the jamb studs supporting the header on each side of the opening and then down into the foundation at locations B and C. The header, jamb studs, and their connections must be designed as elements supporting discontinuous walls, in accordance with Section 12.3.3.3, using the overstrength factor, as applicable. The overturning force at the left side of wall section B also affects design of the hold-down located at the right side of wall section C. In this example, the first-floor pier on the right side of the window does not meet the required aspect ratio to function as a shear wall and therefore cannot be used as part of the lateral-force-resisting system. It is, however, required to support the right side of the discontinuous wall above. Two configurations for the support and tie down for the right side of wall segment B are shown in the figure. If top plates are not installed over the header which can serve as a collector, a tie strap must also attach the header/collector below wall section B to the top plate of wall section C to transfer the lateral sliding force caused by wall section B.

Figure 11.4 shows a similar condition where a perforated shear wall is located over a segmented shear wall of lesser width. In this case, the primary hold-downs are only required at each end of the perforated shear wall above and at each end of wall section C. Since the first-floor full height section on the right side of the opening is too narrow

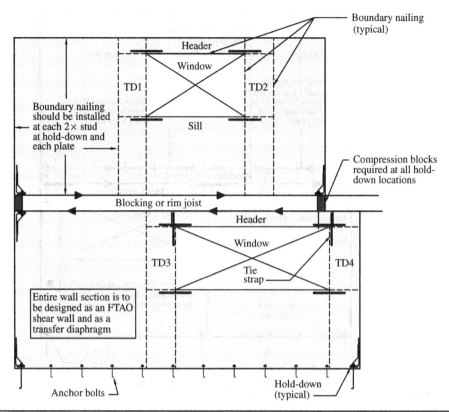

Boundary nailing
(typical)

Header

Window

TD1

TD2

Boundary nailing
should be installed
at each 2× stud
at hold-down and
each plate

Sill

Compression blocks
required at all hold-
down locations

Blocking or rim joist

Header

Window

TD3

Tie
strap

TD4

Entire wall section is to
be designed as an FTAO
shear wall and as a
transfer diaphragm

Anchor bolts

Hold-down
(typical)

Figure 11.5 In-plane offset FTAO shear walls.

to meet the code required aspect ratio, it cannot be used as part of the first-floor shear wall to resist lateral forces. The second-floor wall is vertically discontinuous on the right side of the wall. The end studs of section B are supported at the first floor by multiple studs which must be designed as elements supporting discontinuous walls. These studs must be designed to transfer the tension and compression forces from the end of the shear wall above to the foundation below. When seismic forces are additive with gravity forces, the seismic overstrength load combinations of ASCE 7 Section 2.4.5 are applied. The intermediate uplift anchors at the full height segments, as required by SDPWS[2] Section 4.3.6.4.2.1, can be attached to blocking and into the wall below using tie straps or into a continuous rim joist adequately designed to resist forces from the full height sections and anchored at the ends of the wall into the wall and header below.

Figure 11.5 shows a two-story shear wall, where the wall at each level is designed in accordance with the force transfer around the opening (FTAO) method. The upper wall is shown having a length that is less than the first-floor wall to demonstrate how complicated load paths can become. The overturning force at the right side of the upper wall must be transferred into the lower wall header. The header section and the jamb studs on each side of the opening should be designed as elements supporting discontinuous walls. The figure shows partial strapping placements into the side transfer

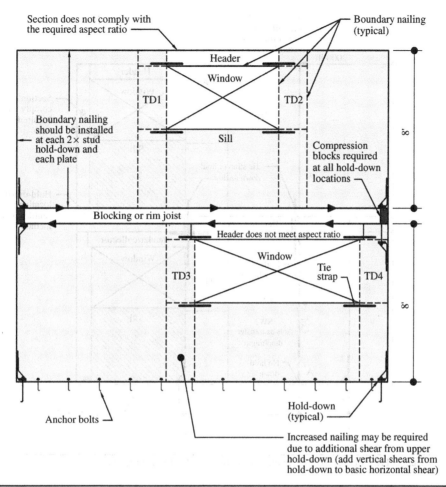

Section does not comply with the required aspect ratio

Boundary nailing (typical)

Header

Window

TD1

TD2

Boundary nailing should be installed at each 2× stud hold-down and each plate

Sill

Compression blocks required at all hold-down locations

8'

Blocking or rim joist

Header does not meet aspect ratio

Window

TD3

Tie strap

TD4

8'

Hold-down (typical)

Anchor bolts

Increased nailing may be required due to additional shear from upper hold-down (add vertical shears from hold-down to basic horizontal shear)

FIGURE 11.6 In-plane offset cantilever shear walls.

diaphragms. The wall top plates or continuous rim joists are continuous and can act as the collector.

Figure 11.6 shows a two-story shear wall where each wall is designed using the cantilever method. The low wall heights prevent the depth of the header sections from complying with code required aspect ratios. Since the upper and lower wall lengths are the same, there is no offset or vertical discontinuity wall condition. The figure shows collector splice issues and possible partial strapping placements. The header or continuous rim joists act as wall boundary members (collectors) that transfer shears from pier to pier.

Example 11.1: In-Plane Offset, Two-Story Shear Segmented Shear Wall—Without Gravity Loads

The wall shown in Fig. 11.7 is a two-story offset segmental shear wall. The headers and full height sections on the right side of the openings on both floors do not comply

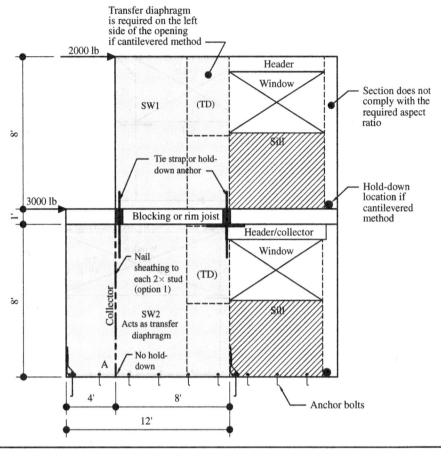

Figure 11.7 In-plane offset segmented shear wall without gravity loads.

with required aspect ratios and are therefore ignored. The upper wall is narrower in length than the lower wall and is offset to the right so that the hold-down anchors on the right side of the walls are in line with the wall below. The overturning force at the left side of SW1 is transferred into SW2 by transfer diaphragm action and must be designed as a wall supporting a discontinuous shear wall above if seismic controls and no hold-down anchor is installed at location A. Although utilizing the lower wall to support the upper wall is not an ideal approach, it is presented here to show how it might be analyzed. As an alternate, the wall lengths could be extended by using the cantilever wall method if the sill sections on the right side of the full height sections are utilized and transfer their forces at the top of the sill sections into the full height wall transfer diaphragms as previously discussed. The headers at the second-floor and first-floor walls can act as collectors to transfer the remaining diaphragm shears into the shear walls, as required by code. Gravity loads are ignored in this example to provide clarity in the method of determining the wall shears. The sill sections below the opening will not be used for this example. Wind controls the design and all loads have been appropriately factored. Because seismic forces do not control, the overturning force to the wall below does not have to be amplified by the overstrength factor. However, because ASCE 7-16 Section 12.3.3.3 applies to SDC B, it is advised to check the seismic overturning force to the wind overturning force to see which one actually controls at the discontinuity.

Basic information (Fig. 11.7):

$$h = 8 \text{ ft } 0 \text{ in, wall height of the first and second-floor walls}$$

$$V_r = 2000 \text{ lb shear force at the roof}$$

$$V_2 = 3000 \text{ lb shear force at the second floor}$$

$$b_{SW1} = 8 \text{ ft } 0 \text{ in}$$

$$b_{SW2} = 12 \text{ ft } 0 \text{ in}$$

$$b_{\text{eff SW1}} = 8 \text{ ft } 0 \text{ in tie straps at ends of wall}$$

$$b_{\text{eff SW2}} = 11 \text{ ft } 6 \text{ in}$$

Basic wall shear, second floor (see Fig. 11.8): The positive sign convention is with respect to the direction of the applied load as shown in the figure.

$$V_{SW1} = \frac{V_r}{b_{SW1}} = \frac{2000}{8} = +250 \text{ plf}$$

$$v_{SW2} = \frac{V_r + V_2}{b_{SW2}} = \frac{2000 + 3000}{12} = +416.67 \text{ plf}$$

Overturning forces, neglecting gravity loads for this example:

$$T_{SW1} = C_{SW1} = \frac{V_r h}{b_{SW1}} = \frac{2000(8)}{8} = 2000 \text{ lb at second floor}$$

$$T_{SW2} = C_{SW2} = [2000(17) + 3000(9)]\frac{1}{11.5} = 5304 \text{ lb at first floor}$$

However, the shear at the end of the wall is

$$V_{SW2} = [2000(17) + 3000(9)]\frac{1}{12} = 5083.3 \text{ lb at first floor}$$

The 2000 lb overturning force from the wall above and the associated overturning force of the rim joist must be translated to the top of SW2. The overturning force from the rim joist is equal to

$$T_{\text{rim}} = C_{\text{rim}} = \frac{\sum V d_{\text{flr}}}{b_{SW1}} = \frac{5000(1)}{12} = 416.67 \text{ lb, see free-body below upper wall}$$

Transfer diaphragm shear, shown at the right of the figure:

$$\text{Depth of transfer diaphragm} = 8 \text{ ft } 0 \text{ in}$$

$$V_{\text{left}} = \frac{F_{o/t} b_{\text{eff SW1}}}{b_{SW2}} = \frac{2000(8)}{12} = 1333.33 \text{ lb}$$

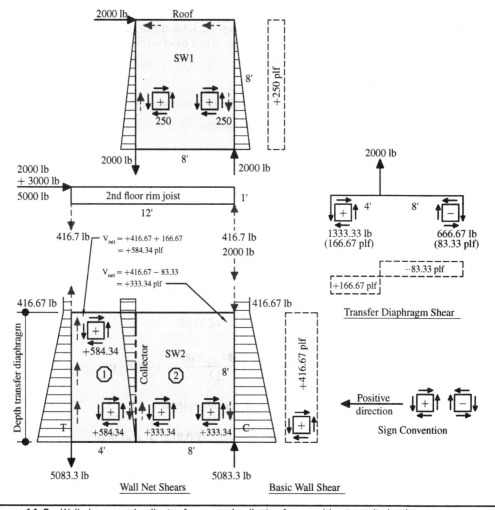

FIGURE 11.8   Wall shears and collector forces and collector forces without gravity loads.

Based on the direction of the shears acting on the sheathing elements:

$$v_{\text{left}} = \frac{V_{\text{left}}}{h_{\text{wall}}} = \frac{1333.33}{8} = +166.67 \text{ plf positive shear}$$

$$V_{\text{right}} = \frac{F_{o/t}(b_{\text{SW2}} - b_{\text{SW1}})}{b_{\text{SW2}}} = \frac{2000(4)}{12} = 666.67 \text{ lb}$$

$$v_{\text{right}} = \frac{V_{\text{right}}}{h_{\text{wall}}} = \frac{666.67}{8} = -83.33 \text{ plf negative shear}$$

Net shears:

$$v_{net} = +416.67 + (166.67) = +584.34 \text{ plf, section 1}$$

$$v_{net} = +416.67 - (83.33) = +333.34 \text{ plf, section 2}$$

The distribution of the overturning force into the lower wall creates varying shears across the wall. Nailing callouts and detailing should reflect these results.

Vertical collector forces: The collector force is equal to the wall shear in section 1 times the wall height plus the overturning force from the rim joist.

$$F_{left} = 584.34(8) + 416.67 = 5083.3 \text{ lb tension, matches previous calculation}$$

$$F_{interior} = (584.34 - 333.34)(8) = 2008 \text{ lb tension} \cong 2000 \text{ lb,} \quad \therefore \text{ OK}$$

The force in the collector on the right side of the wall is equal to the wall shear in section 2 times the wall height plus the overturning force from the rim joist plus the overturning force from SW2.

$$F_{right} = 333.34(8) + 2000 + 416.67 = 5083.3 \text{ lb compression}$$

Therefore, the analysis is confirmed.

Special nailing of the sheathing to the interior multiple stud collectors should be called out and detailed on the drawings. ▲

Example 11.2: In-Plane Offset, Two-Story Segmented Shear Wall—with Gravity Loads and Seismic Controlling

The same wall will be examined with gravity loads added, reference Fig. 11.9. The purpose of this exercise is to show how gravity loads can significantly affect the wall shears. The wall is to be analyzed for the loads shown. Seismic controls the design.

$$\text{ASCE 7-16 Section 2.4.5 (ASD), Load Case 8, } 1.0D + 0.7E_v + 0.7 E_h$$

$$E_v = 0 \text{ per Section 12.4.2.2, exception 2:}$$

$$DL_{roof} = 150 \text{ plf}$$

$$DL_{2nd} = 250 \text{ plf}$$

$$DL_{wall} = 10 \text{ psf}$$

Note: horizontal forces are already factored to $0.7E_h$.

$$V_{roof} = 2000 \text{ lb total shear at the roof}$$

$$V_{2nd} = 3000 \text{ lb total shear at the second floor}$$

$$\Sigma V_{1st} = 5000 \text{ lb total shear at the first floor}$$

Total dead load to wall sections:

Roof, First-floor wall:

$$DL_{SW1} = 150 + 10(8) = 230 \text{ plf}$$

$$DL_{hdr} = 150\left(\frac{6}{2}\right) = 450 \text{ lb}$$

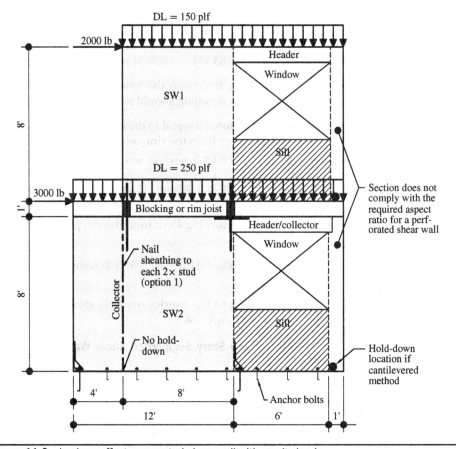

FIGURE 11.9 In-plane offset segmented shear wall with gravity loads.

Second floor, First-floor wall includes roof/floor DL plus wall DL:

$$DL_{SW2} = 250 + 10(8) = 330 \text{ plf}$$

$$DL_{hdr} = [250 + 10(7)](3) = 960 \text{ lb}$$

The header DL includes the floor DL plus the wall DL above not supported by the roof header.

Tension hold-downs:

$$T_{SW1} = [2000(8) - (230)(8)(4)]\frac{1}{8} = 1080 \text{ lb tension at the second floor}$$

Basic wall shear (reference Fig. 11.10):

$$v_{SW1} = \frac{V_r}{L_{SW1}} = \frac{2000}{8} = +250 \text{ plf}$$

The application of the dead loads affects the unit shears as follows:

$$v_{\text{left}} = \frac{1080}{8} = +135 \text{ plf unit shear, chord force divided by the wall height}$$

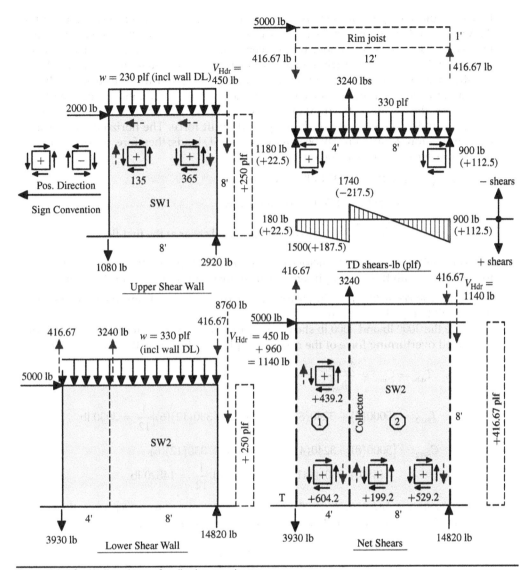

FIGURE 11.10   Wall and transfer diaphragm shears.

$$C_{SW1} = \left[2000(8) + \frac{230(8)^2}{2}\right]\frac{1}{8} = 2920 \text{ lb}$$

$$v_{right} = \frac{V_{right}}{h_{wall}} = \frac{2920}{8} = +365 \text{ plf unit shear, chord force divided by the wall height}$$

$$v_{average} = \frac{135 + 365}{8} = +250 \text{ plf, same as calculated for the basic wall shear}$$

The results show that the calculated shears at the left chord member is less than the basic wall shear, V/L and the shears at the right chord exceed the basic wall shear.

However, the average of these two shears is equal to the basic wall shear. None of the APA tests[3] on diaphragms or shear walls were set up to directly measure the variance in shear across the wall section. It has been assumed and widely accepted that the shears will be uniformly distributed across the wall. The calculations show that this is not the case, especially when gravity loads are applied, and the wall below is being used as a transfer diaphragm or a complex shear wall. Following the discussion of discontinuous shear walls at the front of this section, the wall below and its anchorage are designed under the influence of the amplified uplift force. The horizontal forces and overturning forces have already been reduced to ASD loads; therefore, the overstrength forces transferred to the wall below become

$$\Omega_0 T_{SW1} = (3)(1080) = 3240 \text{ lb to wall below}$$

$$\Omega_0 C_{SW1} = (3)(2920) = 8760 \text{ lb to wall below}$$

$$v_{SW2} = \frac{2000 + 3000}{8} = 416.67 \text{ plf basic wall shear at the first-floor wall}$$

Note that the overstrength factor can be reduced by ½ per footnote (b) of Table 12.2-1 for structures with flexible diaphragms. This reduction is not used for this example.

Overturning forces at first floor (reference Figs. 11.10 and 11.11): Transposing forces from upper wall to top of lower wall.

Since the 2000 lb and 3000 lb shear force must be transposed to the top of SW2, the associated overturning force of the rim joist applied to the top of the wall is equal to

$$T_{rim} = C_{rim} = \frac{\sum V d_{flr}}{b_{SW1}} = \frac{5000(1)}{12} = 416.7 \text{ lb}$$

$$T_{SW2} = [5000(8) + 3240(8) + 416.7(12) - 330(12)(6)]\frac{1}{12} = 3930 \text{ lb}$$

$$C_{SW2} = [5000(8) - 3240(4) + 330(12)(6) + 330(12)(6)$$
$$+ (450 + 960 + 416.7 + 8760)(12)]\frac{1}{12} = 14820 \text{ lb}$$

Transfer diaphragm shear includes the floor dead load with the overturning force from the second-floor wall, SW1. See the free-body diagrams on the right of Fig. 11.10.

$$V_{left} = \left[-\frac{330(12)^2}{2} + 3240(8)\right]\frac{1}{12} = 180 \text{ lb}$$

$$v_{left} = \frac{V_{left}}{h_{wall}} = \frac{180}{8} = 22.5 \text{ plf, from shears acting on shtg. element}$$

$$V_{right} = \left[\frac{330(12)^2}{2} - 3240(4)\right]\frac{1}{12} = -900 \text{ lb}$$

$$v_{right} = \frac{V_{right}}{h_{wall}} = \frac{900}{8} = +112.5 \text{ plf, from shears acting on shtg. element}$$

Construct the resulting transfer diaphragm shear diagram (Fig. 11.10): Note that the direction of the 1260 lb reaction is downward and the uniform gravity loads are acting

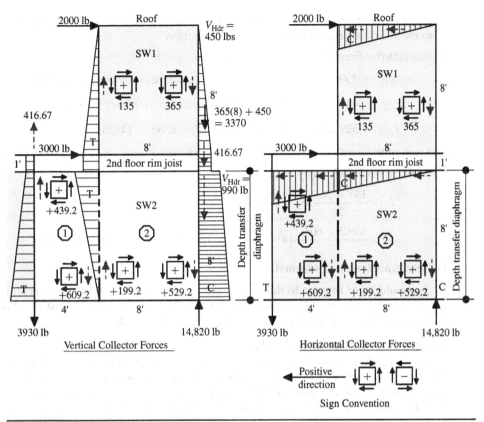

FIGURE 11.11 Collector force diagrams.

downward. The shears in this area are determined by the direction of the shears acting on the sheathing element and are therefore positive below the base line. When the shears drop below the base line, the resulting shears will be positive. Starting on the left:

$$V = -180 - 330(4) = -1500 \text{ lb } (+187.5 \text{ plf by sheathing element symbol}),$$
left side of the interior collector

$$V = -1500 + 3240 = 1740 \text{ lb } (-217.5 \text{ plf}), \text{ right side of the interior collector}$$

$$V = 1740 - 330(8) = -900 \text{ lb } (+112.5 \text{ plf}), \text{ on the right collector}$$

Add or subtract the transfer diaphragm shears to the basic wall shears.

Net shears (nailing requirements):

$$v_{\text{net}} = +416.67 + (22.5) = +439.2 \text{ plf, left side of section 1}$$

$$v_{\text{net}} = +416.67 + (187.5) = +604.2 \text{ plf, right side of section 1}$$

$$v_{\text{net}} = +416.67 - (187.5) = +199.2 \text{ plf, left side of section 2}$$

$$v_{\text{net}} = +416.67 - (112.5) = +529.2 \text{ plf, right side of section 2}$$

Apply the direction of the shears applied into the boundary members and collectors from the sheathing elements onto the wall elevation.

Vertical collector forces (reference Fig. 11.11):

$$F_{\text{left}} = 439.2(8) + 416.67 = 3930.3 \text{ lb, matches previous calculation.}$$

$$F_{\text{interior}} = (604.2 - 199.2)(8) = 3240 \text{ lb, matches o/t force from above.}$$

$$F_{\text{right}} = 529.2(8) + 960 + 450 + 416.67 + 8760 = 14820.3 \text{ lb}$$

The sum of the vertical forces matches the overturning force previously calculated.

Horizontal collector forces (reference Fig. 11.11):

$$F = \frac{135 + 365}{2}(8) = 2000 \text{ lb at roof}$$

$$F = \frac{439.2 + 604.2}{2}(4) + \frac{199.2 + 529.2}{2}(8) = 5000 \text{ lb at second floor}$$

Therefore, analysis is confirmed.

It should be of interest to the reader to note that the cost of the analysis and construction of the transfer wall would be significantly higher than the cost of adding a hold-down at the foundation under the left edge of SW1. ▲

Example 11.3: Partial Shear Panels

Figure 11.12 shows a method of transferring diaphragm shears across floor or roof framing at an interior or exterior shear wall with spaced shear panels or blocking. This is often done as a means to allow the installation of mechanical ducts through the shear wall area within the floor space. Code typically requires that joists be laterally supported at each end and at each support by solid blocking, a header, rim joist, or other means to prevent rotation. This is specifically true with dimensional lumber which is addressed in IBC 2308.4.2.3. IBC Section 2308.4.3 addresses engineered wood products, such as I-joists, and some manufacturers do not require lateral support at intermediate supports in some situations. In many cases, the framing members or blocking that are provided to prevent rotation of the joists and/or transfer bearing forces through a floor also functions as the mechanism to transfer lateral roof or floor diaphragm shears to the shear wall, collector, or diaphragm chord below. Situations may occur where the rotation prevention requirements are met but additional attention is required to provide a complete load path for the transfer of the lateral forces. There have been occasions where it was thought that by reducing the number of shear panels required for the lateral system by staggering them, considerable material costs and labor could be saved. It is important to resist the temptation, where possible, to reduce the number of blocking panels as will be demonstrated in this example. Providing partial shear blocking for only portions of the wall creates a situation that is like the offset shear wall examined in the previous example. Partial blocking places a high demand on a limited number of panels, increasing the chance of an overload to the shear panels. Whenever this approach has been used, it is often assumed that a single note placed on the drawings, requiring the panels to be placed every other bay or every third bay, would provide the necessary information to maintain stability of the joists and effectively transfer the shears into the wall. This approach presumes that the contractor will frame it correctly based solely on the plan note. Contractors are typically familiar with blocking used to prevent rotation

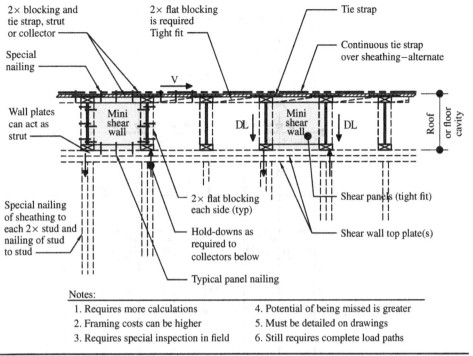

2× blocking and tie strap, strut or collector

Special nailing

Wall plates can act as strut

2× flat blocking is required Tight fit

Tie strap

Continuous tie strap over sheathing—alternate

Mini shear wall

V

DL

Mini shear wall

DL

Roof or floor cavity

Special nailing of sheathing to each 2× stud and nailing of stud to stud

2× flat blocking each side (typ)

Hold-downs as required to collectors below

Typical panel nailing

Shear panels (tight fit)

Shear wall top plate(s)

Notes:
1. Requires more calculations
2. Framing costs can be higher
3. Requires special inspection in field
4. Potential of being missed is greater
5. Must be detailed on drawings
6. Still requires complete load paths

Figure 11.12 Partial shear panel layout.

but are not always aware of the lateral load path requirements. To successfully address this concern, calculations are required to confirm that a complete load path has been established and the design must be fully detailed on the drawings. In addition to building department inspections, structural observation, per IBC Sections 1704.6.2 for seismic, 1704.6.3 for wind, or 1704.6.1(3) when required by the registered design professional, should also be provided to verify that the intent of the design has been met.

The shear panels in the roof or floor cavity must act as mini-shear walls. When the shear panels are spaced, 2× blocking and tie straps are required between the shear panels to act as collectors or drag elements (boundary elements), which transfer the diaphragm shears and collector forces from panel to panel in accordance with ASCE 7-16 Sections 12.10.1 and 12.10.2. The tie straps are required because code does not allow the roof and floor sheathing be used to splice boundary elements. When shear panels are installed between each joist, the vertical shear forces from adjacent shear panels are available to oppose the overturning forces of the panels and cancel out to zero. Because these opposing forces are not present in cases of partial blocking, hold-down connections might be required at the end of each shear panel if the overturning force exceeds the dead load resistive force. The dead load of the trusses or joists can be used to resist overturning of the shear panels; however, a proper connection of the panel to the joist must be made to accomplish this. If the floor or roof framing is parallel to the wall, the opportunity to use dead load is diminished. The distribution of the forces from the shear panels into a wall is shown in Fig. 11.13. The overturning forces at each end of the panels must be tied to a vertical collector in the wall below to distribute the force into the wall. The wall acts as a transfer diaphragm if hold-downs are not installed in

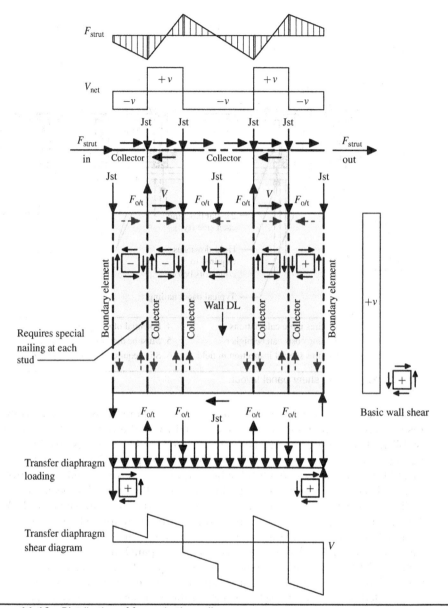

Figure 11.13 Distribution of forces in the wall.

the foundation at each collector location. Special nailing of the sheathing to each 2× stud in the collector is required. Included in the figure is the construction of the strut force diagram that must be checked. Note that additional strut forces are applied at the beginning and end of the shear wall. The dead load of the wall is included in the transfer diaphragm loading diagram. The net wall shears are determined by adding or subtracting the transfer diaphragm shears to the basic wall shears.

The wall shown in Fig. 11.13 has a height of 8 ft and a length of 12 ft. The depth of the roof joists is 24 in with a joist spacing of 2'-0" o.c. Only two shear panels are installed as shown in the figure. The uniform shear from the diaphragm is 175 plf and the collector forces into and out of the wall are 2150 lb and 1250 lb, respectively. The dead load of each joist is equal to 750 lb. The wall dead load is 80 plf. Analyze the wall for the loads shown. The overstrength factor has not been included in this example for simplification.

Sum of shear forces to each shear panel:

$$\Sigma V_{horiz} = 2150 + 1250 + 175(12) = 5500 \text{ lb}$$

$$V_{panel} = \frac{5500}{2} = 2750 \text{ lb per panel}, \quad v_{panel} = \frac{2750}{2} = 1375 \text{ plf}$$

The unit shear in the blocking panels suggests that an additional blocking panel is required, say one at each end plus one in the middle.

Check the overturning of the wall (see Fig. 11.14):

$$T = \left| 5500(10) - 750(2 + 4 + 6 + 8 + 10 + 12) - \frac{80(12)^2}{2} \right| \frac{1}{12} = 1478.3 \text{ lb}$$

$$C = \left| 5500(10) + 750(2 + 4 + 6 + 8 + 10 + 12) + \frac{80(12)^2}{2} \right| \frac{1}{12} = 7688.3 \text{ lb}$$

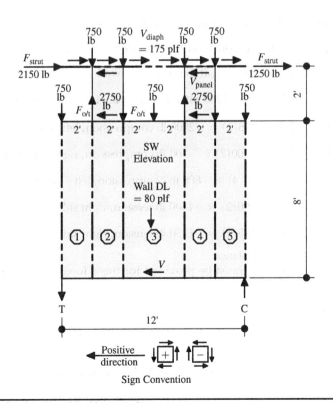

FIGURE 11.14 Partial shear blocking panels.

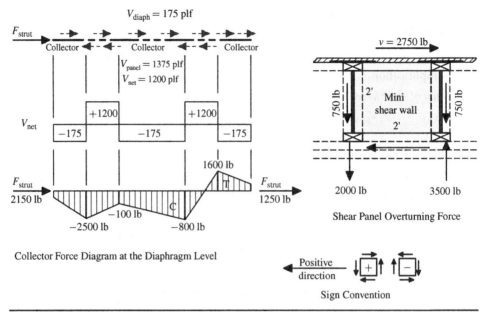

FIGURE 11.15 Collector force diagram at diaphragm level.

Determine the collector forces at the roof diaphragm (reference Fig. 11.15):

F_{strut} = −2150 lb compression, at the start of the wall

F_{strut} = +1250 lb tension, at the end of the wall

V_{net} = 1375 − 175 = 1200 plf at panels (panel shear minus diaphragm shear)

F = −2150 − 175(2) = −2500 lb compression, left side of the first panel

F = −2500 + 1200(2) = −100 lb compression, right side of the first panel

F = −100 − 175(4) = −800 lb compression, left side of the second panel

F = −800 + 1200(2) = +1600 lb tension, right side of the second panel

F = +1600 − 175(2) = +1250 lb tension, right side of the second panel

Therefore, the forces balance.

Splice connections must be provided for these forces, 2500 lb maximum tie strap tension.

Determine the overturning forces of the panels (see Fig. 11.15):

$$T = \left[2750(2) - 750(2)\right]\frac{1}{2} = 2000 \text{ lb tension}$$

$$C = \left[2750(2) + 750(2)\right]\frac{1}{2} = 3500 \text{ lb compression}$$

Determine the transfer diaphragm shears (see Fig. 11.16):

Basic **wall shear:**

$$v_{SW} = \frac{5500}{12} + 458.3 \text{ plf}$$

Transfer diaphragm shear, summing moments about the right end:

$$V_{left} = \left[3500(2 + 8) + \frac{80(12)^2}{2} + 750(6) - 2000(4 + 10) \right]\frac{1}{12} = 1438.33 \text{ lb}$$

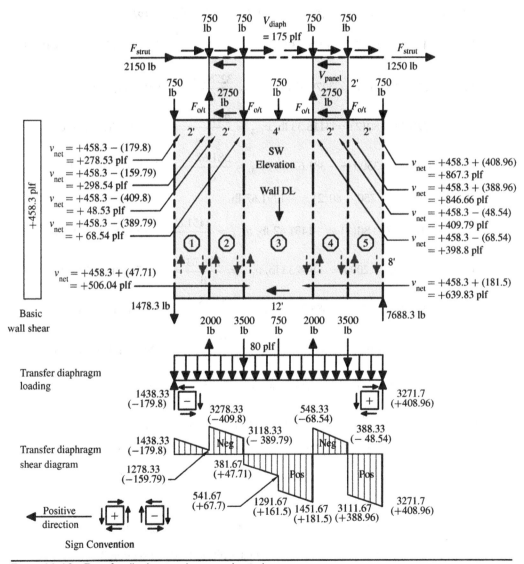

FIGURE 11.16 Transfer diaphragm shears and net shears.

$$v_{\text{left}} = \frac{V_{\text{left}}}{h_{\text{wall}}} = \frac{1438.33}{8} = -179.8 \text{ plf, negative shear from shtg. element}$$

$$V_{\text{right}} = \left[3500(4 + 10) + \frac{80(12)^2}{2} + 750(6) - 2000(2 + 8) \right] \frac{1}{12} = 3271.7 \text{ lb}$$

$$v_{\text{right}} = \frac{V_{\text{right}}}{h_{\text{wall}}} = \frac{3271.7}{8} = +408.96 \text{ plf, positive shear from shtg. element}$$

Construction of the transfer diaphragm shear diagram (see Fig. 11.16):

$$V_{1-2} = 1438.33 - 80(2) = 1278.33 \text{ lb}, \ v_{1-2} = \frac{V_{1-2}}{h_{\text{wall}}} = \frac{1278.33}{8} = -159.79 \text{ plf, left side}$$

$$V_{1-2} = 1278.33 + 2000 = 3278.33 \text{ lb}, \ v_{1-2} = \frac{3278.33}{8} = -409.8 \text{ plf, right side}$$

$$V_{2-3} = 3278.33 - 80(2) = 3118.33 \text{ lb}, \ v_{2-3} = \frac{3118.33}{8} = -389.79 \text{ plf, left side}$$

$$V_{2-3} = 3118.33 - 3500 = -381.67 \text{ lb}, \ v_{2-3} = \frac{381.67}{8} = +47.71 \text{ plf, right side}$$

$$V_{3.5} = -381.67 - 750 - 80(2) = -1291.67 \text{ lb}$$

$$V_{3-4} = -1291.67 - 80(2) = -1451.67 \text{ lb}, \ v_{3-4} = \frac{1451.67}{8} = +181.5 \text{ plf, left side}$$

$$V_{3-4} = -1451.67 + 2000 = +548.33 \text{ lb}, \ v_{3-4} = \frac{548.33}{8} = -68.54 \text{ plf, right side}$$

$$V_{4-5} = 548.33 - 80(2) = +388.33 \text{ lb}, \ v_{4-5} = \frac{388.33}{8} = -48.54 \text{ plf, left side}$$

$$V_{4-5} = 388.33 - 3500 = -3111.67 \text{ lb}, \ v_{2-3} = \frac{3111.67}{8} = +388.96 \text{ plf, right side}$$

$$V_5 = -3111.67 - 80(2) = -3271.7 \text{ lb}, \ v_5 = \frac{3271.7}{8} = +408.96 \text{ plf}$$

Determine wall net shears, basic wall shear +/− transfer diaphragm shear:

$$v_{\text{net}} = +458.3 - (179.8) = +278.53 \text{ plf, left side of section 1}$$

$$v_{\text{net}} = +458.3 - (159.8) = +298.53 \text{ plf, right side of section 1}$$

$$v_{\text{net}} = +458.3 - (409.8) = +48.53 \text{ plf, left side of section 2}$$

$$v_{\text{net}} = +458.3 - (389.8) = +68.54 \text{ plf, right side of section 2}$$

$$v_{\text{net}} = +458.3 + (47.71) = +506.04 \text{ plf, left side of section 3}$$

$$v_{\text{net}} = +458.3 + (181.5) = +639.83 \text{ plf, right side of section 3}$$

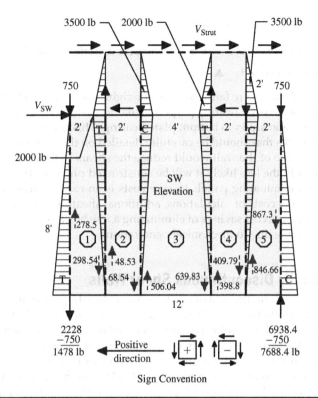

Figure 11.17 Determination of collector forces.

$$v_{net} = +458.3 - (68.54) = +389.8 \text{ plf, left side of section 4}$$

$$v_{net} = +458.3 - (48.54) = +409.79 \text{ plf, right side of section 4}$$

$$v_{net} = +458.3 + (388.96) = +846.66 \text{ plf, left side of section 5}$$

$$v_{net} = +458.3 + (408.96) = +867.3 \text{ plf, right side of section 5}$$

The direction of the shears transferred into the collectors and boundary members are shown in the figure.

Determine the vertical collector forces (see Fig. 11.17):

$$F_{left} = 278.5(8) = 2228 \text{ lb}$$

$$T = F_{left} - DL_{joist} = 2228 - 750 = 1478 \text{ lb reaction in left boundary member}$$

$$F_{1-2} = (298.54 - 48.53)(8) = 2000 \text{ lb, matches overturning force,} \quad \therefore \text{ OK}$$

$$F_{2-3} = (506.04 - 68.54)(8) = 3500 \text{ lb, matches overturning force,} \quad \therefore \text{ OK}$$

$$F_{3-4} = (639.83 - 389.8)(8) = 2000 \text{ lb, matches overturning force,} \quad \therefore \text{ OK}$$

$$F_{4-5} = (846.66 - 409.79)(8) = 3495 \text{ lb, matches overturning force} \pm$$

$$F_{\text{right}} = 867.3(8) = 6938.4 \text{ lb}$$

$$C = 6938.4 + 750 = 7688.4 \text{ lb compression, right boundary member}$$

Reaction is confirmed.   ▲

The wall shears are fairly high on the right side of the wall, requiring sheathing on both sides of the wall. Bearing perpendicular to the grain stresses should be checked for the compression studs to bottom plate bearing. This is a high demand wall with complex load paths that should be carefully detailed on the drawings. One additional shear panel at the top of the wall would reduce the shears in the wall. The more complicated the load path, the less likely it will be constructed properly. It should be obvious to the reader that eliminating panels to save costs is an exercise that is not worth pursuing. The additional costs of calculations, additional sheathing, nailing straps collectors, etc. far outweigh the cost savings of eliminating a few blocking panels. However, it can also be unavoidable due to mechanical penetrations.

11.3   Vertically Discontinuous Shear Walls

Discontinuities in the seismic lateral-force-resisting system occur frequently. For seismic controlled designs, it is important to understand what constitutes a horizontal and vertical irregularity and their related code sections listed in ASCE 7 Tables 12.3-1 and 12.3-2. The two-story interior shear walls at grid line 2 as shown in Fig. 11.18 are in vertical alignment. The arrangement of the walls creates simple lateral load paths to the foundation and splits the diaphragm into two simple rectangular diaphragms. For this condition, there are no irregularities. Figure 11.19 shows the same structure with in-plane offset walls at grid lines 1 and 3, which creates a type 4 vertical offset irregularity. The horizontally offset shear wall at grid line 2 is discontinuous to the foundation, which creates a type 4 horizontal irregularity. Shear wall SW3 at grid line 2 supports the roof diaphragms that span from grid lines 1 and 2 and from 2 to 3. Since the shear wall at grid line 2 does not continue to the foundation, the second-floor diaphragm must span from grid line 1 to 3 and act as a transfer diaphragm, having to comply with ASCE 7 Sections 12.10.1.1 and 12.10.3.3. The lateral force from SW3 becomes a diaphragm transfer force and must be amplified by the overstrength factor of Section 12.4.3 in the design of the second-floor diaphragm. The loading of the second-floor diaphragm is shown in the figure, which consists of the diaphragm inertial uniform load tributary to the second floor plus the concentrated lateral transfer force from SW3 that must be amplified by the overstrength factor before adding it to the diaphragm. Additionally, Section 12.3.3.4 must also be applied for these types of irregularities were noted in Table 12.3-1, which requires a 25 percent increase in the connections of the diaphragm to the vertical elements and collectors; and collectors and their connections, including their connection to the vertical elements of the SFRS. Forces calculated using the seismic load effects including overstrength need not be increased. Shear wall SW5 is a high-demand wall that receives the reaction of the second-floor diaphragm plus the tributary load from the roof diaphragm and must be carefully designed due to the complex load paths within the wall line.

Whenever type 4 horizontal or vertical irregularities exist, elements supporting discontinuous walls or frames must comply with ASCE7-16 Section 12.3.3.3. This

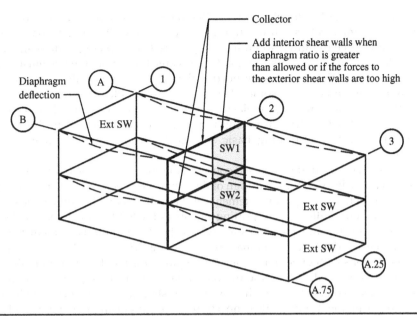

FIGURE **11.18**   Walls in-line and continuous to the foundation.

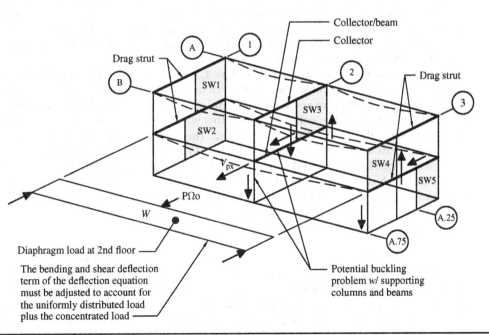

FIGURE **11.19**   Vertically discontinuous and horizontally offset walls.

requirement applies to all structures that have a type 4 vertical or horizontal irregularity designed in SDCs B through F. The purpose of the special load combinations including overstrength is to protect the gravity load carrying system against possible overloads caused by overstrength of the seismic force-resisting system and to prevent the collapse of the elements supporting the discontinuous walls or frames. ASCE7-16 commentary notes that some elements are incapable of safely resisting ground-shaking demands through inelastic behavior. To assure safety, these elements must be designed with sufficient strength to remain elastic. See ASCE7-16 Sections 12.4.3 and 12.4.3.1 for controlling basic seismic load combinations, special notes, and exceptions.

Another issue that has caused considerable confusion is the required load combination that is to be used for the design of the connections between the discontinuous walls and frames to the supporting elements and podium slab. The connections between the discontinuous walls and supporting members for light-framed construction must be adequate to transmit the forces for which the discontinuous element was designed, using the standard load combinations of Sections 2.3 and 2.4. The special seismic load combinations with overstrength are not required for the discontinuous shear wall that is being supported. Where the discontinuous element is required to be designed for seismic load combinations including the overstrength factor, such as columns of steel braced or moment frames, its connection to the supporting member shall be designed to transmit the same force. ASCE 7-16 commentary Figure C12.3-5 shows the requirements applied to wood light-framed shear walls. After the design load of the discontinuous shear wall is transferred out of the shear wall, the beams, columns, and other supporting load path connections and elements to the foundation are designed for combinations with overstrength. Also note that the design of anchorage in the supporting system may require use of load combinations including seismic load effect with overstrength in accordance with the material design standard of the supporting system. For example, when anchoring to concrete, design of the concrete in accordance with ACI 318 may require use of load combinations including seismic load effect with overstrength even through ASCE 7 may not.

The shear wall configuration at grid line 1, shown in Figs. 11.19 and 11.20, is a standard two-story segmented shear wall system. The walls do not extend across the full depth of the diaphragm and therefore require collectors to complete the boundary line of the diaphragm. The collectors along this wall line are required to be designed in accordance with ASCE 7-16 Section 12.10.2. There are two approaches that can be taken for the design of the lower shear wall, SW2. First, the overturning force for shear wall 1 can be transferred down to the foundation and anchored with a hold-down, in which case, there would be no vertical irregularity. The other option is to eliminate the hold-down at the foundation and use SW2 as a transfer wall, where the wall would become an element that supports a discontinuous wall, which may trigger the application of the overstrength factor for its design as previously discussed in Section 11.2.

Shear wall SW3 at grid line 2, as shown in Fig. 11.21, is discontinuous with no shear wall directly beneath. The collector or strut at the roof and its connection to the wall is designed using the same requirements as the collectors at grid line 1. The connection of the wall to the supporting beam must be adequate to transmit the forces for which the discontinuous element was designed. The supporting beam/collector, columns, and their connections are all subject to the special load combinations with overstrength

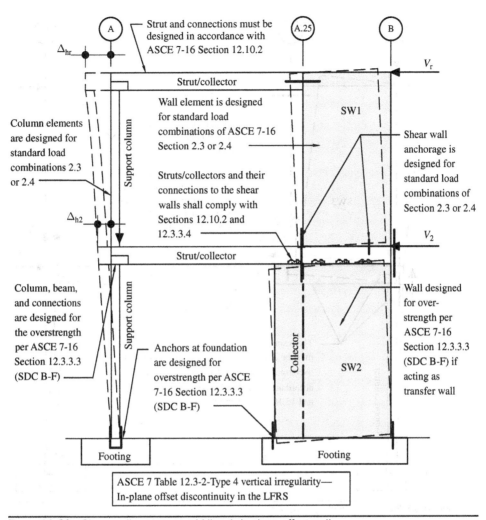

Figure 11.20 Shear wall system at grid line 1, in-plane offset wall.

factor as shown in Fig. 11.21 and ASCE 7-16 Section 12.3.3.3. The columns should also be checked for *P*-delta effects due to the deflection of the diaphragm. Occasionally, the supporting beam is loaded in a manner that creates potential buckling issues as demonstrated in Figs. 11.21 and 11.22. In Fig. 11.22, when the load is applied acting to the left, the top flange of the beam is in compression, which is braced by the diaphragm sheathing and framing, and the bottom flange is in tension as in a typically loaded beam. When the load is applied acting to the right, the bottom flange of the beam is in compression and the top flange is in tension. The support beam must be designed for bending in both directions due to load reversal. If the compression flange at the bottom of the beam is completely unbraced as shown in Fig. 11.23, it will tend to buckle. Steel beams are typically installed with a 2×, or 4× wood plate bolted to the top flange of the steel beam as shown in Fig. 11.23. The floor sheathing is nailed directly to the wood plate,

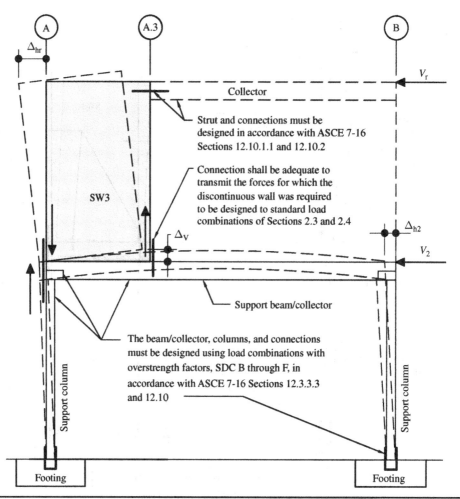

Figure 11.21 Shear wall system at grid line 2.

which braces the beam for flexural buckling forces when the top flange is in compression. Ceiling material or other forms of lateral bracing are rarely connected to the bottom flange. Therefore, if the bottom flange goes into compression due to uplift loads, it can be unbraced along its entire length. For glu-lam beams, the issue lies in the lay-up of the beam as shown in Fig. 11.23. The standard lay-up commonly used for glu-lam beams is combination V3 or V4. These are fabricated with a lower grade material at the top, so if bending reversal occurs, the lower negative bending capacity needs to be checked. For combination 24F-V4, the reference negative bending stress capacity, F_b', is 1850 psi and considerably less than the reference positive bending stress capacity of 2400 psi. For 24F-V3 the difference is 2000 versus 2400 psi. With negative bending under a discontinuous wall under amplified seismic load, the beam needs to be checked, but a V4 or V3 beam can be OK in many cases. Combination V8 is typically used for cantilever and continuous span beams and may be needed for beams supporting discontinuous walls as well.

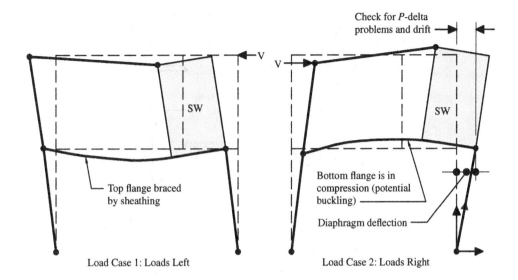

SW

SW

Top flange braced
by sheathing

Bottom flange is in
compression (potential
buckling)

Diaphragm deflection

Load Case 1: Loads Left

Load Case 2: Loads Right

When using glu-lams, if the beam does not have a
cantilever or is part of a continuous span, normal practice
would be to use combination V4. However, glu-lam
beams supporting discontinuous walls should be
combination V8 due to the reversal of stresses.

FIGURE 11.22 Load reversal from discontinuous walls.

The displacement at the top of the discontinuous shear wall will be amplified due to the flexibility of the beam under bending, causing the wall rotation to increase, which must be checked. The displacement of the wall itself consists of two parts, the deflection of the wall and the additional displacement caused by the rotation of the beam in flexure as shown in Fig. 11.24. The additional rotation of the wall due to bending of the beam will increase the total lateral displacement of the wall which will decrease the stiffness of the wall along that line of lateral force-resistance, drawing less load than anticipated. The total lateral displacement of the whole system measured at the top of the wall consists of the two deflections mentioned above plus the deflection of the diaphragm at the wall line.

$$\Delta_{SW} = \frac{8vh^3}{EAb} + \frac{vh}{Gt} + 0.75he_n + d_a\frac{h}{b_{eff}}, \text{ shear wall deflection. Option, use 3-term eq.}$$

$$F_{o/t} = \frac{vh}{b_{eff}}, \text{ overturning force applied to the beam}$$

Deflection of the beam with a concentrated force:

$$\Delta_{Bm} = \frac{Pa^2b^2}{3EIL} = \frac{F_{o/t}a^2b^2}{3EIL} = \Delta_V$$

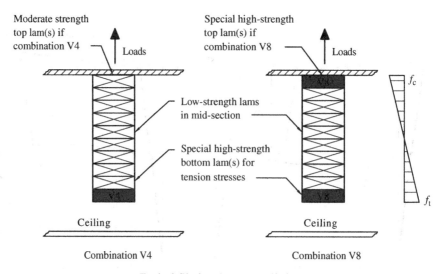

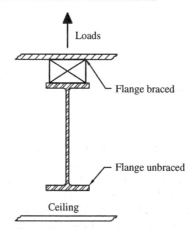

Figure 11.23 Examples of unbraced support beams.

Displacement at the top of the wall due to beam deflection (rotation) (see SEAOC's *Structural/Seismic Design Manual, Vols. 1 and 2* for an example)[4,5]:

$$\tan \theta = \frac{\Delta_{\text{h}}}{h} = \frac{\Delta_{\text{v}}}{b}$$

$$\Delta_{\text{h}} = \frac{h\Delta_{\text{v}}}{b}$$

$$\sum \Delta_{\text{wall}} = \Delta_{\text{h}} + \Delta_{\text{sw}}$$

This must be less than the allowable drift limit.

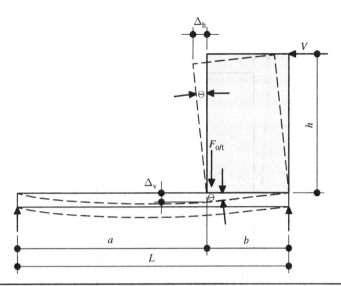

Figure 11.24 Deflection of discontinuous walls.

There are occasions where a discontinuous shear wall is placed perpendicular to the floor framing as shown in Fig. 11.25. This is not an ideal condition and should be avoided whenever possible. Overturning tie straps or hold-downs must be tied to the supporting beam and multiple joists below. The figure clearly shows that control of the displacement at the top of the wall and creating an effective collector would be challenging due to the flexible beam and joists, especially if the wall is located at mid-span of the joist or beam. If the deflection and rotation of the wall become excessive, the wall will become ineffective to resist lateral forces. The forces calculated for that wall would then shift to the adjacent shear walls. Special plan notes and details should be called out on the drawings to notify the joist specifier of the overturning force and the deflection limitations that are required for the design of the supporting joists. Unless the collector is extended into a transfer diaphragm, it would have to extend the full depth of the diaphragm. The figure provides a possible solution for the creation of the collector. The floor framing members in the figure are assumed to be open web joists. Solid blocking is installed between the top chords of the joists. The diaphragm sheathing is nailed to the blocking with special nailing. Multiple $2\times$ members are installed below the top chords to act as the continuous collector. The collector can be nailed directly to the blocking or can have shear clips installed to transfer the shears from the blocking into the collector. The joints in the collector are spliced with tie straps, as required. Hold-down anchors would have to be used to transfer the collector force across the beam.

The cross-section of the wall located at grid line 3 in Fig. 11.19 is shown in Fig. 11.26. The wall line has a type 4 vertical irregularity, which exists when there is an in-plane offset in the lateral-force-resisting elements greater than the length of those elements. The offset shown causes shear wall SW4 to be vertically discontinuous to the foundation. SEAOC noted in their *Structural/Seismic Design Manual, Vol. 1,* that any offset, even those less than or equal to the length of the wall can result in an overturning

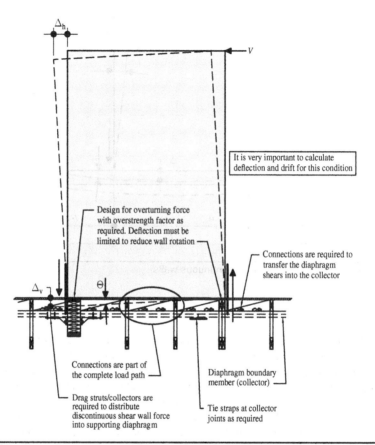

It is very important to calculate deflection and drift for this condition

Design for overturning force with overstrength factor as required. Deflection must be limited to reduce wall rotation

Connections are required to transfer the diaphragm shears into the collector

Connections are part of the complete load path

Diaphragm boundary member (collector)

Drag struts/collectors are required to distribute discontinuous shear wall force into supporting diaphragm

Tie straps at collector joints as required

FIGURE 11.25 Discontinuous shear wall perpendicular to framing.

moment-resisting load transfer discontinuity that requires the application of ASCE7-16 Section 12.3.3.3. The columns, beam, and their connections supporting this wall must be designed using the overstrength factor. When the offset exceeds the length of the resisting element, there is also a shear transfer discontinuity in the collector between the two walls, again requiring the collector to be designed with the overstrength factor. If failure occurs in the connections above SW5 or the collector, SW4 could lose its ability to resist lateral forces.

A good example of complex load paths is the multi-family structure shown in Fig. 11.27 which has multiple irregularities and is in an area of high seismicity. The front wall of the structure at the first floor is set back about 20 ft to allow for parking. This creates a type 4 horizontal irregularity, which is an out-of-plane offset in the shear wall system. The lateral force of the second-floor shear wall must be transferred back to the first-floor shear wall through the diaphragm by cantilever action because the framing at the front on the first floor is not a moment-resisting frame. The second-floor wall is

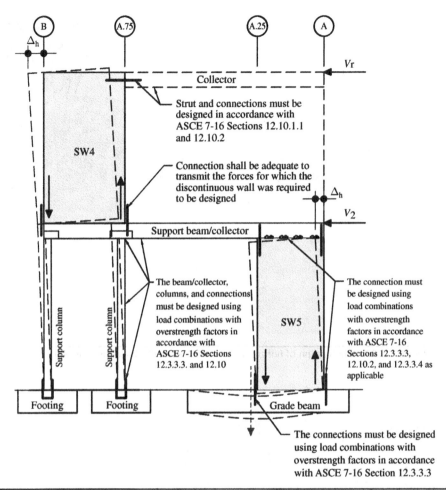

Inside the figure:

Collector

Strut and connections must be designed in accordance with ASCE 7-16 Sections 12.10.1.1 and 12.10.2

Connection shall be adequate to transmit the forces for which the discontinuous wall was required to be designed

SW4

Support beam/collector

The beam/collector, columns, and connections must be designed using load combinations with overstrength factors in accordance with ASCE 7-16 Sections 12.3.3.3. and 12.10

The connection must be designed using load combinations with overstrength factors in accordance with ASCE 7-16 Sections 12.3.3.3, 12.10.2, and 12.3.3.4 as applicable

SW5

Support column

Footing

Grade beam

The connections must be designed using load combinations with overstrength factors in accordance with ASCE 7-16 Section 12.3.3.3

V_r, V_2, Δ_h, B, A.75, A.25, A

Figure 11.26 Shear wall system at grid line 3.

discontinuous to the foundation and must be supported by the beam and columns at the front of the structure. The overturning forces from the wall must be transferred into the beam and columns below. The front support framing, as designed and built, prevents the successful transfer of lateral and overturning forces to the foundation at that location. Figure 11.28 shows some of the unresolved issues that were revealed in the discussion and by on-site observations.

- The second-floor shear wall does not fall directly over the supporting beam. The wall is offset about 2 ft past the beam and is supported on cantilever floor joists.

- Connections were not installed that could transfer the shear forces and overturning forces across the offset.

FIGURE 11.27 Photograph of horizontally offset—vertically discontinuous walls.

FIGURE 11.28 Photograph of close-up of discontinuity.

- The supporting beam was not designed using the overstrength load combinations and has a 10 ft or greater portion of the beam where neither the top nor the bottom lams are laterally braced. The glu-lam beam combination is likely to be combination 24F-V4 and designed for gravity loads alone where only the top flange would be in compression.

- The connections of the side wall to the support beam and the connection of the end of the beam to the column are inadequate (minimal residential connections).

- The supporting columns are not continuous to the foundation. They consist of wood posts connected to the top of masonry piers, which creates a knuckle at approximately mid-height of the column.

- The footings and masonry piers were designed for gravity loads only. Buckling of the column at the knuckle would produce a lateral force at the top of the masonry pier, which in turn would cause an overturning moment at the base of the footing.

Clearly, the front part of the structure is irregular and should have been required to be engineered.

11.4   Two-Stage Podium Structures

The development of mid-rise construction is becoming commonplace and often includes up to six stories of light-framed construction over one or multiple stories of concrete or steel framed podiums. Such structures can benefit from a two-stage seismic analysis. A two-stage analysis is permitted for structures that have a flexible upper portion over a rigid lower portion using the equivalent lateral force procedure, provided the design complies with all of the following:

a. The stiffness of the lower portion shall be at least 10 times stiffer than the upper portion.

b. The period of the entire structure shall not be greater than 1.1 times the period of the upper portion considered as a separate structure supported at the transition of the upper portion to the lower portion.

c. The upper portion shall be designed as a separate structure using the appropriate R and ρ values.

d. The lower portion shall be designed as separate structures using the appropriate R and ρ value. The seismic shear and overturning reactions from the upper portion to the lower portion shall be amplified by the ratio R/ρ of the upper portion over the R/ρ of the lower portion, which shall not be less than 1.0.

e. The upper portion is analyzed using the equivalent lateral force or modal response spectrum procedure, and the lower portion is analyzed with equivalent lateral force procedure.

Questions often arise regarding what amplification factors must be applied to the connections and forces that are transferred at the interface of the structure above to the podium below, as shown in Fig. 11.29. For standard shear walls, the seismic shear, and

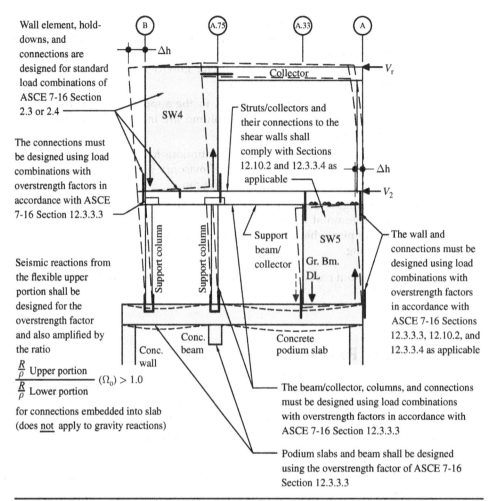

Wall element, hold-downs, and connections are designed for standard load combinations of ASCE 7-16 Section 2.3 or 2.4

The connections must be designed using load combinations with overstrength factors in accordance with ASCE 7-16 Section 12.3.3.3

Seismic reactions from the flexible upper portion shall be designed for the overstrength factor and also amplified by the ratio

$$\dfrac{\dfrac{R}{\rho}\ \text{Upper portion}}{\dfrac{R}{\rho}\ \text{Lower portion}}\ (\Omega_0) > 1.0$$

for connections embedded into slab (does **not** apply to gravity reactions)

Struts/collectors and their connections to the shear walls shall comply with Sections 12.10.2 and 12.3.3.4 as applicable

The wall and connections must be designed using load combinations with overstrength factors in accordance with ASCE 7-16 Sections 12.3.3.3, 12.10.2, and 12.3.3.4 as applicable

The beam/collector, columns, and connections must be designed using load combinations with overstrength factors in accordance with ASCE 7-16 Section 12.3.3.3

Podium slabs and beam shall be designed using the overstrength factor of ASCE 7-16 Section 12.3.3.3

FIGURE 11.29 Discontinuous shear wall at podium.

overturning reactions of the upper portion to the podium slab are required to be amplified by the ratio of R/ρ of the upper structure divided by the ratio R/ρ of the lower structure in accordance with Section 12.2.3.2(d). If a type 4 horizontal or vertical irregularity exists, that amplification ratio must also be added to the overstrength factor for supporting components.

11.5   Problem

Problem 11.1: Three-Story Segmented Shear Walls with In-Plane Offset-Discontinuous Shear Walls

Given: The wall shown in Fig. P11.1 is a three-story shear wall which is offset in-plane at the second floor. The wall is to be analyzed for the loads shown. Seismic controls the design. There are no hold-downs tying the second- and third-floor wall boundary

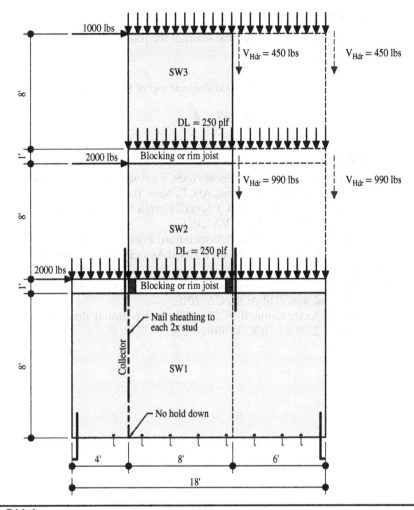

1000 lbs

$V_{Hdr} = 450$ lbs $V_{Hdr} = 450$ lbs

SW3

DL = 250 plf

2000 lbs   Blocking or rim joist

$V_{Hdr} = 990$ lbs $V_{Hdr} = 990$ lbs

SW2

DL = 250 plf

2000 lbs

Blocking or rim joist

Nail sheathing to each 2x stud

Collector

SW1

No hold down

4'   8'   6'

18'

FIGURE P11.1

elements at the foundation. Therefore, the first-floor shear wall is an element that supports a discontinuous shear wall above. The second-floor overturning forces must be amplified for the overstrength factor.

ASCE 7-16 Section 2.4.5 (ASD), Load case 8, $1.0D + 0.7E_v + 0.7E_h$

$\Omega_o = 3.0$ for a bearing wall system, light framed shear walls with wood shtg.

$E_v = 0$ per section 12.4.2.2, exception 2

Assume $b_{eff} = 8$ ft second and third story, $b_{eff} = 18$ ft first story

$DL_{roof} = 150$ plf

$DL_{2nd \text{ and } 3rd} = 250$ plf

$DL_{wall} = 10$ psf

Note: horizontal forces are already factored to $0.7E_h$

$$V_{roof} = 1000 \text{ lb total shear at the roof}$$

$$V_{2nd} \text{ and } V_{3rd} = 2000 \text{ lb}$$

$$\sum V_{1st} = 5000 \text{ lb total shear at top of the first-floor wall}$$

Find: Design the first-floor shear wall. ▲

11.6   References

1. American Society of Civil Engineers (ASCE), *ASCE/SEI 7-16 Minimum Design Loads for Buildings and Other Structures*, ASCE, New York, 2016.
2. American Wood Council (AWC), *Special Design Provisions for Wind and Seismic with Commentary*, AWC, Leesburg, VA, 2021.
3. APA—The Engineered Wood Association, *Plywood Diaphragms, Report 138, APA Form E315H*, APA—The Engineered Wood Association, Engineering Wood Systems, Tacoma, WA, 2000.
4. Structural Engineers Association of California (SEAOC), *2018 IBC Structural/Seismic Design Manual, Vol. 2*, SEAOC, CA, 2018.
5. International Code Council (ICC). 2021. International Building Code, 2021 with commentary. 2021 ed., ICC, Whittier, CA.

CHAPTER **12**

Offset Shear Walls

12.1 Introduction

Several examples of horizontally offset shear walls were presented in Chap. 3. Those examples were included to show the importance in checking diaphragms in both the transverse and longitudinal directions. It is often assumed that loading in the longitudinal direction does not control, especially in areas where the design is controlled by low seismicity or wind. The examples demonstrated that even if the analysis in the longitudinal direction showed that the diaphragm shears are less than those in the transverse direction, the lack of checking drag strut/collector forces and offset shear walls could lead to discontinuous load paths and under-designed shear walls. Further discussion and additional examples are presented in this chapter to help provide a comprehensive coverage on the topic.

12.2 Out-of-Plane Offset Shear Wall Layouts

Four cases of offset shear wall layouts are shown in Fig. 12.1. Case 1 is the classic simple rectangular diaphragm, where all the lateral-force-resisting elements within a line of resistance are in the same plane, no out-of-plane offsets. The load to each wall line is distributed in accordance with tributary area or width of the diaphragm assuming a flexible diaphragm. A brief summary of out-of-plane offset walls was presented in Fig. 2.35 and as depicted in case 2 of Fig. 12.1, which is a common occurrence in many commercial and multi-family structures. Although these walls are assumed to act in the same line of lateral force resistance, these offsets create discontinuities in the diaphragm chords and struts that connect the shear walls together. Case 3 covers horizontally offset shear walls that are assumed to act in the same line of lateral force resistance. The two-story offset shear wall Example 12.2 presented below addresses this type of analysis. Case 4 commonly occurs in multi-family structures, where the shear walls are offset across the corridor. Collectors are often omitted across the corridor, or do not extend across the full depth of the diaphragm at the wall lines. This case will be covered in Sec. 12.4.

12.3 Offset Walls in the Same Line of Lateral Force Resistance

Offsets at exterior walls can require some ingenuity to establish an engineered solution which provides a complete load path across the areas of discontinuity. Not every offset shear wall layout will require the same solution. The approach to solving these types of

445

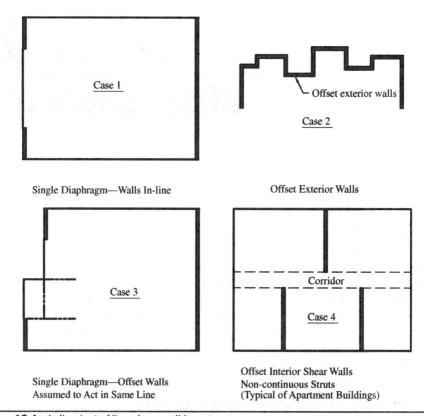

Case 1

Case 2

Offset exterior walls

Single Diaphragm—Walls In-line

Offset Exterior Walls

Case 3

Case 4

Corridor

Single Diaphragm—Offset Walls
Assumed to Act in Same Line

Offset Interior Shear Walls
Non-continuous Struts
(Typical of Apartment Buildings)

Figure 12.1 In-line/out-of-line shear wall layouts.

problems can often require an iterative process. The better the understanding an engineer has for how the forces will be transferred through the structure, the easier it will be to find a solution.

Example 12.1: Offset Exterior Shear Walls

A typical example of case 2 is shown in Fig. 12.2. The structure shown is loaded in the longitudinal direction with a 350 plf uniform wind load. Wind end pressure zones have not been included to simplify the example. Since sheltering is not allowed, the uniform load is applied to all projections at the offsets. The main diaphragm spans 26 ft from grid line C to the corridor walls. SW1 and SW3 are offset 4 ft from grid line C and have a width of 10 ft and 20 ft, respectively. SW2 is offset 8 ft from grid line B with a width of 16 ft. A continuous collector is installed at grid line C full width of the diaphragm, which is 87 ft. The spacing between the walls is shown in the figure.

Before a solution can be determined, it requires some thought as to how the main diaphragm is supported off the exterior shear walls. One way to approach it is to assume that the main diaphragm is supported off the corridor walls and off the continuous collector at grid line C. The diaphragm reaction at grid line C is then distributed

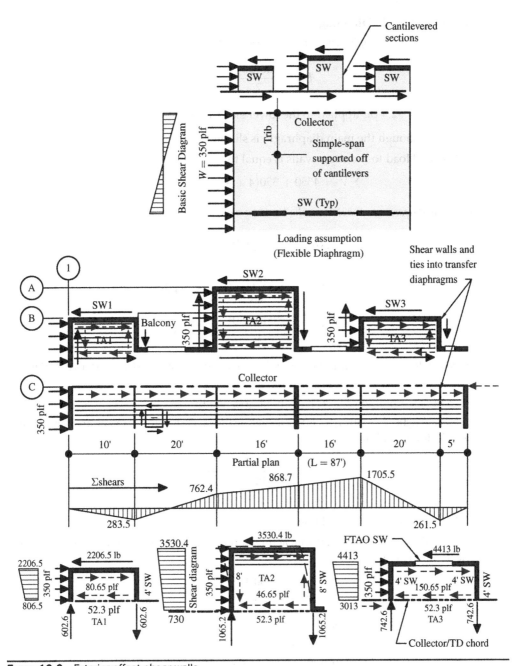

Figure 12.2 Exterior offset shear walls.

into each offset transfer area as a concentrated force, then transferred to the supporting shear walls by cantilever action as shown at the top of Fig. 12.2. There are two approaches that can be taken, one as a flexible diaphragm and the other as a semi-rigid or rigid diaphragm. Both will be presented and discussed here.

Assuming Flexible Diaphragm

Determine the main diaphragm reaction at grid line C:

$$R_B = \frac{350(26)}{2} = 4550 \text{ lb}$$

$v_B = \dfrac{4550}{87} = 52.3$ plf applied to the lower side of the collector at grid line C, acting to the right as though the main diaphragm is sliding off the transfer areas

The total load to the shear walls is equal to

$$\Sigma V = 4550 + 350(4 + 4 + 8) = 10,150 \text{ lb}$$

Determine the forces to the shear walls based on wall length:

$$\text{SW1} = \frac{10(10,150)}{46} = 2206.5 \text{ lb}$$

$$\text{SW2} = \frac{16(10,150)}{46} = 3530.4 \text{ lb}$$

$$\text{SW3} = \frac{20(10,150)}{46} = 4413 \text{ lb}$$

Determine the forces applied to the ends of the cantilever transfer areas (see bottom of Fig. 12.2):

At SW1, summing shear forces about grid line B:

$V_{SW1} = 2206.5 - 350(4) = 806.5$ lb, acting at the end of the transfer area and above the collector at grid line C

$$v_{SW1} = \frac{806.5}{10} = 80.65 \text{ plf}$$

Uniform shear acting on the collector.

At SW2, summing about grid line A:

$V_{SW2} = 3530.4 - 350(8) = 730.4$ lb, acting at the end of the transfer area and above the collector at grid line C

$$v_{SW2} = \frac{730.4}{16} = 45.65 \text{ plf}$$

Uniform shear acting on the collector.

At SW3, summing about grid line B:

$V_{SW3} = 4413 - 350(4) = 3013$ lb, acting at the end of the transfer area and above the collector at grid line C

$$v_{SW3} = \frac{3013}{20} = 150.65 \text{ plf}$$

Uniform shear acting on the collector.

These unit shears will be supporting the collector/strut, so the transfer shears acting on the collector are acting to the left.

Determine the rotational resisting forces at the cantilevers:

The eccentric forces at the ends of the cantilever will be resisted by the transfer area side walls that, in this case, can act as shear walls. Summing about grid lines A and B:

$$F_{SW1} = \frac{806.5(4) + (350(4)(2)}{10} = 602.6 \text{ lb}$$

$$F_{SW2} = \frac{730.4(8) + (350(8)(4)}{16} = 1065.2 \text{ lb}$$

$$F_{SW3} = \frac{3013(4) + (350(4)(2)}{20} = 742.6 \text{ lb}$$

Construction of the force diagram:

Note that the shears transferred into the collector are in opposing directions on each side of the collector at each transfer area and are subtracted from each other to get a net shear value. Starting at grid line 1 and summing shears to the right:

$$F_{10'} = (80.65 - 52.3)(10) = 283.5 \text{ lb}$$

$$F_{30'} = 283.5 - 52.3(20) = -762.5 \text{ lb}$$

$$F_{46'} = -762.5 - (52.3 - 45.65)(16) = -868.7 \text{ lb}$$

$$F_{62'} = -868.7 - 52.3(16) = -1705.5 \text{ lb}$$

$$F_{82'} = -1705.5 + (150.65 - 52.3)(20) = +261.5 \text{ lb}$$

$$F_{87'} = +261.5 - 52.3(5) = 0 \text{ lb}. \qquad \therefore \text{ Diagram closes to zero.}$$

Assuming Semi-Rigid, Rigid Diaphragm

If a semi-rigid or rigid diaphragm exists, a rigid diaphragm analysis, RDA, can be performed to determine the forces that are applied to the exterior shear walls. The transfer area shears and resisting side wall forces can be determined by summing forces about grid lines A and B as before. The shear forces below the collector supporting the main diaphragm can also be determined as before. If the sidewalls cannot act as shear walls, the rotational forces must be distributed into a transfer diaphragm or interior wall within the main diaphragm. ▲

Example 12.2: Two-Story Offset Shear Walls

The structure shown in the partial isometric in Fig. 12.3, and the accompanying floor plans in Figs. 12.4 and 12.5 have two stories of offset shear walls at grid lines 1, 2, a first-floor shear wall at grid line 3, and two-story in-line shear walls at grid line 4. The shear walls at both floors, at grid lines 1 and 2, are assumed to act in the same line of resistance. The first-floor shear wall, SW5, at grid line 3 supports only the second-floor diaphragm and does not continue up to the roof. Shear wall SW1 at the second floor is vertically discontinuous to the first floor and has an in-plane offset from SW3 at the first floor. The roof diaphragm spans from the offset walls at grid lines 1 and 2 to the shear

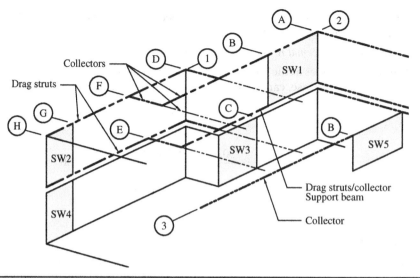

FIGURE 12.3 Two-story offset shear walls—isometric of offset wall line.

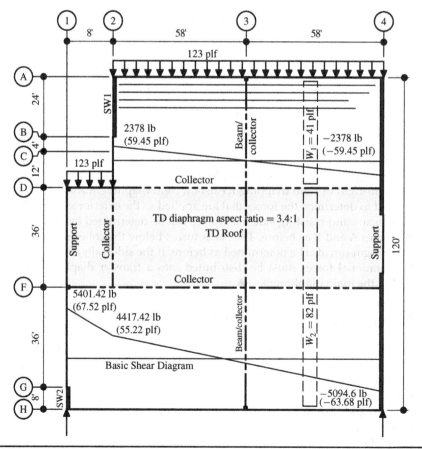

FIGURE 12.4 Roof diaphragm and second-floor walls.

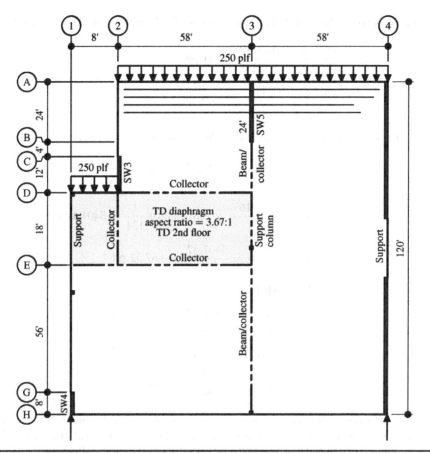

Figure 12.5 Second-floor diaphragm and first-floor walls.

walls at grid line 4. The roof and floor joists are oriented perpendicular to the shear walls. The forces being shared by the second-floor offset walls are transferred through the roof transfer diaphragm located between grid lines D and F, from 1 to 4, as shown in Fig. 12.4. The second-floor diaphragm shown in Fig. 12.5 spans from the offset walls at grid lines 1 and 2 to shear wall SW5 at grid line 3, and from grid line 3 to grid line 4. The forces being shared by the offset walls at the second floor are transferred through the transfer diaphragm located between grid lines D and E, from grid line 1 to 3. The shear wall layout demonstrates the difference in distribution of the forces to the offset walls due to the diaphragm span to offset ratios. This structure could easily have semi-rigid or rigid diaphragms that cause torsional irregularities.

Roof diaphragm (see Fig. 12.4): The uniform load applied to the roof diaphragm is 123 plf. The diaphragm is broken down into two sections between grid lines A and D from 2 to 4, and between grid lines D and H from 1 to 4, as shown in Fig. 12.4. Each section will be treated as separately supported diaphragms. The section between grid lines A and D is supported at its left end by the shear wall and collector that is embedded into the transfer diaphragm. The 123 plf uniform load between grid lines 2 and 4 is distributed to each diaphragm section in accordance with its width.

Construct of the basic shear diagram for the section between A and D from 2 to 4:

$$w_1 = \frac{wW_{A-D}}{W_{A-H}} = \frac{123(40)}{120} = 41 \text{ plf}$$

$$V_2 = \frac{wL_{2-4}}{2} = \frac{41(116)}{2} = 2378 \text{ lb}, \quad v_2 = \frac{V_2}{W_{A-D}} = \frac{2378}{40} = 59.45 \text{ plf}$$

Construct of the basic shear diagram for the section between D and H:

$$w_2 = \frac{wW_{D-H}}{W_{A-H}} = \frac{123(80)}{120} = 82 \text{ plf}$$

$$V_1 = \left[\frac{82(116)^2}{2} + 123(8)(120)\right]\frac{1}{124} = 5401.42 \text{ lb}$$

$$v_1 = \frac{V_1}{W_{D-H}} = \frac{5401.42}{80} = +67.52 \text{ plf}$$

$$V_2 = 5401.42 - 123(8) = 4417.42 \text{ lb}$$

$$v_2 = \frac{V_2}{W_{D-H}} = \frac{4417.42}{80} = +55.22 \text{ plf}$$

$$V_4 = 4417.42 - 82(116) = -5094.6 \text{ lb}$$

$$v_4 = \frac{V_4}{W_{D-H}} = \frac{-5094.6}{80} = -63.68 \text{ plf}$$

Determination of the transfer diaphragm shears (see Fig. 12.6): Before the transfer diaphragm shears can be determined, the force at the end of the drag strut at grid line 2D must be calculated. An initial analysis will show that the transverse collector force diagrams will

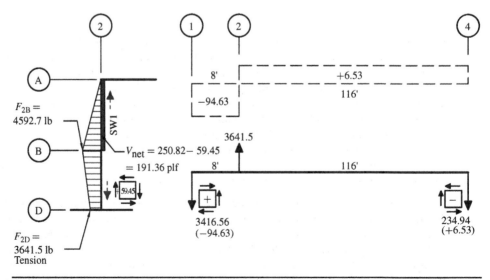

Figure 12.6 Roof transfer diagram shears.

not close because the actual force to the offset wall line is greater than that calculated by using the standard tributary width method as demonstrated in Example 3.3. The increase in the reaction to the offset walls is approximately 5.6 percent, by trial and error or can be calculated by weighted average by shear wall length. The diaphragm shears will remain the same, but the load to the walls will increase by this amount.

Force to transfer diaphragm (adjusted) (reference Fig. 12.6):

$$R_{1,2} = \frac{wL_{1-4}}{2}\text{(calculated increase)} = \frac{123(124)}{2}(1.056) = 8026.5 \text{ lb}$$

$$v_{SW1} = \frac{R_{1,2}}{L_{SW1}} = \frac{8026.5}{32} = 250.82 \text{ plf unit wall shear} = v_{SW2}$$

$$V_{SW1} = 250.82(24) = 6020 \text{ lb total shear to SW1}$$

$$v_{diaph} = +59.45 \text{ plf}$$

$$v_{net} = v_{SW1} - v_{diaph} = 250.82 - 59.45 = 191.36 \text{ plf}$$

$$V_{SW2} = 250.82(8) = 2006.5 \text{ lb}$$

$$v_{diaph} = +67.52 \text{ plf}$$

$$v_{net} = v_{SW1} - v_{diaph} = 250.82 - 67.52 = 183.29 \text{ plf}$$

Starting at grid line A:

$$F_{2B} = v_{net}L_{SW1} = 191.36(24) = 4592.7 \text{ lb}$$

Force to transfer diaphragm:

$$F_{2D} = F_{2B} - v_{diaph}L_{B-D} = 4592.7 - 59.45(16) = 3641.5 \text{ lb, tension}$$

Transfer diaphragm shears (Fig. 12.6):

$$V_1 = \frac{3641.5(116)}{124} = 3416.56 \text{ lb}$$

$$v_1 = \frac{V_1}{W_{TD}} = \frac{3416.5}{36} = -94.63 \text{ plf}$$

$$V_4 = \frac{3641.5(8)}{124} = +234.94 \text{ lb}$$

$$v_4 = \frac{V_4}{W_{TD}} = \frac{234.94}{36} = +6.53 \text{ plf}$$

Determine the net shears at transfer diaphragm (see Fig. 12.7):

Location (Grid Line)	Net Shear (plf)
2 from A to D	previously calculated
1 from D to F	$v = +67.52 - (94.63) = -27.11$ plf
2 from D to F (left)	$v = +55.22 - (94.63) = -39.41$ plf
2 from D to F (right)	$v = +55.22 + (6.53) = +61.75$ plf

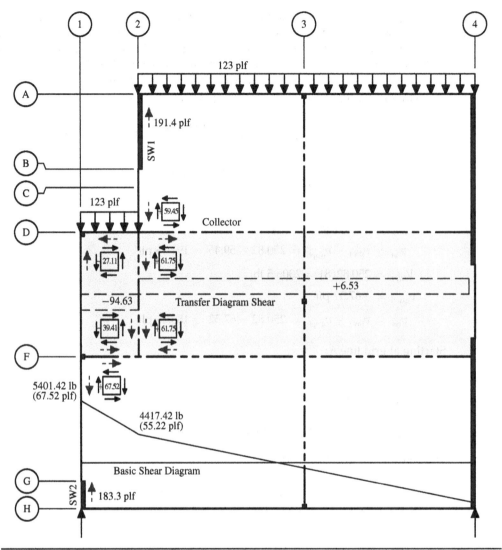

FIGURE 12.7 Roof diaphragm net shears.

Determine the transverse collector/strut forces along grid lines 1 and 2 (see Fig. 12.8):

Forces along grid line 2, starting at grid line A:

Force from A to D previously calculated = 3641.5 lb

$$F_{2F} = 3641.5 - (39.41 + 61.75)(36) = 0 \text{ lb}$$

Forces along grid line 1, starting at grid line D:

$$F_{1F} = 27.11(36) = 975.96 \text{ lb, tension}$$

$$F_{1G} = 975.96 - 67.52(36) = -1454.76 \text{ lb, compression}$$

$$F_{1H} = -1454.76 + 183.29(8) = -11.6 \text{ lb}, \quad \therefore \text{ OK. Close enough.}$$

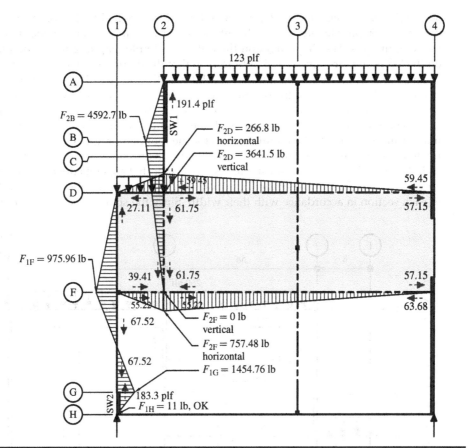

FIGURE 12.8 Roof collector force diagrams.

The 5.6 percent increase in the roof loads to grid lines 1 and 2 is confirmed.

Determine the longitudinal collector/strut forces along grid lines D and E (see Fig. 12.8):

Forces along grid line D:

$$F_{2D} = \frac{(27.11 + 39.41)}{2}(8) = 266.08 \text{ lb, tension}$$

$$F_{4D} = 266.08 - \frac{[(61.75 - 59.45) + (59.45 - 57.15)]}{2}(116) = 0 \text{ lb}$$

Forces along grid line F:

$$F_{2F} = \frac{[(67.52 + 27.11) + (55.22 + 39.41)]}{2}(8) = 757.48 \text{ lb, compression}$$

$$F_{4F} = 757.48 - \frac{[(61.75 - 55.22) + (63.68 - 57.15)]}{2}(116) = 0 \text{ lb}$$

Therefore, all the force diagrams close to zero.

Second floor diaphragm (see Fig. 12.9): The uniform load applied to the second-floor diaphragm is 250 plf. The presence of the interior shear wall at grid line 3 creates two separate diaphragms. The diaphragm on the right is a simple diaphragm. The diaphragm on the left is similar to the roof diaphragm, in that the diaphragm reaction to the offset shear walls at grid lines 1 and 2 is increased due to the offset of the walls. The increase to the first-floor walls from the second-floor diaphragm must be determined before the lateral forces from the second-floor walls can be added. The left diaphragm can be broken down into two sections between grid lines A and D from 2 to 3, and between grid lines D and H from 1 to 3 the same way as for the roof diaphragm. Each section will be treated as separate diaphragms. The section between grid lines A and D is supported at its left end by the drag strut, shear wall, and collector that are embedded into the transfer diaphragm. The portion of the 250 plf uniform load to the diaphragm is distributed to each section in accordance with their width. Since the span of the left diaphragm is

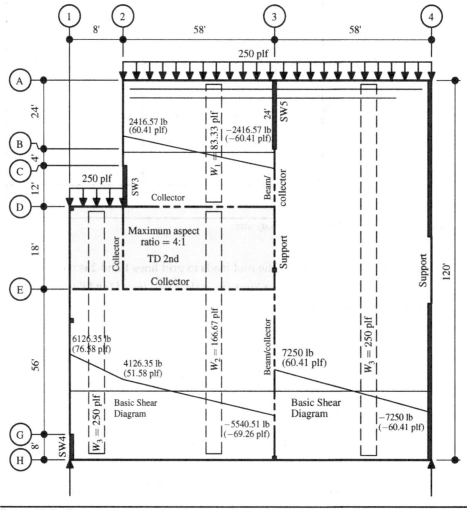

Figure 12.9 Floor diaphragm and basic shear diagram.

less than that of the roof, the width of the second-floor transfer diaphragm can also be reduced. Assume the desired aspect ratio is 4:1.

$$W_{TD} = \frac{66}{4} = 16.4 \text{ ft}$$

Use 18 ft,

$$A.R. = \frac{66}{18} = 3.67 : 1, \text{blocked}$$

Construct of the basic shear diagram for the section between A and D from 2 to 3:

$$w_1 = \frac{wW_{A-D}}{W_{A-H}} = \frac{250(40)}{120} = 83.33 \text{ plf}$$

$$V_2 = \frac{w_1 L_{2-3}}{2} = \frac{83.33(58)}{2} = 2416.57 \text{ lb}, \quad v_2 = \frac{V_2}{W_{A-D}} = \frac{2416.57}{40} = 60.41 \text{ plf}$$

Construct of the basic shear diagram for the section between D and H:

$$w_2 = \frac{wW_{D-H}}{W_{A-H}} = \frac{250(80)}{120} = 166.67 \text{ plf}, \quad w_3 = 250 \text{ plf}$$

$$V_1 = \left[\frac{166.67(58)^2}{2} + 250(8)(62) \right] \frac{1}{66} = 6126.35 \text{ lb, shear (reaction) at grid line 1}$$

$$v_1 = \frac{V_1}{W_{D-H}} = \frac{6126.35}{80} = +76.58 \text{ plf unit shear}$$

$$V_2 = V_1 - wL_{1-2} = 6126.35 - 250(8) = 4126.35 \text{ lb}$$

$$v_2 = \frac{V_2}{W_{D-H}} = \frac{4126.35}{80} = +51.58 \text{ plf unit shear}$$

$$V_4 = V_2 - wL_{2-4} = 4126.35 - 166.67(58) = -5540.51 \text{ lb}$$

$$v_4 = \frac{V_4}{W_{D-H}} = \frac{-5540.51}{80} = -69.26 \text{ plf unit shear}$$

Determination of the transfer diaphragm shears (see Fig. 12.10): Before the transfer diaphragm shears can be determined, the force at the end of the drag strut at grid line 2D must be calculated. The increase in the reaction to the offset walls at the second floor is approximately 8 percent, by trial and error or weighted average. The offset in the walls at the roof and second floor is the same. However, since the span of the left diaphragm is about half that of the roof, it effectively creates a greater offset with respect to the reduced span. The greater the span to offset ratio is, the greater the increase in load to the offset walls.

Force to transfer diaphragm (adjusted):

$$R_{1,2} = \frac{wL_{1-3}}{2}(\text{calculated increase}) = \frac{250(66)}{2}(1.08) = 8910 \text{ lb}$$

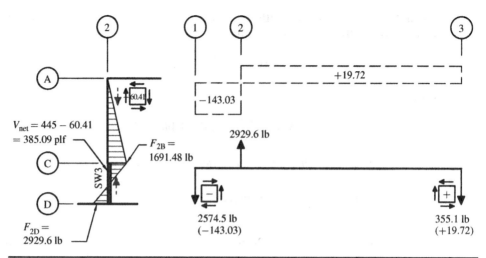

Figure 12.10   Floor transfer diagram shears.

$$v_{SW3} = \frac{R_{1,2}}{\sum L_{SW}} = \frac{8910}{20} = 445.5 \text{ plf} = v_{SW4}$$

$$V_{SW3} = v_{SW3}L_{SW3} = 445.5(12) = 5346 \text{ lb}$$

$$V_{SW4} = v_{SW4}L_{SW4} = 445.5(8) = 3564 \text{ lb}$$

$$v_{SW3} = 445.5 \text{ plf}$$

$$v_{diaph} = +60.14 \text{ plf}$$

$$v_{net} = v_{sw3} - v_{diaph} = 445.5 - 60.41 = 385.09 \text{ plf net shear at SW3}$$

$$v_{SW4} = 445.5 \text{ plf}$$

$$v_{diaph} = +76.58 \text{ plf}$$

$$v_{net} = v_{sw4} - v_{diaph} = 445.5 - 76.58 = 368.92 \text{ plf net shear at SW4}$$

Starting at grid line A:

$$F_{2B} = -60.41(28) = -1691.48 \text{ lb}$$

$$F_{2D} = -1691.48 + 385.09(12) = 2929.6 \text{ lb, tension force to the transfer diaphragm}$$

Transfer diaphragm shears (Fig. 12.10):

$$V_1 = \frac{2929.6(58)}{66} = 2574.5 \text{ lb}$$

$$v_1 = \frac{V_1}{W_{TD}} = \frac{2574.5}{18} = -143.03 \text{ plf}$$

$$V_4 = \frac{2929.6(8)}{66} = +355.1 \text{ lb}$$

$$v_4 = \frac{V_4}{W_{TD}} = \frac{355.1}{18} = +19.72 \text{ plf}$$

Determine the net shears at transfer diaphragm (see Fig. 12.11):

Location (Grid Line)	Net Shear (plf)
2 from A to D	previously calculated
1 from D to E	$v = +76.58 - (143.03) = -66.45$ plf
2 from D to E (left)	$v = +51.58 - (143.03) = -91.45$ plf
2 from D to E (right)	$v = +51.58 + (19.72) = +71.3$ plf
3 from D to E (left)	$v = -69.26 + (19.72) = -49.54$ plf

Transfer the net shears and sheathing elements on the plan.

Determine the transverse collector/strut forces along grid lines 1 and 2 (see Fig. 12.12):

Forces along grid line 2, starting at grid line A:

Forces from A to D were previously calculated $= 2929.6$ lb

$$F_{2E} = 2929.6 - (91.45 + 71.3)(18) = 0 \text{ lb}$$

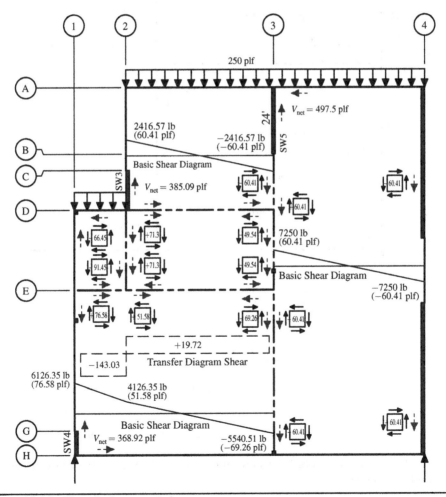

FIGURE 12.11 Floor diaphragm net shears.

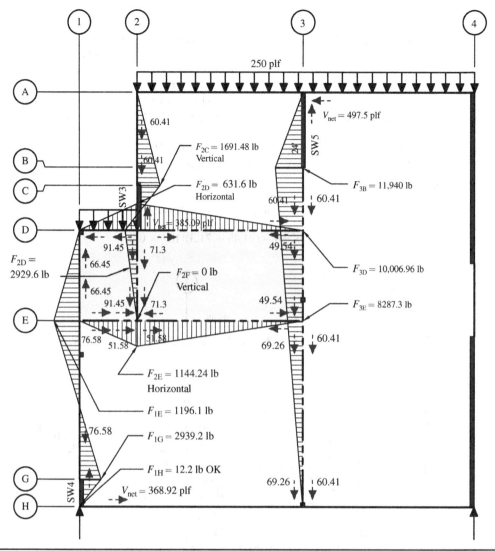

FIGURE 12.12 Floor collector force diagrams.

Forces along grid line 1 starting at grid line D:

$$F_{1E} = 66.45(18) = 1196.1 \text{ lb, tension}$$

$$F_{1G} = 1196.1 - 76.58(54) = -2939.2 \text{ lb, compression}$$

$$F_{1H} = -2939.2 + 368.92(8) = -12.2 \text{ lb}, \quad \therefore \text{ OK}$$

This is close enough. The 8 percent increase in the second-floor loads to grid lines 1 and 2 is confirmed.

Forces along grid line 3 (reference Figs. 12.11 and 12.12): The total load to grid line 3 is equal to the adjusted (92 percent) reaction of the left diaphragm to grid line 3 plus the left reaction of the right diaphragm.

$$V_3 = \frac{250(66)}{2}(0.92) + 7250 = 14,840 \text{ lb}$$

$$v_{SW5} = \frac{V_3}{L_{SW5}} = \frac{14,840}{24} = 618.33 \text{ plf}$$

$$v_{net} = v_{SW5} - v_{diaph} = 618.33 - 60.41 - 60.41 = 497.5 \text{ plf at SW5}$$

Starting at grid line A:

$$F_{3B} = v_{net}L_{SW5} = 497.5(24) = 11,940 \text{ lb}$$

$$F_{3D} = 11,940 - (60.41 + 60.41)(16) = 10,006.96 \text{ lb}$$

$$F_{3E} = 10,006.96 - (49.54 + 60.41)(18) = 8027.9 \text{ lb}$$

$$F_{3G} = 8027.9 - (69.26 + 60.41)(62) = -11.6 \text{ lb}, \quad \therefore \text{ OK}$$

Determine the longitudinal collector/strut forces along grid lines D and E (see Fig. 12.12):

Forces along grid line D starting at grid line 1:

$$F_{2D} = \frac{(66.45 + 91.45)}{2}(8) = 631.6 \text{ lb, tension}$$

$$F_{4D} = 774.64 - \frac{[(71.3 - 60.41) + (60.41 - 49.54)]}{2}(58) = 0 \text{ lb}$$

Forces along grid line E starting at grid line 1:

$$F_{2F} = \frac{[(66.45 + 76.58) + (91.45 + 51.58)]}{2}(8) = 1144.24 \text{ lb, compression}$$

$$F_{4F} = 1144.24 - \frac{[(71.3 - 51.58) + (69.26 - 49.54)]}{2}(58) = 0 \text{ lb}$$

Therefore, all the force diagrams close to zero.

The second-floor shear wall forces can now be added to the second-floor diaphragm for the final forces along the offset wall lines (reference Fig. 12.13). Although shear walls SW2 and SW4 are in vertical alignment, all the shears from SW2 cannot be added directly to SW4 because the walls at the first floor are still designed to share all the forces as a unit.

Final first-floor wall shears:

$$V_{SW1} = 6020 \text{ lb}$$
$$V_{SW2} = 2006.5 \text{ lb}$$
$$V_{SW3} = 5346 \text{ lb}$$
$$V_{SW4} = 3564 \text{ lb}$$

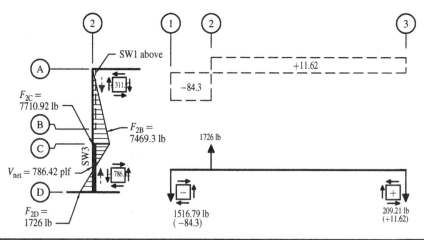

Figure 12.13 Final floor transfer diagram shears.

$$\sum V = 16{,}936.5 \text{ lb}$$

$$V_{SW} = \frac{16{,}936.5}{(12 + 8)} = 846.83 \text{ plf}$$

$$V_{SW3} = 846.83(12) = 10{,}161.96 \text{ lb}$$

$$V_{SW4} = 846.83(8) = 6774.64 \text{ lb}$$

Determine the net shear at shear walls (reference Fig. 12.14):

Shear wall SW3

$$v_{diaph} = +60.14 \text{ plf}$$

$$v_{net} = v_{SW} - v_{diaph} = 846.83 - 60.41 = 786.42 \text{ plf}$$

Shear wall SW4

$$v_{diaph} = +76.58 \text{ plf}$$

$$v_{SW2} = 250.82 \text{ plf, above}$$

$$v_{net} = v_{SW4} - v_{SW2} - v_{diaph} = 846.83 - 250.82 - 76.58 = 519.4 \text{ plf}$$

Determine the net shears between grid lines A to D:

Location (Grid Line)	Net Shear (plf)
V_{SW1} at second floor	$v = 250.82$ plf
2 from A to B	$v = 250.82 + 60.41 = 311.22$ plf
2 from B to C	$v = 60.41$ plf

Force to transfer diaphragm:

$$F_{2B} = 311.22(24) = 7469.3 \text{ lb compression}$$

$$F_{2C} = 7469.3 + 60.41(4) = 7710.92 \text{ lb compression}$$

$$F_{2D} = 7710.92 - 786.42(12) = -1726 \text{ lb tension force to transfer diaphragm}$$

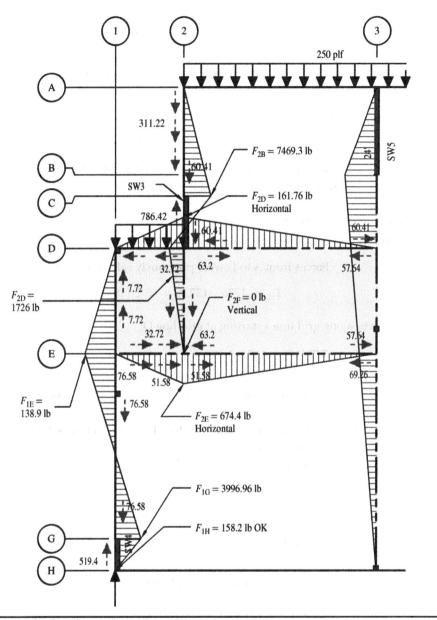

FIGURE 12.14   Final floor collector force diagrams.

Transfer diaphragm shears (Fig. 12.13):

$$V_1 = \frac{1726(58)}{66} = 1516.79 \text{ lb}$$

$$v_1 = \frac{V_1}{W_{TD}} = \frac{1516.79}{18} = -84.3 \text{ plf}$$

$$V_4 = \frac{1726(8)}{66} = +209.21 \text{ lb}$$

$$v_4 = \frac{V_4}{W_{TD}} = \frac{209.21}{18} = +11.62 \text{ plf}$$

Determine the net shears at the transfer diaphragm:

1 from D to E	$v = +76.58 - (84.3) = -7.72 \text{ plf}$
2 from D to E (left)	$v = +51.58 - (84.3) = -32.72 \text{ plf}$
2 from D to E (right)	$v = +51.58 + (11.62) = +63.2 \text{ plf}$
3 from D to E (left)	$v = -69.26 + (11.62) = -57.64 \text{ plf}$

Determine the collector/strut forces along grid lines 1 and 2 (see Fig. 12.14):

Forces along grid line 2 starting at grid line D:

Forces from A to D were previously calculated $= 1726 \text{ lb}$

$$F_{2E} = 1726 - (32.72 + 63.2)(18) = 0 \text{ lb}$$

Forces along grid line 1 starting at grid line D:

$$F_{1E} = 7.72(18) = 138.9 \text{ lb, tension}$$

$$F_{1G} = 138.9 - 76.58(54) = -3996.96 \text{ lb, compression}$$

$$F_{1H} = -3996.96 + 519.4(8) = -158.2 \text{ lb, Therefore OK, close enough}$$

The 8 percent increase in the second-floor loads to grid lines 1 and 2 is confirmed but should be more like 8.5 percent.

Forces along grid line D starting at grid line 1:

$$F_{2D} = \frac{(7.72 + 32.72)}{2}(8) = 161.76 \text{ lb, tension}$$

$$F_{4D} = 161.76 - \frac{[(63.2 - 60.41) + (60.41 - 57.64)]}{2}(58) = 0 \text{ lb}$$

Forces along grid line E starting at grid line 1:

$$F_{2E} = \frac{[(76.58 + 7.72) + (51.58 + 32.72)]}{2}(8) = 674.4 \text{ lb, compression}$$

$$F_{4E} = 674.4 - \frac{[(63.2 - 51.58) + (69.26 - 57.64)]}{2}(58) = 0 \text{ lb}$$

The increase in the shears to the first-floor shear walls is approximately 53 plf due to the offset in the walls at grid lines 1 and 2. If the walls were designed using the conventional tributary width method, the walls and foundations could have been under-designed. ▲

12.4 Interior Offset Shear Walls

Architectural layouts in living units of multi-family structures do not always allow in-line placement of shear walls across the corridor. Offset shear walls in this case need to be connected across the corridor if they are to act as a unit in the same line of lateral force resistance and share the load. Occasionally, shear walls are placed on only one side of the corridor, which would require a full-length collector to complete the diaphragm boundary element.

Figure 12.15 shows five different wall configurations for placement of the transverse shear walls. The most common and preferable arrangement is layout 1. The shear walls in this case are in alignment and are connected across the corridor by joists acting as the collector. The collector and their connections are shown in the typical detail at the lower right of the figure and must be designed for tension and compression forces. In layout 2, the walls are staggered enough so that only one wall can be used. A collector must be installed at the end of the shear wall that extends across the full width of the diaphragm to establish the remaining diaphragm boundary element. Whenever the

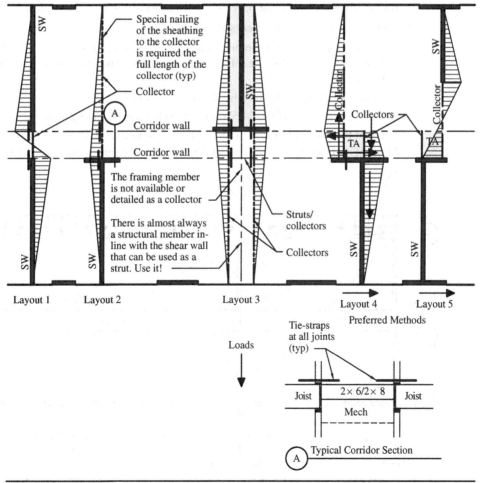

Figure 12.15 Common transverse offset interior wall layouts.

collector force becomes too large, layout 3 can be used to split the loads in half and distribute them into two adjacent members. Their force must be transferred across the offset to the shear wall. A floor framing member does not always line up with the shear wall. As an alternate, the transfer can be made between the corridor walls like a transfer area. Layouts 4 and 5 are preferred methods if offset walls and/or collectors are utilized at the corridor only because of the shorter collector lengths that can be used. The collector must extend across the full width of the diaphragm and lap with the collector of the adjacent shear wall at the corridor. The shear force from the collector must be transferred across the offset into the shear wall by means of a transfer area, which causes an eccentricity. Resisting forces can be transferred into the corridor walls. Four different layouts will be presented to show the difference in collector forces and their connection requirements. The first example is shown in Fig. 12.16, which shows two full-length walls across each side of the corridor that are in alignment. These interior walls create two separate diaphragms. The force diagram can be plotted by the following:

- Calculate each diaphragm reactions on each side of the shear walls at grid line 2.
- Calculate the diaphragm unit shears by dividing the diaphragm reactions by the width of the diaphragm, which will be 156 plf (left) and 244 plf (right).
- Calculate the shear wall shears at grid line 2.
- Before the force diagram can be plotted, the net shear at the shear walls must be determined (e.g., SW shear minus the diaphragm unit shears each side of the wall = 54.55 plf).
- Starting at grid line 2A, sum the shears along grid line 2.
  Force at the end of SW1 = 54.55(22) = 1200 lb tension, acting upward
  Force at the start of SW2 = +1200 − (156 + 244)(6) = −1200 lb
  Force at end of SW2 = −1200 + 54.55(22) = 0 lb. Diagram closes to zero.

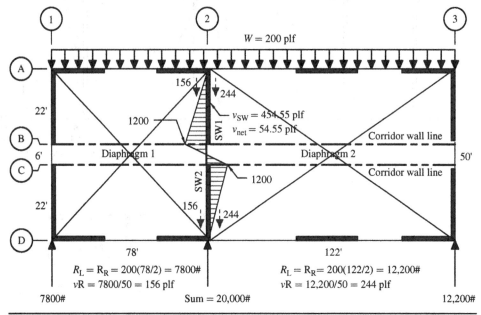

Figure 12.16 Full-length walls aligned.

The following example shows that not connecting the walls will create a discontinuity in the load path and result in an unresolved residual force of 1200 lb. It also shows that a complete load path would not be established by not connecting the walls.

Example 12.3: Full-Length Offset Interior Shear Walls

Figure 12.17 shows a diaphragm that is 200 ft long by 50 ft wide, where two full-length offset interior shear walls are placed on each side of a 6-ft corridor at grid lines 2 and 3 with a 4-ft offset, respectively. This layout requires a little more thought since the diaphragm length is not the same across the corridor. In this case, the diaphragm must be broken down into four diaphragm sections, Diaphragms 1–4, each with boundaries as shown in the figure.

Calculate the uniform load to each diaphragm section: The 200 plf load to each section is determined in accordance with their width.

$$w_{\text{diaph 1}} = w_{\text{diaph 4}} = \frac{(22+6)(200)}{50} = 112 \text{ plf}$$

$$w_{\text{diaph 2}} = w_{\text{diaph 3}} = \frac{(22)(200)}{50} = 88 \text{ plf}$$

$$w_{\text{transfer area}} = \frac{(6)(200)}{50} = 24 \text{ plf.}$$

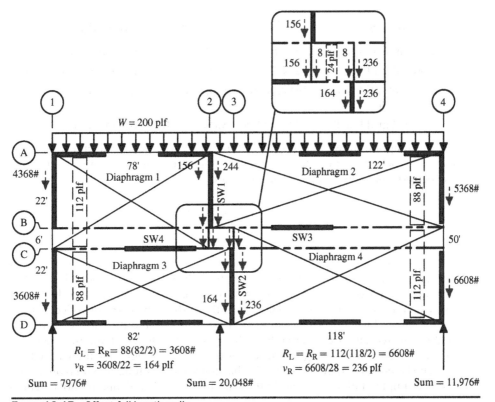

Figure 12.17 Offset full-length walls.

The transfer area is isolated from the other diaphragms and must be included.

$$\text{Shear} = \frac{(24)(4)}{6(2)} = 8 \text{ plf}$$

Calculate the diaphragm reactions and unit shears on each side of the shear walls:

$$R_{R1} = \frac{(112)(78)}{2} = 4368 \text{ lb}, \qquad v_{R1} = \frac{4368}{28} = 156 \text{ plf}$$

$$R_{L2} = \frac{(88)(122)}{2} = 5368 \text{ lb}, \qquad v_{L2} = \frac{5368}{22} = 244 \text{ plf}$$

$$R_{R3} = \frac{(88)(82)}{2} = 3608 \text{ lb}, \qquad v_{R3} = \frac{3608}{22} = 164 \text{ plf}$$

$$R_{L4} = \frac{(112)(118)}{2} = 6608 \text{ lb}, \qquad v_{L4} = \frac{6608}{28} = 236 \text{ plf}$$

Calculate the net shear at the shear walls:

$$v_{SW} = \frac{[4368 + 24(4) + 5368 + 3608 + 6608]}{2(22)} = 455.64 \text{ plf}$$

$$v_{\text{net SW1}} = 455.64 - 156 - 244 = 55.64 \text{ plf}$$

$$v_{\text{net SW2}} = 455.64 - 164 - 236 = 55.64 \text{ plf}$$

Plot the force diagrams at grid line 2 (see Fig. 12.18, left side): Starting at grid line 2A.

Force at end of SW1 = +55.64(22) = 1224 lb tension, acting upward

Force at grid line C = +1224 − (156 + 8)(6) = 240 lb tension, acting upward

Force at grid line C, SW2, working upward from grid line D = +55.64(22) = +1224 lb

Force at grid line B = +1224 − (236 + 8)(6) = −240 lb compression

The diagram shows that a residual force of 240 lb is left at the ends of the corridor collectors, acting in opposite directions, forming a couple. This couple can be resisted by the corridor walls. The resisting forces are

$$F = \frac{(240)(4)}{6} = 160 \text{ lb}$$

All forces have been resolved.

The layout on the right side of Fig. 12.18 has the same diaphragm and layout as the previous example, but the offset wall on the right is only partial length and does not extend up to the corridor. The forces at the ends of the walls are greater than the previous example and that the residual force at the ends of the corridor collector is significantly higher. Likewise, the resisting forces distributed into the corridor walls are also higher. The reason for this is that the collectors are tied together and therefore create in effect a longer collector length. The longer the collector length, the higher the force.

The layout shown in Fig. 12.19 also has the same diaphragm and layout as the previous examples, but in this case, both walls are partial length and do not extend to the

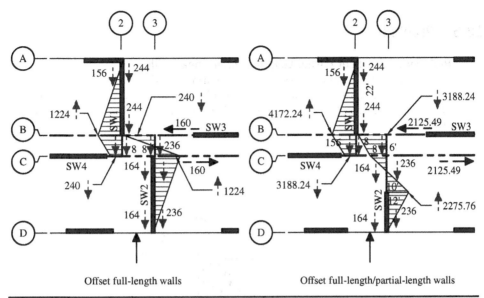

FIGURE 12.18   Offset full-length plus partial length wall.

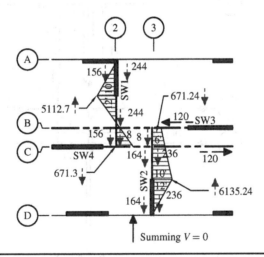

FIGURE 12.19   Both partial-length walls offset.

corridor wall lines. In this case, the forces at the ends of each wall are much greater than the previous layout, but the residual forces at the ends of the collector and the resisting forces have dropped to moderate levels. These comparisons are interesting and provide a powerful statement that the designer should be alert that these conditions should be reviewed as part of the design.

There have been occasions where collectors at transverse shear walls were not installed. The rationale behind this approach is the assumption that by using every transverse wall, the diaphragm forces are reduced to a level where collectors and chords can be ignored. Although this may seem reasonable to some designers, it is not consistent with code, which requires complete load paths. ▲

12.5   Problem

Problem 12.1: Offset Partial Length Interior Shear Walls

Given: Figure P12.1 shows a portion of a diaphragm that is 108 ft long by 58 ft wide, where partial length offset interior shear walls are placed on each side of a 6-ft corridor at grid lines 2 and 3 with a 6-ft offset. This layout is representative of multi-family structures.

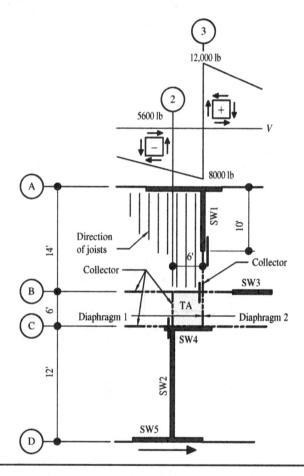

FIGURE P12.1

Find: Calculate the force diagrams at grid lines 2 and 3.   ▲

CHAPTER 13

The Portal Frame

13.1 Introduction

The common configuration of vehicle garages, attached or detached, is to have one or more large door openings with minimal wall sections on either side of the openings. When there is not sufficient wall space for shear walls next to the large openings, this configuration is often addressed with portal frames. The portal frame has undergone substantial change over the past few decades as a result of testing and aesthetic demands. Many home-designers and homeowners felt that pier sections greater than 2 ft 0 in were too large of a wall section to be aesthetically pleasing or fit in the narrow width of the building and parcel allotted to the garage. Marketplace demand has led to the development of additional alternate configurations and methods with narrower panels.

Prior to APA's testing of portal frames, garage door openings with wall segments smaller than the prescriptive 4-ft 0-in braced panel or 2-ft 8-in Alternate Braced Panel (ABP), narrow pier systems have exhibited a lack of capacity for resisting lateral forces. Even single-story structures with garage openings have performed poorly in high wind or seismic events as shown in Fig. 13.1. Whenever a second story is located over a garage opening, as in tuck-under parking complexes, a soft or weak story condition can exist. This type of structure was discussed in ATC-7[1] as having a stiffness irregularity (soft story) and discontinuous load path issues. Structures like these were badly damaged in the San Fernando (1971) and Northridge (1994) earthquakes. Another deficiency was the lack of adequate hold-downs at the piers and/or an inadequate foundation that allowed excessive displacement of the two-story section, thereby causing the collapse at the garage opening. IBC[2] Sections 1604.8 and 1616.3 regarding general structural integrity, and ASCE7-16[3] Section 12.1.3 regarding interconnection ties help resolve some of the issues of building sections not working as a unit.

As aesthetic motivated designs demand floor plans with a greater number of large openings, the market pressure is to reduce the minimum shear wall width to accommodate these large openings. Prescriptive portal framed walls were developed to provide a wall system that allows a minimum wall pier width of 16 in. This creates aspect ratios up to 7.5:1, which exceeds those allowed by code for shear walls. Portal frames are not narrow shear walls because, unlike shear walls, they resist lateral forces by means of a semi-rigid moment-resisting frame and by flexure in horizontal components.

Portal walls are covered under IBC Section 2308.6.5.2. IBC Figure 2308.6.5.2 shown in Fig. 13.2 and APA's Portal-Frame with hold-downs[5] shown in Fig. 13.3 summarize the prescriptive framing requirements for one- and two-story portal frames for both the

471

Figure 13.1   Photograph of portal frame damage.

IRC and the conventional construction Section 2308 of the IBC. Prescriptive options for wall bracing are provided in the code, both IBC and the IRC.

- 2021 IRC Table R602.10.3(3)[4] allows portal frames with hold-downs (PFH), portal frames without hold-downs (PFG) or continuous sheathed portal frames (CS-PF).

  1. Footnote (e) of that table notes that PFG does not apply to SDC D_o or greater.
  2. Footnote (f) notes that PFH and PFG are only allowed in 1-story or the top story of a two-story structure.
  3. Also note that PFG portal frames are limited to pier widths between 24″ and 30″ depending on height, as noted in Table R602.10.6.5.
  4. PFH portal frames have allowable pier widths between 16″ and 24″, and height and SDC restrictions, as noted in Table R602.10.6.5.

- 2021 IBC Table 2308.6.3(1) only allows PFH as an allowable method for portal frames with a maximum wall height equal to 10′-0″, as noted in Figure 2308.6.5.2.

13.2   Testing and Research

The most recent tests conducted by the APA—The Engineered Wood Association's Technical Topic TT-100H[5] are also based on portal framing configurations using hold-downs, PFH. The tests were conducted to determine the strength and stiffness of the walls and to provide recommendations for allowable design loads for engineered solutions. A table is included in the Technical Topic that provided recommended allowable lateral forces that can be applied to each pier section. The paper also noted that the

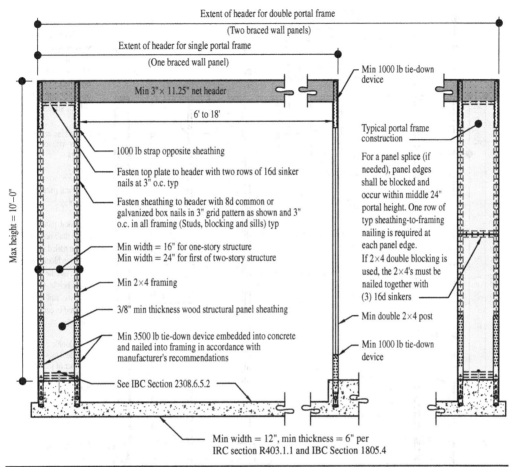

Extent of header for double portal frame
(Two braced wall panels)
Extent of header for single portal frame
(One braced wall panel)

Min 1000 lb tie-down device

Min 3" × 11.25" net header

6' to 18'

Max height = 10'–0"

1000 lb strap opposite sheathing

Fasten top plate to header with two rows of 16d sinker nails at 3" o.c. typ

Fasten sheathing to header with 8d common or galvanized box nails in 3" grid pattern as shown and 3" o.c. in all framing (Studs, blocking and sills) typ

Min width = 16" for one-story structure
Min width = 24" for first of two-story structure

Min 2×4 framing

3/8" min thickness wood structural panel sheathing

Min 3500 lb tie-down device embedded into concrete and nailed into framing in accordance with manufacturer's recommendations

See IBC Section 2308.6.5.2

Typical portal frame construction

For a panel splice (if needed), panel edges shall be blocked and occur within middle 24" portal height. One row of typ sheathing-to-framing nailing is required at each panel edge.
If 2×4 double blocking is used, the 2×4's must be nailed together with (3) 16d sinkers

Min double 2×4 post

Min 1000 lb tie-down device

Min width = 12", min thickness = 6" per IRC section R403.1.1 and IBC Section 1805.4

Figure 13.2 IBC Figure 2308.6.5.2 portal frame with hold-downs (PFH). (*Courtesy, ICC, International Code Council, Washington DC.*)

frame must be attached to a rigid foundation. The table shown in Fig. 13.4 is a summary of APA Technical Topic TT-100H, Table 1. Footnote (c) of that paper notes that the allowable load listed in the table is for a single pier. For multiple piers, the allowable load can be multiplied times the number of piers.

Monotonic and cyclic testing of portal frames has been conducted since mid-1980. The design values listed in Table 1 of APA TT-100H were derived from cyclic test data using a rational procedure that considered both stiffness and strength. The design value derivation procedure assured that the IBC code drift limit and an adequate factor of safety are maintained. In November of 2007, APA conducted monotonic racking tests on 23 full-scale three-dimensional wood-framed houses. The results were published in APA Report T2007-73.[6] The focus of the testing was on IRC 2006 structural panel type wall bracing. A number of variables were also studied, including:

- Hold-down devices in building corners
- Return corners, both 2 ft and 4 ft

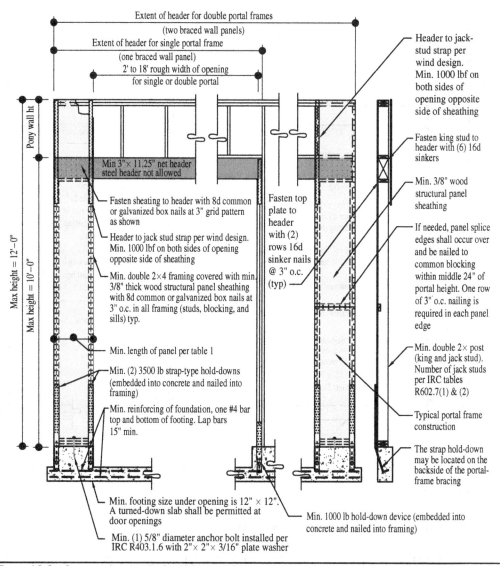

Extent of header for double portal frames
(two braced wall panels)
Extent of header for single portal frame
(one braced wall panel)
2' to 18' rough width of opening
for single or double portal

Pony wall ht

Min 3" × 11.25" net header
steel header not allowed

Fasten sheating to header with 8d common
or galvanized box nails at 3" grid pattern
as shown

Header to jack stud strap per wind design.
Min. 1000 lbf on both sides of opening
opposite side of sheathing

Min. double 2×4 framing covered with min.
3/8" thick wood structural panel sheathing
with 8d common or galvanized box nails at
3" o.c. in all framing (studs, blocking, and
sills) typ.

Min. length of panel per table 1

Min. (2) 3500 lb strap-type hold-downs
(embedded into concrete and nailed into
framing)

Min. reinforcing of foundation, one #4 bar
top and bottom of footing. Lap bars
15" min.

Max height = 12'–0"
Max height = 10'–0"

Fasten top
plate to
header
with (2)
rows 16d
sinker nails
@ 3" o.c.
(typ)

Header to jack-
stud strap per
wind design.
Min. 1000 lbf on
both sides of
opening opposite
side of sheathing

Fasten king stud to
header with (6) 16d
sinkers

Min. 3/8" wood
structural panel
sheathing

If needed, panel splice
edges shall occur over
and be nailed to
common blocking
within middle 24" of
portal height. One row
of 3" o.c. nailing is
required in each panel
edge

Min. double 2× post
(king and jack stud).
Number of jack studs
per IRC tables
R602.7(1) & (2)

Typical portal frame
construction

The strap hold-down
may be located on the
backside of the portal-
frame bracing

Min. footing size under opening is 12" × 12".
A turned-down slab shall be permitted at
door openings

Min. 1000 lb hold-down device (embedded into
concrete and nailed into framing)

Min. (1) 5/8" diameter anchor bolt installed per
IRC R403.1.6 with 2"× 2"× 3/16" plate washer

FIGURE 13.3 Construction details for APA portal frame with hold-downs. (APA—The Engineered Wood Association, Tacoma, WA.)

- Panel placement at corners and away from corners
- Single 1 × 4 let-in bracing
- Isolated wood structural bracing
- Continuous wood structural panel bracing
- Continuous wood structural panel bracing cut around openings
- Walls braced only by 6:1 aspect ratio portal frames
- Wall lines braced with 50 percent 6:1 aspect ratio portal frames mixed with other continuous wood structural panel bracing

RECOMMENDED ALLOWABLE DESIGN VALUES FOR APA PORTAL FRAME USED ON A RIGID-BASE FOUNDATION FOR WIND OR SEISMIC LOADING[a,b,c,d]

Minimum Portal Width (in.)	Maximum Portal Height (ft)	Allowable Design (ASD) Values per Frame Segment		Load Factor
		Shear[e,f] (lbf)	Deflection (in)	
16	8	850	0.33	3.09
	10	625	0.44	2.97
24	8	1675	0.38	2.88
	10	1125	0.51	3.42

a. Design values are based on the use of Douglas-Fir or Southern pine framing. For other species of framing, multiply the above shear design value by the specific gravity adjustment factor $= (1 - (0.5 - SG))$, where SG = specific gravity of the actual framing. This adjustment shall not be greater than 1.0.

b. For construction as shown in Figure 1.

c. Values are for a single portal-frame segment (one vertical leg and a portion of the header). For multiple portal-frame segments, the allowable shear design values are permitted to be multiplied by the number of frame segments (e.g., two $= 2\times$, three $= 3\times$, etc.).

d. Interpolation of design values for heights between 8 and 10 feet, and for portal widths between 16 and 24 inches, is permitted.

e. The allowable shear design value is permitted to be multiplied by a factor of 1.4 for wind design.

f. If story drift is not a design consideration, the tabulated design shear values are permitted to be multiplied by a factor of 1.15. This factor is permitted to be used cumulatively with the wind-design adjustment factor in Footnote (e) above.

Table 1: Construction Details for APA Portal-Frame with Hold-Downs

FIGURE 13.4 APA Technical Bulletin TT-100H Table 1. (*APA—The Engineered Wood Association, Tacoma, WA.*)

Refer to the report for the complete results of the tests. The bases of the portal frame pier sections did not have the embedded tie strap anchors shown in IBC Figure 2308.6.5.2 or IRC Figure R602.10.6.2. Several pier bases were connected to the test frame by only a single bolt. At locations where bolts did not occur within the pier, Simpson LPT4 clips were installed at each end of the piers. The results of the tests showed a significant increase in strength of the portal frames over other IRC wall bracing systems. The increase in strength of equal amounts of portal frame sections to 4-ft-wide wall bracing panels was attributed to the following:

- The semi-rigid moment resisting connections at the top of each narrow wall segment
- 4.5 times more studs
- 1.2 times more sheathing
- 10 times more nailing of the sheathing to the framing
- 2 times more nailing of the sheathing to the top and bottom plates
- 3 or more times greater bottom-of-wall attachment requirements

Another paper worthy of noting is "Principles of Mechanics Model for Wood Structural Panel Portal Frames."[7] This paper provides a "principles of mechanics" model to predict the in-plane racking strength of a wood portal frame, as shown in Fig. 13.5. The model was compared to 10 tested portal frames and showed that the model accurately predicted the strength to within about 5 percent on average. The frames tested resembled the portal frame shown in IBC Figure 2308.6.5.2 with pier widths of 16 in and 24 in, and semi-rigid nailed only header to pier and pier base to sill joints. The model does not consider racking deflection of gravity loads applied to the header.

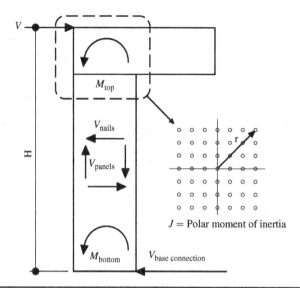

FIGURE 13.5 APA portal frame analysis method.

The maximum lateral racking shear strength, V, of the pier sections is equal to the minimum of

$$V_{\text{moment couples}} = \frac{M_{\text{top}} + M_{\text{bottom}}}{H_{\text{wall}}}, \text{ or}$$

$$V_{\text{shear strength}} = v_{\text{panel}}, v_{\text{nails}}, v_{\text{base connection}}$$

where M_{top} = Minimum of the sheathing to header fastener moment capacity plus the moment capacity of the steel straps, or the sheathing bending strength plus the moment capacity of the steel straps.

M_{bottom} = The tie-down strap capacity times the pier width plus the sheathing to sill plate fastening moment capacity.

V_{panel} = Shear through the thickness strength.

v_{nails} = Wood structural panel to framing shear capacity.

$v_{\text{base connection}}$ = Shear capacity due to base of wall connections to supporting structure.

The semi-rigid joints between the header and pier section are designed using the polar moment of inertia of the nailing pattern at the joint.

13.3 Foundation Issues

The IBC and APA require the portal wall panels to be supported on a foundation that is continuous across the entire length of the braced wall line, similar to that shown in Fig. 13.6. The code section notes:

> The foundation shall be reinforced with not less than (1) no. 4 bar top and bottom. Where the continuous foundation is required to have a depth greater than 12″, a minimum 12″ by 12″

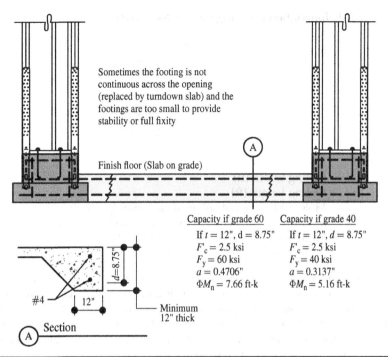

Sometimes the footing is not continuous across the opening (replaced by turndown slab) and the footings are too small to provide stability or full fixity

Ⓐ

Finish floor (Slab on grade)

Capacity if grade 60

If $t = 12"$, $d = 8.75"$
$F'_c = 2.5$ ksi
$F_y = 60$ ksi
$a = 0.4706"$
$\Phi M_n = 7.66$ ft-k

Capacity if grade 40

If $t = 12"$, $d = 8.75"$
$F'_c = 2.5$ ksi
$F_y = 40$ ksi
$a = 0.3137"$
$\Phi M_n = 5.16$ ft-k

#4   12"
$d = 8.75"$

Minimum 12" thick

Ⓐ Section

FIGURE 13.6 Foundation rigidity review.

continuous footing or turn down slab edge is permitted at door openings in the braced wall line. This continuous footing or turn down slab edge shall be reinforced with not less than (1) no. 4 bar top and bottom. This reinforcement shall be lapped not less than 15″ with the reinforcement required in the continuous foundation located directly under the braced wall line.

Tests on portal frames have been conducted with and without tie strap anchors at the base of the piers. The latest tests and current code approved portal frame configurations include base anchors, which provide fixity at the base of the portal frame. APA TT-100H states "Since design values are based on testing conducted with the portal frame attached to a rigid test frame using embedded strap-type hold-downs, design values should be limited to portal frames constructed on similar rigid based foundations such as a concrete foundation, stem wall, or turndown slab, which uses a similar embedded strap-type hold-down." It is important to create a reasonably rigid foundation so that the design of the frame matches the tests. To provide full fixity at the base of the portal pier, it is customarily assumed that the foundation should be 10× stiffer than the pier section.

The approximate rigidity of a foundation can be determined as follows[8]:

Rigid foundation (bending is not influenced as much by K_s):

$$\lambda L < \frac{\pi}{4}$$

Flexible foundation (bending heavily localized):

$$\lambda L > \pi$$

$$\lambda = \sqrt[4]{\frac{K'_S}{4EI}}$$

$$\lambda L = \sqrt[4]{\frac{L^4 K'_S}{4EI}}$$

where $K'_S = K_S B$
B = footing width (ft)
L = footing length (ft)
K_s = MSR = Modulus of subgrade reaction (pcf, kcf)
E = Modulus of elasticity of concrete section (psi, ksi)
I = Moment of inertia of concrete section (in^4, ft^4)
λ = A dimensionless parameter that quantifies the rigidity of a foundation

Check the strength of the turndown footing to resist the applied moment from the pier (see Fig. 13.6): Assume a turndown slab or footing extension occurs across the opening. Check this condition for strength and stiffness. The resulting factored moment capacity of the hold-down anchors at the base of the piers, assuming full fixity, is 6.34 ft-k (strength) for the 16-in pier and 10.42 ft-k (strength) for the 24-in pier. Check footing capacity for both 40 grade. And 60 gr. reinforcement.

$$\text{Width} = 12 \text{ in}, \qquad \text{thickness} = 12 \text{ in}$$

$$F'_C = 2.5 \text{ ksi} \qquad F_y = 40 \text{ ksi}$$

Use (1) #4, at both top and bottom

$$d = 12 - 3 - \frac{0.5}{2} = 8.756''$$

$$a = \frac{A_s F_y}{0.85 b F'_C} = \frac{0.2(40)}{0.85(12)(2.5)} = 0.3137''$$

$$\phi M_n = \phi A_s F_y \left(d - \frac{a}{2} \right) = 0.9(0.2)(40)\left(8.75 - \frac{0.3137}{2} \right)\frac{1}{12} = 5.16 \text{ ft-k}$$

If $F'_C = 2.5 \text{ ksi}, \qquad F_y = 60 \text{ ksi}$

$$\phi M_n = 7.66 \text{ ft-k}$$

Note: The 24-in pier foundation capacity would be questionable in resisting the applied moment. Therefore, a standard stem wall footing, or deeper turndown might be a better choice. The minimum reinforcement and the shear capacity for the section must also be checked in accordance with ACI 318-19 Chapters 9 and 13.

Check the relative stiffness of the walls versus foundation (see Fig. 13.7): In addition to the previous check for the foundation strength, the stiffness of the foundation needs to be checked as rigid relative to the wall pier. As only an approximation is needed, using

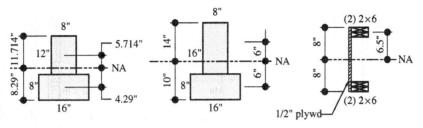

FIGURE 13.7   Foundation sections.

the gross, non-cracked, moment of inertia for the foundation should be acceptable. However, this should be based on engineering judgment.

Wood wall (see Fig. 13.7):

$$C_1 = C_2 = 8 \text{ in}$$

$$I_G = 2(2)(1.5)(5.5)(6.5)^2 + \frac{0.5(16)^3}{12} = 1564.92 \text{ in}^4$$

$$E_W = 1.7 \times 10^6 \text{ psi}$$
$$E_W I_G = 2.66 \times 10^9 \text{ lb in}^2$$

Foundation (see Fig. 13.7):

$$\text{12-in thickness, } I_G = \frac{12(12)^3}{12} = 1728 \text{ in}^4$$

$$E_C = 57,000\sqrt{F'_C} = 57,000\sqrt{2500} = 2.85 \times 10^6 \text{ psi}$$

$$E_C I_G = 4.92 \times 10^9 \text{ lb in}^2$$

$$\frac{E_C I_G}{E_W I_G} = \frac{4.92 \times 10^9}{2.66 \times 10^9} = 1.85 < 10$$

Therefore, N.G.

By inspection, the turndown slabs do not provide enough stiffness to fix the base of the portal pier sections.

Try an 8-in-deep by 16-in-wide footing with an 8-in-thick by 12-in-deep stem wall (see Fig. 13.7):

$$\bar{y} = \frac{8(12)(6) + 8(16)(16)}{8(12) + 8(16)} = 11.714 \text{ in}, \ C_2 = 8.29 \text{ in}$$

$$I_G = 8(12)(5.714)^2 + 8(16)(4.29)^2 + \frac{8(12)^3}{12} + \frac{16(8)^3}{12} = 7320 \text{ in}^4$$

$$E_C I_G = 2.086 \times 10^{10} \text{ lb in}^2$$

$$\frac{E_C I_G}{E_W I_G} = \frac{2.086 \times 10^{10}}{2.66 \times 10^9} = 7.84 < 10$$

Therefore, N.G.

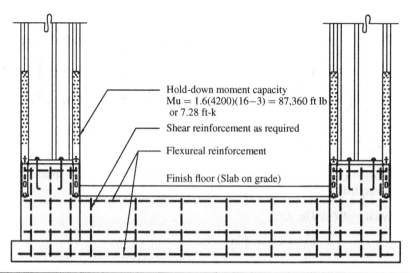

FIGURE 13.8 Required foundation details.

Try an 8-in-deep by 16-in-wide footing with an 8-in-thick by 16-in-deep stem wall (see Fig. 13.7):

$$\bar{y} = \frac{8(16)(8) + 8(16)(20)}{8(16) + 8(16)} = 14'', \ C_2 = 10''$$

$$I_G = 8(16)(6)^2 + 8(16)(6)^2 + \frac{8(16)^3}{12} + \frac{16(8)^3}{12} = 12629 \ \text{in}^4$$

$$E_c I_G = 3.6 \times 10^{10} \ \text{lb in}^2$$

$$\frac{E_c I_G}{E_W I_G} = \frac{3.6 \times 10^{10}}{2.66 \times 10^9} = 13.52 > 10$$

Therefore, OK.

The foundation should be similar to the 16-in-deep foundation walls shown in Figs. 13.7 and 13.8 to provide the necessary stiffness to create the effect of full fixity for the piers.

13.4 Alternate Frame Systems

Alternate frame systems have become available on the commercial market where greater lateral forces, smaller wall pier widths, or more elaborate wall layouts are required. Among them are the Simpson Strong-Tie Strong-Walls and Strong-Frames, and MiTek Shear Wall Systems and Hardy Moment Frames, similar to the configurations illustrated in Fig. 13.9. These lateral resisting systems can be used for prescriptive or engineered solutions for wall bracing applications in single- or two-story structures. Simpson Strong-Tie provides three types of "Strong-Walls," LSL solid wall panels, cold formed steel panels (CFS), and steel sheet panel walls with wood boundary elements. These lateral resisting systems have been tested and can be used for an engineered design. The

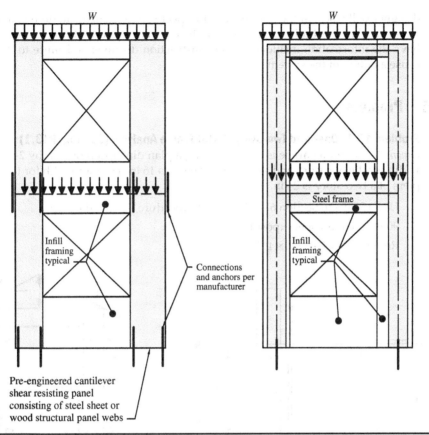

Infill
framing
typical

Connections
and anchors per
manufacturer

Steel frame

Infill
framing
typical

Pre-engineered cantilever
shear resisting panel
consisting of steel sheet or
wood structural panel webs

Figure 13.9 Pre-engineered or tested systems.

Strong-Frames are available as ordinary, intermediate, or special steel moment resisting frames, which are designed in accordance with the latest edition of AISC. General guidelines for the foundation are provided by the manufacturer, which are based only on the foundation size required to develop the full capacity of the anchors. All wall and frame systems are required to be installed on a rigid concrete foundation.

MiTek Industries, Inc. provides two types of Hardy Shear-Walls, which consist of steel sheet panels with cold formed steel boundary elements and diagonal braced strap CFS wall panels. Two moment frame options are available, one is the Hardy CFS Moment Frame and the other is a steel Hardy Special Moment Resisting Frame.

The lateral resisting systems of both manufacturers have been tested and can be used for an engineered design. The product manufacturer will have the required testing performed by an approved testing agency who should publish a product evaluation report that can be submitted to the building official for review and approval. All wall systems are required to be installed on a rigid concrete foundation.

As with any system or component discussed in this book, the engineer is responsible for establishing a complete load path and for assuring compliance with all the manufacturer's requirements and all applicable codes and standards, and to provide the necessary information on the construction documents. Typically, the engineer or designer will research the manufacturers' specific requirements to do the analysis but, to ensure that it is built per the design, the contractor and building inspector need to have the information included in the construction documents. Failure to do so could cause a failure of the system.

13.5 Problem

Problem 13.1: One- and Two-Story Portal Frame Analysis (See Fig. P13.1)

Given: A plan for a garage is shown having a plan dimension of 24 ft by 24 ft. The pier width for the one story and second-floor piers is 16 in. The pier width for the first-floor piers of a two-story is 24 in.

Use ASCE7-16 Section 6.4, Simplified Procedure, for wind design.

90 mph wind zone, exposure C

Roof pitch = 4:12 = 18.43°

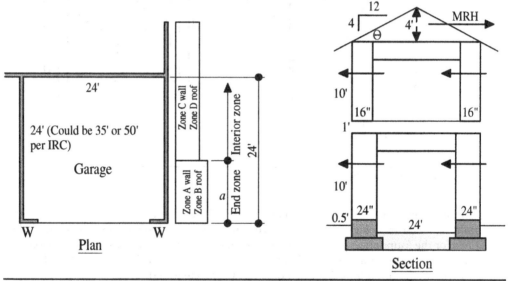

FIGURE P13.1

Find: Calculate the forces to the portal frame and determine if the forces meet or exceed the allowable forces per pier section in accordance with APA TT 100-G.

Check the following configurations for wind loading only.

- One story with an 8-ft wall height and a 4:12 roof pitch.
- One story with a 10-ft wall height and a 4:12 roof pitch.
- A two-story structure with a 10-ft wall height at the first floor and an 8-ft wall height at the second floor and a 4:12 roof pitch.
- A two-story structure with a 10-ft wall height at the first floor and second floor and a 4:12 roof pitch. ▲

13.6 References

1. Applied Technology Council (ATC), *Guidelines for Design of Horizontal Wood Diaphragms, ATC-7*. ATC, Redwood, CA, 1981.
2. International Code Council (ICC), *International Building Code, 2021 with Commentary*, ICC, Whittier, CA, 2021.
3. American Society of Civil Engineers (ASCE), *ASCE/SEI 7-16 Minimum Design Loads for Buildings and Other Structures*, ASCE, New York, 2016.
4. International Council of Building Officials (ICBO), *International Residential Code, 2021 with Commentary*, ICBO, Whittier, CA, 2021.
5. APA—The Engineered Wood Association, *A Portal Frame with Hold Downs for Engineered Applications, APA Form TT-100H*, APA—The Engineered Wood Association, Engineering Wood Systems, Tacoma, WA, 2007.
6. APA—The Engineered Wood Association, *APA Report T2007-73, Full-scale 3D Wall Bracing Tests*, APA Report No. T2007-73, APA—The Engineered Wood Association, Engineering Wood Systems, Tacoma, WA, 2007.
7. Martin, Z., Skaggs, T., Keith, E., and APA, "Principles of Mechanics Model for Wood Structural Panel Portal Frames," Structures Congress Proceedings, ASCE, 2008.
8. Bowles, J. E., *Foundation Analysis and Design*, 3rd ed., McGraw-Hill, New York, NY, 1982.

CHAPTER 14

CLT Diaphragms

14.1 Introduction

Cross-laminated timber (CLT) is a relatively new engineered wood product which consists of multiple wood laminations manufactured into large, flat panel products. CLT commonly varies from 3 to 12 in thick, with up to 20 in maximum possible. Panels can be fabricated from 4- to 12-ft wide and up to 64-ft long and can be made from three or more layers of solid sawn or structural composite lumber, for example laminated veneer lumber (LVL), which are glued and pressed together in alternating orthogonal directions into the final panel size, see Fig. 14.1. CLT products were first developed in Europe the 1990s and have been used in North America since the early 2000s. A CLT product standard for use in North America, ANSI/APA PRG 320 Standard for Performance-Rated Cross-Laminated Timber (PRG 320),[1] was first issued in 2011. Building code recognition of CLT as a structural component first appeared in the 2015 edition of the International Building Code.[2] The structural design of CLT is covered by AWC's National Design Standard for Wood Construction (NDS) starting with the 2015 edition.[3] Mass timber construction using CLT and other large timber products can be used in any IBC construction type allowing combustible construction, including construction types III and V. Starting with the 2015 IBC, CLT, with minimum thickness, has also been recognized as an allowable heavy timber material, and useable in construction type IV-HT and anywhere heavy timber components are permitted.

With the 2021 IBC, three new construction types, IV-A, IV-B, and IV-C, allow mass timber construction, including CLT, to be used in buildings much larger and taller than the historic combustible construction types.[4] With this update to the model building code, residential and business occupancies can be built of mass timber up to 18 stories tall. As jurisdictions adopt these newer editions of the IBC and the industry gains more experience with these new construction types, there may be large increase in the use of CLT; especially in urban development projects where steel and concrete have historically been the primary material options. Owners, designers, and builders are exploring the use of mass timber and CLT for multiple reasons including the speed of construction, aesthetics, and low carbon footprint. Timber is a sustainable and renewable resource grown while sequestering carbon for the life of the timber products. Policy makers, owners, and designers are looking to increase the use of timber products in the next generation of energy efficient and responsibly constructed buildings.

Being assembled of wood fiber, an anisotropic material, and assembled in layers running in alternating directions, CLT is a composite non-homogeneous component. Structurally, it can be modeled as a composite system built on the material properties of

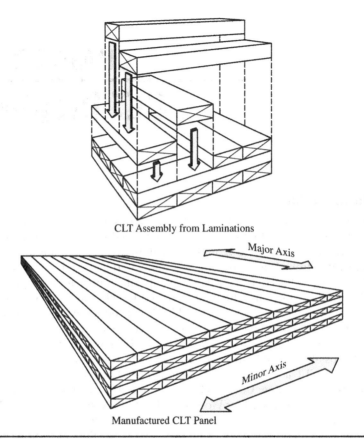

CLT Assembly from Laminations

Manufactured CLT Panel

Figure 14.1 CLT panel construction.

the different layers. For structural engineering of building systems, a simpler ortho-tropic plate and shell view of the component is more common. The structural properties of the composite plate are commonly provided as "flatwise" properties, which pertain to the out-of-plane behaviors, such as a panel supporting gravity loads spanning between bearing walls or beams, and the "edgewise" properties, which pertain to the in-plane behaviors, such as shear in a diaphragm or the bending of a wall section spanning over a window opening as a header. When manufactured of solid sawn lumber laminations, the layup of the panels is usually symmetric from top to bottom with an odd number of alternating layers. This layup, and the use of higher grades of lumber in the laminations running in the direction of the outer laminations, creates a major (stronger) strength direction and a minor strength direction in the panel. Similar to plywood and OSB, the major strength direction is typically in the long direction of the panels. Following the notation of the ANSI/APA PRG 320 standard, the primary flatwise structural properties of CLT are shown in Table 14.1.

The type of laminations used to make the CLT panel define the Grade of CLT. PRG 320 defines a handful of example grades which manufacturer may choose to make. Similarly, the PRG 320 defines several example layups, of which the most common is the 3-ply, 4 1/8-in-thick and 5-ply 6 7/8-in-thick layups. These layups are made from

Property	Notation		U.S. Units
	Major Direction	**Minor Direction**	
Flexural Strength	$F_b S_{eff,0}$	$F_b S_{eff,90}$	lbf-ft/ft
Flexural Stiffness	$EI_{eff,f,0}$	$EI_{eff,f,90}$	lbf-in^2/ft
Shear Strength	$V_{s,0}$	$V_{s,90}$	lbf/ft
Shear Stiffness	$GA_{eff,f,0}$	$GA_{eff,f,90}$	lbf/ft

Table 14.1 Primary Flatwise Structural Properties of CLT Panel

laminations which are 1 3/8 in thick resulting from planning the surface of a nominal "2 by" board, which is approximately 1.5 in thick, before gluing up into the CLT panel.

The example grades and layups in the PRG 320 standard are labeled "Basic" grades and layups. Manufacturers may choose to make other grades and layups of CLT. Regardless of what the grades and layups of panels are, manufacturing according to the PRG 320 product standard requires third-party verification of published design values via testing and ongoing quality control of the manufacturing process.

The in-plane shear capacity of CLT panels of any grade and layup is not found directly in PRG 320. The PRG 320 standard defines a protocol by which the in-plane shear capacity is determined from specific product testing. The protocol uses a thin strip of CLT tested in a beam-like configuration to determine two values $F_{v,e,0}$ and $F_{v,e,90}$ based on the orientation of the outer laminations in the beam test. For large panels in more of a pure shear configuration, such as in CLT diaphragms and shear walls, the lower of the two values from testing can be used. Alternatively, a method to calculate the in-plane shear capacity of CLT is found in the Commentary of NDS 2018.

In many of building projects in the United States using CLT, CLT is used as floor or roof decking supported on a framework of column supported beams or bearing walls. CLT panels are strong and stiff in-plane, and it is a natural fit for CLT to be the primary structural component of a horizontal diaphragm (see Fig. 14.2). Other options for

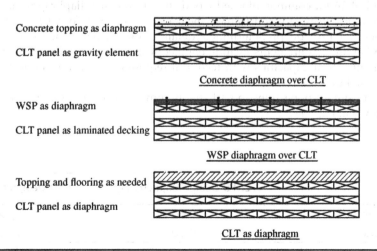

Figure 14.2 Diaphragm options with CLT.

horizontal diaphragms with CLT decking are to use a structural topping layer over the CLT as the diaphragm. Such configurations can use wood structural panels (WSP) or concrete as the topping layers. The option to provide a concrete diaphragm as topping over the CLT panels is possible; however, it is not an easy or efficient construction method. The minimum concrete depth needed to provide sufficient cover of the structural reinforcement can add significant weight and depth to the assembly. Furthermore, the detailing to transfer forces into and out of the concrete diaphragm can be challenging to design and construct. This is especially true with the structural framing beneath the CLT deck serving as diaphragm chords and collectors. When a WSP diaphragm is provided over the CLT, the SDPWS 2015[5] Section 4.2.7.1 is used for code recognition which allows sheathing over laminated decking.

However, direct use of the CLT as the structural diaphragm is a more direct and efficient detailing strategy. SDPWS 2021, referenced from the 2021 IBC, now includes design requirements for CLT diaphragm in Section 4.5.[6]

14.2 SDPWS 2021 Requirements

New with SDPWS 2021, Section 4.1.4 defines a single nominal shear capacity, v_n, to be used for wind and seismic design. From this nominal shear capacity, the ASD and LRFD wind and seismic design capacities can be determined from Table 14.2.

This design capacity is required to be greater than or equal to the induced unit shear wind or seismic load which can be written as follows:

For ASD:
$$\frac{v_n}{\Omega_D} \geq v_{d,ASD}$$

For LRFD:
$$\phi_D v_n \geq v_{d,LRFD}$$

For WSP shear walls and diaphragms, the typical design process uses the nominal shear capacity tables published in the SDPWS for predetermined details such as SDPWS Table 4.2A for common blocked wood structural panel diaphragms and Table 4.3A for common sheathed wood-framed shear walls. However, for CLT diaphragms, SDPWS does not provide predetermined details with tabulated nominal shear capacities. Rather, the requirements include engineering methods to determine the nominal diaphragm shear capacities with technical requirements on the connection details to be verified by engineering calculations.

Example connections in CLT diaphragms transferring diaphragm shear forces are shown in Fig. 14.3.

	LRFD Design Capacity $\phi_D v_n$	ASD Design Capacity $\dfrac{v_n}{\Omega_D}$
Seismic	0.50 v_n	$v_n/2.8$
Wind	0.80 v_n	$v_n/2.0$

TABLE 14.2 Design Shear Capacity from Nominal Capacity in 2021 SDPWS

Top-side spline connection

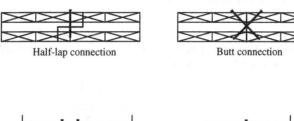

Half-lap connection   Butt connection

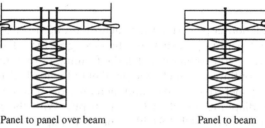

Panel to panel over beam   Panel to beam

FIGURE 14.3 CLT diaphragm shear connections.

For connections in CLT diaphragms transferring diaphragm shear forces between CLT panels to the boundary elements, as shown in Fig. 14.3, SDPWS Section 4.5.4 item 1 requires the forces to be transferred using dowel type fasteners in shear. In the NDS, dowel type fasteners include nails, wood screws, lag screws, and bolts. In practice, nails and proprietary self-tapping screws are used in the CLT diaphragm connections. While some of these types of fasteners can be used to provide tension and compression capacities, such arrangements are not permitted to be used to transfer CLT diaphragm shear forces. The capacity of the dowel type fasteners in shear, Z, is to be calculated using the yield mode equations of NDS Section 12.3.1, and only connections where Mode IIIs or Mode IV controls the capacity are permitted. These yield modes are known to provide consistently ductile failure modes, which are advantageous over more brittle behaviors when overloaded beyond the design load. These modes also have well understood minimum overstrength capacities which are useful in achieving a minimum ultimate strength.

Provided the reference design capacity, Z, of a fastener selected for a CLT diaphragm connection is controlled by Mode IIIs or Mode IV, the nominal capacity of the connection is found by calculating an adjusted design capacity Z^*, which is similar to an adjusted design capacity Z', except that the ASD and LRFD specific adjustment factors, C_D, K_F, ϕ, and λ, are not applied.

$$Z^* = Z \times C_M\, C_t\, C_g\, C_\Delta\, C_{eg}\, C_{di}\, C_{tn}$$

For application of CLT diaphragms within the envelope of a building, C_M and C_t are 1.0. For typical CLT diaphragm connection with fasteners installed perpendicular to the face of the CLT, C_{eg} and C_{tn} are 1.0. Using fasteners with diameters larger than 1/4 in triggers evaluation of C_Δ, C_{eg}, and consideration of wood grain directions in calculation of the bearing capacity of the timber. For dowel type fasteners with a diameter less than 1/4 in, C_Δ and C_{eg} are 1.0. For CLT diaphragms, $C_{di} > 1.0$ does not apply, even if using nailed connections.

For CLT diaphragm connections, SDPWS Section 4.5.4 item 1 defines the nominal shear capacity per fastener as follows:

$$V_n = 4.5Z^*$$

For fasteners at regular spacing, S, in inches, the nominal diaphragm shear capacity of a *connection* is calculated as follows:

$$v_n = 4.5Z^* \ \frac{12 \ \frac{in}{ft}}{S}$$

This calculation is performed for each diaphragm shear connection detail in the system. In WSP diaphragms, such as when using SDPWS Table 4.2A, the table defines the fasteners (e.g., 10d common nails), the framing (2-in nominal width), and the nail spacing along continuous panel edge parallel to the load (4″ o.c.), and the nail spacing along all other panel edges (6″ o.c.) and then provides a nominal unit shear capacity for this complete diaphragm system (e.g., 1190 plf). With the prescriptive definition of these components in the SDPWS tables, the WSP diaphragm system has a single nominal diaphragm shear capacity. With CLT diaphragms, each different connection detail in the system can have unique diaphragm shear capacities calculated using the procedure above. Because of this, it is appropriate to consider the diaphragm shear capacities of the different connections relative to their demands. If a single diaphragm capacity is useful, then this capacity can be found as the lowest shear capacity of all connections in the diaphragm system.

To illustrate common components of a CLT diaphragm, Fig. 14.4 shows framing components in a regular post-and-beam supported floor system. This framing layout is common in office and similar open floor plan occupancies. The left half of the plan shows a 30-ft by 30-ft column grid supporting girders running east-west between columns, and secondary beams or purlins spaced at 10 ft on center running north-south between girders. The CLT panels span east-west supported by beams at 10 ft on-center. As CLT panels are typically much longer than 10 ft, in this configuration panels from 20 to 60 ft long may be used depending on the manufacturing and shipping capabilities of the CLT manufacturer. Figure 14.5 shows an enlarged partial plan of the northwest corner of the building with diaphragm details labeled. At the left-most edge of the diaphragm, there are three connection details, each of which may limit the diaphragm capacity—detail "c" is the CLT panel to collector which transfers the shear from the CLT to the collector line detail "a" is the panel-to-panel shear connection and detail "d" is the panel-to-chord connection. Each of these details is required by the SDPWS to use dowel type fasteners in shear controlled by yield mode IIIs or IV. Each of these three connections participates in transferring the applied loads through the diaphragm system and could be the detail limiting the shear capacity at the left edge of the diaphragm system.

The connections transferring diaphragm shear (details "a," "b," "c," and "d" in Fig. 14.5) each need to meet the requirements of SDPWS Section 4.5.4 item 1 with a nominal shear capacity calculated as described above using 4.5Z*. The nominal shear capacity is adjusted to a design capacity that must be greater than or equal to the code prescribed diaphragm design load.

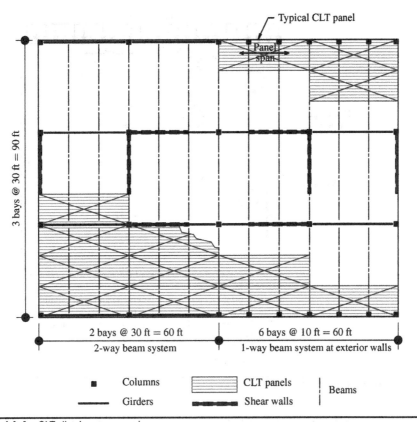

Typical CLT panel

FIGURE 14.4 CLT diaphragm overview.

For the rest of the CLT diaphragm system, SDPWS uses a different approach to define the design capacity and a force amplification factor applies to the demand. The force amplification factor varies based on the component and type of loading (i.e., wind vs. seismic). The components of the diaphragm following this approach include:

- CLT panels
- Chords and their splices and connections
- Collectors and their splices and connections
- Shear transfer elements of panel-to-panel connections (splines, etc.)

In Fig. 14.5, the shear in the CLT panel (e), the collectors and connections (y), and the chords and connections (z) all are designed using the force amplification approach. For these components, the capacity is calculated directly following the provisions of the appropriate materials standard and are *not* calculated using the provisions of SDPWS Section 4.1 and the reduction factors therein. For wood components following the NDS, a reference design value, generically labeled R, is calculated using the applicable provisions and adjusted to the ASD or LRFD adjusted design capacity, R'_{ASD} and R'_{LRFD}.

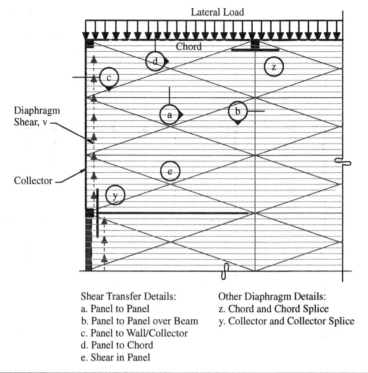

Shear Transfer Details:
a. Panel to Panel
b. Panel to Panel over Beam
c. Panel to Wall/Collector
d. Panel to Chord
e. Shear in Panel

Other Diaphragm Details:
z. Chord and Chord Splice
y. Collector and Collector Splice

FIGURE 14.5   Common CLT diaphragm details.

For example, the ASD adjusted shear capacity of a plywood spline used in a panel-to-panel connection would be found using NDS Table 9.3.1 as follows:

$$F_v t_v{}'_{ASD} = F_v t_v \, C_D \, C_M \, C_t$$

where $C_D = 1.6$ and $C_M = C_t = 1.0$ for interior dry locations.

On the demand side, the force amplification factors for different components and loading types are found in SDPWS Section 4.5.4 item 3 including Exceptions 1 and 2. The load force amplification factors are summarized in Table 14.3.

Component	Force Amplification Factor γ_D	
	Seismic	**Wind**
Chord splice connections between wood elements where the connection is using fasteners in shear controlled by Yield Mode IIIs or Mode IV	1.5	1.0
Wood elements including other connections between wood elements	2.0	1.5
Steel elements including connections between steel elements	2.0	2.0

TABLE 14.3   Summary of Force Amplification Factors for CLT Diaphragms

Concrete materials and connections are not explicitly covered; however, using the 2.0 force amplification factor appears appropriate. The adjusted capacity of the component must be equal to or greater than the amplified design force. In equation form, the required capacity check for the components subject to the force amplification factor approach can be written for ASD as follows:

$$R'_{ASD} \geq \gamma_D{}^*F_{ASD}$$

14.3   Connection Detailing

CLT panels used to carry gravity loads in floors and roofs are very stiff and very strong in-plane relative to the shear connections at the edges of the panels. There are several categories of panel shear connections.

The first category of CLT panel connection transferring diaphragm shear forces is where adjacent panels meet and are not located over any framing below. This commonly occurs at the CLT panel edges running perpendicular to the gravity framing. Here the primary purpose of the panel-to-panel connection is to transfer diaphragm shear forces between the panels. Some nominal vertical shear capacity is valuable to provide "bridging" between parallel spans of adjacent CLT panels to minimize differential deflections and cracking of finishes at the connection line. Figure 14.3 shows the common top-side spline connection and the half-lap connection.

The common top-side spline connection uses a wood spline recessed into the top side of the two adjacent panels with nails or screws fastening the spline to the panels. Figure 14.6 shows this connection style in more detail. The spline is frequently plywood or LVL. Commonly used plywood ranges from 23/32 in thick to 1 1/8 in thick. Using Struct-I plywood provides higher capacities; however, care should be given to specify a plywood panel which is available in the region of the construction. A convenient splice width is to cut a 4-ft-wide plywood panel into 8 splines, each 5 7/8 in wide, allowing for a 1/8-in saw kerf. Solid sawn lumber should not be used for the spline because lumber has a low tension perpendicular to the grain capacity.

For an economical connection, consider collated nails installed with a nail gun. In the past, collated nails were not available in "common" nail sizes and were typically "sinker" nail sizes. However, in recent years, manufacturers have begun producing common nail sizes for use in nail guns. As always when designing nailed connections for structural purposes, the required diameter and length of the nail should be specified, not just a name such as "10d nails." Tightly spaced nails in a row can potentially cause localized splitting in face lamination of solid sawn CLT similar to conditions in other wood framing products. The perpendicular layers in CLT can help reduce splitting relative to solid sawn lumber; however, the triggers for staggering tightly spaced nails of SDPWS 2021 Section 4.2.8 are good guidance for nails in CLT of solid sawn laminations. Any nails spaced at 2.5" o.c. or tighter should be staggered, and nails with a diameter of 0.148 in or larger, such as a 10d common nail, should be staggered when spaced at 3" o.c. or tighter. When nails are used as the main fastener transferring the diaphragm shear through the connection, it is a good practice to first install partially threaded screws widely spaced at 24" o.c. or similar. The screws hold the connection tightly together while the nails are placed to avoid any permanent vertical offset between the panels and create a stronger vertical connection for the bridging behavior of the connection.

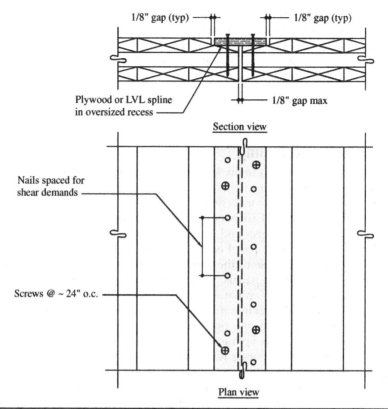

1/8" gap (typ) — 1/8" gap (typ)

Plywood or LVL spline
in oversized recess

1/8" gap max

Section view

Nails spaced for
shear demands

Screws @ ~ 24" o.c.

Plan view

FIGURE 14.6   CLT panel-to-panel spline connections.

A spline with all screws can be stronger than a nailed spline given the fact that larger diameter screw fasteners are available. However, small diameter collated screws can be an economical connection option. Some such screws are categorized as decking screws and have specialized installation tools which can be used while the installer is standing upright. Consider the relative effort between installing collated decking screws while standing to installing one handheld screw at a time while kneeling or bending over with a handheld driver.

Another consideration of the spline detail is the range of acceptable gaps between CLT panels and splines and between the CLT panels themselves. Based on informal testing observations, compression between panels can potentially be detrimental to the ultimate connection performance if the CLT panels close on the spline before the CLT panels bear on each other as shown in Fig. 14.7.

To mitigate the potential impact of panel-to-panel gap closure on the ultimate performance, gaps such as those shown in Fig. 14.6 are recommended. The objective is to have the typical gap between the CLT recess and the spline at least as large as the allowable gap between CLT panels. In practice, the spline will not be perfectly centered in the recess and the sum of the gaps on each side of the spline are more important than the specific gap on one side. For a CLT diaphragm system with competent boundary elements, this gap closure behavior should not be a significant concern, but attention to this detail may take the system performance from good to great.

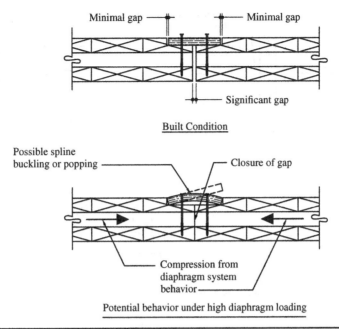

FIGURE 14.7 Potential gap impacts on spline connection.

Another recommendation is to detail spline connections so each plywood or LVL spline element only connects two CLT panels. This recommendation is to avoid unnecessary secondary stresses being induced in the spline if it were to cross over panel-to-panel connections perpendicular to the spline where the multiple corners of panels come together.

The half-lap panel-to-panel connection shown in Fig. 14.3 appears simpler with only two CLT panels and one row of screws. The half-lap connection can be significantly stronger and stiffer than the spline detail. However, it is typically more complicated and expensive to build. With the overlapping panels, more area of panel needs to be manufactured. With a top-side spline, three 10-ft-wide panels can cover a 30-ft-wide floor. With a half-lap connection, this would need four panels from a supplier with a 10-ft maximum panel width. The larger cuts into the panels required more fabrication effort. Also, the computer numerical controlled (CNC) equipment available at some manufacturers is only configured to work from one side of the panel. To fabricate the half-lap connection on such equipment, the panels will need to be flipped over to install the recess on both sides, significantly impacting fabrication efficiency. Because the half-lap has a top and a bottom panel, the connections need to be carefully coordinated with the construction sequence, as the orientation of the half-lap detail implies a certain installation sequence of the panels. Given all these considerations, the half-lap connection is used much less frequently than the top-side spline. The additional considerations and costs of the half-lap connection may be worthwhile if the top-side spline connection cannot provide the needed capacity of stiffness.

Panel-to-panel shear connections on the short end of the panels typically occur over a gravity support element perpendicular to the panel. Common connection details for CLT panels bearing on gravity supports at the end of panels are shown in Fig. 14.8.

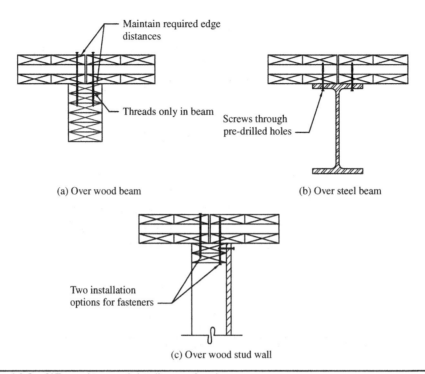

(a) Over wood beam

(b) Over steel beam

(c) Over wood stud wall

FIGURE 14.8   CLT panel-to-panel details over framing.

In these details, the gravity framing is used to transfer diaphragm shear from one panel to the other panel. In many cases the shear demand through the two rows of fasteners is equal. However, if the framing element is used as a diaphragm chord or collector, then the shear demands on the left and right side of the detail may not be equal and different fastener spacing may be more efficient. When designing such details, the minimum required edge distance may control the geometries. It is also a good detailing practice to select partially threaded screws where in the fully installed condition, the threads are only located within the component in which the tip of the screw is located. Partially threaded screws pull the two components tight together, otherwise any gap present when the screw is installed will be permanent. In mass timber details using long screws, proprietary self-tapping screws should almost always be used instead of lag screws. The installation requirements for lag screws, shown in NDS Section 12.1.4, require two pre-drilled holes of different diameters and different depths—one for the threaded portion of the screw and one for the shank. If the construction quality control process of a project is stringent enough that these holes are properly drilled thousands of times on a CLT diaphragm project, the installation costs while following this process will be multiple times the installation costs of using proprietary self-tapping screws. As implied by the name, self-tapping screws do not always require pre-drilled lead or clearance holes. There are cases where a single lead hole is recommended, so consultation with the manufacturer's installation instructions is still needed. As proprietary products, the capacities are commonly found using methods defined in third-party evaluation reports, such as ICC-ES reports.

The possible details at the edges of the diaphragms over framing location are similar to the interior panel-to-panel connections as shown in Fig. 14.9.

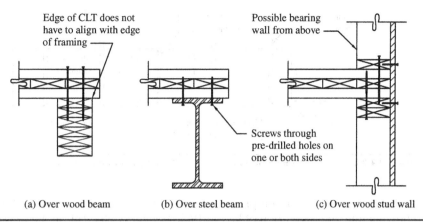

(a) Over wood beam   (b) Over steel beam   (c) Over wood stud wall

FIGURE 14.9   CLT panel to perimeter framing.

14.4   Boundary Element Detailing

In CLT diaphragms, several boundary element detailing styles are possible. One style is to use the gravity support element below the panel as the chord or collector. Referring to Fig. 14.4, this could occur at the chords and collectors in the left half of the plan and at the right/east edge of the diaphragm. For CLT diaphragm panels resting on bearing walls, the continuous boundary element can be the top plates of a conventional wood stud wall or a support angle in a CMU wall. For CLT diaphragm panels resting on beams, the beam closest to edge of the diaphragm can be selected to act as the boundary element as shown in the left portion of Fig. 14.10. Another option is to use the CLT panels to carry the boundary-element-forces directly. CLT panels directly in contact with

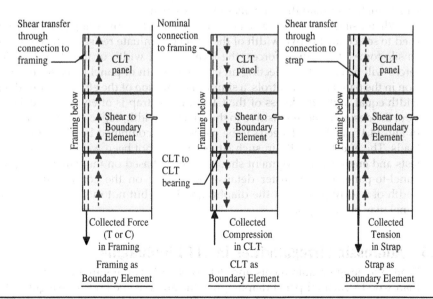

FIGURE 14.10   Boundary element options at framing below (plan view).

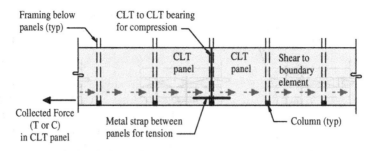

FIGURE 14.11 Boundary element options with no framing below (plan view).

each other can transfer compression forces through bearing between the panels as shown in the middle portion of Fig. 14.10. In boundary elements along the short direction of the panel, the tension forces are commonly carried by light gauge metal or steel straps on the top side of the CLT panels.

Where the long direction of the CLT panels are aligned with needed boundary elements it is also possible to use the CLT panel itself as the boundary element as shown in Fig. 14.11. This works best where the exterior wall does not require a beam or bearing wall at the edge of the diaphragm, as shown in the right half of the plan in Fig. 14.4. With this configuration, the CLT spanning between vertical load-bearing elements will act as both the gravity framing and as the boundary element carrying tension and compression loads. As such, the CLT will need to be checked in combined gravity (out-of-plane) and wind or seismic (compression and tension) loads as a beam-column element. At the ends of the CLT panels, short steel straps can provide continuity of tensile forces from panel to panel. Direct panel-to-panel contact at the ends of the panel can provide a load path for compression forces. A framing element may be needed below the CLT panel along the long direction of the panel for gravity loads or boundary element forces which are higher than the capacity of the CLT panel.

When using CLT panels as a boundary element, engineering judgment needs to be used to select an effective width of the panel to evaluate resisting the compression and tensile forces. For tensile forces, the location and width of the strap transferring the forces will be critical in selecting an effective width of panel resisting tension. As tension in the panel rarely controls, a simple assumption of the effective boundary element width equal to the thickness of the CLT at each strap is often adequate. For compression, the design is more sensitive to the effective width, especially when used as shown in Fig. 14.11, as the CLT needs to be checked for combined bending and compression loads. The effective width in such situations has not been well established via system tests and engineering judgment should be used based on the panel configuration and panel-to-panel force transfer details. Depending on the configuration, an effective width of 5 to 10 percent of the diaphragm depth but not more than the width of one panel may be appropriate.

14.5 Addressing Irregularities in CLT Diaphragms

While the design requirements and details of CLT diaphragm are significantly different than wood structural panel diaphragms, the analysis methods of irregular shaped diaphragms presented earlier in this book are still applicable as shown in Figs. 14-12 and 14-13.

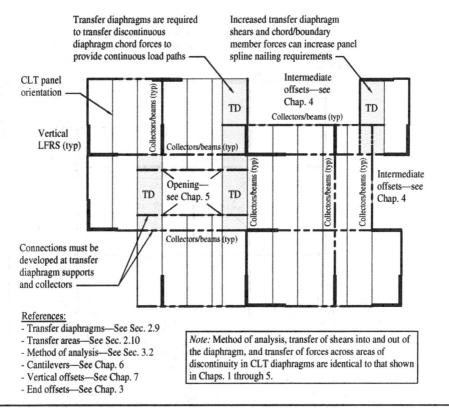

Transfer diaphragms are required to transfer discontinuous diaphragm chord forces to provide continuous load paths

Increased transfer diaphragm shears and chord/boundary member forces can increase panel spline nailing requirements

CLT panel orientation

Intermediate offsets—see Chap. 4

Vertical LFRS (typ)

Intermediate offsets—see Chap. 4

Connections must be developed at transfer diaphragm supports and collectors

References:
- Transfer diaphragms—See Sec. 2.9
- Transfer areas—See Sec. 2.10
- Method of analysis—See Sec. 3.2
- Cantilevers—See Chap. 6
- Vertical offsets—See Chap. 7
- End offsets—See Chap. 3

Note: Method of analysis, transfer of shears into and out of the diaphragm, and transfer of forces across areas of discontinuity in CLT diaphragms are identical to that shown in Chaps. 1 through 5.

FIGURE **14.12** Irregular CLT diaphragms with panels oriented north-south.

Example 14.1: Notched CLT Diaphragm, Analysis in the Transverse Direction

To illustrate the design and detailing aspects which are important to CLT diaphragms, Fig. 14.14 shows a diaphragm with a simple exterior notch (offset) in the diaphragm. This example is very similar to Example 3.1, which has a similar diaphragm with an offset chord. The analysis of this diaphragm follows the analysis procedure of the disrupted chord and transfer diaphragm shown in Figs. 3.9 to 3.11.

This example will evaluate the diaphragm with a 300 plf wind load applied in the transverse (N/S) direction using a transfer diaphragm between grid lines 2 and 3 for the disrupted chord along the south edge of the diaphragm as shown in Figs. 14.14 and 14.15. The initial diaphragm analysis can be made independent of the CLT panel layout. However, locating the internal transfer diaphragm, chords, and collectors should be performed in consultation with the framing and panel layout selected. For this, Figs. 14.12 and 14.13 can be used as guidance on possibilities.

Following the steps detailed in Chap. 3 and Example 3.1, the basic and transfer diaphragm shears and net diaphragm shears are developed in Fig. 14.16 based on a flexible diaphragm assumption. The diaphragm shears at the supports are 300 plf at grid line 1 and 250 plf at grid line 4. A higher unit shear of 350 plf occurs in the transfer area of the transfer diaphragm between grids C and D along grid line 2. If a single unit shear force is used throughout the diaphragm except at the transfer diaphragm, the design unit shear force of 300 plf controls the typical detailing. The connection details

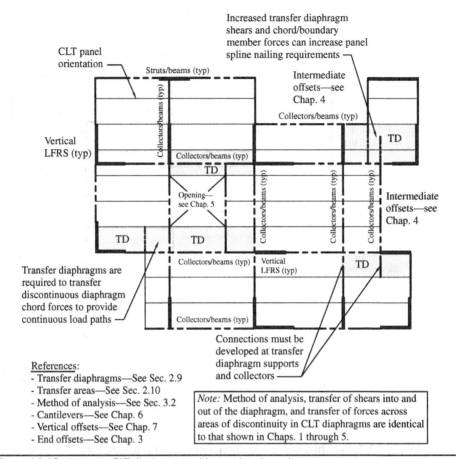

FIGURE 14.13 Irregular CLT diaphragms with panels oriented east-west.

and fastener spacing used in the transfer area of the transfer diaphragm will need to be checked against the higher unit shear demands in this area. This could result in the requirement for tighter fastener spacing. The corresponding forces within the diaphragm chords and collectors, including the transfer diaphragm, are developed in Fig. 14.17.

These diaphragm forces are now used to demonstrate design and detailing considerations of the boundary elements of the diaphragm. Figure 14.18 shows the location of four details near the notch under consideration.

Detail A is representative of a typical chord splice which occurs at each column location along Grids A and D where the chord forces are aligned along the long length of the CLT panels. One possible approach of chord and collector design is to use the CLT panels themselves as the primary boundary element handling axial forces. Metal straps attached on the top of the panels can transfer tension forces from one panel to the next as shown in Fig. 14.19. At the ends of the panels, direct bearing between adjacent CLT panels is used to transfer compression forces. Where compression forces are to be transferred, large gaps between the panels can be filled with metal shims, non-shrink grout, and/or plywood with an area large enough to transfer the compression loads.

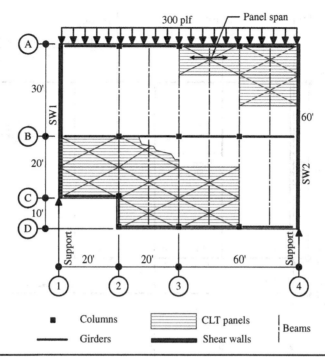

FIGURE 14.14 Notched CLT diaphragm example plan.

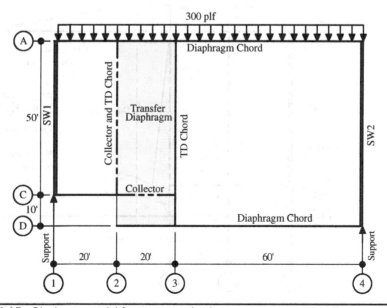

FIGURE 14.15 Diaphragm model for transverse load.

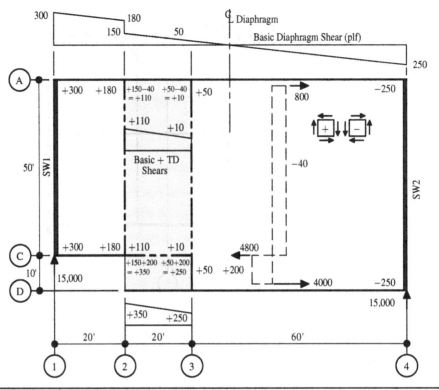

Figure 14.16 Diaphragm shears for transverse load.

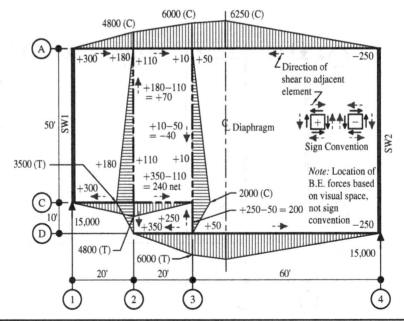

Figure 14.17 Chord and collector forces for transverse load.

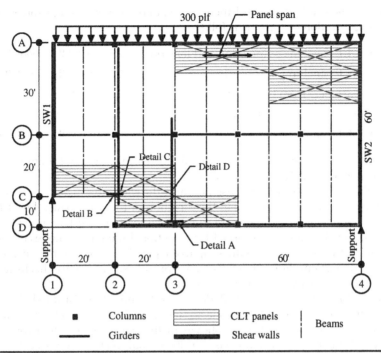

FIGURE 14.18 Details of interest at example diaphragm notch.

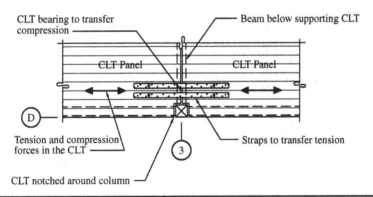

FIGURE 14.19 Detail A of notched diaphragm.

When installing non-shrink grout, care must be taken to damn the flow during construction, but also detailing provided such that the grout or other filler material will not fall out over time or during a seismic event.

As discussed in Sec. 14.2 of this chapter, the 2021 SDPWS requirements for CLT diaphragm design includes required force amplification factor for chord elements. The calculated chord load at grid line D/3 for the 300 plf wind load is 6000 lb. The load path components at this detail and their corresponding wind design forces including the load amplification factors are as follows:

- CLT panel in compression and tension: $1.5 \times 6000 \text{ lb} = 9000 \text{ lb}$ (T or C)
- CLT panel-to-panel bearing: $1.5 \times 6000 \text{ lb} = 9000 \text{ lb}$ (C)
- Fasteners from CLT to straps: $1.0 \times 6000 \text{ lb} = 6000 \text{ lb}$

  The 1.0 factor applies if fasteners are controlled by Yield Mode IIIs or IV

- Metal straps $2.0 \times 6000 \text{ lb} = 12,000 \text{ lb}$ (T)

If the design forces are from seismic loads, the 1.5 and 1.0 amplification factors above increase to 2.0 and 1.5, respectively.

If the CLT panels are being used to carry the chord forces and are also used as gravity framing elements spanning in the direction of the axial loads, the CLT needs to be checked under the combination of the vertical loading and lateral loading using 2018 NDS Section 3.9. At the strap-to-CLT connection, local failure modes of group and row tear-out of the fasteners from the strap to the panel may need to be considered at highly loaded connections. Further details on performing such capacity checks can be found in WoodWorks' *CLT Diaphragm Design Guide*.[7]

Detail B, shown in Fig. 14.20, demonstrates how the disrupted chord at grid line C/2 is transferred through the collector along grid line C into the transfer diaphragm between grid lines 2 and 3. As shown in the boundary element force diagrams in Fig. 14.17, the design chord force is 4800 lb. Similar to Detail A, the east-west chord and

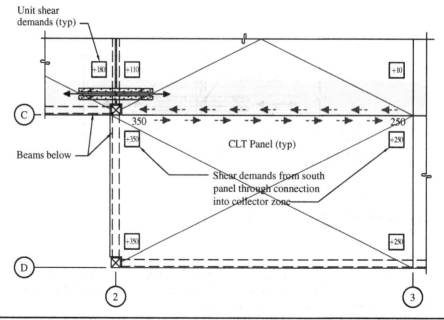

FIGURE 14.20   Detail B of notched diaphragm.

collector forces are primarily being carried by the CLT panel, metal straps transfer the shear between the panels, and panel-to-panel bearing transfers the compression forces. The required wind design forces for these components, including the SDPWS force amplification factors are as follows:

- CLT panel in compression and tension: 1.5×4800 lb $= 7200$ lb (T or C)
- CLT panel-to-panel bearing: 1.5×4800 lb $= 7200$ lb (C)
- Fasteners from CLT to straps: 1.0×4800 lb $= 4800$ lb

  The 1.0 factor applies if fasteners are controlled by Yield Mode IIIs or IV

- Metal straps 2.0×4800 lb $= 9600$ lb (T)

The idealized diaphragm analysis considered the chords and collectors located at the grid lines; however, the specific force demands on different components of the boundary elements and connections depend on the configuration of the detail. In Detail B, the collector zone between grid lines 2 and 3 is in the panel north of grid line C. Therefore, diaphragm shear south of grid line C needs to be transferred through the panel-to-panel connection and into the collector zone north of grid line C. In this case, a unit shear of 350 plf near grid line 2 and a unit shear of 250 plf near grid line 3 need to be transferred through the panel-to-panel connection along grid line C. As 350 plf is larger than the maximum shear at the support, special detailing such as tighter fastener spacing may be needed at this location.

If the panel layout is known or precisely specified, then the unit shear forces along the connection of two adjoining panels can be averaged over the length of the connection. If the panel-to-panel connection along grid line C, between grid lines 2 and 3, has only one panel north of the connection and one panel south of the connection, the panel-to-panel connection can be checked against the average 300 plf, instead of the peak unit diaphragm shear of 350 plf. This is possible because the CLT panels are much more rigid than the connection between them and this rigid behavior is very efficient in distributing the forces along the length of the panel and connection. However, this averaging method should only be used where the panel layout is very certain. Otherwise, the peak diaphragm shear values should be used to design and verify connection details.

Detail C, shown in Fig. 14.21, is the collector and transfer diaphragm chord in the N/S direction along grid line 2. The detailing approach taken in this example is to use a strap on the top of the CLT for tension loads and the CLT panels in bearing for compression loads. Along this line, the CLT panels are vertically attached to gravity beams below which restrain the panels from buckling out-of-plane under compression loads. While the full width of the building isn't shown, the strap at Detail C is installed along grid line 2 from grid lines A to D.

Considering the design forces on the components, the strap from grid lines C to D is installed for only 9 ft length instead of the full 10-ft grid spacing. This increases the nominal 350 plf unit shear to 3500 lb/9 ft $= 389$ plf, so that the full 3500 lb transfer diaphragm chord force can be developed over the length of strap provided. North of grid line B, the strap and chord zone are east of grid line 2, so the panel-to-panel connection along grid line 2 is designed for the 180 plf demand coming from the west side. The net lineal shear force transferred to the strap north of grid line C is 180 plf $-$ 110 plf $= 70$ plf.

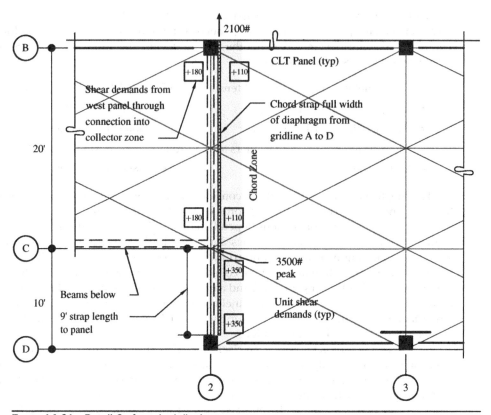

Figure 14.21 Detail C of notched diaphragm.

The wind design forces on the components of this detail, including the force amplification factors of SDPWS are as follows:

- Fasteners from CLT panel to 9 ft strap from B to C: 389 plf

  To be controlled by Yield Mode IIIs or IV

- CLT panel-to-panel bearing @ C/2: 1.5×3500 lb = 5250 lb (C)
- Metal strap @ C/2: 2.0×3500 lb = 7000 lb (T)
- Fasteners from strap to CLT from A to C: 70 plf

  To be controlled by Yield Mode IIIs or IV

Detail D, shown in Fig. 14.22, is the transfer diaphragm chord in the N/S direction along grid line 3. Unlike Detail C, Detail D does not use a chord zone in the CLT panels, but rather uses the framing below as the primary chord element. This detailing approach has the advantage of needing less strap length; however, there are different challenges to address. Starting at the south end of the chord, the unit diaphragm shears of 250 plf from the left and 50 plf from the right need to be transferred to the beam below from grid lines D to C. If the column significantly reduces the bearing length of the panels on the beam below, the design shear loads should be increased to handle the total shear

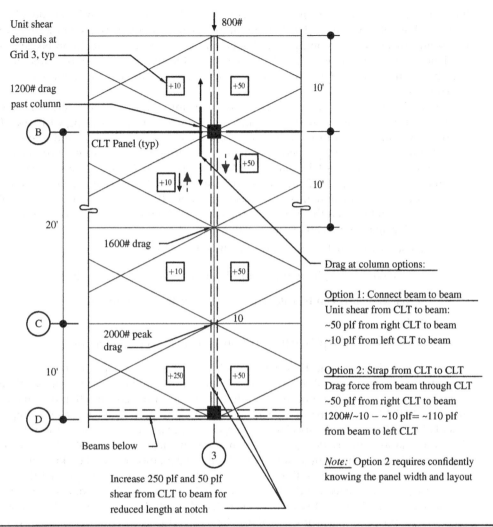

Unit shear demands at Grid 3, typ

1200# drag past column

CLT Panel (typ)

800#

+10   +50   10'

B

+10   +50   10'

20'

1600# drag

+10   +50

C   10

2000# peak drag

10'   +250   +50

D

Beams below

3

Increase 250 plf and 50 plf shear from CLT to beam for reduced length at notch

Drag at column options:

Option 1: Connect beam to beam
Unit shear from CLT to beam:
~50 plf from right CLT to beam
~10 plf from left CLT to beam

Option 2: Strap from CLT to CLT
Drag force from beam through CLT
~50 plf from right CLT to beam
1200#/~10 − ~10 plf= ~110 plf
from beam to left CLT

Note: Option 2 requires confidently knowing the panel width and layout

FIGURE 14.22 Detail D of notched diaphragm.

transfer of 2500 lb (250 × 10) from the left and 500 lb (50 × 10) from the right. The peak force in the chord of 2000 lb occurs at grid line C and this force used to check the beam spanning from grid lines B to D under combined bending from vertical loads and axial chord loads.

From grid lines C to B, the nominal diaphragm design shear loads are 10 plf from the left and 50 plf from the right and the total drag force in the chord element at grid line B is 1200 lb. Because the column at grid line B/3 interrupts the framing serving at the chord, detailing to transfer the compression and tension forces past the column is needed.

Option 1 is to transfer the chord forces directly from beam to beam on the adjacent sides of the column. If the beams and column are steel or concrete, then direct transfer of the chord forces through the column can be readily accomplished. With steel framing, load transfer or stiffener plates can be detailed through the web of the column. If the beams and column are timber elements, then this approach can be challenging. Often timber columns are wider than the timber beams, making straps on the sides of the beams infeasible. For compression loads, direct timber beam-to-column bearing is challenging or impossible to detail in a constructible manner considering tolerance gaps between the components and the compression perpendicular to grain capacity of the column.

Option 2 is to transfer the collected chord loads into the CLT panel nearest the column, then providing a chord splice connection similar to Detail A using CLT to CLT bearing for compression loads and straps for tension loads. This is a valid approach provided the forces can be properly detailed, considering the concentration of transfer forces from the beam to the CLT panel near the column. With the strap located left of the column as shown in Fig. 14.22, the chord force at grid line B is 1200 lb. The chord force located at the panel break half-way between grid lines B and C is 1600 lb. The critical shear connection between the beam and the CLT panel left of grid line 3 can be calculated considering the sign conventions as follows:

Force left CLT panel south of grid line B and beam below:

$$1200 \text{ lb} - 10 \text{ plf} (10 \text{ ft}) = 1100 \text{ lb, or } 110 \text{ plf over 10 ft length.}$$

Alternatively, this can be calculated starting at the panel-to-panel break further south:

$$1600 \text{ lb} - 50 \text{ plf} (10 \text{ ft}) = 1100 \text{ lb, or } 110 \text{ plf over 10 ft length.}$$

The CLT-to-beam connection detail needs to provide this capacity from the left panel to the beam. If special detailing is needed to provide the capacity, such as tight fastener spacing, this should be specifically called out on the construction documents along with field inspections or structural observations performed to ensure proper installation. In this example, special detailing would not be needed if the connections of the entire diaphragm are designed to the maximum shear demands of 300 plf at the supports. It is very important to note that this option 2 method is based on knowing the width of the panels adjacent to the chord break at the column. For example, if 4-ft-wide panels are provided instead of the 10-ft-wide panels, the CLT to chord connection needs to transfer 1160 lb. If the 4-ft-wide panel is notched 6 in to fit around the column, then the shear from the CLT to beam is 1160 lb/(3.5 ft) = 331 plf—a factor of 3 increase over the shear force of the 10-ft-wide panel! Project construction costs often benefit from flexibility in CLT panel sizes. While possible, the option 2 method requires extra design, detailing and inspection efforts. It may be more practical to not use the framing below as a primary boundary element unless the splices at columns can remain directly attached to the beams. It is often preferable to use the option 1 shown in Detail D with steel or concrete framing and a detailing approach similar to shown in Detail C with timber framing.

The wind design forces on the components of this detail, including the force amplification factors of SDPWS are as follows:

- Fasteners from left CLT panel to beam from D to C: 250 plf

  To be controlled by Yield Mode IIIs or IV

- Fasteners from right CLT panels to beam from D to A: 50 plf

To be controlled by Yield Mode IIIs or IV

- Axial force on timber beam as chord between B and D: $1.5 \times 2000\ \text{lb} = 3500\ \text{lb}$
- Axial force on steel beam as chord between B and D: $2.0 \times 2000\ \text{lb} = 4000\ \text{lb}$

Option 1:

- Chord force through steel column @ B/3: $2.0 \times 1200\ \text{lb} = 2400\ \text{lb}$

Option 2:

- Fasteners from 10-ft-wide CLT to beam N/S of B $110\ \text{plf}$
  To be controlled by Yield Mode IIIs or IV
- Metal strap adjacent to B/3: $2.0 \times 1200\ \text{lb} = 2400\ \text{lb (T)}$
- CLT to CLT bearing adjacent to @ B/3: $1.5 \times 1200\ \text{lb} = 1800\ \text{lb (C)}$ ▲

14.6 Diaphragm Stiffness and Deflections

One of the first questions that arises when designing a diaphragm is if the diaphragm can be idealized as flexible or idealized as rigid. For seismic design, ASCE 7-16[8] Section 12.3.1 provides conditions in which diaphragms can be categorically idealized as flexible or rigid. This does not include conditions which apply to CLT diaphragms. ASCE 7-16 Section 12.3.1.3 provides that diaphragms can be idealized as flexible by calculation when

$$\frac{\delta_{MDD}}{\Delta_{ADVE}} > 2$$

where δ_{MDD} is the maximum in-plane diaphragm deflection and Δ_{ADVE} is the average deflection of adjoining vertical elements of the lateral force resisting system. Similarly, IBC Section 1604.4 provides a condition by which a diaphragm can be idealized as rigid effectively when

$$\delta_{MDD} \leq 2\ \Delta_{ADVE}$$

SDPWS 2021 Section 4.1.7 has a similar provision as the IBC to justify a rigid diaphragm by calculation and includes further clarification on how to apply to the provisions to cantilevered diaphragms.

Many CLT diaphragms supported by sheathed wood frame shear walls and steel moment frames can be categorized as rigid diaphragms. Some CLT diaphragms supported by concrete shear walls and steel braced frames can be categorized as flexible; however, some can be categorized as rigid. In buildings with multiple diaphragm spans, it is often the case that the entire building cannot be categorized as either rigid or flexible. In such cases, an envelope solution can take the highest demands on each component from a rigid and a flexible diaphragm analysis. Alternatively, a semi-rigid diaphragm analysis can be performed to provide a more precise distribution of the lateral loads through the structure.

The calculations to justify a diaphragm as flexible or as rigid and semi-rigid diaphragm analysis both require a method to calculate the deflections and stiffness of the diaphragm system. Chapter 2 presented the equations recognized in the SDPWS for calculating deflection for WSP diaphragms; however, these equations do not apply to CLT diaphragms. For CLT diaphragms, SDPWS 2021 does not provide similar standardized deflection equations. SDPWS 2021 does provide direction that CLT diaphragm deflections are to be calculated using "principles of engineering mechanics."

An ambitious design effort could undertake finite element modeling of a complete CLT diaphragm system as shown in Breneman et al.[9] and Kneer et al.[10] Such efforts are beyond the means of typical building design projects which can benefit from simpler methods.

The derivations of the deflection equations for WSP diaphragms found in the SDPWS are based on basic principles of mechanics as documented in ATC-7.[11] Several assumptions made in the derivation of the sheathed wood-frame diaphragm equations are not generally applicable to CLT diaphragms, notably:

1. 4 ft × 8 ft panels, typical
2. Equal stiffness to shear slip at all edges of the panels
3. Panel to adjoining panel shear connections occur through a third-framing element.

All of these assumptions are used in the derivation of the third term of the SDPWS equation for uniformly loaded simply span WSP diaphragms discussed in Chap. 2.

$$\Delta = \frac{5vL^3}{8EAW} + \frac{vL}{4Gt} + 0.188Le_n + \frac{\Sigma(\Delta_c X)}{2W} \qquad \text{SDPWS Eq. C4.3.2-1}$$

The third term of this equation calculates the contribution to mid-span diaphragm deflections from the nail slip at the support and the span of the diaphragm. Assumptions number 1 to 3 are all built into the term $0.188Le_n$. Also built into the 0.188 term is a unit conversion from L in feet to the deflection Δ in inches.

As CLT panels are not universally 4 ft by 8 ft in dimensions, assumption number 1 is obviously not valid. Spickler et al.[12] addressed this by extracting from the derivations in ATC-7 the more general form of the third term, δ_{slip} as follows:

$$\delta_{slip} = \frac{L\,e_f}{2}\left(\frac{1}{P_{\perp}} + \frac{1}{P_{\parallel}}\right)$$

where $P_{\perp}$ is panel length perpendicular to the applied load, $P_{\parallel}$ is the panel length parallel to the applied load, and e_f is the fastener slip, recognizing the fasteners are not all nails.

Referring to the common CLT diaphragm connections shown in Fig. 14.3, it is apparent that different connections in the same diaphragm may have very different fasteners and connection types. Panel-to-panel connections along the long edge of the CLT may have closely spaced nails through a plywood spline. Panel-to-panel connections over beams may have larger diameter, long self-tapping screws. These two connections in the same diaphragm may have very different stiffness values and not be aligned with assumption number 2. Both connection styles use a third-framing element to transfer the shear forces. In the top-side spline connection, the spline is the shear element connected to the two panels with two different rows of fasteners through the spline to the two panels. The panel-to-panel shear connection over the beam also has two rows of fasteners with the beam providing the shear transfer from one panel to the other. The half-lap style of the panel-to-panel connection is different with one necessary row of fasteners connecting the panels directly to each other.

To relax assumptions 2 and 3, John Lawson of California Polytechnic State University, Marco Lo Ricco of the USDA Forest Products Lab, and Scott Breneman have developed a more general form of the fasteners slip term of the deflection as follows:

$$\delta_{slip} = \frac{L}{4}\left(\frac{n_{\parallel}e_{f\parallel}}{P_{\perp}} + \frac{n_{\perp}e_{f\perp}}{P_{\parallel}}\right)$$

where $e_{f\parallel}$ is the slip of the fasteners in the panel-to-panel connections parallel to the loading, $n_{\parallel}$ is the number of slip planes at the parallel connection, $e_{f\perp}$ is the slip of the fasteners in the panel-to-panel connection perpendicular to the loading, and $n_{\perp}$ is the number of slip planes at the perpendicular connection. For the calculation of both fastener slip terms, the maximum diaphragm shear load at the supports, v, is used. Figure 14.23 illustrates the location of the slip planes in the different types of connections. One way to determine how many slip planes there are in a connection is to consider how many screws or nails in shear a single pound of force goes through to be transferred from one panel to the next. The typical WSP connection, the top-side spline CLT connection and the typical CLT-to-beam-to-CLT connection all have two slip planes, while the half-lap CLT-to-CLT connection has one slip plane.

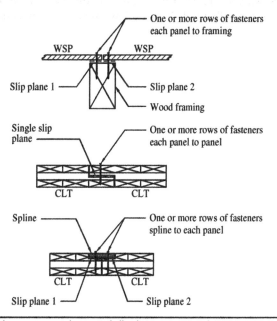

FIGURE 14.23 Slip planes at panel-to-panel diaphragm connections.

14.7 References

1. APA—The Engineered Wood Association (APA), *ANSI/APA PRG 320-2019 Standard for Performance-Rated Cross-Laminated Timber*, APA, Tacoma, WA, 2020.
2. International Code Council (ICC), *2015 International Building Code*, ICC, Whittier, CA, 2014.
3. American Wood Council (AWC), *National Design Specification (NDS) for Wood Construction with Commentary 2015 Edition*, AWC, Leesburg, VA, 2014.
4. ICC, *2021 International Building Code*, ICC, Whittier, CA, 2020.
5. AWC, *Special Design Provisions for Wind and Seismic with Commentary 2015 Edition*, AWC, Leesburg, VA, 2015.
6. AWC, *Special Design Provisions for Wind and Seismic 2021 Edition*, AWC, Leesburg, VA, 2020.
7. WoodWorks—Wood Products Council, *CLT Diaphragm Design Guide*, WoodWorks, Washington, DC, 2022.
8. American Society of Civil Engineers (ASCE), *ASCE/SEI 7-16 Minimum Design Loads for Buildings and Other Structures*, ASCE, New York, 2016.
9. Breneman, S., McDonnell, E., and Zimmerman, R. B., *An Approach to CLT Diaphragm Modeling for Seismic Design with Application to a US High-Rise Project.* Proceedings of 2016 World Conference on Timber Engineering. Vienna, Austria, 2016.
10. Kneer, E., Houston, J., Breneman S, *Analytical modeling of cross-laminated timber (CLT) diaphragms.* Proceeding of the 2016 SEAOC Convention. SEAOC, Sacramento, CA, 2016.
11. Applied Technology Council (ATC), *Guidelines for the Design of Horizontal Wood Diaphragms, ATC-7*, ATC, Berkeley, CA, 1981.
12. Spickler, K., Closen, M., Line, P., and Pohil, M, *CLT Horizontal Diaphragm Design Example*, Informally published, 2015. www.structurlam.com" www.structurlam.com.

CHAPTER **15**

CLT Shear Walls

15.1   Introduction

CLT panels are very strong and stiff for in-plane forces which makes them good candidates for use in shear walls. The structural panel properties described in Chap. 14 are valid for CLT used in shear walls. The use of CLT as structural walls has not grown as widespread as CLT panels used in floor and roof systems. However, this use is increasing in popularity and is currently a subject of significant discussion and research.

Many of the early applications of CLT as structural elements in Europe used the CLT panels as the primary structural element for the floors, walls, and roofs. These early applications of systems fully framed with CLT primarily were in Austria and Switzerland, followed by expanded use in the United Kingdom. Researchers in the higher seismic regions of Italy and Slovenia investigated CLT and connection behaviors related to earthquake response and this work has been built upon by North American researchers over the past decade.

Engineers in the U.S. reference ASCE 7-16[1] Table 12.2-1 for a library of code recognized vertical lateral-force-resisting systems available for both seismic and wind applications. For light-frame wood bearing-wall construction, there is currently one main all-wood entry—the wood structural panel (WSP) sheathed wood-framed shear walls. This differs from steel and concrete systems where engineers are accustomed to a hierarchy of seismic systems with varying levels of detailing requirements. The benefit of choosing a system with higher levels of seismic detailing requirements is increased seismic performance as reflected by a higher seismic reduction coefficient, R, and permissible use in higher seismic design categories. A familiar pattern can be found in steel moment frames, concrete moment frames, reinforced concrete shear walls, and masonry shear walls. Each of these materials has a hierarchy of "special," "intermediate," and "ordinary" seismic systems in ASCE Table 12.2-1.

The 2021 SDPWS[2] now provides definition of a CLT shear wall system which is a large and significant step forward for standardization of lateral-force-resisting systems for CLT. This step puts forward one seismic system using CLT as a structural component which has been through the standardization process. Other systems are not yet standardized and continue to be in research and development. In some cases, these non-standardized lateral systems have been constructed in buildings in the United States through an alternative means and methods process using performance-based design. Looking forward, the future in the CLT industry may be more like steel than conventional light wood framing, where a variety of defined systems with corresponding seismic design parameters are available for the engineers to use in design projects.

513

15.2 Range of Possible CLT Shear Wall Systems

This section provides an overview of a range of CLT shear wall systems, both those standardized in the 2021 SDPWS and those being considered by researchers and innovative practicing engineers. For this discussion a categorization of possible CLT shear walls systems include:

- CLT shear walls with no special seismic detailing requirements
- Seismically detailed platform-framed CLT shear walls
- Seismically detailed balloon-framed CLT shear walls
- CLT rocking shear walls
- Hybrid CLT shear wall systems

These categories are approximately ordered from lower performing seismic systems to higher performing seismic systems.

15.2.1 CLT Shear Walls with No Special Seismic Detailing Requirements

This category of CLT shear walls is for CLT shear walls with no special seismic detailing provisions *required* by the building code and referenced standards. It may seem unusual to start the discussion with this system; however, this category will likely be used in many applications where appropriate in the United States. While there is no formal definition of this category in standards, it is conceptually similar to the "Ordinary" or "Plain" seismic systems of other materials. For brevity, ordinary CLT shear walls will be used in this chapter to refer to CLT shear walls with no special seismic detailing requirements.

Structurally, such shear walls are designed for the lateral design forces of wind, seismic, and general structural integrity as applicable per the building code. CLT panels can be of any size feasible, which is a primary benefit of this category of CLT shear walls. Larger panels are typically more efficient than smaller panels for construction efficiency and total installed costs. The high strength and stiffness of CLT can create a very efficient lateral system for wind loads. Connections between diaphragms and wall panels, between adjoining wall panels, and between the wall panels and the foundation are selected based on the required strength, stiffness, economy, and constructability.

Both balloon-framed and platform-framed ordinary CLT shear wall configurations are possible. Figure 15.1 shows the schematic of a platform-framed ordinary CLT shear wall system. Such a system may have CLT panels with the width of the panel matching the clear story height and the length of the panels running long in the plan direction of the walls. The chosen ceiling height determines the required width of the panels. Each CLT manufacturing plant has a maximum panel width and length capacity. Such a design should be selected with panel manufacturing capability in mind. Widths available include 8 ft, 10 ft, and 12 ft, leading to efficiencies in manufacturing when the ceiling height matches. With a platform-framed layout, it can be efficient to pre-fabricate the opening for windows and other penetrations into large wall CLT panels, as shown in Fig. 15.1. For larger openings, the panel layout can include separate header pieces over openings to reduce material waste. Header pieces can be CLT panels or glu-lam or structural composite lumber beams. Such headers can bear on notches in the CLT panels, provided the bearing area of the beam and CLT wall have the capacity to support

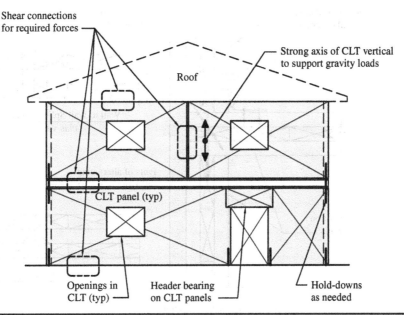

Shear connections for required forces

Strong axis of CLT vertical to support gravity loads

Roof

CLT panel (typ)

Openings in CLT (typ)

Header bearing on CLT panels

Hold-downs as needed

Figure 15.1 Platform-framed ordinary CLT shear walls.

the loading. When used in a wall application, CLT is often used with the major strength axis of the panels oriented vertically, which is 8 to 12 ft width of the panel. This orientation of the major strength axis makes the most efficient use of the CLT to support gravity compression loads and out-of-plane wind loads. As floor and roof panels commonly have the strength axis oriented with the long direction of the panel, care needs to be taken in specifying the orientation of major strength direction of the panels and verifying these in shop drawing review process. As such systems are designed without any required seismic detailing, the panel layout can provide the best fit for the project and manufacturing limitations.

Figure 15.2 shows some of the more common connection details which may be used in platform-framed ordinary CLT shear walls. Such connections can be detailed similar to other types of CLT shear walls and CLT diaphragms, such as using screws and nail primarily loaded in shear to transfer lateral forces, but this is not a requirement. The selection of fasteners and connection hardware needs to be coordinated with the architectural plan to cover the CLT in the interior spaces or leave it visibly exposed for aesthetic purposes. For example, an efficient connection such as using light-gage angle connectors nailed between the wall and floor panels may not be acceptable to the aesthetic plan of the project.

Another possible configuration of ordinary CLT shear walls is using CLT at the primary framing of core wall systems as shown in Fig. 15.3. Such cores can form elevator and stair shafts in either mass timber or conventional wood frame buildings. One construction method for CLT cores is to assemble the core quickly in a balloon frame style installing site-built or manufactured stair systems before the remaining of the structure is built around the cores. Such construction can be much quicker than site-built masonry or concrete cores. This method benefits from using CLT panels which can be multiple stories tall. Another construction method for CLT cores is to build the core

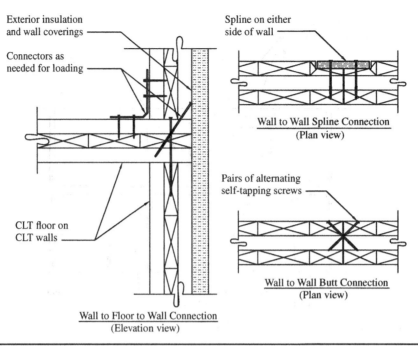

Exterior insulation and wall coverings

Connectors as needed for loading

CLT floor on CLT walls

Wall to Floor to Wall Connection
(Elevation view)

Spline on either side of wall

Wall to Wall Spline Connection
(Plan view)

Pairs of alternating self-tapping screws

Wall to Wall Butt Connection
(Plan view)

Figure 15.2 Platform-framed ordinary CLT shear wall details.

concurrently with the rest of the structure. This second method typically uses CLT panels which are one story in height at a time.

For warehouse and box-store type buildings, CLT framing can also be used in a configuration similar to concrete tilt-up construction. CLT panels can form the exterior wall framing, running the full height of the exterior wall, creating high-ceiling spaces. The CLT panels can extend past the roof line to form the parapet wall when required. Alternatively, the CLT panels can stop at the roof line and site-built framing used for the parapet. When running the CLT panels vertically, the panels can be used as both vertical load bearing walls and shear walls. For relatively short buildings with short roof framing spans, roof trusses or beams can sometimes be designed to bear directly on the CLT walls. For higher roof levels and longer framing spans, the roof girders will need to be supported by columns, typically located on the inside face of the CLT panels in the exterior wall. Alternatively, the outside face of the column can be aligned with the outside face of the CLT panels, which limits the size of the panel to match the spacing between columns. Considering the maximum slenderness ratio of 50 for wood elements in compression per NDS Section 3.7, the maximum possible load bearing height for 4.125 in 3-ply is about 17 ft. For 6.875 in 5-ply, this is a little over 28 ft. The load bearing capacity of such walls would need to be verified to handle the applied loads for the project.

When approaching the load bearing limits with vertically oriented CLT, an alternative framing style to consider is to have regularly spaced columns supporting both the roof framing and install CLT panels horizontally in a girt-style arrangement between columns. For out-of-plane wind-loads, the CLT panels span horizontally between the columns while the CLT panels can be detailed to serve as shear walls from the roof to foundation.

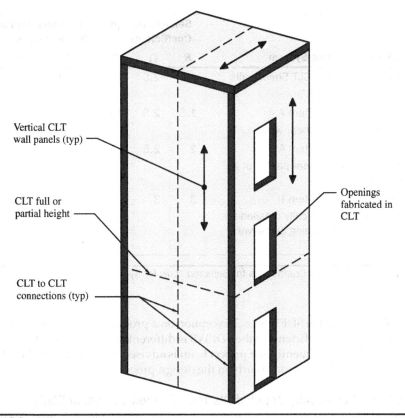

Figure 15.3 CLT core wall system.

For these CLT shear walls with no special seismic detailing requirements, SDPWS Section 4.6.3 provides a route to code acceptance via the following:

Approved CLT shear wall systems, other than those in accordance with Appendix B, shall be permitted to resist wind forces, and to resist seismic forces in Seismic Design Category A and in Seismic Design Category B where R = 1.5, C_d = 1.5 and Ω_o = 2.5 and with structural height limit of 65', unless other values are approved.

This language provides a low-seismic catch-all system similar to several other lateral systems in ASCE 7-16 Table 12.1-1, such as the $R = 3$ catch-all system for "steel system not specifically detailed for seismic resistance." The seismic design coefficients for several systems are shown in Table 15.1.

Comparing the value assigned to the different systems in Table 15.1, it is apparent the values assigned to ordinary CLT shear walls via the SDPWS 4.6.3 Exception are the lowest assigned to systems which are not cantilevered column systems. The ordinary CLT shear wall system has the same, low, seismic design coefficients as an ordinary plain (unreinforced) concrete shear wall and the height limitations are even more restrictive. As a relatively new system without a long track-record in the field, it is natural to use conservative values; however, with more experience and careful studies, a reasonable prediction is that higher seismic design values will be justified for ordinary CLT shear walls in the future.

Seismic Force-Resisting System	Seismic Design Coefficients			Structural Height Limit for Seismic Design Category		
	R	Ω_o	C_d	B	C	D
SDPWS 4.6.3 Exception CLT Shear walls "Ordinary CLT shear walls"	1.5	2.5	1.5	65 ft	NP	NP
ASCE 7-16 Table 12.2-1 item A.4 Ordinary plain concrete shear walls	1.5	2.5	1.5	NL	NP	NP
ASCE 7-16 Table 12.2-1 item A.17 Light-Frame walls with shear panels of all other materials	2	2.5	2	NL	NL	35
ASCE 7-16 Table 12.2-1 item H Steel systems not specifically detailed for seismic resistance excluding cantilever column systems	3	3	3	NL	NL	NP

TABLE 15.1   Seismic Design Coefficients for Selected "Low R" Systems

When using the SDPWS 4.6.3 Exception on a project, the code path to recognition of seismic design coefficients in the SDPWS is different path than those located in ASCE 7 Chapter 12. Consequently, the project team is advised to consult with the AHJ to obtain acceptance of this approach early in the design process.

15.2.2   Seismically Detailed Platform-Framed CLT Shear Walls

Using large CLT panels along the length of a wall as shown in Fig. 15.1 results in a strong and stiff system. However, it is a system which may not have a large deformation capacity nor dissipate significant energy in response to a high-level seismic event. A natural strategy to increase the deformation capacity and ductile behavior of a CLT shear wall system is to construct the wall from many smaller panels as shown in Fig. 15.4. In such a configuration, under shear loads, each panel can shift and rotate a small amount relative to each other, while maintaining a competent lateral load path.

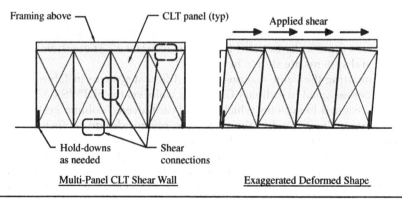

FIGURE 15.4   Multi-panel CLT shear wall construction.

The connections of the panels at the top and bottom of wall and at adjoining edges of panels can be detailed to provide ductile behavior under the shear forces.

A common detailing strategy to provide ductile behavior at the CLT shear wall connections is to use dowel-type fasteners in shear as discussed in Chap. 14 for CLT diaphragms. Based on research performed by FPInnovations,[3] the US CLT Handbook[4] published in 2012 recommends that CLT systems resisting seismic forces using fasteners loaded in shear are designed such that fastener Yield Mode III or Mode IV from NDS[5] controls the capacity of the connection. The US CLT Handbook also recommends designing all of the wood limit states to design forces matching the nominal connection capacity. The US CLT Handbook provided an early recommendation of using $R = 2$ for CLT shear walls following the detailing rules of the handbook.

These concepts are the foundation behind the CLT shear wall requirements in Section 4.6 of SDPWS. Section 4.6 is the beginning of the SDPWS CLT shear wall requirements with most of the content located in SDPWS Appendix B. Important characteristics of using CLT shear walls following Appendix B include:

- Specific connection components for shear transfer at the top and bottom of the wall and between adjoining wall panels are required.

- The range of allowable wall panel aspect ratio h/b_s is from 2 to 4.

- Hold-down design forces are determined using amplified *capacities* of shear walls.

The defined nomenclature for this system in SDPWS is simply "CLT shear walls." SDPWS also defines a special case of this system when using only panels with an aspect ratio h/b_s of 4. The SDPWS refers to this system as "CLT shear walls with shear resistance provided by high aspect ratio panels only." Further discussion of the SDPWS 2021 CLT Shear Walls is found in Sec. 15.3 of this chapter.

15.2.3 Seismically Detailed Balloon-Framed CLT Shear Walls

In addition to any explicit height limits applied to the CLT shear wall systems in the SDPWS and ASCE 7, platform-framed CLT walls have a practical limit of application. The perpendicular to grain compression strength and total shrinkage of the CLT floor panels in the vertical load path may limit the application of platform framing. Taller buildings using mass timber elements are increasingly common and it may be advantageous in such buildings to have an all-timber gravity and lateral load carrying structural system. Seismically detailed balloon-framed CLT shear walls, as shown in Fig. 15.5, are one possible configuration under current investigation.[6] Simply described, such systems combine balloon frame CLT shear walls configured in core applications like shown in Fig. 15.3 or in isolated balloon frame multi-story walls with design details for them to be high performing seismic systems. Unlike the SDPWS CLT Shear Walls, there are not yet specific standardized requirements developed for this system. Detailing approaches for such systems combined high-aspect ratio CLT wall panels connected with dowel type fasteners in shear using hold-downs similar to those used in conventional light-frame construction.

One design challenge for balloon-framed CLT shear walls, including the specialized rocking-wall systems described in the next section, is the floor framing to wall connections. Lateral and sometimes vertical forces need to be transferred from the floor framing to the balloon frame shear wall system. This connectivity needs to be detailed in a

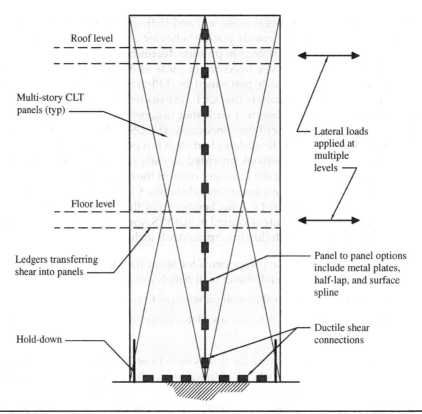

Figure 15.5 Seismically detailed balloon frame CLT shear walls.

manner to competently transfer the needed gravity forces, yet not impede the seismic performance of the shear wall system, or adversely impact the floor framing system. A steel or timber ledger attached to the wall panels and supporting floor panels is a straight-forward connection. Testing has indicated that this ledger configuration is adequate for seismically detailed balloon frame CLT shear walls.[7]

15.2.4 CLT Rocking Shear Wall Systems

CLT rocking shear walls are constructed as high-aspect-ratio balloon-framed walls with special detailing which allows the walls to physically lift up off their supports at the tension end of each wall under lateral seismic loading. As the wall cycles back and forth under lateral seismic loading, each end of the wall alternates being in contact with the foundation and lifting off the foundation. While similar behavior is observed in the prior systems discussed at a small scale, the CLT rocking shear wall system is detailed such that this behavior occurs while the wall and main connections respond elastically—without yielding or significant damage occurring in the wall or main connections. Unique detailing aspects of this system are discussed in the following.

Several detailing aspects particularly to CLT rocking shear wall systems are shown in Fig. 15.6.

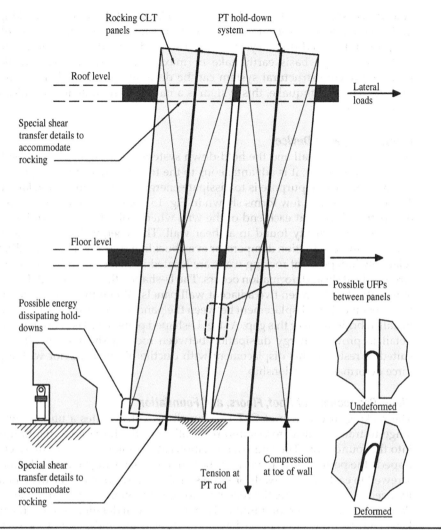

Roof level

Lateral loads

Rocking CLT panels

PT hold-down system

Special shear transfer details to accommodate rocking

Floor level

Possible energy dissipating hold-downs

Possible UFPs between panels

Undeformed

Special shear transfer details to accommodate rocking

Tension at PT rod

Compression at toe of wall

Deformed

FIGURE 15.6 CLT rocking shear wall systems.

Post-Tensioned Hold-Down

The rocking wall system frequently includes a hold-down, commonly a continuous steel rod system, placed in a cavity within the center of the wall or adjacent to the center of the wall.[8] This system is designed to provide a restoring force which pulls the wall system back to center (plumb) when pushed laterally by a seismic event. Generally applicable to high-seismic locations, the hold-down system is also detailed such that rocking and uplift behavior do not occur during service level wind events. To create a significant restoring force at even minor rocking angles, the hold-down rod system can be post-tensioned to increase the restoring moment associated with a specific rocking angle. As with other post-tensioned structural system, the immediate loss of tension when applying the tension forces and long-term loss from creep deformations need to be considered in the long-term performance of the system. The tensioned hold-down system is designed to remain elastic while allowing rocking behavior under moderate-level seismic excitations. Depending on the performance objectives of the project, the hold-down system may also be designed to have an elastic response under the design basis earthquake or maximum considered earthquake loading. Because the core structural system can be designed to target damage-free behavior during a large earthquake, this system is a natural candidate for seismically resilient structural design.

Energy Dissipative Device

Because both the wall and the hold-down systems are designed to respond elastically to the seismic event, it is advantageous to the total behavior to add specific structural components whose purpose is to dissipate energy and to dampen the lateral response. These can come in a few forms shown in Fig. 15.6. One technique is to connect energy dissipating devices at each end of the wall where hold-downs providing overturning restraint are normally found in a shear wall. This energy dissipative devices can be viscous dampers, friction dampers, or even buckling restrained brace (BRB). Where two relatively independent rocking walls are located in-plane and adjacent to each other, a large vertical shear deformation occurs. The U-shaped flexural plate (UFP) is one mechanism attached between two adjacent wall panels.[9] When the walls rock, this creates a large relative shear displacement between the panels. Energy dissipative devices can be installed bridging over this gap. A steel U-shaped plate (UFP) is a low-tech and effective detail to provide energy dissipation between rocking shear walls. The UFP is well suited to resisting this displacement with ductile plastic behavior with a predictable force deformation relationship.

Shear Connections at Roof, Floors, and Foundation

The rocking mechanism of CLT rocking shear walls creates a unique detailing challenge to transfer shear forces into the wall from the horizontal diaphragms and out into the foundation. The large deformations at the intersections of these elements can impede the performance of normal shear details like using bent metal plates nailed or screwed to each surface. Nails or screws in shear may not have the deformation capacity needed. Furthermore, the rocking and uplift motion of the wall panels should not damage the floor and roof framing. Therefore, special detailing such as the use of shear keys and shear pins to transfer lateral forces near the mid-point of the rocking can be used. The rocking behavior creates uplift displacement even at the mid-point of the walls, so such shear keys need to be detailed to allow for this uplift while transferring horizontal shear forces. To minimize negative interactions, rocking shear walls are sometimes detailed to be independent of the vertical gravity load path of the surrounding structure.

Several projects have been built in North America using CLT rocking shear walls. The Catalyst project designed by KPFF Engineers in Spokane, Washington, has CLT multi-story shear walls with BRBs as energy dissipating hold-downs. The Peavy Hall replacement building at Oregon State University in Corvallis, Oregon, uses multiple CLT rocking walls with UFP connectors designed by Equilibrium Consulting of Vancouver, British Columbia. Such innovative projects have been designed using advanced performance-based seismic design methods. There is ongoing research and development of the design method to simplify the design process and work toward standardization.[10,11]

15.2.5 Hybrid CLT Shear Wall Systems

Hybrid systems combining CLT shear wall systems with other lateral-force-resisting elements include many possibilities. A range of hybrid coupled wall systems has been proposed. Coupled wall systems combine tall shear wall components with horizontal spandrel beams in a system which looks like a cross between a shear wall and a moment frame. The combination of the shear wall as a column and spandrel beam follows the familiar strong-column/weak-beam philosophy common in seismic moment-frame systems. As an example of such a system which has been standardized, ASCE 7-22[12] added the system reinforced concrete ductile coupled walls in Table 12.2-1 with $R = 8$. This system has the highest R value of any bearing wall system and matches the highest assigned to any seismic-force-resisting system in ASCE 7-22. For more information on this system, refer to S. K. Ghosh (2019).[13]

Coupled CLT shear wall systems are similar to the balloon frame CLT wall and CLT rocking wall with multiple adjacent walls. However, instead of having wall piers located within inches of each other, coupled wall systems move the walls apart and interconnect them with spandrel elements. These spandrels provide ductile shear transfer of forces between the walls, similar to a UFP in the rocking wall system. Figure 15.7 shows several possible coupled wall systems.

In *The Case for Tall Wood Buildings*,[14] Equilibrium Consulting presented a hybrid system with mass timber walls interconnected with steel beams. The components of this system are proportioned so that as the walls sway, the steel beams form plastic hinges similar to beams in a steel moment frame. In Skidmore, Owings, and Merrill's

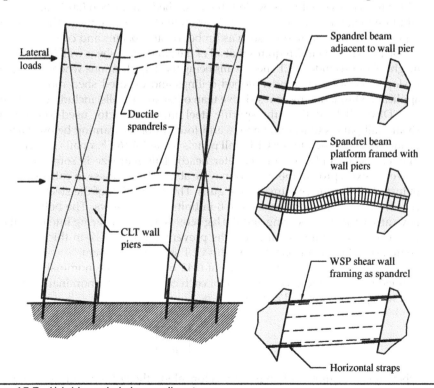

FIGURE 15.7 Hybrid coupled shear wall systems.

Timber Tower Research Project,[15] a similar system was explored using reinforced concrete spandrel beams instead of steel spandrel beams. Either of these combinations has the potential to achieve high seismic performance exemplified in the new reinforced ductile concrete coupled wall system of ASCE 7-22. As both systems are constructed only of mass timber and non-combustible structural elements, they have potential application in the tall Type IV construction types first added in the 2021 IBC.[16]

A similar coupled wall system with combined vertical CLT walls with horizontal spandrels constructed as WSP shear walls has also been envisioned. The use of such a system will be limited in buildings where combustible light frame construction is permitted; however, this system has the potential of being a high capacity, high performing seismic force resisting supporting buildings up to 85 ft tall—the current IBC limit for light frame combustible construction.

15.3   SDPWS Seismically Detailed Platform-Framed CLT Shear Walls

The 2021 SDPWS CLT shear wall system has very detailed and prescriptive requirements contained in SDPWS Appendix B worth further discussion.

SDPWS B.2 has application requirements, including that construction be platform-framed and that CLT floors panels bear on the CLT shear walls. This requirement and the accompanying commentary make it clear that compliant CLT shear walls in multistory applications only be used with CLT floor panels bearing on the walls. This requirement does not apply to the roof over the CLT shear wall. At the roof, framing other than CLT panels is acceptable, provided the prescribed connection hardware can be used. In addition to having CLT floor panels bear on the CLT wall panel, other gravity framing is permitted in the structure, such as timber or steel beams and columns.

SDPWS B.3.1 item 2 requires that all of the panels in a shear wall have the same panel height. This precludes CLT shear walls being used applications where the top of the wall is sloped as can occur under sloped ceilings and exposed shed roofs. This creates an application limitation more restrictive than other shear walls, including WSP shear walls.

SDPWS B.3.2 defines the specific steel angle connector used to connect the CLT shear wall panels to framing above and foundation or framing below. Each bent steel connector is fastened to the CLT wall panels using (8) 16d *box* nails (3.5 in long, 0.135-in diameter shaft and 0.344-in diameter head). This nail size is somewhat uncommon; however, extra procurement effort may be able to locate these in collated forms for ease of installation using pneumatic nail guns. The connection of the steel angle to the adjacent framing or foundation uses 5/8-in bolts or lag screws. The bolts are required to have at least 4½ in of bearing and the lag screws 5⅞ in of bearing into the adjacent framing. These minimum bearing lengths prevent common 4⅛-in-thick 3-ply CLT from being used at the floor panels with this CLT shear wall system.

For CLT with a specific gravity of 0.42 or greater, the nominal shear capacity is defined based on the shear capacity of each connector. The nominal shear capacity, v_n, of a wall panel is defined as follows:

$$v_n = n\left(\frac{2605}{b_s}\right)C_G$$

where n is the number of angle connectors along the bottom and top of the panel, b_s is the individual panel length, and C_G is an adjustment factor for CLT with a specific gravity less than 0.42. A higher nominal capacity is not recognized by the SDPWS when using CLT made from lumber with a specific gravity greater than 0.42, such as from Douglas Fir-Larch, or southern pine lumber.

For convenience during design, the nominal shear capacity of *each angle connector* at the bottom of the wall can be calculated using:

$$V_n = C_G \, 2605 \text{ lb}$$

The ASD seismic capacity of each angle connector is

$$V_{s,ASD} = C_G \, 2605/2.8 = C_G \, 930 \text{ lb}$$

The ASD wind capacity of each angle connector is

$$V_{s,ASD} = C_G \, 2605/2.0 = C_G \, 1302.5 \text{ lb}$$

Using linear interpolation on the specific gravity, G, of the CLT, as provided by the SDPWS, C_G can be written as follows:

$$C_G = 2G + 0.16 \le 1.0$$

In SDPWS B.3.2 at least two angle connectors are required at the top and bottom of the panel and within 12 in of each end of the panel. The numbers of connectors at the top and bottom of the panel are to be equal. The ASD and LRFD design capacities are calculated from the nominal shear capacity using the provisions of the 2021 SDPWS Section 4.1.4, as described in Sec. 14.2 of this book.

The adjoining CLT panel edges within the shear wall are connected to each other with flat steel plates defined in SDPWS B.3.3. Each flat plate is attached to each panel with (8) 16d box nails just as the angle plates at the top and bottom of the wall are attached to the CLT. The number of such plates required in SDPWS B.3.3 item 3 equals the number of angle connectors at the bottom of the wall, n, times h/b_s and rounded up if not an integer.

In SDPWS B.3.1 item 2, all of the panels in the shear wall are required to have an aspect ratio h/b_s at least 2 and not greater than 4, as shown in Fig. 15.8.

SDPWS B.3.1 item 2 also requires that all CLT panel forming a multi-panel shear wall shall have the same height, h, and the same individual panel length, b_s. This requirement can have a significant impact on the construction of an SDPWS CLT shear wall. Consider a 12-ft-long shear wall is designed using (3) 4-ft-wide panels. How will the design team respond if the plans change so that only 11 ft is available for the wall? Per the SDPWS requirement, it is not acceptable to cut one panel down by 12 in and use (2) 4-ft-wide panels and (1) 3-ft-wide panel. Instead, to meet the requirements (3) 3-ft 8-in wide panels would need to be used. If such a change occurs during construction, it could be problematic if the CLT panels are already fabricated and shipped to the construction site.

There are no aspect ratio limits for the entire CLT shear wall stated in the 2021 SDPWS, as the maximum panel aspect ratio of 4 in SDPWS B.3.1 item 2 creates a limit on the maximum wall aspect ratio form from a single CLT panel.

The hold-downs at the ends of the CLT shear wall are not prescriptively detailed; however, hold-downs are required to use threaded anchor rods for the tension forces and nails, screws, or bolts to attach to the CLT. This allows for use of many hold-downs developed for conventional wood construction provided they meet the requirements.

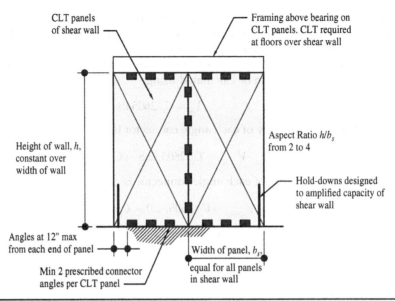

CLT panels of shear wall

Framing above bearing on CLT panels. CLT required at floors over shear wall

Height of wall, h, constant over width of wall

Aspect Ratio h/b_s from 2 to 4

Hold-downs designed to amplified capacity of shear wall

Angles at 12" max from each end of panel

Width of panel, b_s, equal for all panels in shear wall

Min 2 prescribed connector angles per CLT panel

Figure 15.8 SDPWS 2021 CLT shear wall geometric limitations.

Strap hold-downs which do not transfer tension forces via threaded rods are not permitted. The required design forces on the hold-downs are unique to CLT shear walls. SDPWS B.3.4 item 3 requires the hold-down system have a capacity not less than 2.0 times the forces associated with the seismic design shear *capacity* and not less than 1.5 times the forces associated with the wind design *capacity*. This requires the hold-downs to be designed using a factored-up *capacity* of the shear wall, which will be significantly higher than the required *design forces* on the shear wall.

Because of applicable reduction and amplification factors, the required wind shear forces are consistently larger than the required seismic forces for the design of the hold-downs. If the hold-downs considered have different seismic and wind capacities, then calculating both the required capacities is useful.

15.3.1 CLT Walls Not Part of the Seismic-Force-Resisting System

SDPWS B.2 items 3 & 4 provide requirements on how to consider CLT walls which are not part of the designated seismic-force-resisting system. Because of the complexities of these requirements, it is best to avoid using a combination of CLT walls which are considered shear walls with CLT walls which are not considered shear walls.

However, if CLT walls not part of the designated seismic-force-resisting system are used, there are considerations required by the SDPWS. SDPWS B.2 item 3 requires the walls to be detailed in the same manner as the CLT shear walls with the same limits on panel aspect ratio using the same connectors. The number of connectors is not prescribed nor are hold-downs required. Using CLT walls which are not part of the seismic-force-resisting system is not a means to using larger wall panels or using different connection hardware. SDPWS B.2 item 4.a requires that the designer perform two analyses with different assumptions to distribute lateral forces through the structure: one considering the designated seismic-force-resisting system only, and one considering all CLT walls as resisting the seismic forces. The gravity load carrying system, the seismic-force-resisting system, and the diaphragms are all to be designed to the most critical of these two analyses. In addition, the presence of irregularities defined in ASCE 7 12.3.2 must also be checked using both analyses. If a design project explores this route, an important modeling decision, without guidance in the SDPWS, is how to determine the stiffness of the CLT walls which are not part of the seismic-force-resisting system which do not have hold-down devices installed.

15.3.2   Other Load Path Connections

A requirement of SDPWS B.3.6 is the angle connector at the top of the SDPWS CLT shear wall shall not be considered in the design of out-of-plane wall forces. Additional connections to the CLT may be needed to handle out-of-plane wind or seismic forces and possible uplift forces from framing above. SDPWS B.3.6 requires any additional connections to use dowel-type fasteners design to develop Mode IIIs or Mode IV yielding. Additionally, screws are not to be used in connections at the top and bottom of the walls for additional load path considerations. Meeting these requirements when addressing out-of-plane loading on the wall or uplift from roof panels above can require some creative detailing. Examples of such details using toenails and long drift pins can be found in SDPWS Table C-B.2.

15.3.3   CLT Shear Walls with Resistance Provided
by High-Aspect-Ratio Panels Only

This long moniker is the name of a defined variant of the SDPWS CLT shear wall system which only uses panels with aspect ratios equal to 4 in all CLT walls in the system. This special case provides smaller walls and more connectors which results in a higher drift capacity and more ductile, energy dissipating behavior during a seismic event. The higher level of energy dissipation with such detailing leads to this system having a higher seismic R coefficient. The commentary provides some allowable variation from this exact aspect ratio by allowing the ratio of $4 +/- 2.5$ percent.

15.3.4   Seismic Design Coefficients

Regarding seismic design coefficients, besides the recognition of $R = 1.5$ for CLT shear walls not seismically detailed, the R values for the CLT shear wall system in SDPWS Appendix B are not found in the SDPWS. Rather these are found in Table 12.2-1 of ASCE 7-22 which references the 2021 SDPWS. The CLT shear wall system seismic design coefficients are shown in Table 15.2 alongside the SDPWS 4.6.3 Exception CLT Shear Walls and the WSP sheathed shear wall entry.

The SDPWS Appendix B CLT shear wall system is granted an R value of 3. Limiting the aspect ratio of the CLT panels to 4 increases the R value to 4. The limit on the structural height for the Appendix B CLT shear walls is 65 ft matching that for WSP shear walls in SDCs D and above. For SDC B and C, the 65-ft limit still applies to the Appendix B CLT shear walls while it doesn't for the WSP shear walls. In projects where combustible light frame construction is permitted, the trade-off between using CLT shear walls and WSP shear is a worthwhile evaluation process. In many cases, the answer may be that WSP shear walls are a better solution for the project. Prefabricated panelized light frame walls are often a good fit for use with CLT floor and roof diaphragm panels. In terms of the IBC construction types, light frame WSP shear walls are

Seismic-Force-Resisting System	Seismic Design Coefficients			Structural Height Limit for Seismic Design Category		
	R	Ω_0	C_d	B	C	D
SDPWS 4.6.3 Exception CLT Shear walls (Ordinary CLT shear walls)	1.5	2.5	1.5	65 ft	NP	NP
ASCE 7-22 Table 12.2-1 item A.20 Cross-laminated timber shear walls (SDPWS Appendix B CLT shear walls)	3	3	3	65 ft	65 ft	65 ft
ASCE 7-22 Table 12.2-1 item A.21 Cross-laminated timber shear walls with shear resistance provided by high-aspect-ratio panels only. (SDPWS Appendix B CLT shear walls with A.R. = 4)	4	3	4	65 ft	65 ft	65 ft
ASCE 7-22 Table 12.2-1 item A.16 Light-frame (Wood) walls sheathed with wood structural panels rated for shear resistance	6.5	3	4	NL	NL	65 ft

TABLE **15.2**   Seismic Design Coefficients for SDPWS CLT Shear Walls

permitted in Type III, Type IV-HT, and Type V construction. This includes the exterior walls of Type III and IV-HT, provided fire-retardant-treated wood is used. Construction Types IV-A, IV-B, and IV-C in IBC 2021 do allow CLT walls where light frame wood walls are not allowed. Buildings using these construction types are typically going to be taller than 65 ft; however, there could be certain projects below 65 ft tall using Type IV-A, IV-B, or IV-C because of unusually large areas or occupancies besides business and residential. If you encounter a project looking to use CLT shear walls, prepare yourself and your client on their detailing limitations.

Example 15.1: SDPWS Appendix B CLT Shear Wall

Consider a 9-ft-tall, 12-ft-long wall to be designed as a 2021 SDPWS Appendix B CLT shear wall. Design and detail the wall for ASD shear demands of 300 plf wind and 500 plf seismic demand corresponding to the $R = 3$ system.

Select (3) 4-ft-wide, $4\frac{1}{8}$-in-thick panels of PRG 320 V4 Grade CLT for the walls.

Check the aspect ratio of the walls:

$h/b_s = 9$ ft$/4$ ft $= 2.25 \geq 2$ and ≤ 4; therefore, panel aspect ratio is acceptable.

The ASD seismic shear demand to each 4-ft panel is (4 ft) (500 plf) = 2000 lb.

V4 CLT is made from spruce-pine-fir (south) visually graded lumber with $G = 0.36$. The capacity reduction factor for the specific gravity, based on linear interpolation is

$$C_G = 2\,G + 0.16 \leq 1.0$$

$$C_G = 2\,(0.36) + 0.16 = 0.88$$

The ASD seismic shear capacity per connector:

$$V_{s,ASD} = C_G\,(930\text{ lb}) = 0.88\,(930\text{ lb}) = 818\text{ lb per connector}$$

The required number of connectors at the bottom and top of the panel is

$$n = (2000\text{ lb/panel})/(818\text{ lb/connector}) = 2.44\text{ connectors/panel, use 3}$$

The required number of connectors at vertical adjoining edges of the panels is

$$n(h/b_s) = 3(9\text{ ft}/4\text{ ft}) = 6.75,\text{ use 7}$$

The nominal shear capacity of this CLT shear wall is

$$v_n = n\left(\frac{2605}{b_s}\right)C_G = 3\left(\frac{2605\text{ lb}}{4\text{ ft}}\right)(0.88) = 1719\text{ plf}$$

The LRFD and ASD wind and seismic design capacities are subsequently calculated as

	LRFD Design Capacity $\phi_D v_n$	ASD Design Capacity $\dfrac{v_n}{\Omega_D}$
Seismic	0.50 (1719) = 860 plf	(1719)/2.8 = 614 plf
Wind	0.80 (1719) = 1375 plf	(1719)/2.0 = 860 plf

The hold-down design forces are based on the amplified seismic and wind shear capacity. For ASD design, the equivalent unit shear demands on the wall for the hold-down design are as follows:

$$\text{Seismic: } v = 2.0 \times 614\text{ plf} = 1218\text{ plf (ASD)}$$

$$\text{Wind: } v = 1.5 \times 860\text{ plf} = 1290\text{ plf (ASD)}$$

Notice the wind shear force for hold-down design is greater than the seismic shear force for hold-down design, even though the code-prescribed seismic forces are larger than the wind forces.

If a 200 plf dead load is supported by the wall which can be used to reduce the hold-down design forces, SDPWS commentary Eq. C-B.2 can be used to calculate the tension force in the hold-down. This only uses the dead load on the last CLT panel to reduce the uplift load. If the location of the hold-down decreases the moment arm of the restraining forces by 3 in, $b_{eff} = 3.75$ ft, and the resulting hold-down design forces for seismic and wind are

$$\text{Seismic: } T = \frac{1218\text{ plf }(4\text{ ft})(9\text{ ft}) - 200\text{ plf }(4\text{ ft})(4\text{ ft}/2)}{3.5\text{ ft}} = 12{,}070\text{ lb}$$

$$\text{Wind: } T = \frac{1290\text{ plf }(4\text{ ft})(9\text{ ft}) - 200\text{ plf }(4\text{ ft})(4\text{ ft}/2)}{3.5\text{ ft}} = 12{,}811\text{ lb}$$

A hold-down needs to be selected with ASD capacities of at least these values. The resulting CLT shear wall configuration is shown in Fig. 15.9.

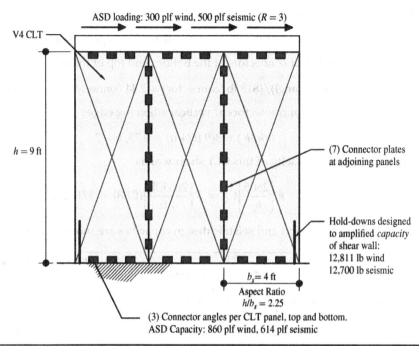

ASD loading: 300 plf wind, 500 plf seismic ($R = 3$)

V4 CLT

$h = 9$ ft

(7) Connector plates at adjoining panels

Hold-downs designed to amplified *capacity* of shear wall:
12,811 lb wind
12,700 lb seismic

$b_s = 4$ ft

Aspect Ratio
$h/b_s = 2.25$

(3) Connector angles per CLT panel, top and bottom.
ASD Capacity: 860 plf wind, 614 plf seismic

Figure 15.9 Example—Designed 2021 SDPWS CLT shear wall.

As an alternative, consider using the high aspect ratio, $R = 4$ CLT shear wall system and designing the corresponding wall. Without changing the wall height ($h = 9$ ft), the CLT panel width is required to be

$$b_s = 9 \text{ ft}/4 = 2.25 \text{ ft } (2'3'')$$

Using panels of this width cannot result in a 12-ft-long wall, so five panels of 2.25 ft width will be used for a wall length of 11.25 ft. Assuming the same distribution of loading from the rest of the structure, and adjusting for the different seismic R value, the loading onto this alternative configuration is

$$\text{Wind: 300 plf } (12 \text{ ft}/11.25 \text{ ft}) = 320 \text{ plf}$$

$$\text{Seismic: 500 plf } (12 \text{ ft}/11.25 \text{ ft}) (3/4) = 400 \text{ plf}$$

The ASD seismic shear demand to each 2.25-ft panel is (2.25 ft) (400 plf) = 900 lb. The required number of connectors at the bottom and top of the panel is

$$n = (900 \text{ lb/panel})/(818 \text{ lb/connector}) = 1.10 \text{ connectors/panel, use 2}$$

The required number of connectors at vertical adjoining edges of the panels is

$$n \, (h/b_s) = 2 \, (9 \text{ ft}/2.25 \text{ ft}) = 8.0, \text{ use 8}$$

The nominal shear capacity of this CLT shear wall is

$$v_n = n \left(\frac{2605}{b_s} \right) C_G = 2 \left(\frac{2605 \text{ lb}}{2.25 \text{ ft}} \right) (0.88) = 2038 \text{ plf}$$

The LRFD and ASD wind and seismic design capacities are calculated as

	LRFD Design Capacity $\phi_D v_n$	ASD Design Capacity $\frac{v_n}{\Omega_D}$
Seismic	0.50 (2308) = 1019 plf	(2308)/2.8 = 728 plf
Wind	0.80 (2308) = 1630 plf	(2308)/2.0 = 1019 plf

The hold-down design forces are based on the amplified seismic and wind shear capacity. For ASD design, the equivalent unit shear demands on the wall for the hold-down design are

$$\text{Seismic: } v = 2.0 \times 728 \text{ plf} = 1455 \text{ plf (ASD)}$$

$$\text{Wind: } v = 1.5 \times 1019 \text{ plf} = 1528 \text{ plf (ASD)}$$

Using the same 3-in reduction in the moment arm resisting overturning, $b_{eff} = 2.0$ ft, and the hold-down design forces using SDPWS commentary Eq. C-B.2 for seismic and wind are

$$\text{Seismic: } T = \frac{1455 \text{ plf } (2.25 \text{ ft})(9 \text{ ft}) - 200 \text{ plf } (2.25 \text{ ft})(2.25 \text{ ft}/2)}{2.0 \text{ ft}} = 14{,}478 \text{ lb}$$

$$\text{Wind: } T = \frac{1528 \text{ plf } (2.25 \text{ ft})(9 \text{ ft}) - 200 \text{ plf } (2.25 \text{ ft})(2.25 \text{ ft}/2)}{2.0 \text{ ft}} = 15{,}219 \text{ lb}$$

A hold-down needs to be selected with ASD capacities of at least these values. ▲

15.4 References

1. American Society of Civil Engineers (ASCE), *ASCE/SEI 7-16 Minimum Design Loads for Buildings and Other Structures*, ASCE, New York, 2016.
2. AWC, *Special Design Provisions for Wind and Seismic 2021 Edition*, AWC, Leesburg, VA, 2020.
3. Gagnon, S. and Ciprian, P. (eds.), *CLT Handbook: Canadian Edition*, FPInnovations, Pointe-Claire, QC, Canada, 2011.
4. Karacabeyli, E. and Douglas, B. (eds.), *CLT Handbook: US Edition*, FPInnovations, Pointe-Claire, QC, Canada, 2012.
5. American Wood Council (AWC), *National Design Specification (NDS) for Wood Construction with Commentary, 2018 Edition*, AWC, Leesburg, VA, 2017.
6. Chen, Z. and Popovski, P., Mechanics-based analytical models for balloon-type cross-laminated timber (CLT) shear walls under lateral loads. *Engineering Structures*, Vol. 208, 2020.
7. Shahnewaz, M., Dickhof, C., and Tannert, T., Seismic behavior of balloon frame CLT shear walls with different ledgers. *Journal of Structural Engineering*, Vol. 147, ASCE, Reston, VA, 2021.
8. Zimmerman, R. and McDonnell, E., *Framework—Innovation in Re-Centering Mass Timber Wall Buildings*. In 11th U.S. National Conference on Earthquake Engineering, Los Angeles, California, 2018.

9. Wilson, A., Christopher, J. M., Phillips, A. R., and Dolan, J. D., Seismic response of post-tensioned cross-laminated timber rocking wall buildings. *Journal of Structural Engineering*, Vol. 146, ASCE Reston, VA, 2020.

10. Jin, Z., Pei, S., Blomgren, H., and Powers, J., Simplified mechanistic model for seismic response prediction of coupled cross-laminated timber rocking walls. *Journal of Engineering Structures*, Vol. 145, ASCE, Reston, VA, 2019.

11. Chen, Z., Popovski, M., and Iqbal, A., Structural performance of post-tensions CLT shear walls with energy dissipators. *Journal of Structural Engineering*, Vol. 146, ASCE, Reston, VA, 2020.

12. American Society of Civil Engineers (ASCE), *ASCE/SEI 7-22 Minimum Design Loads and Associated Criteria for Buildings and Other Structures*, ASCE, Reston, VA, 2021.

13. Ghosh, S. K., *Ductile Coupled Reinforced Concrete Shear Walls and Coupled Composite Steel Plate Shear Walls as Distinct Seismic Force-Resisting Systems in ASCE 7*, in 2019 SEAOC Convention Proceedings, Structural Engineers Association of California, Sacramento, CA, 2019.

14. mgb Architecture + Design, and Equilibrium Consulting, *The Case for Tall Wood Buildings, mgb Architecture and Design*, Vancouver, BC, Canada, 2022.

15. Skidmore, Owings, & Merrill (SOM), *Timber Tower Research Project*, SOM, Chicago, IL, 2013.

16. International Code Council (ICC), *2021 International Building Code*, ICC, Whittier, CA, 2020.

Index

Note: Page numbers referencing figures are followed by an "*f*".